Numerical Modelling and Simulation of Metal Processing

Numerical Modelling and Simulation of Metal Processing

Editor

Christof Sommitsch

MDPI • Basel • Beijing • Wuhan • Barcelona • Belgrade • Manchester • Tokyo • Cluj • Tianjin

Editor
Christof Sommitsch
Graz University of Technology
Austria

Editorial Office
MDPI
St. Alban-Anlage 66
4052 Basel, Switzerland

This is a reprint of articles from the Special Issue published online in the open access journal *Metals* (ISSN 2075-4701) (available at: https://www.mdpi.com/journal/metals/special_issues/modelling_simulation_metal_processing).

For citation purposes, cite each article independently as indicated on the article page online and as indicated below:

LastName, A.A.; LastName, B.B.; LastName, C.C. Article Title. *Journal Name* **Year**, *Volume Number*, Page Range.

ISBN 978-3-0365-1080-4 (Hbk)
ISBN 978-3-0365-1081-1 (PDF)

Contents

About the Editor

Christof Sommitsch is Austrian and graduated as Diplom-Ingeniuer for Materials Science at Montanuniversität Leoben. He received his PhD at Graz University of Technology with the thesis *Theory and Model of the Microstructure Development of Nickel-Based Superalloys during Hot Forming.* After 5 years in industry, he returned to Montanuniversität Leoben at Chair of Metal Forming. In 2009, he was appointed as Professor of Materials Science and Welding at Graz University of Technology. From 2012 to 2015, he was Dean of Faculty of Mechanical Engineering and Economic Sciences. Christof Sommitsch is a member of several scientific and industrial boards and has published close to 500 scientific contributions.

Preface to "Numerical Modelling and Simulation of Metal Processing"

The service properties of materials strongly depend on their chemical composition, as well as their processing conditions. The specific applications and the optimization of thermal and thermomechanical processes enable the conservation of processes, the treatment of new alloy variants, and the realization of materials with special properties. The numerical simulation of metal processing, coupled with the modelling of the structural evolution, reduces the need for time- and cost-expensive tests at lab and industrial scale. Here, multiscale modelling is considered as an appropriate means to describe the development of the nano-, micro-, and macrostructure and, hence, to determine the local material properties.

Christof Sommitsch
Editor

Technical Note

Performance of Optimization Algorithms in the Model Fitting of the Multi-Scale Numerical Simulation of Ductile Iron Solidification

Eva Anglada [1], Antton Meléndez [2], Alejandro Obregón [2], Ester Villanueva [2] and Iñaki Garmendia [3],*

[1] TECNALIA, Basque Research and Technology Alliance (BRTA), Mikeletegi Pasealekua, 2, E-20009 Donostia–San Sebastián, Spain; eva.anglada@tecnalia.com

[2] TECNALIA, Basque Research and Technology Alliance (BRTA), Astondo Bidea, Edificio 700, E-48160 Derio, Spain; antton.melendez@tecnalia.com (A.M.); alejandro.obregon@tecnalia.com (A.O.); ester.villanueva@tecnalia.com (E.V.)

[3] Mechanical Engineering Department, University of the Basque Country UPV/EHU, Engineering School of Gipuzkoa, Plaza de Europa, 1, E-20018 Donostia–San Sebastián, Spain

*   Correspondence: inaki.garmendia@ehu.es; Tel.:+34-43-018630

Received: 1 July 2020; Accepted: 5 August 2020; Published: 8 August 2020

Abstract: The use of optimization algorithms to adjust the numerical models with experimental values has been applied in other fields, but the efforts done in metal casting sector are much more limited. The advances in this area may contribute to get metal casting adjusted models in less time improving the confidence in their predictions and contributing to reduce tests at laboratory scale. This work compares the performance of four algorithms (compass search, NEWUOA, genetic algorithm (GA) and particle swarm optimization (PSO)) in the adjustment of the metal casting simulation models. The case study used in the comparison is the multiscale simulation of the hypereutectic ductile iron (SGI) casting solidification. The model fitting criteria is the value of the tensile strength. Four different situations have been studied: model fitting based in 2, 3, 6 and 10 variables. Compass search and PSO have succeeded in reaching the error target in the four cases studied, while NEWUOA and GA have failed in some cases. In the case of the deterministic algorithms, compass search and NEWUOA, the use of a multiple random initial guess has been clearly beneficious.

Keywords: model fitting; optimization; FEM; metal casting; SGI; numerical simulation; compass search; NEWUOA; genetic algorithm; particle swarm optimization

---

1. Introduction

Today, numerical simulation of metal casting processes at macro scale is a well-known technology. Although new approaches based for example in dynamic mesh techniques are being currently investigated [1], the use of commercial programs based on the finite element method (FEM) or on the finite volume method (FVM) is quite widespread in the industry. However, the agreement between the obtained predictions and the actual results obtained at the foundry plant, is not always the desired one. This agreement is more difficult even in the case of multiscale simulation. For this reason, the model fitting is of high interest to improve the confidence in the predictions. Even more in the present context where, in one side the integrated computational materials engineering and in other side the hybrid prognostic models combining numerical simulation and data-driven models applied to the Industry 4.0 movement of the manufacturing industry, appear as two of the most promising and challenging research trends [2].

The obtaining of reliable predictions in metal casting processes is not a trivial task. The results do not only depend on numeric codes in which they are based. The use of the appropriate values of the thermophysical parameters along with the proper establishment of boundary conditions, is essential to avoid poor results or even erroneous ones. In fact, the ASM Committee [3] remarks that each manufacturing process has unique boundary conditions that can be even equipment specific, which must be identified and characterized for the specific application being simulated.

The characterization of the material properties for the whole range of temperatures involved in the metal casting processes is very difficult. Additionally, the determination of the values of some boundary conditions, as for example the heat transfer coefficients (HTC) between the alloy and the mold, is even more complex. The limitations of bibliographic data and the extreme difficulty of the direct determination of these values have inspired the application of inverse methods to avoid these difficulties.

The inverse methods can be considered a subset of the field of numerical optimization where the objective is to minimize one objective function, which represents the error between the simulation results and the experimental measurements. The challenge is that the corresponding objective function can be highly nonlinear or non-monotonic, may have a very complex form or its analytical expression may be unknown. Moreover, in the case of metal casting inverse problems, the effect of changes in boundary conditions are normally damped or lagged, i.e., the varying magnitude of the interior temperature profile lags behind the changes in boundary conditions and is generally of lesser magnitude. Therefore, such a problem would be a typically ill-posed and would normally be sensitive to measurement errors. Additionally, the uniqueness and stability of the solution are not generally guaranteed [4–6].

This type of optimization problems is usually solved using iterative strategies. The methodology consists basically in an iterative process where several parameters of the model are modified until the simulation predictions agree with the experimental measurements. Different strategies can be applied to advance in the iterative process depending on the selected optimization method.

In general, it is not possible to establish which optimization methods are the best between the high number of alternatives that exist. The performance of the optimization methods is completely related with the problem type to be tackled. Therefore, for each problem type, the most appropriate method should be searched.

The use of optimization algorithms to correlate the numerical models with experimental values has been applied in other fields different from the metal casting. The sector where possibly more efforts have been done in this sense, is the space industry. The correlation of the spacecraft thermal mathematical models with the results of the thermal tests is mandatory, which has encouraged the evaluation of different solutions. Some authors have explored the use of deterministic methods, as Klement, who used several quasi-newton type algorithms based on the Broyden methods [7,8] or Torralbo et al. who used a generalized iterative pseudo-inverse algorithm [9,10]. But most of them have studied and compared different stochastics methods. For example Dudon [11] use the Latin hypercube sampling (LHS), the self-adaptive evolution (SAE) and the branch and bound methods; Beck [12] use the adaptive particle swarm optimization (APSO); Van Zijl [13] use the Monte Carlo, the genetic algorithm (GA) and the APSO methods; Trinoga and Frey et al. [14,15] use the simulated annealing (SA), the threshold accepting (TA) and the GA; Anglada et al. work mainly with GA [16,17] but also made some comparisons with Klement algorithms [18] and with the deterministic algorithms TOLMIN, NEWUOA, BOBYQA and LINCOA [19]. In addition, works about the use of hybrid algorithms combining deterministic and stochastic methods can also be found, as the works of De Palo [20] where the downhill simplex is combined with the LHS or the work of Cheng et al. [21] where the Broyden–Fletcher–Goldfarb–Shanno (BFGS) is combined with the Monte Carlo method.

In contrast, the efforts done in the field of metal casting are much more limited. The number of bibliographic references devoted to the application of optimization algorithms to the adjustment of metal casting simulation models is quite reduced. Most of them [22–25] use the maximum A posteriori (MAP) algorithm, which is a variation of the least-square technique explained in reference [26]. Others perform the adjustment manually [27] or combining the manual adjustment with the MAP algorithm [28–32]. Moreover, some authors have tested different methods as Ilkhchy et al. [33] which use a nonlinear estimation technique similar to that stated by Beck in [34]; Dong et al. [35] that use a combined function specification and regularization method; Zhang et al. [36] with a neural network or Zhang et al. [37] that use a globally convergent method (GCM).

The authors of the present study believe that a deeper study in the performance of different optimization methods may contribute to give a step forward in the state of art of the model fitting applied to the metal casting. It would make possible to obtain adjusted models in less time and moreover will open the door to the future integration of automatic model fitting algorithms in prognostic models to be used in the framework of the industry 4.0.

The work presented henceforward compares the performance of several algorithms, deterministic and stochastic, in the adjustment of the multiscale simulation model of the solidification of a ductile iron casting, with the objective of contributing to the development of this research line in the search of a methodology solid enough to be used in the adjustment of metal casting simulation models.

2. Methodology

The methodology used for the adjustment of the simulation model is summarized in Figure 1.

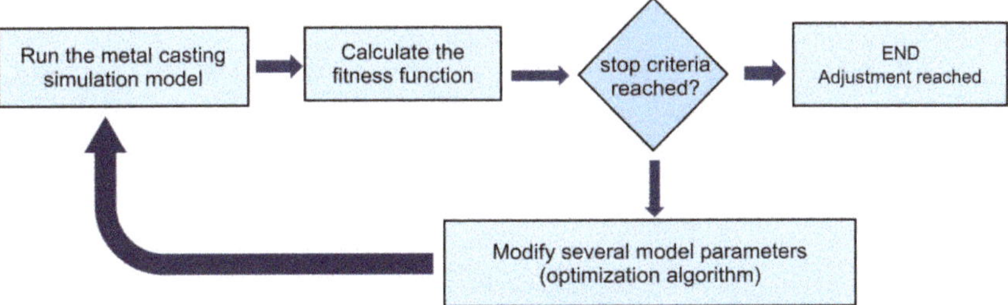

Figure 1. Flowchart of the model adjustment procedure.

The multiscale simulation model of the metal casting process was done with the commercial finite element software ProCAST 2018.0 (ESI Group, Paris, FRANCE), focused on metal casting simulation [38]. The fitness function represents the error between the simulation model predictions and the experimental results, that in this case are the alloy tensile strength. The parameters modification is guided by the optimization method used in each case, which was implemented by means of the PyGMO (Parallel Global Multiobjective Optimizer) library. PyGMO is a scientific library for massively parallel optimization developed in the context of the European Space Agency to evolve interplanetary spacecraft trajectories as well as designs for spacecraft parts and more [39].

2.1. Simulation Model

2.1.1. The Physics

Different approximations have been studied for the prediction of the tensile strength of cast alloys, as the use of machine learning algorithms [40] or analytical approximations as the Equation (1) stated by [41]. In this case, this last approximation, Equation (1), has been used by the simulation model to predict the tensile strength at room temperature (σ_{ULT}), where f_g is the graphite fraction, n is the shape factor for graphite nodules which is related with their sphericity (maximum value, that is a perfect sphere, is unity), f_α is the ferrite fraction, f_p the perlite fraction, and $(dT/dt)_{850}$ is the cooling rate at 850 °C. Therefore, the microstructure and the cooling rate must be previously calculated.

$$\sigma_{ULT} = \left(1 - f_g^n\right)\left(482.2 f_\alpha + 991.5 f_p\right) + 50[(dT/dt)_{850} - 0.5], \tag{1}$$

In cast irons the carbon and silicon contents are high, compared with steels, which origins a rich carbon content phase in their structure. When the alloy graphitization potential is high, which is determined by its composition and melt treatment, the iron solidifies according to the stable Fe-graphite system and the rich carbon phase is graphite, a hexagonal-close-pack form of carbon. In addition, the presence of magnesium makes the graphite structure be spheroidal instead of lamellar. One of the main melt treatments is the inoculation, which consists in the deliberate addition of elements that promote more active graphite nucleation sites (usually ferrosilicon enriched with Sr, Ba, Al, Zr, Ca and with very reduced and well calculated quantities of rare earths). The matrix structure is essentially determined by the composition and the cooling rate through the eutectoid temperature range. The eutectoid reaction leads to the decomposition of austenite into ferrite and graphite for the case of the stable eutectoid and to pearlite for the metastable eutectoid transformation. Slower cooling rates result in more stable eutectoid structure promoting a more ferritic matrix. If the complete transformation of austenite is not achieved when the metastable temperature is reached, pearlite forms and grows in competition with ferrite.

In the metal casting model used in this case, the influence of the chemical composition is managed by means of the thermodynamic database for multicomponent iron-based alloys developed by CompuTherm, PanFe 2017 (CompuTherm LLC, Middleton, WI, USA) [42]. This thermodynamic database is based on the calculation phase diagrams (CALPHAD) methodology, also called computer coupling of phase diagrams. CALPHAD is a materials genome method which plays a central and fundamental role in materials design, see reference [43] for more information. This method develops models to represent thermodynamic properties for various phases. The thermodynamic properties of each phase are described through the Gibbs free energy, based on the experimental and theoretical information available on phase equilibria and thermochemical properties in a system. Following this, it is possible to recalculate the phase diagram, as well as the thermodynamic properties, of all the phases and the system as a whole. It allows to reliably predict the set of stable phases and their thermodynamic properties in regions without experimental information and for metastable states during simulations of phase transformations together with the properties of multicomponent systems from those of binary and ternary subsystems. This is accomplished by considering the physical and chemical properties of the system in the thermodynamic model. In this case, the "lever" micro-segregation model, which assumes a complete mixing of the solute in the solid, was applied.

The influence of the melt treatment is managed by means of the nucleation parameters which must be specified/adjusted for each case, as they are not an intrinsic property of the material.

- The nucleation of the primary dendritic phase, austenite, is modeled following the gaussian distribution model proposed by [44]. This model defines the relationship between the number of nuclei and the undercooling (ΔT) following Equation (2), where n_{max} is the maximum grain nuclei density, ΔT_δ the undercooling standard deviation and ΔT_n the average undercooling (see Figure 2):

$$n(\Delta T(t)) = \frac{n_{max}}{\sqrt{2\pi} \cdot \Delta T_\delta} \int_0^{\Delta T(t)} exp\left(-\frac{(\Delta T(t) - \Delta T_n)^2}{2\Delta T_\delta^2}\right) d(\Delta T(t)), \tag{2}$$

- The nucleation of the graphite nodules is calculated as a power law of the undercooling following Equations (3) and (4) of the model proposed by [45], where A_e and n are the nucleation constants. This model assumes bulk heterogeneous nucleation at foreign sites which are already present within melt or intentionally added to the melt by inoculation:

$$N_{eut} = A_e(\Delta T)^n, \tag{3}$$

$$\frac{dN_{eut}}{dt} = -nA_e(\Delta T)^{n-1}\frac{dT}{dt}, \tag{4}$$

- The graphite nodules growth is calculated with a quadratic power of the undercooling following Equation (5), where μ_e is the eutectic growth coefficient:

$$\frac{dR_g}{dt} = \mu_e(\Delta T)^2,\tag{5}$$

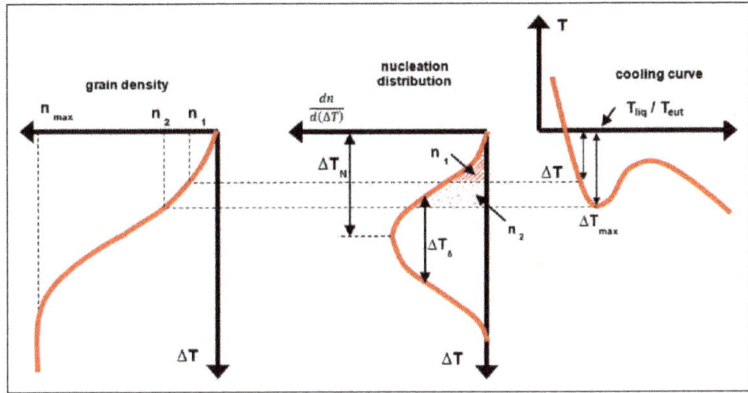

Figure 2. Undercooling definition.

Relating to the cooling rates, the governing equations of the physics involved in the mold filling, and in the alloy cooling and solidification, are the energy conservation, Equation (6), the mass conservation or continuity, Equation (7), and the Boussinesq form of the Navier–Stokes equation for incompressible Newtonian fluids, Equation (8), where ρ is the density; h is the specific enthalpy; t is the time; v is the velocity; k is the thermal conductivity; $\dot{R}_q$ is the heat generation per unit mass; ρ_0 is density at reference temperature and pressure; $\hat{p}$ is the modified pressure ($\hat{p} = p + \rho_0 gh$); μ_l is the shear viscosity; g is the gravity; β_T is the volumetric thermal expansion coefficient; T is the temperature and T_0 is the boundary temperature. More detailed information about them can be found at references [46,47].

$$\rho\frac{\partial h}{\partial t} + \rho v\cdot\nabla h = \nabla\cdot(k\nabla T) + \rho\dot{R}_q,\tag{6}$$

$$\frac{\partial \rho}{\partial t} + \nabla\cdot(\rho v) = 0\tag{7}$$

$$\rho_0\left(\frac{\partial v}{\partial t} + v\cdot\nabla v\right) = -\nabla\hat{p} + \mu_l\nabla^2 v - \rho_0 g\beta_T(T - T_0)\tag{8}$$

2.1.2. Model Reduction

A preliminary detailed 3D model has been developed. As can be observed in Figure 3a, it represents the complete geometry of the cast parts. The two similar, but different cast parts, the feeding systems formed by two risers, one for each cast part, the filling system and the mold. This 3D model is formed by 1,148,327 tetrahedral elements and 199,031 nodes. In this first attempt the complete simulation has been performed, that is, the part filling and the cooling and solidification process together with the microstructural calculations. The time step has taken values between 1×10^{-3} (during filling) and 50 s (during the last cooling) depending on the calculation evolution. The resolution of all the equations, shown in last section, for a transient analysis from pouring temperature (1410 °C) down to values below the eutectoid transformation temperature (738 °C), with the mentioned mesh and time steps, requires a high CPU time. In this case, the simulation has required about 5 h using a computer with 64 Gb of RAM using 8 cores Intel-Xeon CPU E5-2630 v3 @ 2.4 GHz (Dell Inc., Round Rock, TX, USA).

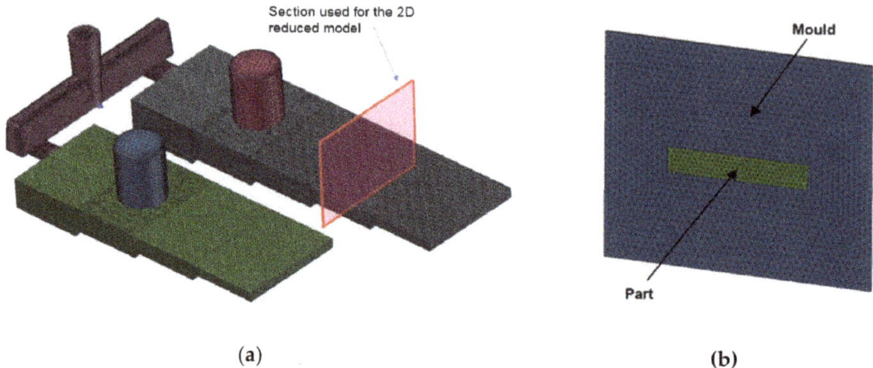

(**a**) (**b**)

Figure 3. (**a**) Mesh used in the 3D simulation model; (**b**) mesh used in the pseudo-2D model.

As it has been previously introduced, the model adjustment is an iterative procedure where the simulation model must be solved each time to get the predictions needed to calculate the fitness function. Therefore, it is highly advised to employ simulation models whose calculation times are as reduced as possible.

For this reason, due to the long calculation times of this model, a model reduction strategy was followed to reduce the high CPU time for the fitness function calculation. In first place the mold filling was suppressed from the calculation. As in this case the mold used was manufactured in resin bonded sand, the temperature drop during filling is not very pronounced. Therefore, the assumption of a mold completely filled of alloy at uniform temperature at the initial step of the simulation, is admissible. In second place the 3D geometry was replaced by a slice of the central area where the tensile test bar was extracted (see Figure 3b and Figure 5). This pseudo-2D geometry assumes that the heat flows mainly across the mold walls.

This model reduction presents some differences in the cooling rate (see Figure 4) which implies an error of 3.6% in the tensile strength prediction (433.64 MPa in the full 3D model and 418.14 MPa in the 2D model). However, the CPU time required to solve this 2D model is of only 29 s (620 times less). Therefore, regarding the CPU time saving, the error level is considered acceptable.

2.2. Fitness Function

The fitness function—also called cost function or objective function—is the mathematical representation of the aspect under evaluation which will be minimized. Therefore, it is the function that represents the difference between the hypothesis (model predictions) and the real results (experimental measurements), that is, the error. The expression used in this case, is the relative error between the predicted tensile strength ($\sigma_{predicted}$) and the measured tensile strength ($\sigma_{reference}$), following Equation (9). Note that absolute value of the difference is used.

$$fitness = \frac{\left|\sigma_{predicted} - \sigma_{reference}\right|}{\sigma_{reference}}, \tag{9}$$

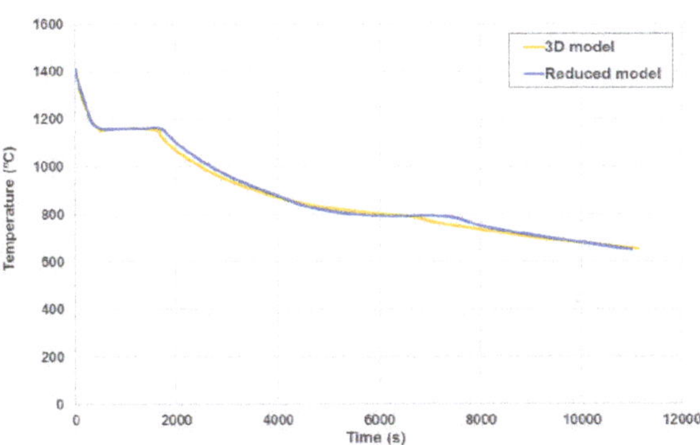

Figure 4. Cooling curve comparison between the pseudo-2D model and the 3D model.

2.3. Variables Selection

The variables selection that is the parameters of the model that will be modified by the algorithm to adjust the model, was selected in a supervised mode based on the expert's knowledge about the physics involved.

It is advised to select those variables which have more influence in the fitness function. Moreover, among them, those whose values are more uncertain. As it has been previously introduced, the tensile strength prediction is mainly related with the chemical composition of the alloy, the melt treatment and the cooling rate.

In this case the influence of the chemical composition is managed by the thermodynamic database, which calculates the material properties and the different phases that appears during the solidification. CALPHAD is a well-known methodology widely used in computational materials engineering, therefore, although new updated databases are continuously developed, we can assume its predictions as reliable.

The melt treatment can be modeled by means of the parameters that control the nucleation of the austenite: the maximum grain nuclei density n_{max}, the undercooling standard deviation ΔT_δ and the average undercooling ΔT_n; those that control the nucleation of the graphite nodules: A_e and n; and, finally, the parameter that controls the graphite nodules growth, the eutectic growth coefficient μ_e. All these parameters are very difficult to estimate, so they are good candidates to be included in the parameters group to be modified by the optimization method.

The cooling rate is mainly determined by the material properties of the alloy and the mold and for the heat transfer coefficient (HTC) between them. As said, the material properties of the alloy are calculated by the thermodynamic database, so they can be considered reliable. Resin bonded sand molds are manufactured mixing together clean and dry sand, with a binder and a catalyst in a continuous mixer. Different granulometries and types of sand can be used together with different binders and catalysts from several manufacturers, which can be also mixed in different proportions, so the resin bonded sand molds may present differences in their properties. Therefore, they can be also included in the parameters group to be modified by the optimization method. The heat transfer coefficient (HTC) between the alloy and the mold is difficult to estimate. When the metal enters in the mold, the macroscopic contact between the alloy and the mold is good because of the conformance of the molten metal. In this period the HTC presents high values, typically between 100 and 3000 W/m^2 K depending on different factors (surface roughness, pressure, wettability, etc.). Once a solidified layer with enough strength has grown, the contraction suffered by the alloy, together with the mold distortion caused by the mold heating, are likely to start a gap between them. In this phase, the HTC will mainly depend on the gas conductivity and the gap size, and its values may drop to values 10 times lower (10–100 W/m^2 K). In the ductile iron case, the expansion that takes place in the alloy during the graphite precipitation tends to reduce that gap but makes even more difficult the estimation of the HTC value. Therefore, the HTC is also a good candidate to be modified during the adjustment. However, in order to facilitate the implementation of the algorithm designed to perform the model fitting, a constant value was used for the HTC.

2.4. Optimization Methods

In the model fitting procedure followed, the parameters modification is guided by the optimization method. The performance of four different methods was compared. Two of them are deterministic methods, the compass search and the NEWUOA. The other two are heuristic methods, the particle swarm optimization (PSO) and the genetic algorithm (GA).

Compass search is a deterministic local optimization algorithm. It uses a simple and slow, but very sure procedure. The algorithm advances varying one of the parameters that form the solution vector at a time by steps of the same magnitude and when no such increase or decrease in any one parameter further improves the fit to the experimental data, they halve the step size and the process is repeated until the steps are deemed sufficiently small. For example, in a minimization problem with only two variables, the algorithm can be summarized as follows: Try steps to the East, West, North and South. If one of these steps yields a reduction in the function, the improved point becomes the new, iterate. If none of these steps yields improvement, try again with steps half as long. More details about this algorithm can be found in [48].

NEWUOA is also a deterministic local optimization algorithm. It seeks the least value of a function F without constraints and without derivatives. A quadratic approximation Q of the F function is used to obtain a fast convergence ratio, using a trust region strategy (see Figure 5). In this type of strategy, a model which represents a behavior similar to the objective function in a point x_k is generated. The search of the optimum is restricted to the surroundings to that point, which is called trust region. As the optimization progress, successive approximations of Q are used. More details about this algorithm can be found in [49].

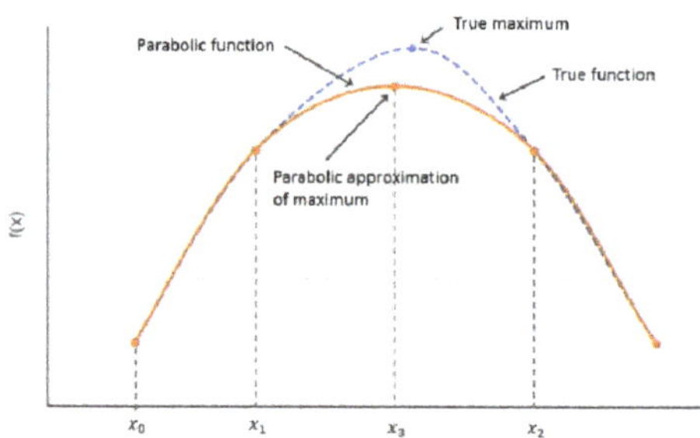

Figure 5. Quadratic approximation of a function

Particle swarm optimization (PSO) is a heuristic bioinspired algorithm for global optimization. A number of particles are placed in the search space of some problem or function and each evaluates the objective function at its current location. Each particle then determines its movement through the search space by combining some aspects of the history of its own current and best (best-fitness) locations with those of one or more members of the swarm, with some random perturbations. The next iteration takes place after all particles have been moved. Each individual in the particle swarm is composed of three D-dimensional vectors, where D is the dimensionality of the search space. The first vector is the current position x_i, which can be considered as a set of coordinates describing a point in space that is evaluated as a problem solution. Second vector is formed by the coordinates of previous best position p_i. The value of the best global position so far is stored in a variable for comparison on later iterations. The third vector is the velocity v_i, which can effectively be seen as a step size. In this way, new points are chosen by adding v_i coordinates to x_i and the algorithm operates by adjusting v_i. Each particle communicates with some other particles and is affected by the best point found by any member of its neighborhood. More details about this type of algorithm can be found in reference [50].

Genetic algorithms (GA) are heuristic search algorithms of general purpose inspired in the genetic processes that take place in biologic organisms and in the principles of the natural evolution of the populations. The basic idea consists in the generation of a random population of individuals or chromosomes and the advance to better individuals by means of the application of genetic operators, which are based in the genetic processes that takes place in the nature. Each of the individuals represents a possible solution to the problem, so the population represents the solutions search space. During the successive iterations, called generations, the individuals of the population are evaluated depending on their adaptation as solution. To do that, each one has associated a fitness value which measures how of good or bad is the possible solution that represents. The new population of individuals is created by selection mechanisms and by crossover and mutation operators. In this way, with the successive iterations or generations the algorithm converges to better solutions until reaching the optimum or the maximum number of iterations. More information about this type of algorithms can be found in reference [51].

3. Case Study

The case study is a part with a stepped geometry, cast in a sand mold bonded with resin where two similar parts are cast simultaneously. The reference value used for the model fitting is the tensile strength measured experimentally. Three tensile test bars were extracted from the central area of the 35-mm high step of the longer part, shown in Figure 6. The mean tensile strength obtained is equal to 552 MPa and the standard deviation equal to 4 MPa. The material is ductile iron, also called spheroidal gray iron (SGI), whose slightly hypereutectic composition is shown in Table 1, together with the tensile strength value measured experimentally. The mold was manufactured with silica sand mixed with a phenolic urethane resin (Pep-Set) and the corresponding catalyst. Figure 7 shows the manufactured cast part.

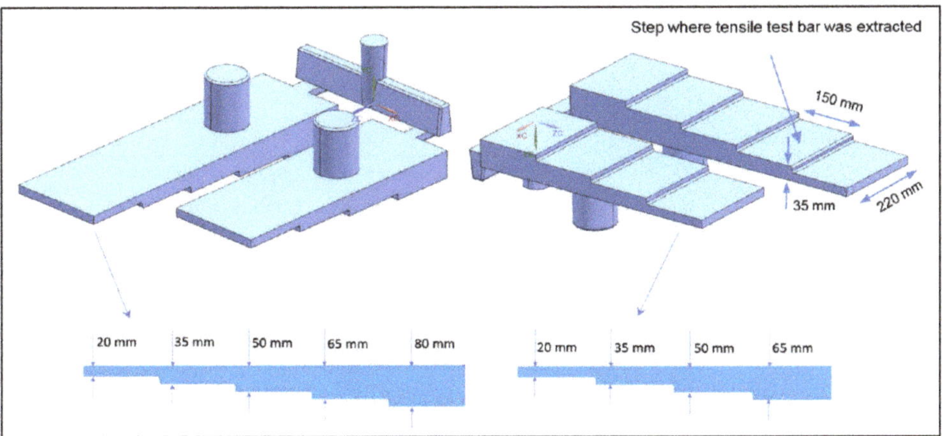

Figure 6. Top and bottom view of the part geometry.

Table 1. Ductile iron chemical composition.

C (wt%)	Si (wt%)	Mn (wt%)	Cr (wt%)	Cu (wt%)	Mg (wt%)	CE	Tensile Strength (MPa)
3.53	2.65	0.27	0.037	0.37	0.04	4.404	552

Figure 7. Manufactured cast part.

One important aspect in the performance of optimization algorithms is the dimensionality problem—that is, the number of variables that can be modified by the algorithm. Typically, classical optimization algorithms, as compass search or NEWUOA in this case, performs better in low dimensional problems, but when the number of variables increases it may require too long calculation times. Instead, the advantage of the heuristic algorithms, as PSO and GA in this case, takes place when the problem dimensionality is high. For this reason, the performance of the four optimization algorithms was evaluated for different numbers of variables included in the model fitting. The cases of 2, 3, 6 and 10 variables were studied. The corresponding variables considered in each case are detailed in Table 2. Table 3 summarize the default values used for these variables together with their ranges.

Table 2. Variables included in the model fitting cases.

Number of Variables	Variables
2	Parameters that control the nucleation of the graphite nodules: A_e and n.
3	Parameters that control the nucleation of the graphite nodules: A_e and n. HTC between the alloy and the mold
6	Parameters that control the nucleation of the graphite nodules: A_e and n. HTC between the alloy and the mold. Density, thermal conductivity and specific heat of the mold
10	Parameters that control the nucleation of the graphite nodules: A_e and n. HTC between the alloy and the mold. Density, thermal conductivity and specific heat of the mold. Parameters that control the nucleation of the austenite: Maximum grain nuclei density n_{max}, undercooling standard deviation ΔT_δ and average undercooling ΔT_n. Parameters that control the graphite nodules growth (eutectic growth coefficient μ_e)

Table 3. Default values and ranges of the variables included in model fitting.

Parameter	Default Value	Minimum Value	Maximum Value
A_e	1.0×10^3	71.5	1500
n	2.5	1.0	3.0
$\text{HTC}_{\text{alloy-mold}}$ (W/m^2 K)	300	30	3000
ρ_{mold} (kg/m^3)	1590	1113	2067
k_{mold} (W/mK)	0.621	0.435	0.808
$\left(C_p\right)_{mold}$ (J/kgK)	914.0	640.0	1188.0
n_{max}	1.0×10^3	100	1.0×10^4
ΔT_δ (°C)	1.0	0.5	5.0
ΔT_n (°C)	6.0	3.0	20.0
μ_e	3.87×10^{-6}	1.5×10^{-6}	5.0×10^{-6}

The optimization algorithms were used as they are implemented in the PyGMO library previously mentioned, using quite standard values for their internal parameters. Compass search was defined with a start range equal to 0.1, a stop range equal to 0.001 and a reduction coefficient equal to 0.5. In the case of NEWUOA, no additional parameters were setup. In order to compare both algorithms in similar conditions, the same seed were used in both cases.

Compass search and NEWUOA are local optimization algorithms, for this reason the selection of the initial guess may have influence in the final solution of the algorithm. For this reason, two different configurations were used for both. In the first configuration the algorithms are initialized with only one random guess. In the second one, they are initialized with 20 random guesses. In this last case, after the first iteration with those 20 individuals, the best of them is selected and the algorithms continue iterating from that point.

PSO and GA are stochastic population-based algorithms. In both cases a population size of 20 individuals were used. The canonical version of PSO based on the constriction coefficient velocity update formula was used. In addition, in this case quite standard values were used: constriction factor equal to 0.7298, social component equal to 2.05, a cognitive component equal to 2.05, maximum allowed particle velocities normalized respect to the bounds width equal to 0.5 and 4 individuals considered as neighbors. In the GA case, an exponential crossover operator with a probability equal to 0.90, a polynomial mutation operator with a probability equal to 0.02 and a tournament selection operator were used.

More details about the algorithm parameters can be found in [39].

4. Results and Discussion

The simulation model solved for the default parameters shown in Table 3, predicts a tensile strength value equal to 418.14 MPa, which corresponds to a relative error equal to 24% (see Equation (10)). The objective is to evaluate the ability of the algorithms to reduce that error to values below 0.1%, modifying the variables shown in Table 2.

$$\text{Error} = \frac{|418.14 - 552|}{552} = 0.2425, \tag{10}$$

Figure 8 collects the convergence evolution of each algorithm for the different cases studied. The behavior of the deterministic algorithms was uneven. The performance of the compass search was very good, reducing the error to values below the 0.1% in all cases with a reasonable number of calculations. The use of 20 initial random points has improved the algorithm performance reducing the number of calculations, especially in the 10 variables case where the difference is clear.

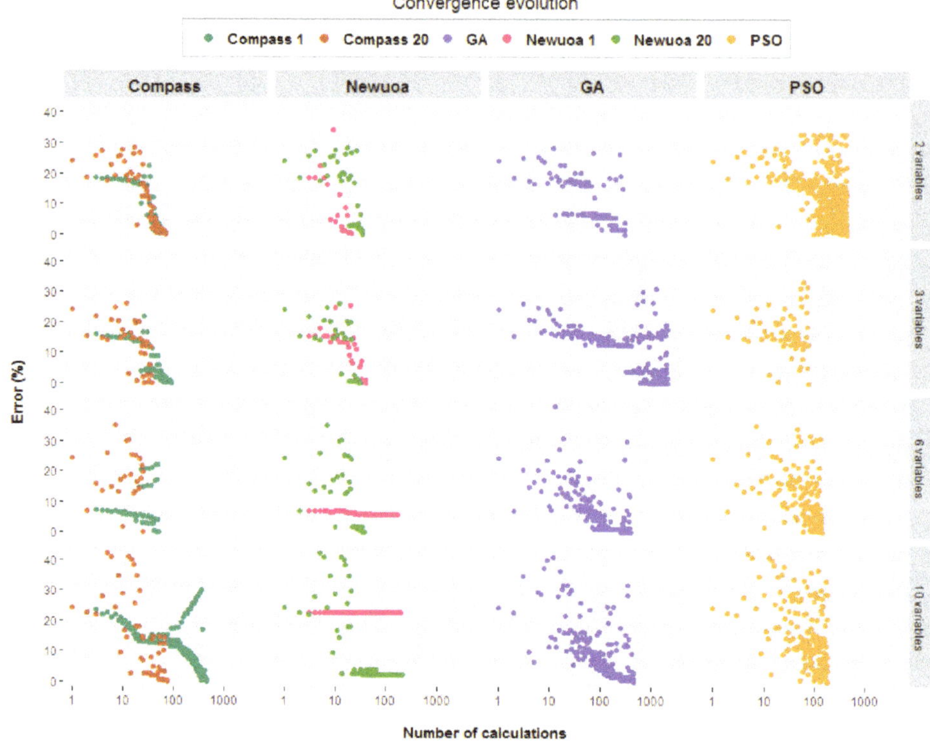

Figure 8. Convergence evolution.

NEWUOA algorithm instead, failed in reaching the error target in several cases. Using an initialization based on only one point, the algorithm has failed in the 6 and 10 variables cases with error values equal to 22.62% and 5.77%, respectively. The use of 20 random initial points clearly has improved the behavior of the algorithm making possible to reach the target in the 6 variables case and to reduce the error value to 2.07% in the 10 variables case.

The heuristic algorithms, GA and PSO, have succeeded in reaching the error target in all the cases, with the exception of the GA in the 3 variables case, where the minimum error value was equal to 0.124%. The number of calculations needed to reach the target error was lower for PSO than for GA, except for the 2 variables case.

Therefore, the best options were the compass search algorithm with 20 random initial points, followed by the PSO algorithm.

A priori, it can be thought that as higher the number of variables, larger the search domain and higher the difficulty of the algorithm to reach the target. However, a higher number of variables may also imply a bigger number of solutions that fulfil the target value, making easy to find one of them. On the other side, a reduced number of variables may be too restrictive limiting the number of possible solutions and making difficult to reach the target. The results obtained are not clear in this sense. There is a tendency in the deterministic algorithms to find more difficulties in cases with higher number of variables in contrast with the heuristics, which have found more difficulties in cases with a lower number of variables. However, the obtained results are not conclusive in this sense.

Relating to the values assigned to the variables, as it was explained in the introduction, it does not exist a unique solution of the problem. Therefore, the model may be fitted assigning different combinations of the values assigned to the variables. The risk of this fact is that sometimes, the values assigned to the variables by the algorithm may cause a certain loss in the physical sense of the model although the results are accurate from the mathematical point of view. However, this is a common difficulty of this type of approximation that has not been solved by now.

Figure 9 shows the distribution of the values assigned to the different variables in boxplot form. Black points represent the values assigned for each variable jittered, the blue line corresponds to the value assigned for each variable in the initial model and red lines to the bounds fixed for each of them (In the case of NEWUOA no bounds are used). As can be observed the distribution is quite different depending on the adjusted variable.

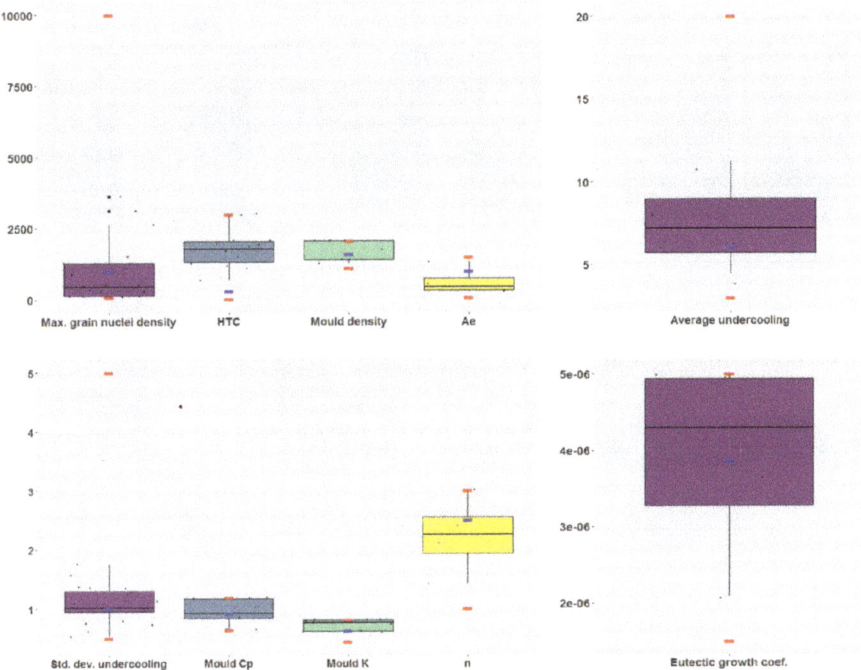

Figure 9. Distribution of the values assigned to the different variables in boxplot form.

Figure 10 collects the values assigned grouped by the number of variables included in the fitting and colored by the algorithm used in each case. Colored bars represent the values assigned to the variables by the different optimization algorithms, the dotted red lines represent the bounds fixed for the corresponding variable and the dotted blue line the value initially assigned.

As expected, the higher the number of variables included in the adjustment, the higher the variability in the values assigned. This can be observed in the cases of the maximum grain nuclei density, the undercooling standard deviation and the heat transfer coefficient. Otherwise, the variability of the values assigned to the variable seems also related with the algorithm used. For example, compass search presents a bigger tendency to adjust the variable to higher values. However, as the "true" value of each variable is actually unknown, it is difficult to evaluate if any of the algorithms has a better performance in assigning more reliable values to the variables.

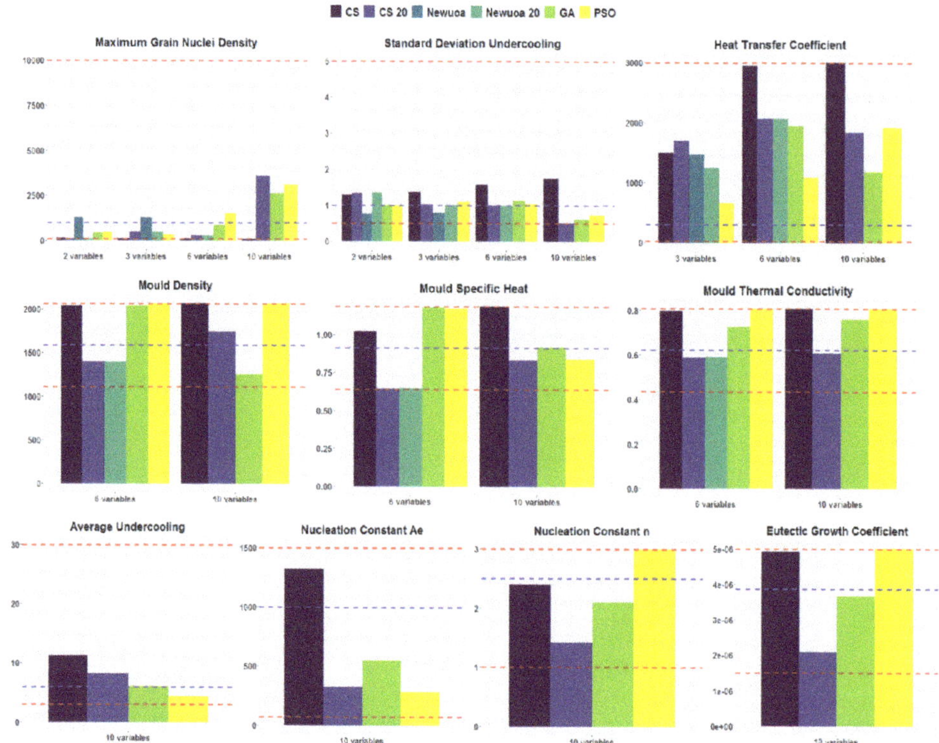

Figure 10. Values assigned to variables grouped by the number of variables included in the fitting and colored by the algorithm.

5. Conclusions

Only compass search and PSO have succeeded in reaching the error target in the four cases studied (2, 3, 6 and 10 variables). The NEWUOA and GA algorithms have failed in some case.

In the case of the deterministic algorithms, compass search and NEWUOA, the use of a multiple random initial guess was clearly beneficious.

Compass search with an initial 20 random points has presented the best performance, reaching the error target in all cases with the lower number of calculations.

The known problem related with the absence of a unique solution was found, obtaining different combinations of the values assigned to the variables. However, as the 'true' value of each variable is actually unknown, it is difficult to evaluate if any of the algorithms has a better performance in assigning more reliable values to the variables.

Future works. The authors believe that the research line focused in the fitting of metal casting models by means of optimization algorithms is very interesting, as the advances in this area may contribute to reduce the time needed to adjust the metal casting models improving the confidence in their predictions and contributing to reduce tests at laboratory scale. In fact, some of the future works planned are:

- Model fitting based on several properties, for example ultimate and yield strengths. It would be a very interesting improvement in this type of technique, although it would imply to tackle a multi-objective optimization, much more challenging that the classical optimization presented in this work;

- Evaluate the results obtained in this study for a completely different type of material (for example some aluminum alloy or superalloy) and/or for a different manufacturing casting process (for example die casting or investment casting);
- Include the possibility of using temperature dependent variables in the model fitting.

Author Contributions: Conceptualization, E.A.; data curation, A.O. and E.V.; formal analysis, E.A.; investigation, E.A., A.M., A.O. and E.V.; methodology, E.A.; resources, A.M.; software, E.A.; validation, A.M.; writing—original draft, E.A.; writing—review & editing, E.A., A.M. and I.G. All authors have read and agreed to the published version of the manuscript.

Funding: This research was funded by the Basque Government under the ELKARTEK Program (ARGIA Project, ELKARTEK KK-2019/00068) and by the HAZITEK Program (CASTMART Project, HAZITEK ZL-2019/00562).

Conflicts of Interest: The authors declare no conflict of interest.

Nomenclature

CE	Carbon equivalent	n_{max}	Maximum grain nuclei density
FEM	Finite element method	p_i	Previous position vector in PSO algorithm
FVM	Finite volume method	$\hat{p}$	Modified pressure (Pa)
HTC	Heat transfer coefficient (W/m^2 K)	Q	Quadratic approximation of F function
SGI	Spheroidal graphite iron / ductile iron	$\dot{R}_q$	Heat generation per unit mass (W/kg)
BFGS	Broyden–Fletcher–Goldfarb–Shanno	R_g	Graphite radius
GA	Genetic algorithm	T	Temperature (°C)
GCM	Globally convergent method	T_0	Boundary temperature (°C)
LHS	Latin hypercube sampling	t	Time (s)
MAP	Maximum a posteriori	v_i	Velocity vector in PSO algorithm
PSO	Particle swarm optimization	v	Velocity (m/s)
SAE	Self-adaptive evolution	x_k	X coordinate of a point
A_e	Graphite nodules nucleation constant	x_i	Current position vector in PSO algorithm
C_p	Specific heat (J/kgK)	ΔT_δ	Standard deviation of the undercooling (°C)
fitness	Fitness function	ΔT_n	Average undercooling (°C)
F	Generic function	β_T	Volumetric thermal expansion coefficient (°C^{-1})
f_α	Ferrite fraction	ρ	Density (kg/m^3)
f_g	Graphite fraction	ρ_0	Density at reference temperature (kg/m^3)
f_p	Perlite fraction	μ_e	Eutectic growth coefficient
g	Gravity (m/s^2)	μ_l	Shear viscosity (Pa·s)
h	Specific enthalpy (J/kg)	$\sigma_{predicted}$	Predicted tensile strength (Pa)
k	Thermal conductivity (W/mK)	$\sigma_{reference}$	Measured tensile strength (Pa)
n	Graphite nodules nucleation constant	σ_{ULT}	Ultimate strength (Pa)

References

1. Horr, A.M.; Kronsteiner, J. On numerical simulation of casting in new foundries: Dynamic process simulations. *Metals* **2020**, *10*, 886. [CrossRef]
2. Diez-Olivan, A.; Del Ser, J.; Galar, D.; Serra, B. Data fusion and machine learning for industrial prognosis: Trends and perspectives towards Industry 4.0. *Inf. Fusion* **2019**, *50*, 92–111. [CrossRef]
3. ASM. *International Handbook Committee ASM Handbook, Metals Process Simulation*, 1st ed.; Furrer, D.U., Semiatin, S.L., Eds.; ASM International: Novelty, OH, USA, 2010; Volume 22B, ISBN 978 1 61503 005 7.
4. Gadala, M.S.; Vakili, S. Assesment of Various Methods in Solving Inverse Heat Conduction Problems. In *Heat Conduction—Basic Research*; Vikhrenko, V., Ed.; InTech: Rijeka, Croatia, 2011; pp. 37–62. ISBN 978-953-307-404-7.
5. Beck, J.V.; Blackwell, B.; St. Clair, C.R. *Inverse Heat Conduction: Ill-Posed Problems*; John Wiley & Sons Inc: Hoboken, NJ, USA, 1985; ISBN 0-471-08319-4.
6. Özisik, M.N.; Orlande, H.R.B. *Inverse Heat Transfer: Fundamentals and Applications*; Taylor & Francis: New York, NY, USA, 2000; ISBN 1-56032-838-X.

7. Klement, J. On using quasi Newton algorithms of the Broyden class for model-to-test correlation. In Proceedings of the 28th European Space Thermal Analysis Workshop, Noordwijk, The Netherlands, 14–15 October 2014; ESA, Ed.; ESA Publications Division: Noordwijk, The Netherlands, 2014; pp. 213–228.

8. Klement, J. On Using Quasi-Newton Algorithms of the Broyden Class for Model-to-Test Correlation. *J. Aerosp. Technol. Manag.* **2014**, *6*, 407–414. [CrossRef]

9. Torralbo, I.; Perez-Grande, I.; Sanz-Andrés, A.; Piqueras, J. Correlation of Thermal Mathematical Models to test data using Jacobian matrix formulation. In Proceedings of the 48th International Conference on Environmental Systems ICES-2018, Alburquerque, NM, USA, 8–12 July 2018.

10. Torralbo, I.; Perez-Grande, I.; Sanz-Andres, A.; Piqueras, J. Correlation of spacecraft thermal mathematical models to reference data. *Acta Astronaut.* **2017**. [CrossRef]

11. Dudon, J.P.; Pasquier, H.M. Evaluation of stochastic & statistics methods for spacecraft thermal analysis. In Proceedings of the 25th European Workshop on Thermal and ECLS Software, Noordwijk, The Netherlands, 8–9 November 2011; ESA, Ed.; ESA Publications Division: Noordwijk, The Netherlands, 2011; pp. 319–335.

12. Beck, T.; Bieler, A.; Thomas, N. Numerical thermal mathematical model correlation to thermal balance test using adaptive particle swarm optimization (APSO). *Appl. Therm. Eng.* **2012**, *38*, 168–174. [CrossRef]

13. Van Zijl, N.; Zandbergen, B.; Benthem, B. Correlating thermal balance test results with a thermal mathematical model using evolutionary algorithms. In Proceedings of the 27th European Space Thermal Analysis Workshop, Noordwijk, The Netherlands, 3–4 December 2013; ESA, Ed.; ESA Publications Division: Noordwijk, The Netherlands, 2013; pp. 89–108.

14. Trinoga, M. Development of an automated thermal model correlation tool. In Proceedings of the 28th European Space Thermal Analysis Workshop, Noordwijk, The Netherlands, 14–15 October 2014; ESA, Ed.; ESA Publications Division: Noordwijk, The Netherlands, 2014; pp. 201–212.

15. Frey, B.; Trinoga, M.; Hoppe, M.; Ebeling, W.D. Development of an Automated Thermal Model Correlation Method and Tool. In Proceedings of the 45th International Conference on Environmental Systems, ICES 2015, Washington, DC, USA, 12–16 July 2015; ICES STEERING COMMITTEE, Ed.; Texas Tech University: Lubbock, TX, USA, 2015; pp. 1–14.

16. Anglada, E.; Garmendia, I. Correlation of thermal mathematical models for thermal control of space vehicles by means of genetic algorithms. *Acta Astronaut.* **2015**, *108*, 1–17. [CrossRef]

17. Garmendia, I.; Anglada, E. Thermal mathematical model correlation through genetic algorithms of an experiment conducted on board the International Space Station. *Acta Astronaut.* **2016**, *122*, 63–75. [CrossRef]

18. Klement, J.; Anglada, E.; Garmendia, I. Advances in automatic thermal model to test correlation in space industry. In Proceedings of the 46th International Conference on Environmental Systems, ICES 2016, Vienna, Austria, 10–14 July 2016; ICES STEERING COMMITTEE, Ed.; Texas Tech University: Lubbock, TX, USA, 2016; pp. 1–11.

19. Anglada, E.; Martinez-Jimenez, L.; Garmendia, I. Performance of Gradient-Based Solutions versus Genetic Algorithms in the Correlation of Thermal Mathematical Models of Spacecrafts. *Int. J. Aerosp. Eng.* **2017**, *2017*, 1–12. [CrossRef]

20. De Palo, S.; Malosti, T.; Filiddani, G. Thermal Correlation of BepiColombo MOSIF 10 Solar Constants Simulation Test. In Proceedings of the 25th European Workshop on Thermal and ECLS Software, Noordwijk, The Netherlands, 8–9 November 2011; ESA, Ed.; ESA Publications Division: Noordwijk, The Netherlands, 2011; pp. 271–284.

21. Cheng, W.; Liu, N.; Li, Z.; Zhong, Q.; Wang, A.; Zhang, Z.; He, Z. Application study of a correction method for a spacecraft thermal model with a Monte-Carlo hybrid algorithm. *Chin. Sci. Bull.* **2011**, *56*, 1407–1412. [CrossRef]

22. Opstelten, I.J.; Rabenberg, J.M. Determination of the thermal boundary conditions during aluminium DC casting from experimental data using inverse modelling. In *Essential Readings in Light Metals*; Grandfield, J.F., Eskin, D.G., Eds.; Springer: Berlin/Heidelberg, Germany, 2016; p. 7.

23. Leder, M.O.; Gorina, A.V.; Kornilova, M.A.; Tarenkova, N.Y.; Kondrashov, E.N. Determination of the Thermophysical Properties of Titanium Alloys from Liquid Bath Profiles. *Russ. Metall.* **2015**, *12*, 964–969. [CrossRef]

24. Torroba, A.J.; Koeser, O.; Calba, L.; Maestro, L.; Carreño-Morelli, E.; Rahimian, M.; Milenkovic, S.; Sabirov, I.; LLorca, J. Investment casting of nozzle guide vanes from nickel-based superalloys: Part I—Thermal calibration and porosity prediction. *Integr. Mater. Manuf. Innov.* **2014**, 3–25. [CrossRef]

25. Jin, H.; Li, J.; Pan, D. Application of inverse method to estimation of boundary conditions during investment casting simulation. *Acta Metall. Sin. (Engl. Lett.)* **2009**, *22*. [CrossRef]
26. Rappaz, M.; Bellet, M.; Deville, M. Inverse Methods. In *Numerical Modeling in Materials Science and Engineering*; Springer Series in Computational Mathematics; Springer: Berlin/Heidelberg, Germany, 2010; pp. 447–477.
27. Long, A.; Thornhill, D.; Armstrong, C.; Watson, D. Determination of the heat transfer coefficient at the metal-die interface for high pressure die cast AlSi9Cu3Fe. *Appl. Therm. Eng.* **2011**, *31*, 3996–4006. [CrossRef]
28. Anglada, E.; Meléndez, A.; Vicario, I.; Arratibel, E.; Aguillo, I. Adjustment of a High Pressure Die Casting Simulation Model against Experimental Data. *Procedia Eng.* **2015**, *132*, 966–973. [CrossRef]
29. Meléndez, A.; Anglada, E.; Maestro, L.; Dominguez, I. More robust processes and more added value for foundries based on inverse modelling and tailor-made software tools. In Proceedings of the 71st World Foundry Congress—Advanced Sustainable Foundry, Bilbao, Spain, 19–21 May 2014; Tabira Foundry Institute, WFO, IK4 Azterlan: Bilbao, Spain, 2014.
30. Meléndez, A.; Anglada, E.; Rodriguez, P.P.; Armenteros, A. Investment Casting Moulds Optimisation by means of Advanced Simulation and Instrumented Fibre-wrapped Moulds. In Proceedings of the European Investment Casters' Federation (EICF). Technical Workshop for Foundry Engineers—Method Engineering & Process Modelling, Rotherham, UK, 12–13 May 2015.
31. Anglada, E.; Meléndez, A.; Maestro, L.; Dominguez, I. Adjustment of Numerical Simulation Model to the Investment Casting Process. *Procedia Eng.* **2013**, *63*, 75–83. [CrossRef]
32. Anglada, E.; Meléndez, A.; Maestro, L.; Dominguez, I. Finite Element Model Correlation of an Investment Casting Process. *Mater. Sci. Forum. Adv. Mater. Process. Technol. MESIC V* **2014**, *797*, 105–110. [CrossRef]
33. Ilkhchy, A.F.; Jabbari, M.; Davami, P. Effect of pressure on heat transfer coefficient at the metal/mold interface of A356 aluminum alloy. Int. Commun. *Heat Mass Transf.* **2012**, *39*, 705–712. [CrossRef]
34. Beck, J. V Nonlinear estimation applied to the nonlinear inverse heat conduction problem. *Int. J. Heat Mass Transf.* **1970**, *13*, 703–716. [CrossRef]
35. Dong, Y.; Bu, K.; Dou, Y.; Zhang, D. Determination of interfacial heat-transfer coefficient during investment-casting process of single-crystal blades. *J. Mater. Process. Technol.* **2011**, *211*, 2123–2131. [CrossRef]
36. Zhang, L.; Li, L.; Ju, H.; Zhu, B. Inverse identification of interfacial heat transfer coefficient between the casting and metal mold using neural network. *Energy Convers. Manag.* **2010**, *51*, 1898–1904. [CrossRef]
37. Zhang, W.; Xie, G.; Zhang, D. Application of an optimization method and experiment in inverse determination of interfacial heat transfer coefficients in the blade casting process. *Exp. Therm. Fluid Sci.* **2010**, *34*, 1068–1076. [CrossRef]
38. ProCAST. Version 2018.0. ESI Group: Paris, France. Available online: https://www.esi-group.com/products/casting (accessed on 1 July 2020).
39. Biscani, F.; Izzo, D. Esa/Pagmo2: Pygmo 2.15.0 (Version v2.15.0). Zenodo. Available online: https://ui.adsabs.harvard.edu/abs/2019zndo...2529931B/abstract (accessed on 2 April 2020).
40. Fragassa, C.; Babic, M.; Perez Bergman, C.; Minak, G. Predicting the Tensile Behaviour of Cast Alloys by a Pattern Recognition Analysis on Experimental Data. *Metals* **2019**, *9*, 557. [CrossRef]
41. Stefanescu, D.M.; Catalina, A.V.; Guo, X.; Chuzhoy, L.; Pershing, M.A.; Biltgen, G.L. Prediction of room temperature microestructure and mechanical properties in ductile iron castings. In Proceedings of the 4th Decennial International Conference on Solidification Processing, Sheffield, UK, 7–10 July 1997; p. 7.
42. PanFe. Version 2017. CompuTherm LLC: Middleton, WI, USA. Available online: https://computherm.com (accessed on 1 July 2020).
43. Chen, Q. CALPHAD and Beyond—The True Story of Materials Genome. In Proceedings of the International Forum on Advanced Materials, IFAM 2016, Nanjing, China, 24–26 September 2016; p. 22.
44. Rappaz, M.; Thevoz, P.H. Solute diffusion model for equiaxed dendritic growth: Analytical solution. *Acta Metall.* **1987**, *35*, 2929–2933. [CrossRef]
45. Oldfield, W. Freezing of Cast Irons. *ASM Trans.* **1966**, *59*, 945–959.
46. Dantzig, J.A.; Rappaz, M. *Solidification*; Dantzig, J.A., Rappaz, M., Eds.; CRC Press: Boca Raton, FL, USA, 2009; ISBN 978 0 8493 8238 3.
47. Lewis, R.W.; Ravindran, K. Finite element simulation of metal casting. *Int. J. Numer. Methods Eng.* **2000**, *47*, 29–59. [CrossRef]

48. Kolda, T.G.; Lewis, R.M.; Torczon, V. Optimization by Direct Search: New Perspectives on Some Classical and Modern Methods. *SIAM Rev.* **2003**, *45*, 385–482. [CrossRef]
49. Powell, M.J.D. The NEWUOA software for unconstrained optimization with derivatives. In *Large-Scale Nonlinear Optimization*; Di Pillo, G., Roma, M., Eds.; Springer: New York, NY, USA, 2006; pp. 255–297. ISBN 978-0387-30063-4.
50. Poli, R.; Kennedy, J.; Blackwell, T. Particle swarm optimization. *Swarm Intell.* **2007**, *1*, 33–57. [CrossRef]
51. Herrera, F.; Lozano, M.; Verdegay, J.L. Tackling Real-Coded Genetic Algorithms: Operators and Tools for Behavioural Analysis. *Artif. Intell. Rev.* **1998**, *12*, 265–319. [CrossRef]

Article

Analysis of Melt-Pool Behaviors during Selective Laser Melting of AISI 304 Stainless-Steel Composites

Daniyal Abolhasani [1,2], S. M. Hossein Seyedkashi [2], Namhyun Kang [3], Yang Jin Kim [1], Young Yun Woo [1] and Young Hoon Moon [1,*]

[1] School of Mechanical Engineering, Pusan National University, Busan 46241, Korea
[2] Department of Mechanical Engineering, University of Birjand, Birjand 97175-376, Iran
[3] School of Material Science and Engineering, Pusan National University, Busan 46241, Korea
* Correspondence: yhmoon@pusan.ac.kr; Tel.: +82-51-510-2472

Received: 29 June 2019; Accepted: 6 August 2019; Published: 8 August 2019

Abstract: The melt-pool behaviors during selective laser melting (SLM) of Al_2O_3-reinforced and a eutectic mixture of Al_2O_3-ZrO_2-reinforced AISI 304 stainless-steel composites were numerically analyzed and experimentally validated. The thermal analysis results show that the geometry of the melt pool is significantly dependent on reinforcing particles, owing to the variations in the melting point and the thermal conductivity of the powder mixture. With the use of a eutectic mixture of Al_2O_3-ZrO_2 instead of an Al_2O_3 reinforcing particle, the maximum temperature of the melt pool was increased. Meanwhile, a negligible corresponding relationship was observed between the cooling rate of both reinforcements. Therefore, it was identified that the liquid lifetime of the melt pool has the effect on the melting behavior, rather than the cooling rate, and the liquid lifetime increases with the eutectic ratio of Al_2O_3-ZrO_2 reinforcement. The temperature gradient at the top surface reduces with the use of an Al_2O_3-ZrO_2 reinforcement particle due to the wider melt pool. Inversely, the temperature gradient in the thickness direction increases with the use of an Al_2O_3-ZrO_2 reinforcement particle. The results of melt-pool behaviors will provide a deep understanding of the effect of reinforcing particles on the dimensional accuracies and properties of fabricated AISI 304 stainless-steel composites.

Keywords: selective laser melting; additive manufacturing; SLM; FEM; Al_2O_3; reinforced; Al_2O_3-ZrO_2; 304; stainless; composite

---

1. Introduction

Austenitic stainless steels can find more applications if their strength is improved. Hence, some studies have investigated the reinforcement of stainless steel matrices via the selective laser melting (SLM) process [1,2]. Reinforcing can be achieved through the use of Al_2O_3 to produce a metal matrix composite that exhibits high mechanical properties, in addition to high corrosion and wear resistance. However, Al_2O_3 particle has very high melting temperatures and very low thermal conductivities, which result in a very high thermal gradient during the process, causing high local stresses and crack formation [3]. The use of a eutectic mixture of Al_2O_3 and ZrO_2 powders decreases the melting temperature of Al_2O_3 from 2313 K to about 2133 K [3,4]. If an optimum volume can be identified for the samples reinforced with a eutectic mixture of $Al2O_3$-ZrO_2, it can be a good alternative to strengthen the AISI 304 stainless steel.

To provide a deep understanding in relation to composites reinforced with alumina particles, this paper presents an innovative numerical study dealing with melt-pool behaviors during selective laser melting (SLM) of Al_2O_3-reinforced and a eutectic mixture of Al_2O_3-ZrO_2-reinforced AISI 304 stainless-steel composites. As is well known, the dimensional accuracies and properties of fabricated composite parts are significantly dependent on the melt-pool behaviors [2–4].

SLM is a manufacturing technique to construct three-dimensional parts in which a high-power density laser is used to melt and fuse metallic powders [4–7]. Compared with other traditional techniques, laser processing typically does not require mechanical tooling and therefore exhibits high flexibility [8–12]. In SLM, the energy density should be sufficient to ensure both the melting of the powder and the bonding of the underlying substrate. If the bonds between the scan tracks and layers are weak, defects such as cracks may be generated [13–16]. The trapped gas in the melt pool can increase the number of pores in SLM-fabricated parts. Additionally, the reduction in solubility of the element during the rapid melting and cooling process can cause detrimental defects [17]. Schleifenbaum et al. [18] found that with the increase in the laser power, the rate of material evaporation and the incidence of spattering increased. These defects should be prevented by identifying appropriate process parameters of the SLM process.

The prediction of the temperature evolution in SLM is commonly performed using the finite-element (FE) method or statistical approaches. The temperature and stress fields in samples of 90W-7Ni-3F and 316L stainless steels were predicted by Zhang et al. [19] and Hussein et al. [20]. Dai and Shaw [21] simulated the transient temperature, as well as the thermal and residual stress fields. Kundakcioglu et al. [22] performed transient thermal modeling of laser-based Additive Manufacturing (AM) for 3D freeform structures.

In the case of composites, AlMangour et al. [23] presented a simulation model for predicting the temperature evolution of the melt pool of TiC/SUS316L samples. Shi et al. [24] investigated the effects of the laser power and scan speed on the melting and solidification mechanisms during SLM of TiC/Inconel718 via a simulation approach. Li et al. [1] investigated the microstructural and mechanical properties of Al_2O_3/316L stainless-steel metal-matrix composites (MMCs) fabricated via SLM through experimental and numerical methods.

The aforementioned studies focused on a single reinforcement particle in the case of SLM of a metal matrix composite. They did not consider an SLM composite with binary-phase reinforcement, such as a eutectic mixture. The thermal cycle in the SLM process increases the complexity of the material melting and thermal behavior, if new combinations of the reinforcements are applied [25]. The melting temperature of the powder mixture is an important microstructural feature in the thermal process, as it affects the final temperature of the surface during the heating or cooling of the melt pool. No detailed analysis involving the 3D FE modeling of a selective-laser-melted part, including the metal matrix and binary-phase reinforcement and its relationship with the geometric features of the melt pool during the process, has been performed.

In this study, a eutectic mixture of Al_2O_3 and ZrO_2 powder particles was added to AISI 304 stainless steel as the metal-matrix media in a set of numerical simulation runs, incorporating experimental runs as validation tests. The use of a eutectic mixture of Al_2O_3 and ZrO_2 powders reduces the melting temperature of Al_2O_3 particles. Hence, for a better understanding of the results, both sets of parts modeled with various weight percentages of Al_2O_3 and Al_2O_3-ZrO_2 particles were considered.

For this investigation, the melt-pool dimensions and the thermal evolution of AISI 304-Al_2O_3 and AISI 304-Al_2O_3-ZrO_2 SLM composites were simulated, and the predicted thermal results were described.

2. Materials and SLM System

In this study, the eutectic mixture was prepared using powders containing 58.5 wt% Al_2O_3 and 41.5 wt% ZrO_2, corresponding to the eutectic point of the Al_2O_3-ZrO_2 binary phase diagram [4], with the reinforcement content within the AISI 304 stainless-steel powder as the metal matrix. An experiment was performed using an SLM system with a continuous-wave, ytterbium fiber laser (IPG YLR-200, IPG Photonics, Burbach, Germany), as shown in Figure 1. The maximum available power of 200 W at 6 A was used, and the laser scanning was controlled using a scanner (hurrySCAN®20, SCANLAB, Puchheim, Germany). Argon gas was used as a shielding gas to prevent oxidation in the melt pool. The layering bar spread the powder, and the build cylinder moved vertically to control the powder-bed height. During the numerical and experimental runs, a 30-μm-thick layer of powder mixture was used. The laser spot diameter was fixed at 80 µm. The laser power and scanning velocity were selected after conducting a large number of preliminary experiments associated with the single-line formation tests.

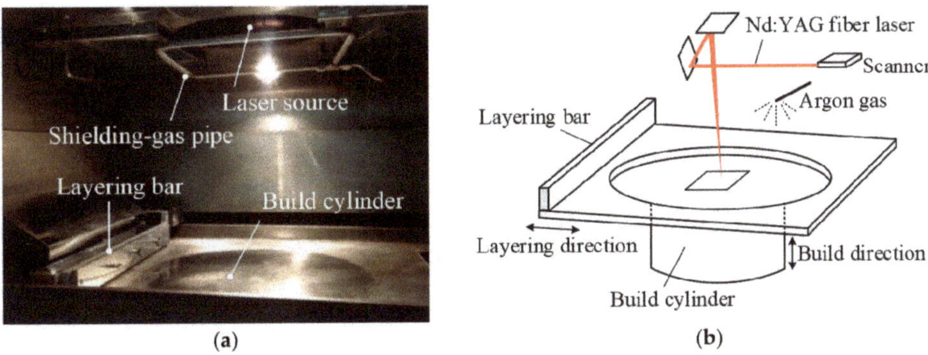

Figure 1. Selective laser melting (SLM) system: (**a**) Experimental setup; (**b**) Schematic illustration.

The laser power and scan speed were 200 W and 732 mm/s, respectively. For comprehensive comparison, various weight percentages of Al_2O_3 and the eutectic mixture of Al_2O_3-ZrO_2 were employed in the reinforcing experiments, as shown in Table 1.

Table 1. Numerical and experimental runs.

Ex. No.	wt% of Al_2O_3	wt% of Al_2O_3-ZrO_2	wt% of 304SS
1	0	0	100%
2	3%	-	97%
3	5%	-	95%
4	7%	-	93%
5	-	3%	97%
6	-	5%	95%
7	-	7%	93%

3. FE Modeling

3.1. Physical Description of Model

A 3D model was developed, and the ABAQUS^{TM} commercial software (version 6-14, Dassault Systems, Providence, RI, USA) was utilized to simulate volumetric laser energy deposition with a Gaussian distribution. A schematic of the SLM process, which shows the interactions between the laser beam and powder bed, as well as the melting-pool dimensions and solidification, is presented in Figure 2a. A square composite powder layer with dimensions of 7 mm × 7 mm × 0.030 mm was placed on a stainless-steel substrate. The powder bed was meshed with eight-node linear hexahedral elements with a size of 70×10^{-6} m that were distributed uniformly throughout the powder-bed model, as shown in Figure 2b. The sweep method was used to mesh the substrate and accurately capture the flux distribution of the moving laser beam.

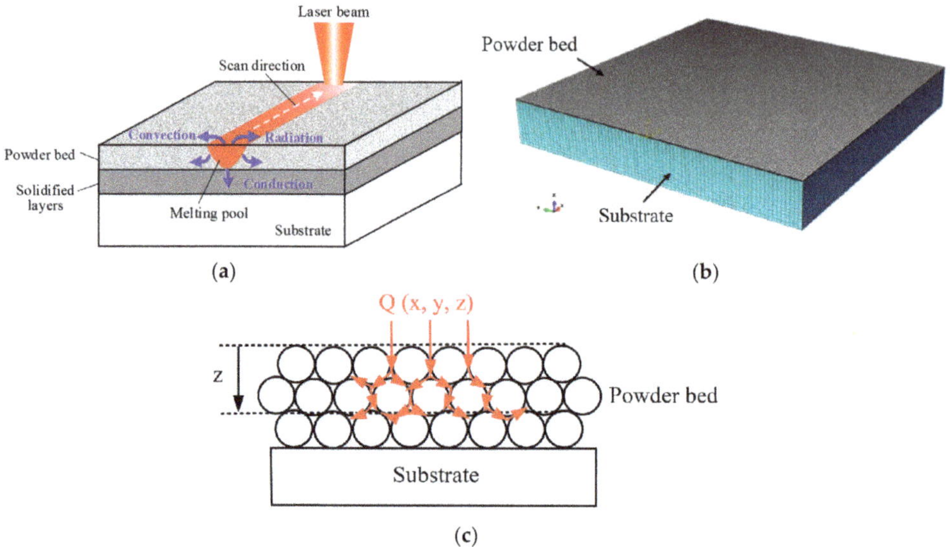

(a)

(b)

(c)

Figure 2. (a) Schematic showing the thermal behavior of the powder bed under laser irradiation. **(b)** 3D finite-element (FE) model. **(c)** Schematic showing the transfer of laser radiation into the powder bed.

3.2. Initial Governing Equation

In the powder-bed additive layer manufacturing system with a moving laser source, the heat-transport equation for 3D Cartesian coordinate systems is expressed as follows [22]:

$$\rho{\cdot}C_p{\cdot}\frac{\partial T}{\partial t} = \frac{\partial}{\partial x}(k{\cdot}\frac{\partial T}{\partial x}) + \frac{\partial}{\partial y}(k{\cdot}\frac{\partial T}{\partial y}) + \frac{\partial}{\partial z}(k{\cdot}\frac{\partial T}{\partial z}) + \rho{\cdot}C_p{\cdot}v{\cdot}\frac{\partial T}{\partial x} + Q, \tag{1}$$

where T is the temperature (K), ρ is the density (kg/m^3), C_p is the specific heat (J/(kg·K)), k is the thermal conductivity (W/m K), and ∇ is the differential or gradient operator. Q is the rate of internal energy conversion per unit volume (referred to as the source term, W/m^3).

The corresponding boundary condition was [20]:

$$-\frac{\partial T}{\partial z} = \frac{h}{k}(T - T_0) + \frac{\varepsilon\, 6}{k}(T^4 - T_0{}^4), \tag{2}$$

where T is the surface temperature of the powder bed, z is the axis perpendicular to the powder surface, $h = 200$ [26] is the forced-convection heat-transfer coefficient (W/m^2 K) due to argon gas, $\varepsilon = 0.8$ is the emissivity of the heated surface, and $6 = 5.6703 \times 10^{-8}$ W/m^{2}K^4 is the Stefan–Boltzmann constant.

3.3. Heat-Source Model

The travel distance of the laser beam into the powder media (z), which is schematically presented in Figure 2c, can be modeled using a volumetric heat source ($Q(x, y, z)$) [27]:

$$Q(x,\, y,\, z) = (1 - R)\, n\, A\, \frac{P}{\pi r_0{}^2}\, exp(\frac{-n(x^2 + y^2)}{r_0{}^2})\, exp(-\frac{1}{d}\, z), \tag{3}$$

where R is the surface reflectivity of the powder mixture, and n is a shape parameter of the heat-flux distribution, $n = 2$. A is the laser-absorption coefficient of the irradiated surface and it was modified during numerical validation. P is the laser power (W), r_0 is the laser-beam radius (m), x and y are the

Cartesian coordinates at the surface, and z is the Cartesian coordinate perpendicular to the powder bed. d is the penetration depth equal to 40 µm for the SLM process [27].

The surface reflectivity of the powder mixture (R) is expressed as follows [25,28]:

$$R = \frac{\sum_{i=1}^{n} \frac{R_{(i)}\, \beta_{(i)}\, w_{(i)}}{\rho_{(i)}}}{\sum_{i=1}^{n} \frac{\beta_{(i)}\, w_{(i)}}{\rho_{(i)}}},$$ (4)

Here, in accordance with the three components utilized in this study (AISI 304, Al_2O_3, and ZrO_2), n is 3; i is a subscript referring to a specific mixture component; $R_{(i)}$ is the surface reflectance of component i; $\beta_{(i)}$ is the extinction coefficient of component i, $\rho_{(i)}$ is the density of component i, and $w_{(i)}$ is the weight percentage of component i in the mixture.

The extinction coefficient β is defined as [29]:

$$\beta = \frac{3\,(1-\phi)}{2\,\phi\, D},$$ (5)

where ϕ is the porosity of the bed, and D is the average particle diameter of the powder. The powder density was calculated using the following equation [26]:

$$\rho_{powder} = (1 - \phi)\, \rho_{bulk},$$ (6)

The laser-absorption coefficient of a powder bed with a Gaussian distribution (A_{powder}) was estimated using the absorption coefficient of the bulk material (A_{bulk}) [30]:

$$A_{powder} = 0.0413 + 2.89A_{bulk} - 5.36A^2{}_{bulk} + 4.50A^3{}_{bulk},$$ (7)

The values of A in Equation (3) were calculated according to the linear rule of mixtures [31]:

$$A = w_1 \cdot A_1 + w_2 \cdot A_2 + w_3 \cdot A_3,$$ (8)

where w_1, w_2, and w_3 represent the volumetric contents of the three components in the mixture, and A_1, A_2, and A_3 represent the absorption coefficients of the components ($\sum_{i=1}^{3} w_i = 1, 0 < A < 1$). The final values of A applied in Equation (3) were obtained through numerical validations, as discussed in Section 4.1.

The parameters presented in Table 2 were obtained from the material supplier and previous works [26,32], as well as an online source. Then, the corresponding values were calculated using Equations (4)–(8).

Table 2. Parameters used for obtaining the values of the surface reflectivity (R) and absorption coefficient (A) of the powder mixture.

Powder	$R_{(i)}$	$D_{(i)}$ (m)	$\phi_{(i)}$	ρ_{bulk} (kg/m^3)	ρ_{powder} (kg/m^3)	A_{powder}
AISI 304	0.46	20×10^{-6}	0.25	7900	5861.8	0.604
Al_2O_3	0.79	3×10^{-6}	0.65	3970	1580	0.173
ZrO_2	0.82	1×10^{-6}	0.21	6000	4740	0.266

3.4. Physical Properties of Materials

The values of k are expressed by the following equations [29,31,33], where T is the temperature and the superscripts s and m correspond to "solidus" and "melting", respectively.

$$k = \begin{cases} \text{for AISI 304, } T < T_s \; : k_{powder} = k_{bulk}(1-\phi)^n, \; n = 4 \\ \text{for Al}_2\text{O}_3 \text{ and ZrO}_2, \; T < T_s \; : \\ \dfrac{k_{powder}}{k_{atm}} = 1 - \sqrt{1-\phi}\left(1 + \dfrac{\phi\, k_{rad}}{k_{atm}}\right) + \sqrt{1-\phi}\left(\dfrac{2}{1-\frac{k_{atm}}{k_{bulk}}}\left(\dfrac{1}{1-\frac{k_{atm}}{k_{bulk}}}\ln\left(\dfrac{k_{bulk}}{k_{atm}}\right) - 1\right) + \dfrac{k_{rad}}{k_{atm}}\right), \\ \text{for all components, } T_s < T < T_m \; : \\ k = \left(\dfrac{k_{0\ bulk}\ (T_m) - k_{0\ powder}\ (T_s)}{T_m - T_s}\ (T - T_s) + k_{0\ powder}\ (T_s)\right) \times 10^{-3} \\ \text{for mixture, } T < T_s, T_s < T < T_m \; : k_{mixture} = w_1 k_1 + w_2 k_2 + w_3 k_3 \\ \text{for mixture, } T \geq T_m \; : \\ k_{mixture} = k_1\left(\dfrac{k_2\,(1+2w_2) - k_1(2w_2-2)}{k_1(2+w_2) - k_2(1-w_2)}\right), \; w_2 < w_1\,, k_1 = k_{1\ bulk}, k_2 = k_{2\ bulk} \end{cases} \tag{9}$$

Here, K_{atm} is the thermal conductivity of the ambient atmosphere and is 0.018 for argon gas; ϕ is the porosity of the powder bed; k_0 is the thermal conductivity at the ambient temperature; K_{rad} is the thermal conductivity due to the radiation among the particles in the powder bed and is evaluated as in [29]:

$$K_{rad} = 4F\cdot\sigma T^3 \cdot D, \tag{10}$$

where $F = 1/3$ is the view factor; σ is the Stefan–Boltzmann constant; T is the temperature; and D is the average particle diameter of the powder.

$$C_p = \begin{cases} \text{for all components, } T_s < T < T_m \; : C_p = C_{p0} + \left(\dfrac{1}{(T_m - T_s)\sqrt{\pi}}\, e^{-\left(\frac{T - T_m}{T_m - T_s}\right)}\right) \times L \\ \text{for mixture, } T < T_s, \; T_s < T < T_m \; : C_{p\,mixture} = w_1 C_{p1} + w_2 C_{p2} + w_3 C_{p3} \\ \text{for mixture, } T \geq T_m; \; C_p \text{ of mixture was not available} \end{cases} \tag{11}$$

Here, C_{p0} is the specific heat capacity at the ambient temperature, L is the latent heat of melting, and C_{p1}, C_{p2}, and C_{p3} are the specific heat capacities of the three components [33].

$$\rho = \begin{cases} \text{for AISI 304, } T < T_s \; : \rho = \rho_1 \\ \text{for Al}_2\text{O}_3 \text{ and ZrO}_2, \; T < T_s \; : \rho = \rho_0 \\ \text{for all components, } T_s < T < T_m \; : \rho = \left(1 - \dfrac{T - T_s}{T_m - T_s}\right)\rho_0 + \left(\dfrac{T - T_s}{T_m - T_s}\right)\rho_{0\ bulk}, \\ \text{for mixture, } T < T_s, \; T_s < T < T_m \; : \rho_{mixture} = w_1 \rho_1 + w_2\, \rho_2 + w_3\, \rho_3 \\ \text{for mixture, } T \geq T_m \; : \rho_{mixture} = w_1\, \rho_{bulk\ 1} + w_2\, \rho_{bulk\ 2} + w_3\, \rho_{bulk\ 3} \end{cases} \tag{12}$$

Here, ρ_1, ρ_2 and ρ_3 are the densities of the 304 stainless-steel, Al$_2$O$_3$ and ZrO$_2$ powder components, respectively. ρ_0 is the density of the powder component at the ambient temperature [31,33,34]. Note that $\rho_{powder} = \rho_{bulk}$ when $\phi = 0$ at the melting point.

The phase change was included in the model by using the definition of the latent heat of melting. The T_s of the component with the lowest value (1615 K for AISI 304 stainless steel) and the T_m of the component with the highest value were used for the mixture when simulating composite parts. The solution-dependent state variables of the three fields (powder, solid, and liquid) were based on the density of the mixture (ρ) using a User Defined Field (USDFLD) subroutine.

4. Results and Discussion

To analyze melt-pool behaviors, cross-sectional views of the melt pools obtained from experimental and numerical models are compared first. Additional simulation results are then presented to exhibit the temperature effects on the liquid lifetime of the melt pools. The effect of temperature gradient on microstructural features is also discussed.

4.1. Numerical Validation and Melt-Pool Characterization

To reduce the simulation time, the single track of conditions presented in Table 1 was modeled. The numerical model was validated using the calculated dimensions of the melt pool and compared with the experimental results of the single-line formation test. Image processing was performed on the cross-sectional optical micrograph to evaluate the melt-pool morphology, as shown in Figure 3.

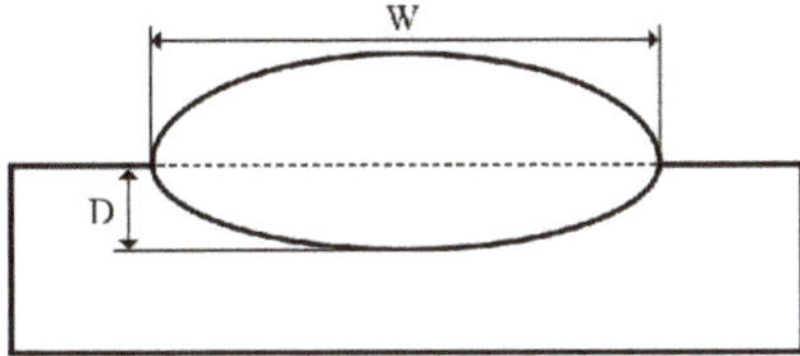

Figure 3. Schematic showing the dimensions of the melt pool. W and D represent the width (μm) and depth (μm), respectively.

The melt-pool boundary was distinguished by the melting-temperature line of the numerical-simulation thermal field. Thus, the width and depth of the numerical melt pool were compared with the average width and depth of the experimental melt pool.

Through numerical validation, the best results for the melt-pool dimensions were obtained by changing the absorption coefficient (*A*) of the mixture. The experimental and numerical cross-sections of three samples—pure AISI 304, 3 wt% Al_2O_3, and 3 wt% Al_2O_3-ZrO_2 (eutectic mixture)—are shown in Figure 4a–f. The black lines in the cross-section of the simulated conditions represent the melt-pool boundaries, which are the melting temperatures, i.e., 1670 K for pure AISI 304, 2313 K for Al_2O_3 reinforced, and 2133 K for the eutectic mixture reinforced samples [4].

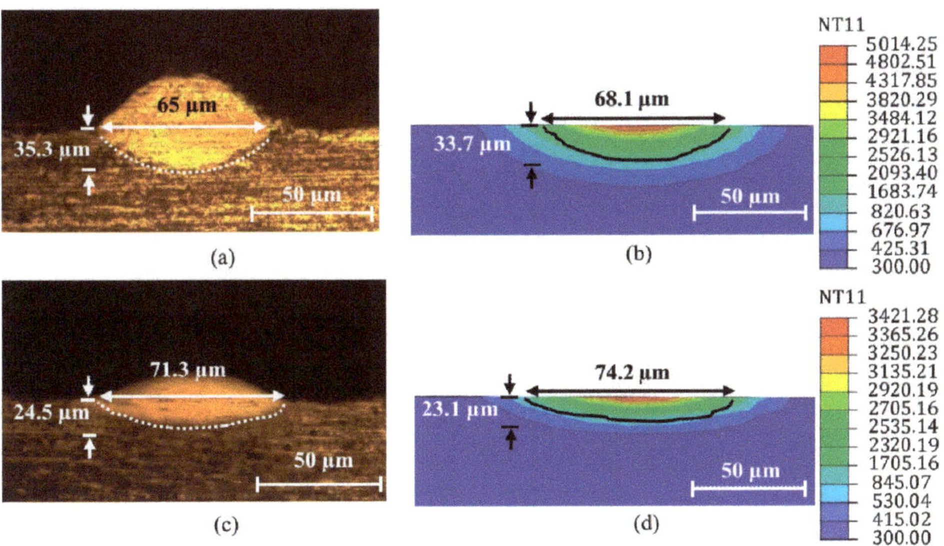

Figure 4. *Cont.*

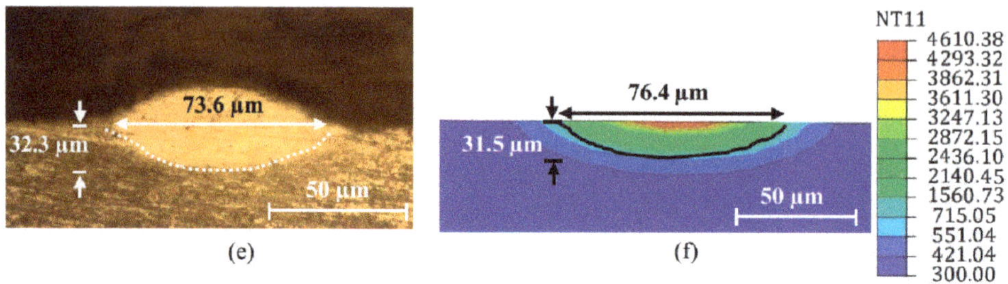

Figure 4. Cross-sectional views of the experimental and numerical models: (**a,b**) pure AISI 304; (**c,d**) 3 wt% Al$_2$O$_3$; (**e,f**) 3 wt% Al$_2$O$_3$-ZrO$_2$. (NT11: Nodal Temperature).

The experimental and numerical melt pools exhibited similar shapes. The width and depth of the melt pools under the conditions listed in Table 1 are plotted in Figure 5a,b, respectively. The two reinforced composites exhibited slight differences in width, and the width of the melt pool was smaller for the pure AISI 304 sample. As thermal conduction is the most influential factor for the melt pool [29], the results are attributed to the lower thermal conductivity in the reinforced samples compared with the pure AISI 304, as shown in Figure 5c at 800 K. The Al$_2$O$_3$-reinforced composite exhibited the smallest melt-pool depth among the samples, as it had the highest melting point, making it difficult for the molten powders to penetrate deep inside the melting pool. The larger melt-pool width and depth for the eutectic-reinforced sample compared with the Al$_2$O$_3$-reinforced sample at each weight percentage are attributed to the reduction in the melting point, along with the reduction in the thermal conductivity.

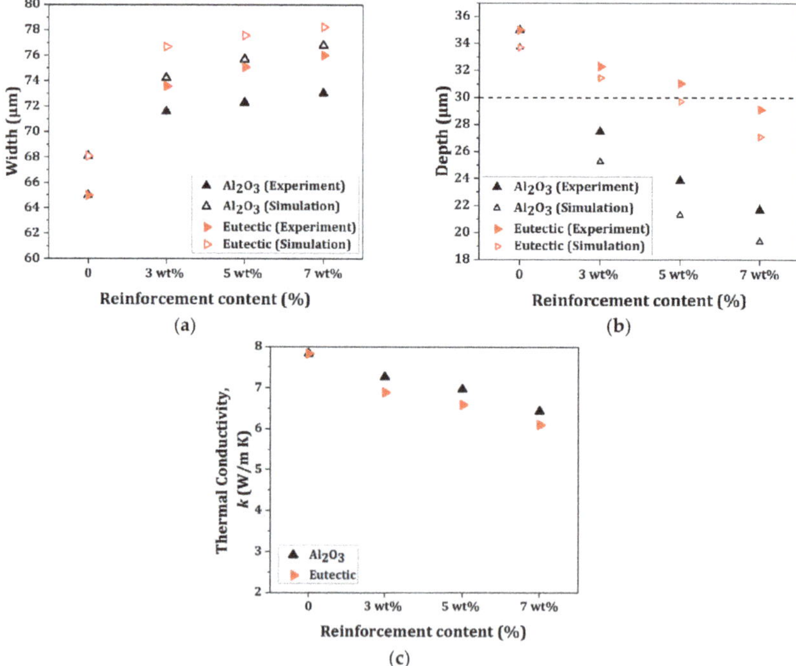

Figure 5. (**a**) Cross-section width and (**b**) depth of the experimental and numerical models. (**c**) Thermal conductivity (*k*) of the mixtures at 800 K.

4.2. Temperature Distribution and Liquid Lifetime

The variation of the temperature with the time needed for scanning a single path during the process, with different weight percentages of reinforcement contents, is explored. The center point of the top surface shown in Figure 6a is considered for plotting the corresponding temperature-time profiles shown in Figure 6b. Figure 6c,d also exhibit the maximum temperature of the top surface and melt-pool lifetime reaching to 300 K for each condition, respectively. As the eutectic mixture of Al_2O_3-ZrO_2 is replaced with the Al_2O_3 reinforcing particle, the maximum temperature is increased significantly due to the reduction in the thermal conductivity and latent heat of the mixture. Also according to Figure 6b, concerning the slope of the cooling curve as the cooling rate, this rate decreased especially in the cases with 7 wt% of reinforcement particles. The simulation results show that as the reinforcement content increases, the cooling rate decreases. The main factor affecting the maximum temperature in various weight percentages of a specific reinforcement in the simulations is identified to be the absorption coefficient. The absorption coefficient of the mixtures was calculated using Equation (8) and exhibited a decreasing trend when increasing a reinforcement content, as is shown in Figure 6e. It is important to note that the cooling rate, i.e., the slope of the cooling curve of Al_2O_3 and Al_2O_3-ZrO_2 systems upon each weight percentage of the reinforcing particle, showed a negligible discrepancy. From the cooling step in simulation runs, the liquid lifetime of the melt pool is distinguished. From Figure 6d, it can be seen that the liquid lifetime increases with the use of a eutectic ratio of Al_2O_3-ZrO_2 reinforcement. Meanwhile, it also rises as the reinforcement content increases. Because the gas atoms are released from the lattice in the heat-affected zone, yielding a longer liquid lifetime in samples with a high weight percentage of reinforcement, e.g., 7 wt%, may result in the formation of a higher amount of gas. It is reported that this phenomenon may cause the formation of porosity in SLM-produced parts [35]. This was explored by multi-line formation tests shown in Figure 7, where the presence of cracks is evident in the reinforced sample with 7 wt% Al_2O_3-ZrO_2, seen in Figure 7f. On the other hand, a short liquid lifetime of the melt pool combined with a lower temperature, observed in the sample with 3 wt% Al_2O_3, is detrimental for wetting the farther gaps of powders, which generated some inter-gaps, as observed in Figure 7a. This can also be observed in Figure 5b, in which in the sample with 3 wt% Al_2O_3 is shown, and the depth of the melt pool cannot reach to the powder bed thickness indicated by the dashed line, i.e., 30 µm. This can be expanded to other reinforced samples with Al_2O_3 particles, where lower temperatures and shorter liquid lifetimes are seen compared with reinforced samples with eutectic mixtures. An appropriate weight percentage of the reinforcement particle of Al_2O_3 or the eutectic mixture of Al_2O_3-ZrO_2 plays an important role in the SLM process of AISI 304 stainless-steel composites. In the case of 3 wt% Al_2O_3-ZrO_2, the formation of an uneven surface, as observed in Figure 7d, is the reason for which it is not regarded as an optimum condition. A proper temperature to melt the particles with appropriate liquid lifetime and melt-pool depth was achieved using a 3 wt% eutectic mixture. Considering a more moderate condition for liquid lifetime, however, the sample of 5 wt% eutectic mixture shows a reasonable trend, as indicated by Figures 6b–d and 7e. From the above observation and discussion, it is evident that the melt-pool lifetime has an effect on the melting behavior, rather than the cooling rate.

4.3. Temperature Gradient

As SLM is an unsteady-state process, temperature gradients along surface direction and thickness direction are unavoidable. The heating and solidification during the process will alter the temperature gradient. This affects the microstructural features, such as grain morphology, grain size and its size distribution, as well as residual stress accumulation due to a large temperature gradient. In general, the unstable material flow, warpage and delamination between fabricated layers are detrimental results of process-induced stresses [36–38].

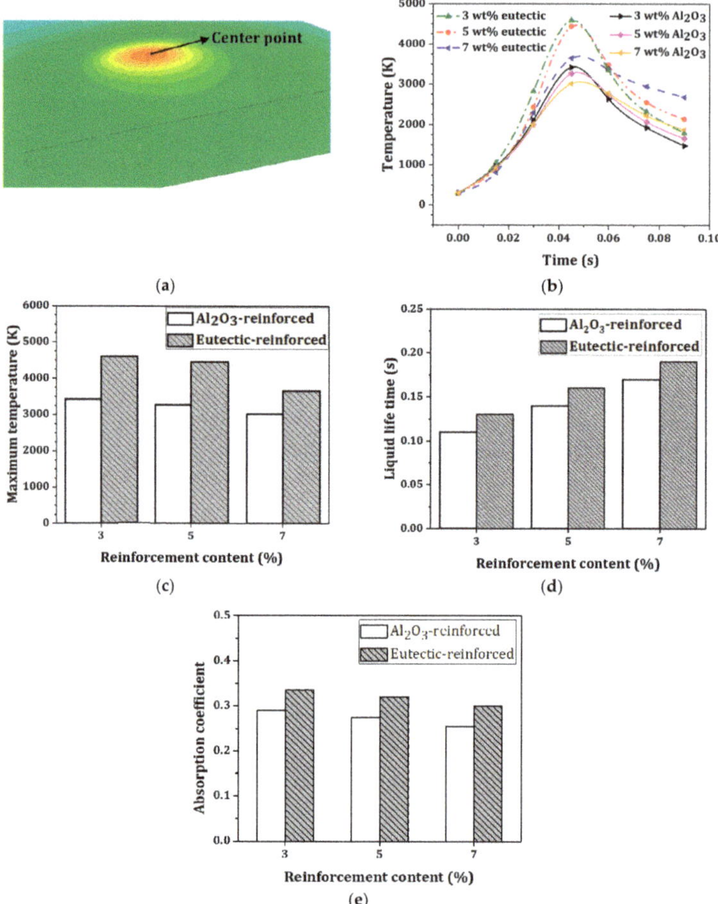

Figure 6. (**a**) Center point on the top surface chosen for plotting the (**b**) Temperature-time curves. (**c**) Results for the maximum temperatures of the reinforced samples. (**d**) Liquid lifetimes. (**e**) The final estimations of the absorption coefficients.

The slope of the curve of temperature distribution presents the temperature gradient of the material [20]. Figure 8a,b show the effects of reinforcement contents on the temperature distribution along the Y-direction at the top surface and Z-direction, or thickness direction, during SLM of Al_2O_3/Al_2O_3-ZrO_2, AISI 304 systems. With the use of an Al_2O_3-ZrO_2 reinforcement system, the temperature gradient at the top surface reduces. This can be recognized in Figure 8a by reducing the slope of the curve of temperature distribution compared with those of Al_2O_3 systems with similar weight percentages of reinforcement content. The generation of a wider melt pool, seen in Figure 5a, or the observation of a larger temperature distribution, seen in Figure 6b, are responsible for this observation. In Figure 8b, the temperature gradient in the thickness direction (Z-direction) of the melt pool is larger in Al_2O_3-ZrO_2-reinforced composites than those of Al_2O_3-reinforced composites. The temperature at the bottom area of the powder bed, shown on a vertical dashed line in Figure 8b, yields higher magnitudes in Al_2O_3-ZrO_2 reinforced samples, but the slope of the plot or temperature gradient in the powder bed is higher compared with Al_2O_3 samples. When a large temperature gradient in the thickness direction is accompanied by a larger depth of the melt pool, as seen in

Figure 5b, this may imply a larger dissipation of laser energy through the pre-fabricated layers or metal substrate. So, it can exert a larger liquid flow in the melt pool in Al_2O_3-ZrO_2 samples.

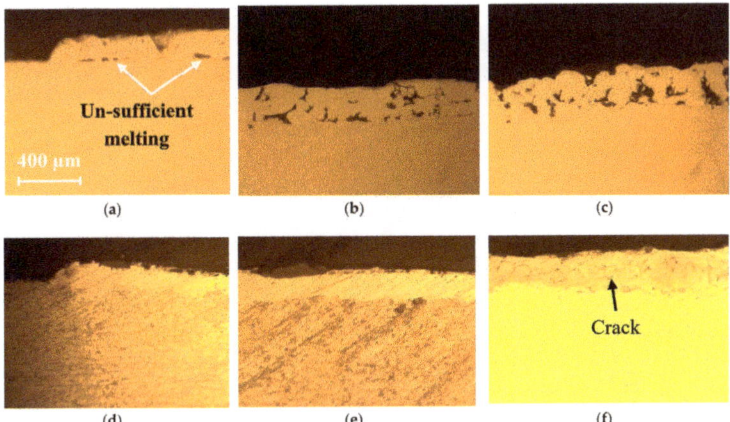

(a)   (b)   (c)   (d)   (e)   (f)

Figure 7. Optical micrographs showing the multi-line formation test results. Al_2O_3-reinforced samples of: (**a**) 3 wt%., (**b**) 5 wt%., and (**c**) 7 wt%. Al_2O_3-ZrO_2 reinforced samples of: (**d**) 3 wt%., (**e**) 5 wt%., and (**f**) 7 wt%.

Liquid flow in the melt pool is an important issue in processes related to molten metals. The free surface energy, which is changed by local heating or cooling, is used to drive liquid metal. However, the term "free surface energy" is not commonly used in regard to liquids and refers to the "gradient in the surface tension" [39]. Thus, the thermal creep in different directions can be obtained by applying a temperature gradient to the surfaces, because heating a surface causes the surface tension to decrease or increase. Inhomogeneities in the gradient of the temperature of a liquid surface generate forces related to the Marangoni effect. This effect can typically be defined as a dimensionless number (M) in the characterization of flows, as indicated by Equation (13) [39].

$$M = \left| \frac{d\gamma}{dT} \right| \times \left| \frac{dT}{di} \right| \times \frac{L^2}{\mu \, \alpha}, \tag{13}$$

Here, $d\gamma/dT$ is the surface-tension gradient or surface-tension temperature coefficient (N m^{-1} K^{-1}), dT/di is the temperature gradient (K m^{-1}) along a specific direction of i, L is the characteristic length (m), μ is the dynamic viscosity (N m^{-2} s), and α is the thermal diffusivity (m^2 s^{-1}). The characteristic length (L) can be considered in processes involving melting, such as welding and AM, for generating deep penetration of surfactants, joints, or additives. For evaluating the convection of the minimum length between the highest temperature (T_{max}) and the lowest temperature ($T_{min} = 300$) along the melt-pool depth, as schematically shown in Figure 8c, the horizontal dashed line drawn at 300 K in Figure 8b can illustrate this length. Following the points of minimum temperatures in the Z-direction on the horizontal dashed line in this figure, for all the models the distance between the points of the maximum and minimum temperatures (L) decreased with the increase in the reinforcement content within the metal matrix. The decrease in the maximum temperature at the top surface observed in Figure 6c may be related to the decrease in the distance between the points of the maximum and minimum temperatures, i.e., L. However, a significant discrepancy is observed between, e.g., 3 wt% Al_2O_3-reinforced and 3 wt% eutectic-reinforced samples and so on. From a numerical viewpoint, this may be due to the higher melting temperature and hence the smaller melt-pool depth in the Al_2O_3-reinforced sample, as previously discussed. Because the thermocapillary (Marangoni) convection always flows from a lower-surface tension region to a higher-surface tension region [39], the larger temperature gradient in each direction causes a larger Marangoni convection in that direction. As a result, when applying the Al_2O_3 reinforcement particle, the Marangoni flow at the top surface increases, which may benefit processes such as surface alloying/hardening. While using the Al_2O_3-ZrO_2 reinforcement particle, the Marangoni flow at the thickness direction increases, which can benefit processes such as 3D-printing. The micro-hardness measurements of the multi-layered samples were investigated at different locations, as shown in Figure 8d, based on the American Society for Testing and Materials (ASTM) E384-16 standard. Figure 8e illustrates that with the use of reinforcement particles, an improvement in the micro-hardness of the fabricated parts is achieved. However, it can be seen that the discrepancy between the micro-hardness of Al_2O_3-reinforced samples and the eutectic ratio of Al_2O_3-ZrO_2 reinforced samples becomes narrower at the top layer, indicating an improvement in reinforcement particles' distribution towards the surface in Al_2O_3-reinforced samples.

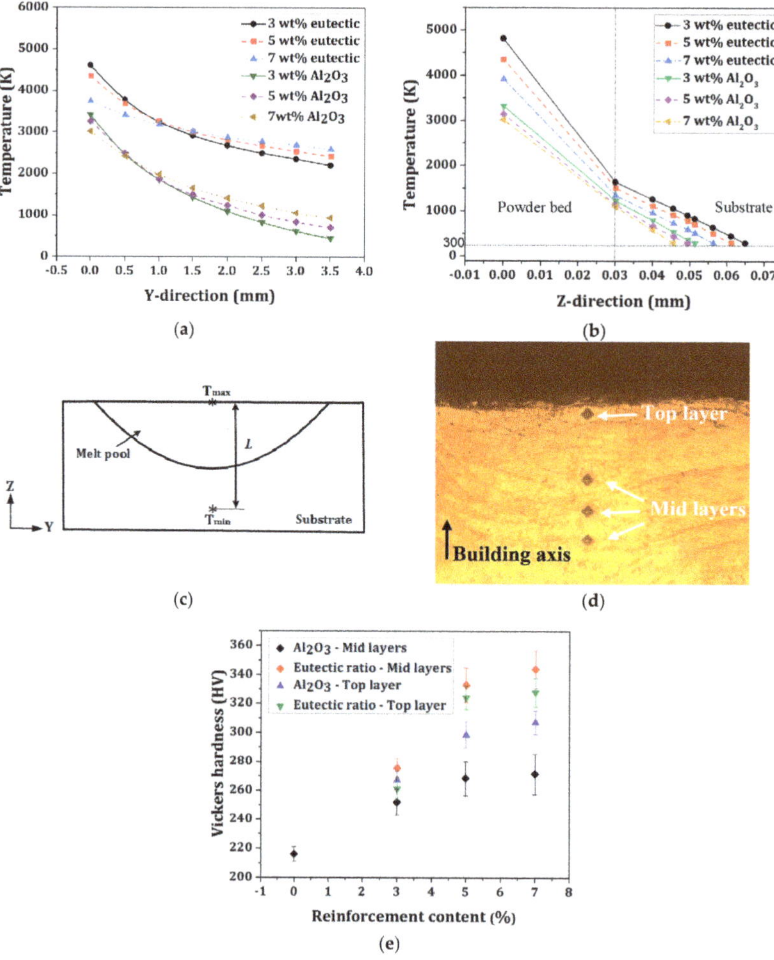

Figure 8. Temperature distribution. (**a**) Along the Y- direction at the top surface. (**b**) Along the Z-direction, i.e., thickness direction. (**c**) Definition of the distance between the points of the maximum and minimum temperatures (L). (**d**) The locations for measuring the micro-hardness. (**e**) Measurements of micro-hardness at different conditions and locations.

5. Conclusions

Melt-pool behaviors during selective laser melting (SLM) of Al_2O_3-reinfored and a eutectic mixture of Al_2O_3-ZrO_2-reinforced AISI 304 stainless-steel composites were analyzed both numerically and experimentally, and the following conclusions are drawn.

(1) A 3D FE model for the SLM of Al_2O_3-reinforced and eutectic Al_2O_3-ZrO_2-reinforced AISI 304 steel composite powders was developed and successfully employed to compare the effects of the reinforcing materials on the melt-pool behaviors.

(2) The width and depth of the melt pool were larger for the eutectic-reinforced sample, which is mainly attributed to the reduction in the melting point and thermal conductivity in this sample. With the use of the eutectic Al_2O_3-ZrO_2 instead of the Al_2O_3 reinforcing particle, the maximum temperature is increased due to the reduction in the thermal conductivity and latent heat of the mixture.

(3) As the reinforcement content increases, the cooling rate decreases. The liquid lifetime of the melt pool has the effect on the melting behavior, rather than the cooling rate, and the liquid lifetime increases with the use of a eutectic ratio of Al_2O_3-ZrO_2 reinforcement. An average and moderate condition for the liquid lifetime was identified to be 5 wt% for the eutectic mixture.

(4) With the use of a eutectic Al_2O_3-ZrO_2 reinforcing particle, the temperature gradient at the top surface reduces compared with the Al_2O_3-reinforced sample, due to a wider melt pool and a larger temperature distribution. This led to a narrower discrepancy between the micro-hardness of Al_2O_3-reinforced samples and the eutectic ratio of Al_2O_3-ZrO_2-reinforced samples at the top layer, indicating an improvement in reinforcement particles' distribution towards the surface in Al_2O_3-reinforced samples. The molten-pool behaviors and the thermal evolution of AISI 304 stainless-steel composites during the selective laser melting process will provide a deep understanding of the effect of reinforcing particles on the shape accuracies and properties of fabricated products.

Author Contributions: Conceptualization, Y.J.K.; Investigation, D.A. and Y.Y.W.; Methodology, S.M.H.S.; Supervision, Y.H.M.; Validation, N.K.

Funding: This work was supported by a National Research Foundation of Korea (NRF) grant (No. 2012R1A5A1048294) funded by the Korean Government (Ministry of Science, ICT and future Planning).

References

1. Li, X.; Willy, H.J.; Chang, S.; Lu, W.; Herng, T.S.; Ding, J. Selective laser melting of stainless steel and alumina composite: Experimental and simulation studies on processing parameters, microstructure and mechanical properties. *Mater. Des.* **2018**, *145*, 1–10. [CrossRef]

2. AlMangour, B.; Grzesiak, D.; Yang, J.M. Selective laser melting of TiC reinforced 316L stainless steel matrix nanocomposites: Influence of starting TiC particle size and volume content. *Mater. Des.* **2016**, *104*, 141–151. [CrossRef]

3. Yves-Christian, H.; Jan, W.; Wilhelm, M.; Konrad, W.; Reinhart, P. Net shaped high performance oxide ceramic parts by selective laser melting. *Physics Procedia* **2010**, *5*, 587–594. [CrossRef]

4. Wilkes, J.; Hagedorn, Y.C.; Meiners, W.; Wissenbach, K. Additive manufacturing of ZrO_2-Al_2O_3 ceramic components by selective laser melting. *Rapid Prototyp. J.* **2013**, *19*, 51–57. [CrossRef]

5. Jang, J.H.; Joo, B.D.; Van Tyne, C.J.; Moon, Y.H. Characterization of deposited layer fabricated by direct laser melting process. *Met. Mater. Int.* **2013**, *19*, 497–506. [CrossRef]

6. Hwang, T.W.; Woo, Y.Y.; Han, S.W.; Moon, Y.H. Fabrication of mesh patterns using a selective laser-melting process. *Appl. Sci.* **2019**, *9*, 1922. [CrossRef]

7. Joo, B.D.; Jang, J.H.; Lee, J.H.; Son, Y.M.; Moon, Y.H. Selective laser melting of Fe-Ni-Cr layer on AISI H13 tool steel. *Trans. Nonferrous Met. Soc. China* **2009**, *19*, 921–924. [CrossRef]

8. Altan, T.; Lilly, B.; Yen, Y.C.; Altan, T. Manufacturing of dies and molds. *CIRP Ann.* **2001**, *50*, 404–422. [CrossRef]
9. Kim, S.Y.; Joo, B.D.; Shin, S.; Van Tyne, C.J.; Moon, Y.H. Discrete layer hydroforming of three-layered tubes. *Int. J. Mach. Tools Manuf.* **2013**, *68*, 56–62. [CrossRef]
10. Karnati, S.; Zhang, Y.; Liou, F.F.; Newkirk, J.W. On the feasibility of tailoring copper–nickel functionally graded materials fabricated through laser metal deposition. *Metals* **2019**, *9*, 287. [CrossRef]
11. Han, S.W.; Woo, Y.Y.; Hwang, T.W.; Oh, I.Y.; Moon, Y.H. Tailor layered tube hydroforming for fabricating tubular parts with dissimilar thickness. *Int. J. Mach. Tools Manuf.* **2019**, *138*, 51–65. [CrossRef]
12. Abazari, H.D.; Seyedkashi, S.M.H.; Gollo, M.H.; Moon, Y.H. Evolution of microstructure and mechanical properties of SUS430/C11000/SUS430 composites during the laser-forming process. *Met. Mater. Int.* **2017**, *23*, 865–876. [CrossRef]
13. Zhao, J.R.; Hung, F.Y.; Lui, T.S.; Wu, Y.L. The relationship of fracture mechanism between high temperature tensile mechanical properties and particle erosion resistance of selective laser melting Ti-6Al-4V alloy. *Metals* **2019**, *9*, 501. [CrossRef]
14. Hwang, T.W.; Woo, Y.Y.; Han, S.W.; Moon, Y.H. Functionally graded properties in directed-energy-deposition titanium parts. *Opt. Laser Technol.* **2018**, *105*, 80–88. [CrossRef]
15. Jang, J.H.; Lee, J.H.; Joo, B.D.; Moon, Y.H. Flow characteristics of aluminum coated boron steel in hot press forming. *T. Nonferr. Metal. Soc.* **2009**, *19*, 913–916. [CrossRef]
16. Seo, D.M.; Hwang, T.W.; Moon, Y.H. Carbonitriding of Ti-6Al-4V alloy via laser irradiation of pure graphite powder in nitrogen environment. *Surf. Coat. Technol.* **2019**, *363*, 244–254. [CrossRef]
17. King, W.; Anderson, A.T.; Ferencz, R.M.; Hodge, N.E.; Kamath, C.; Khairallah, S.A. Overview of modelling and simulation of metal powder bed fusion process at Lawrence Livermore National Laboratory. *Mater. Sci. Technol.* **2015**, *31*, 957–968. [CrossRef]
18. Schleifenbaum, H.; Meiners, W.; Wissenbach, K.; Hinke, C. Individualized production by means of high power selective laser melting. *CIRP J. Manuf. Sci. Technol.* **2010**, *2*, 161–169. [CrossRef]
19. Zhang, D.Q.; Cai, Q.Z.; Liu, J.H.; Zhang, L.; Li, R.D. Select laser melting of W–Ni–Fe powders: Simulation and experimental study. *Int. J. Adv. Manuf. Technol.* **2010**, *51*, 649–658. [CrossRef]
20. Hussein, A.; Hao, L.; Yan, C.; Everson, R. Finite element simulation of the temperature and stress fields in single layers built without-support in selective laser melting. *Mater. Des.* **2013**, *52*, 638–647. [CrossRef]
21. Dai, K.; Shaw, L. Thermal and mechanical finite element modeling of laser forming from metal and ceramic powders. *Acta Mater.* **2004**, *52*, 69–80. [CrossRef]
22. Kundakcioglu, E.; Lazoglu, I.; Rawal, S. Transient thermal modeling of laser-based additive manufacturing for 3D freeform structures. *Int. J. Adv. Manuf. Technol.* **2016**, *85*, 493–501. [CrossRef]
23. AlMangour, B.; Grzesiak, D.; Cheng, J.; Ertas, Y. Thermal behavior of the molten pool, microstructural evolution, and tribological performance during selective laser melting of TiC/316L stainless steel nanocomposites: Experimental and simulation methods. *J. Mater. Process. Technol.* **2018**, *257*, 288–301. [CrossRef]
24. Shi, Q.; Gu, D.; Xia, M.; Cao, S.; Rong, T. Effects of laser processing parameters on thermal behavior and melting/solidification mechanism during selective laser melting of TiC/Inconel 718 composites. *Opt. Laser Technol.* **2016**, *84*, 9–22. [CrossRef]
25. Wang, L.; Jue, J.; Xia, M.; Guo, L.; Yan, B.; Gu, D. Effect of the Thermodynamic behavior of selective laser melting on the formation of in situ oxide dispersion-strengthened aluminum-based composites. *Metals* **2016**, *6*, 286. [CrossRef]
26. Fan, Z.; Lu, M.; Huang, H. Selective laser melting of alumina: A single track study. *Ceram. Int.* **2018**, *44*, 9484–9493. [CrossRef]
27. Vastola, G.; Zhang, G.; Pei, X.Q.; Zhang, Y.W. Modeling the microstructure evolution during additive manufacturing of Ti6Al4V: A comparison between electron beam melting and selective laser melting. *JOM* **2016**, *68*, 1370–1375. [CrossRef]
28. Gusarov, A.V.; Kruth, J.P. Modelling of radiation transfer in metallic powders at laser treatment. *Int. J. Heat Mass Transfer* **2005**, *48*, 3423–3434. [CrossRef]
29. Sih, S.S.; Barlow, J.W. The prediction of the emissivity and thermal conductivity of powder beds. *Part. Sci. Technol.* **2004**, *22*, 427–440. [CrossRef]

30. Boley, C.D.; Mitchell, S.C.; Rubenchik, A.M.; Wu, S.S.Q. Metal powder absorptivity: Modeling and experiment. *Appl. Optics.* **2016**, *55*, 6496. [CrossRef]

31. Angle, J.P.; Wang, Z.; Dames, C.; Mecartney, M.L. Comparison of two-phase thermal conductivity models with experiments on dilute ceramic composites. *J. Am. Ceram. Soc.* **2013**, *96*, 2935–2942. [CrossRef]

32. Spierings, A.B.; Dawson, K.; Heeling, T.; Uggowitzer, P.J.; Schäublin, R.; Palm, F.; Wegener, K. Microstructural features of Sc-and Zr-modified Al-Mg alloys processed by selective laser melting. *Mater. Des.* **2017**, *115*, 52–63. [CrossRef]

33. Loh, L.E.; Chua, C.K.; Yeong, W.Y.; Song, J.; Mapar, M.; Sing, S.L.; Liu, Z.H.; Zhang, D.Q. Numerical investigation and an effective modelling on the Selective Laser Melting (SLM) process with aluminium alloy 6061. *Int. J. Heat Mass Transfer* **2015**, *80*, 288–300. [CrossRef]

34. Mishra, A.K.; Aggarwal, A.; Kumar, A.; Sinha, N. Identification of a suitable volumetric heat source for modelling of selective laser melting of Ti6Al4V powder using numerical and experimental validation approach. *Int. J. Adv. Manuf. Technol.* **2018**, *99*, 2257–2270. [CrossRef]

35. Jeon, T.J.; Hwang, T.W.; Yun, H.J.; Van Tyne, C.J.; Moon, Y.H. Control of porosity in parts produced by a direct laser melting process. *Appl. Sci.* **2018**, *8*, 2573. [CrossRef]

36. Jeon, C.H.; Han, S.W.; Joo, B.D.; Van Tyne, C.J.; Moon, Y.H. Deformation analysis for cold rolling of Al-Cu double layered sheet by physical modeling and finite element method. *Met. Mater. Int.* **2013**, *19*, 1069–1076. [CrossRef]

37. Gordon, W.A.; Van Tyne, C.J.; Moon, Y.H. Axisymmetric extrusion through adaptable dies—Part 1: Flexible velocity fields and power terms. *Int. J. Mech. Sci.* **2007**, *49*, 86–95. [CrossRef]

38. Gordon, W.A.; Van Tyne, C.J.; Moon, Y.H. Axisymmetric extrusion through adaptable dies—Part 3: Minimum pressure streamlined die shapes. *Int. J. Mech. Sci.* **2007**, *49*, 104–115. [CrossRef]

39. AlMangour, B.; Grzesiak, D.; Borkar, T.; Yang, J.M. Densification behavior, microstructural evolution, and mechanical properties of TiC/316L stainless steel nanocomposites fabricated by selective laser melting. *Mater. Des.* **2018**, *138*, 119–128. [CrossRef]

Article

Atomistic Simulations of Pure Tin Based on a New Modified Embedded-Atom Method Interatomic Potential

Won-Seok Ko [1],*, Dong-Hyun Kim [2],*, Yong-Jai Kwon [1] and Min Hyung Lee [3]

[1]   School of Materials Science and Engineering, University of Ulsan, Ulsan 44610, Korea;
     yongjaikwon@ulsan.ac.kr
[2]   Advanced Manufacturing Process R&D Group, Korea Institute of Industrial Technology (KITECH),
     Ulsan 44413, Korea
[3]   Surface R&D Group, Korea Institute of Industrial Technology (KITECH), Incheon 21999, Korea;
     minhyung@kitech.re.kr
*   Correspondence: wonsko@ulsan.ac.kr (W.-S.K.); dhk@kitech.re.kr (D.-H.K.);
     Tel.: +82-52-712-8068 (W.-S.K.); +82-52-980-6711 (D.-H.K.)

Received: 16 October 2018; Accepted: 31 October 2018; Published: 3 November 2018

Abstract: A new interatomic potential for the pure tin (Sn) system is developed on the basis of the second-nearest-neighbor modified embedded-atom-method formalism. The potential parameters were optimized based on the force-matching method utilizing the density functional theory (DFT) database of energies and forces of atomic configurations under various conditions. The developed potential significantly improves the reproducibility of many fundamental physical properties compared to previously reported modified embedded-atom method (MEAM) potentials, especially properties of the β phase that is stable at the ambient condition. Subsequent free energy calculations based on the quasiharmonic approximation and molecular-dynamics simulations verify that the developed potential can be successfully applied to study the allotropic phase transformation between α and β phases and diffusion phenomena of pure tin.

Keywords: tin alloy; modified embedded-atom method; molecular dynamics simulation; phase transformation; diffusion

---

1. Introduction

The materials systems with tin (Sn) have received great attention owing to their importance in many modern technologies. In the electronics industry, soldering is the most important technique of connecting the substrate and electronic devices. Among various kinds of solders, Sn-Pb alloys are the most popular and traditional solders. Recently, Sn-based lead-free solders such as Sn-Ag-Cu alloys have replaced the traditional solders [1]. In the electrochemical industry, Sn-Li alloys are notable materials due to their applications to electrodes [2]. Moreover, liquid Sn-Li alloys are currently being considered as candidates for plasmafacing materials in tokamak fusion reactors [3,4].

For the understanding of detailed phenomena in Sn and its alloys, atomic scale simulations are effective to complement experiments providing detailed insights into the atomic scale processes. Among those simulations, density functional theory (DFT) calculation is a widely used method with many advantages such as a straightforward selectivity of alloy components and the accuracy. However, the DFT calculation is computationally demanding and cannot be applied in all applications. If the material phenomena are related to a comparably large system which cannot be covered by the DFT calculation, large-scale atomistic simulations such as molecular dynamics (MD) and Monte Carlo (MC) are highly desirable. Because the predictive accuracy of such atomistic simulations is critically dependent on the accuracy of interatomic potentials, the availability of reliable interatomic potentials of target systems is of crucial importance. In particular, a relevant description of pure Sn should take precedence because it can be a basis for the development of interatomic potentials for concerned Sn-based multi-component alloy systems.

Until now, the development of an accurate interatomic potential for pure Sn has been a difficult task because of the complexity of the system due to the presence of allotropic phase transformation. In the periodic table, Sn is located at the borderline between covalent and metallic bonding elements. Above 286 K, Sn crystallizes into a body-centered tetragonal crystal structure (β-Sn) having a characteristic of metallic bonding [5]. Below 286 K, Sn crystallizes into a diamond cubic structure (α-Sn) having a characteristic of covalent bonding [5]. So far, several interatomic potentials for pure Sn have been developed in previous works. There are available potentials based on pair potential models [6–8], but the practical utilization of these potentials is greatly limited because these models ignore many-body bonding characteristics of pure Sn. There is an available many-body interatomic potential [9] based on embedded-atom method (EAM) model [10], but this potential cannot predict the β-Sn phase as a stable phase at the ambient condition because target properties of this potential are confined to phases at a high-pressure. Alternatively, there are available many-body interatomic potentials based on the modified embedded-atom method (MEAM) model [11] by Ravelo and Baskes [5], by Vella et al. [12], and by Etesami et al. [13]. It seems reasonable to apply the MEAM model to the Sn system since this model was originally devised to well describe various types of atomic bonds, including the metallic and covalent bonds, in a single formalism. However, these MEAM potentials were developed mostly focusing on properties of liquid phase and showed deficiencies in reproducing physical properties of solid phases, especially for properties of the β-Sn phase.

In the present study, we provide a new interatomic potential based on the second nearest-neighbor modified embedded-atom method (2NN MEAM) model [14–16]. The 2NN MEAM is a model that overcomes several shortcomings of the original MEAM model by partially considering 2NN interactions of atoms. In addition, the present study considered a relevant method of the parameter optimization to improve the reliability of the developed potential. In previous works for MEAM potentials [5,12,13], potential parameters were optimized focusing on specific physical properties obtained by experimental and density-functional theory information. In contrast, the present optimization was performed based on the force-matching method proposed by Ercolessi and Adams [17]. This method considers not a physical property itself, but forces and energies related to various atomic configurations including configurations at finite temperatures expected from the DFT calculation. It has been confirmed that this method can greatly improve the accuracy and transferability of developed interatomic potentials [18–20].

The remainder of the present article is organized as follows. Section 2 describes detailed processes of the DFT calculations and optimization of potential parameters. In Section 3, the accuracy and transferability of the developed potential is presented with suitable examples. The conclusion is finally drawn in Section 4.

2. Methods

The optimization of the present interatomic potential was performed based on the force-matching method which considers both the structural energy and the forces acting on each atom as a fitting target. The overall procedure follows previous works [20,21] in which 2NN MEAM potentials for pure Ni, Ti and Li were similarly determined based on the same method in a systematic manner. First, a database of atomic energies and forces of various atomic configurations was established based on the DFT calculation. Then, an optimum set of potential parameters was determined by minimizing errors between the expectation by potential parameters and the DFT database.

2.1. Construction of a DFT Database

A series of DFT calculations were performed using Vienna Ab initio Simulation Package (VASP) code [22–24] based on the projector augmented wave (PAW) method [25]. For the exchange-correlation functional, the Perdew-Burke-Ernzerhof generalized-gradient approximation (GGA) [26] was used.

A plane-wave kinetic-energy cutoff of 400 eV and the Methfessel-Paxton smearing method with a width of 0.1 eV were used. All calculations were performed with a density of the k-point mesh equivalent to a 21 × 21 × 21 mesh for the face-centered cubic (fcc) primitive unit cell, and the corresponding similar density of the k-point mesh were employed for other unit cells and supercells. In the present DFT calculation, atomic configurations resulting from various conditions were considered for the construction of the DFT database to ensure the sufficient transferability of the developed interatomic potential to various possible applications. Detailed information on cell structures and corresponding k-point meshes is summarized in Table 1.

Table 1. Atomic configurations considered in the present density functional theory (DFT) calculations. In the Stability column, "stable" indicates that the structure is reported in an equilibrium phase diagram. Other phases are labeled as "hypothetical". The Strain column indicates the strain applied to the supercells, where H, O, and M denote hydrostatic, orthorhombic, and monoclinic strains, respectively.

Structure	$N_{Sn\ atoms}$	$N_{vacancies}$	Stability	Temp. (K)	Strain (%)	k-Point
	64	0	low T stable	0, 300	0, H (± 5, ± 10)	5 × 5 × 5
diamond	64	0	low T stable	0	O (± 5, ± 10), M (± 4, ± 8)	5 × 5 × 5
cubic (α)	63	1	low T stable	0, 300	0, H (± 5, ± 10)	5 × 5 × 5
	62	2	low T stable	0	0	5 × 5 × 5
body-centered	64	0	high T stable	0, 300	0, H (± 5, ± 10)	5 × 5 × 5
tetragonal	64	0	high T stable	0	O (± 10), M (± 7)	5 × 5 × 5
(β)	63	1	high T stable	0, 300	0, H (± 5, ± 10)	5 × 5 × 5
	62	2	high T stable	0	0	5 × 5 × 5
fcc	108	0	hypothetical	0	0	4 × 4 × 4
bcc	128	0	hypothetical	0	0	4 × 4 × 4
hcp	96	0	hypothetical	0	0	4 × 4 × 5
liquid	64	0	stable	400, 500, 600	0	5 × 5 × 5

The equilibrium lattice constants and bulk modulus at 0 K were calculated by employing the Birch-Murnaghan equation of state [27,28]. For the calculation of properties involved with defects, positions of each atom were relaxed at a constant volume and cell shape. The vacancy migration energy was calculated with a suitable saddle-point configuration utilizing the nudged elastic band (NEB) method [29,30]. For the calculation of the surface energy, rectangular cells with a stacking of 12 Å to 13 Å thick slab and a vacuum region of 10 Å were employed without the relaxation of cell dimensions parallel to free surfaces. The convergence criteria for energies and forces of all defect calculations were 10^{-6} eV/atom and 10^{-2} eV/Å, respectively. Phonon calculations were performed using the "Phonopy" code [31,32] based on the direct force constant approach [33]. Supercells of 64 atoms were used for the phonon calculation for α and β phases with the convergence criteria for energy and forces of 10^{-8} eV and 10^{-4} eV/Å, respectively. To calculate the cohesive energy, a reference energy with a single atom was obtained by considering the spin-polarized calculation.

To obtain a sufficient number of effective force data at finite temperatures, two-step DFT calculations were conducted. First, ab initio MD simulations [22] were performed for a total of 2000 steps with a timestep of 1.5 fs using relatively low convergence parameters such as a single k-point and a default value of the cutoff energy. To determine accurate energies and forces of each configuration, well-converged calculations with a higher cutoff energy (400 eV) and denser k-point mesh were then followed using randomly extracted configurations from ab initio MD simulations.

2.2. Optimization of Potential Parameters

For the development of a unary potential based on the 2NN MEAM formalism, an optimization of 14 independent potential parameters is required. Four parameters [the cohesive energy (E_c), the equilibrium nearest-neighbor distance (r_e), the bulk modulus (B) of the reference structure and the adjustable parameter d] are involved with to the universal equation of state. Seven parameters [the decay lengths ($\beta^{(0)}$, $\beta^{(1)}$, $\beta^{(2)}$, and $\beta^{(3)}$) and the weighting factors ($t^{(1)}$, $t^{(2)}$, and $t^{(3)}$)] are involved with the electron density. The parameter A is required for the embedding function and two parameters [C_{min} and C_{max}] are required for the many-body screening. Detailed explanations on these potential parameters can be found in literature [14–16].

The target configurations in the DFT database consist of configurations of stable phases (α, β, and liquid) and hypothetical phases (fcc, bcc, and hcp) at various temperatures. In particular, the inclusion of configurations at finite temperatures is indispensable for the force-matching because these configurations can provide effective atomic force data for the optimization. A total of 63 configurations resulting from various temperatures and strain conditions as well as various defect configurations were considered for the fitting. Detailed information on atomic configurations used for the fitting process is listed in Table 1.

The optimization was performed by comparing energies of target configurations and forces on each atom expected by a candidate set of potential parameters and those expected by the DFT calculation. The optimization started with specifying a reference structure of the potential parameters, a radial cutoff distance, and fitting weights of each configuration. An optimizer based on the genetic algorithm then adjusted the candidate set of parameters to reduce the sum of energy errors of each configuration and force errors of each atom. If the derived set of potential parameters does not satisfactorily reproduce overall physical properties, another optimization was performed with a different set of the reference structure, radial cutoff value, and fitting weights of configurations. The trial reference structures were diamond cubic, fcc and bcc structures, and the trial radial cutoff values were 4.5, 5.0, 5.5 and 6.0 Å. During the optimization process, we realized that the consideration of similar weighting between configurations of α and β phases results in a general worsening of the reproducibility of various physical properties. Considering the importance of the β phase stable at ambient condition, the final optimization was performed using increased fitting weights for atomic configurations of the β phase.

Figure 1 shows results of the final optimization represented by statistical correlations between the calculated energies and forces from the developed potential and the DFT calculation. Each correlation was obtained using atomic configurations expected by the DFT calculation without further atomic relaxations. The resultant root mean square (RMS) errors of energies and forces are 0.044 eV/atom and 0.097 eV/Å, respectively. These values are significantly higher than the results of previous optimization for pure Li [21]. It seems that this is because the handling of phases with different types of atomic bonding (covalent and metallic) simultaneously is significantly more difficult than the handling of phases with single type of atomic bonding (metallic) based on a single potential formalism.

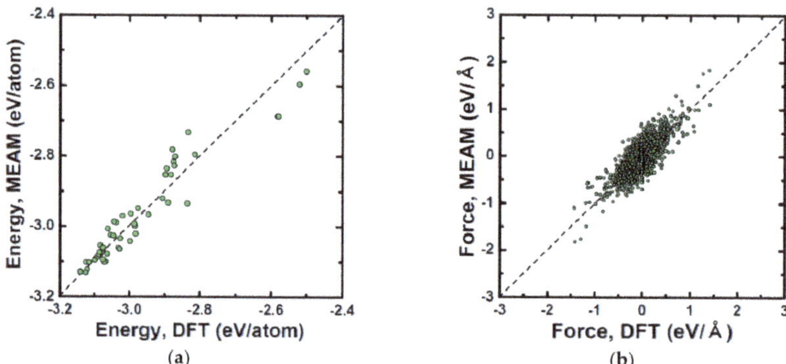

Figure 1. Scatter plots for (**a**) energies and (**b**) forces of pure Sn with respect to the DFT database. The dashed lines indicate a perfect correlation between the present 2NN modified embedded-atom method (MEAM) and the DFT values.

We finally confirmed that a reference structure of fcc provides the optimum result in reproducing various physical properties of pure Sn. A cutoff value of 5.0 Å was confirmed to be large enough to reproduce various physical properties with an acceptable computational efficiency. In the subsequent section, all calculations based on the new MEAM potential were performed using this radial cutoff distance. Table 2 lists the final set of potential parameters for pure Sn.

Table 2. Optimized 2NN MEAM potential parameter set for the pure Sn system. The following properties are dimensionful: the cohesive energy E_c (eV/atom), the equilibrium nearest-neighbor distance r_e (Å), and the bulk modulus B (10^{12} dyne/cm^2). The reference structure is fcc Sn.

E_c	r_e	B	A	$\beta^{(0)}$	$\beta^{(1)}$	$\beta^{(2)}$	$\beta^{(3)}$	$t^{(1)}$	$t^{(2)}$	$t^{(3)}$	C_{min}	C_{max}	d
3.05	3.480	0.6088	1.05	5.50	5.10	4.50	4.30	1.30	3.60	−0.90	1.29	4.43	0.02

3. Results and Discussion

In this section, results of atomistic simulations performed by the developed potential are presented. The results are compared to corresponding experiments and DFT calculations to evaluate the accuracy and transferability of the developed potential. In addition, the performance of the developed potential was compared with previous MEAM potentials by Ravelo and Baskes [5], by Vella et al. [12], and by Etesami et al. [13]. We used the potential of Ravelo and Baskes [5] in a form adopted by Vella et al. [12]. For the clear comparisons between the present potential and the previous potentials, all physical properties were recalculated in the same manner, whether or not some of properties were already reported in previous studies [5,12,13]. All atomistic simulations were performed based on the LAMMPS code [34]. If not specially designated as MD simulations, obtained properties represent results of molecular statics simulations at 0 K using atomic configurations of at least 4000 atoms. All MD runs were performed starting from initial configurations optimized by corresponding molecular statics simulations at 0 K using a timestep of 0.002 ps. Nosé-Hoover thermostat and barostat [35,36] were used for controlling temperature and pressure, respectively.

In Table 3, calculated bulk, elastic and defect properties of pure Sn at 0 K using the present potential are compared with experimental data, DFT calculations and results using previous MEAM potentials. The experimental cohesive energy, which is also in good agreement with the DFT expectation, is well reproduced by the present potential. The ground state structure of pure Sn expected by the DFT calculation is the α structure. The present potential well reproduces such tendency and the structural energy differences between various stable and hypothetical solid phases as well. Among structural energy differences, the difference between the low temperature stable α phase and the high temperature stable β phase is important for the expectation of the allotropic transformation. The present potential accurately reproduces this trend while previous potentials by Vella et al. [12] and by Etesami et al. [13] show significantly higher energy differences. The structural parameters such as lattice constants of the α phase and lattice constants and c/a ratio of the β phase are also well reproduced by the present potential in closer agreement with experimental data.

In the elastic properties, all MEAM potentials generally present a difficulty in reproducing the C_{44} of the β phase. Especially, previous potentials by Ravelo and Baskes [5] and by Etesami et al. [13] indicate the deviation more than one order of magnitude. The present potential significantly improves the reproducibility of the C_{44} and other elastic details as well. For example, it was reported that the β phase lattice is stiffer in the c-direction compared to a- or b-direction ($C_{33} > C_{11}$). The present potential correctly reproduces this trend as well as the absolute values of each elastic constant.

The defect properties of the β phase such as the properties related to the vacancy and the free surface are also examined and listed in Table 3. In the present study, the activation energy of vacancy diffusion (Q^{vac}) is defined as a sum of the vacancy formation energy (E_f^{vac}) and the vacancy migration energy (E_m^{vac}) considering the reported substitutional diffusion mechanism of the β phase [37]. Based on previous experimental study [38] and the present DFT calculation, the self-diffusivity of the β

phase is anisotropic along crystal directions, and the activation energy of vacancy diffusion along the c-direction ($\parallel c$) is higher than that along the a- or b-direction ($\perp c$). The present potential successfully reproduces this trend. In the calculation of the vacancy migration energy and resultant activation energy of vacancy diffusion, we realized that the calculation along the a- or b-direction by previous potentials indicates unphysical results such as the negative migration energy and the instability of the saddle point position. These values are thus not presented in Table 3. The surface energies of the β phase are also correctly reproduced only by the present potential while previous potentials show a significant overestimation.

Table 3. Calculated bulk, elastic and defect properties of pure Sn using the present 2NN MEAM potential, in comparison with experimental data, DFT data, and the calculation results using previous MEAM potentials by Ravelo and Baskes [5], Vella et al. [12], and Etesami et al. [13]. The following quantities are listed: the cohesive energy E_c (eV/atom), the lattice constant a (Å), the bulk modulus B and the elastic constants C_{11}, C_{12}, C_{13}, C_{33}, C_{44} and C_{66} (10^{12} dyne/cm^2), the structural energy differences ΔE (eV/atom), the vacancy formation energy E_f^{vac} (eV), the vacancy migration energy E_m^{vac} (eV), the activation energy of vacancy diffusion Q^{vac} (eV), and the surface energies E_{surf} (erg/cm^2) for the orientations indicated by the superscript. For the E_m^{vac} and Q^{vac}, values considering diffusional paths along the a- or b-direction ($\perp c$) and along the c-direction ($\parallel c$) of the β phase are presented. The DFT and MEAM calculations were performed at 0 K while the experimental data were obtained at finite temperatures.

Phase	Property	Exp.	DFT [f]	MEAM [g,h] [Ravelo]	MEAM [g] [Vella]	MEAM [g] [Etesami]	2NN MEAM [This Work]
	E_c	3.140 [a]	3.154	3.140	3.220	3.209	3.135
	a	6.483 [b]	6.658	6.483	6.304	6.430	6.581
	B	0.426 [c]	0.358	0.422	0.442	0.436	0.406
	C_{11}	0.691 [c]	-	0.704	0.649	0.819	0.504
diamond	C_{12}	0.213 [c]	-	0.281	0.339	0.244	0.357
cubic (α)	C_{44}	0.426 [c]	-	0.367	0.426	0.949	0.105
	$\Delta E_{\alpha \to \beta}$	-	0.041	0.055	0.105	0.118	0.033
	$\Delta E_{\alpha \to \mathrm{fcc}}$	-	0.065	0.060	0.160	0.129	0.052
	$\Delta E_{\alpha \to \mathrm{bcc}}$	-	0.075	0.060	0.144	0.129	0.053
	$\Delta E_{\alpha \to \mathrm{hcp}}$	-	0.063	0.059	0.160	0.128	0.050
	E_c	3.10 [d]	3.113	3.085	3.115	3.091	3.102
	a	5.831 [b]	5.938	5.920	5.682	5.914	5.859
	c	3.184 [b]	3.224	3.235	3.334	3.237	3.206
	c/a	0.546 [b]	0.543	0.546	0.587	0.547	0.547
	B	0.570 [c]	0.479	0.645	0.656	0.647	0.571
	C_{11}	0.734 [c]	-	1.093	1.233	1.329	0.897
	C_{12}	0.599 [c]	-	0.625	0.343	0.469	0.467
	C_{13}	0.391 [c]	-	0.244	0.540	0.166	0.369
body-centered	C_{33}	0.907 [c]	-	1.396	0.592	1.571	0.937
tetragonal (β)	C_{44}	0.220 [c]	-	0.007	0.042	0.012	0.079
	C_{66}	0.239 [c]	-	0.225	0.215	0.281	0.106
	E_f^{vac}	-	0.734	1.112	1.545	1.617	0.848
	$E_m^{\mathrm{vac}}(\perp c)$	-	0.082	-	-	-	0.235
	$E_m^{\mathrm{vac}}(\parallel c)$	-	0.290	0.257	0.268	0.439	0.291
	$Q^{\mathrm{vac}}(\perp c)$	1.089 [e]	0.816	-	-	-	1.083
	$Q^{\mathrm{vac}}(\parallel c)$	1.111 [e]	1.023	1.369	1.813	2.057	1.139
	$E_{\mathrm{surf}}^{(100)}$	-	378	725	813	1375	345
	$E_{\mathrm{surf}}^{(001)}$	-	359	889	792	1300	393

[a] Ref. [39]. [b] Ref. [40]. [c] Ref. [41]. [d] Ref. [42]. [e] Ref. [38]. [f] Present DFT calculation. [g] Present calculation using the reported potential parameters. [h] The potential parameters adopted by Vella et al. [12] are used.

We further examined the transferability of the developed potential by comparing a group of properties at finite temperatures. Table 4 lists the thermal properties of the β phase (the thermal expansion coefficient and the specific heat), in comparison with experimental data. These properties were obtained based on the MD simulations using an isobaric-isothermal (*NPT*) ensemble at the target temperature and zero pressure. For the thermal expansion coefficient of the β phase, the potential by Ravelo and Baskes [5] shows a better agreement with the experiment than other potentials. Instead, other potentials show a better reproducibility than the potential by Ravelo and Baskes [5] in the case of the specific heat of the β phase.

Table 4. Calculated thermal properties of pure Sn using the present 2NN MEAM potential, in comparison with experimental data and the calculation results using previous MEAM potentials by Ravelo and Baskes [5], Vella et al. [12], and Etesami et al. [13]. The listed quantities correspond to the thermal expansion coefficient ε (10^{-6}/K), the heat capacity at constant pressure C_P (J/mol K), the melting temperature T_m (K), the enthalpy of melting ΔH_m (kJ/mol), and the volume change upon melting $\Delta V_m / V_{\text{solid}}$ (%).

Property	Exp.	MEAM [d,e] [Ravelo]	MEAM [d] [Vella]	MEAM [d] [Etesami]	MEAM [This Work]
ε (300 K)	23.5 [a]	23.0	20.3	19.3	18.6
C_P (295 K)	26.5 [b]	27.2	26.6	26.5	26.1
T_m	505 [c]	-	435	502	368
ΔH_m	7.0 [c]	-	4.2	4.1	3.1
$\Delta V_m / V_{\text{solid}}$	2.3 [c]	-	2.6	2.5	4.4

[a] Ref. [41]. [b] Ref. [43]. [c] Ref. [44]. [d] Present calculation using the reported potential parameters. [e] The potential parameters adopted by Vella et al. [12] are used.

Figure 2 shows the temperature dependence of the atomic volume of pure Sn. Initially, the β phase with 16,224 atoms was equilibrated at 5 K, and the temperature was then gradually increased 1000 K with a heating rate of 1.0 K/ps using the *NPT* ensemble at zero pressure. Each potential indicates a discontinuous jump in the volume at a certain temperature. This jump indicates the occurrence of the melting ($\beta \rightarrow$ liquid). Because the heating simulation was performed without any heterogeneous nucleation sites, the melting occurs at a temperature much higher than the equilibrium melting temperature. Therefore, this temperature can be regarded as an overheated melting point. The equilibrium melting temperature at zero pressure was further calculated using the interface method [45,46] employing a simulation cell consisting of liquid and solid phases and listed in Table 4. The table also lists the enthalpy change and the volume change upon melting which were obtained at the calculated equilibrium melting temperature. For the previous potential by Ravelo and Baskes [5], it was reported that the liquid part of the interface simulation is crystallized into a structure different from the β phase [12], and thus the equilibrium melting point and resultant enthalpy and volume changes are not presented. The developed potential shows discrepancies in the properties associated with the melting while the potential by Etesami et al. [13] indicates the best agreement in the melting point with the experiment. This can be interpreted by the fact that the potential [13] was developed focusing mostly on the melting phenomenon without consideration to the reproducibility of many other important properties.

As listed in Table 4, the present potential is somewhat deficient in describing properties related to the melting compared to previous potentials. The present potential underestimates the enthalpy of melting and overestimates the volume change upon melting compared to experimental values. Considering that the present potential satisfactory describes physical properties of β phase, it is necessary to further investigate the properties associated with the structure of the liquid phase. Figure 3 shows the calculated radial distribution function of liquid structures at different temperatures. Although the present potential shows a generally acceptable quality to reproduce the height and position of the first peak, it shows a deficiency in reproducing characteristics of other peaks. We attribute these deficiencies to the use of increased fitting weights of the β phase compared to other phases during the fitting process.

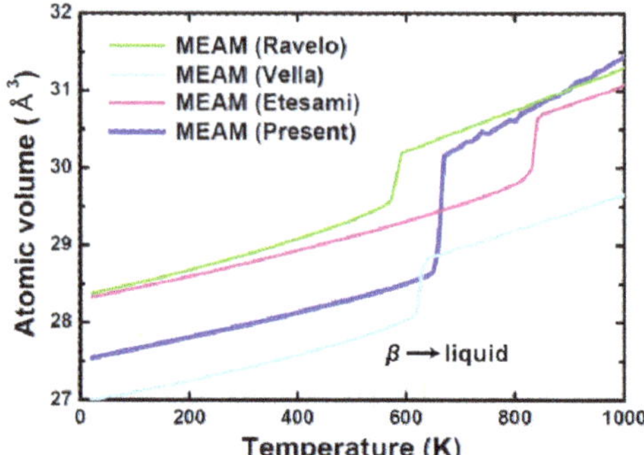

Figure 2. Atomic volume dependence of pure Sn, calculated using the present 2NN MEAM potential, in comparison with the calculation results using previous MEAM potentials by Ravelo and Baskes [5], Vella et al. [12], and Etesami et al. [13]. The heating simulations were started with the β phase at 5 K. The discrete jumps represent the occurrence of the melting.

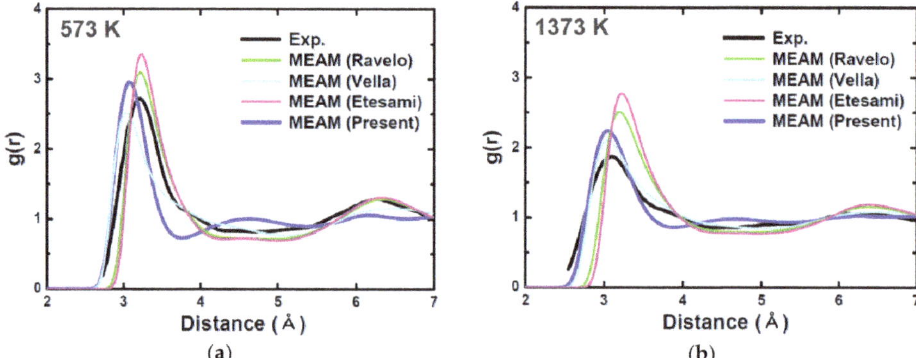

Figure 3. Calculated radial distribution function of liquid Sn at (**a**) 573 K and (**b**) 1373 K using the present 2NN MEAM potential, in comparison with experimental data (Ref. [7]) and the calculation results using previous MEAM potentials by Ravelo and Baskes [5], Vella et al. [12], and Etesami et al. [13].

We further examined the reproducibility of phonon properties and resultant properties at finite temperature based on the harmonic approximation (HA) and quasiharmonic (QHA) approximation. Figure 4 shows the phonon density of states (DOS) of α and β phases expected by the present and previous potentials compared to the DFT expectation. As shown in Figure 4a, the phonon DOS of the α phase is closely reproduced by the present potential. In contrast, previous potentials indicate a significant overestimation of the phonon DOS at high frequencies compared to the DFT expectation. A similar trend is also presented for the phonon DOS of the β phase as shown in Figure 4b. Moreover, previous potentials cannot reproduce the stability of the β phase under perturbative forces. As shown in the magnified figure of the phonon DOS of the β phase (Figure 4c), previous potentials indicate dynamical instability (imaginary phonon frequencies) in contrast to expectations by the present potential and DFT calculations. Even though this problem of previous potentials is expected to interfere with various possible applications of previous potentials, the present potential is free from this problem.

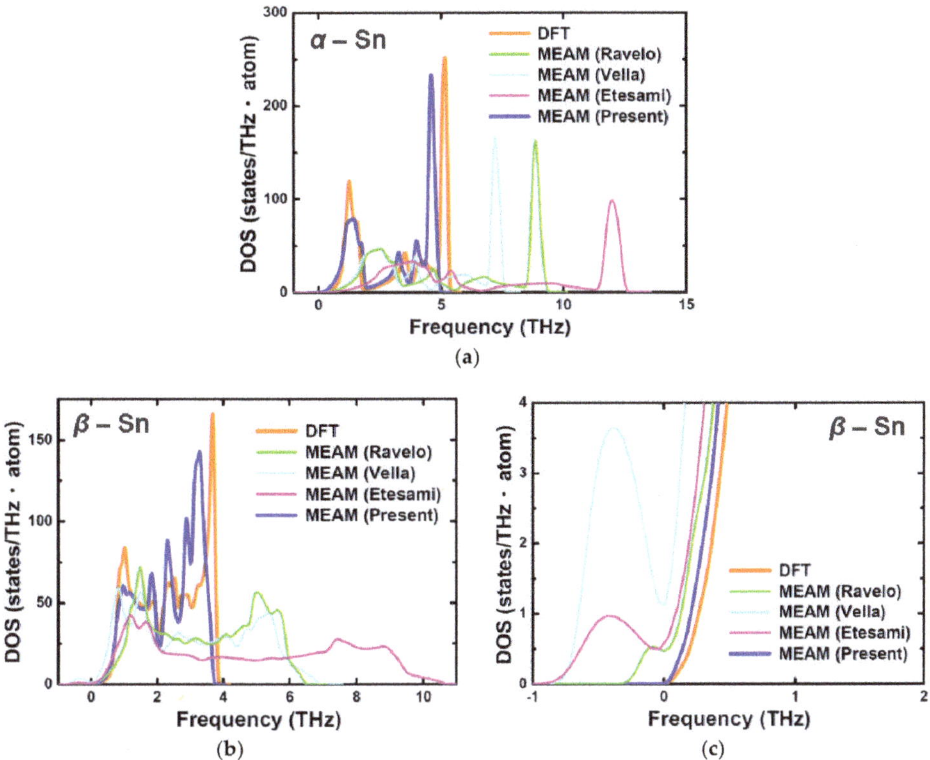

Figure 4. Phonon density of states (DOS) of (**a**) α phase and (**b**) β phase, calculated based on the harmonic approximation (HA). The results using the present 2NN MEAM potential are compared with the present DFT results and the results using previous MEAM potentials by Ravelo and Baskes [5], Vella et al. [12], and Etesami et al. [13]. (**c**) A magnified figure of Figure (**b**) is presented to clarify the existence of imaginary phonon modes.

Therefore, only the present potential can be further utilized to expect the physical properties at finite temperatures based on the QHA. One simple example is calculations of the thermal expansion coefficient and the heat capacity as presented in Figure 5. The present potential exhibits an acceptable quality comparable to the DFT calculation when reproducing such properties. Another important example is the expectation of the allotropic phase transformation between α and β phases. Because the present potential can accurately reproduce the phonon properties without suffering from the imaginary phonon modes, a vibrational contribution on the free energy can be calculated based on the QHA. The Gibbs free energy at the given temperature T and pressure P can be obtained from the Helmholtz free energy $F(T, V)$ at the given temperature and volume V through the transformation,

$$G(T, P) = \min[F(T, V) + PV],\tag{1}$$

where the right-hand side of this equation represents a value when the minimum value is found in square brackets by varying the volume. In the present study, the $F(T, V)$ was computed by a sum of the DFT total energy $E_{DFT}(V)$ at the given volume, and the vibrational contribution to the free energy $F_{vib}(T, V)$ at the given temperature and volume.

$$F(T, V) = E_{DFT}(V) + F_{vib}(T, V)\tag{2}$$

A detailed explanation on the determination of the free energy based on the phonon calculation is given in Ref. [47].

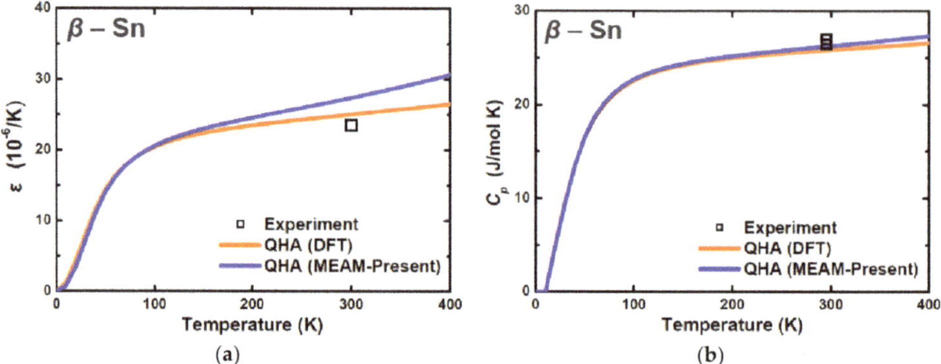

Figure 5. (a) Thermal expansion coefficient (ε) and (b) heat capacity (C_P) of β-Sn, calculated based on the quasiharmonic approximation (QHA). The results using the present 2NN MEAM potential are compared with those using the present DFT calculation and experimental data (Refs. [43,48]).

Figure 6 shows the calculated temperature dependence of Gibbs free energies of α and β phases when the reference state is set to the β phase at 0 K. As stated, both DFT and the present MEAM potential expect that the α phase is a ground state of pure Sn. The Gibbs free energy of each phase decreases with increasing temperature due to the increasing contribution of the vibrational entropy. Because the vibrational contribution is more significant in the β phase than in the α phase, the β phase starts to become more stable than the α phase at a certain temperature. We define this temperature as the phase transformation temperature between α and β phases. The calculated transformation temperature by the present MEAM potential (460 ± 5 K) agrees well with that by the present DFT calculation (470 ± 5 K) while these values are higher than experimental data (286 K) [5]. Although this discrepancy seems to be caused by the limitation of the QHA, the present potential at least reproduces the DFT expectation.

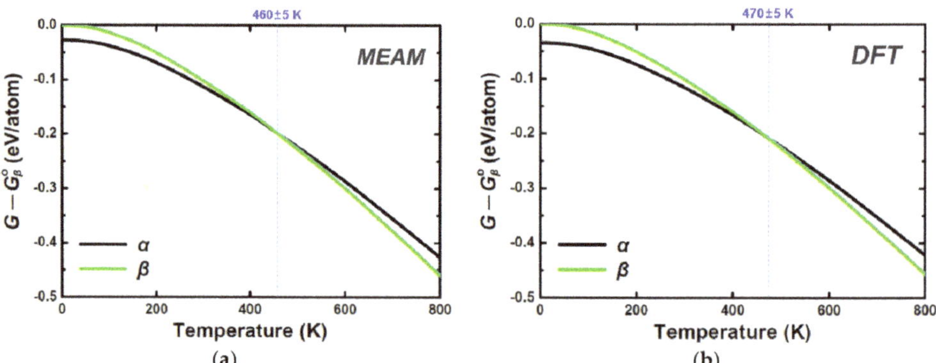

Figure 6. Relative Gibbs free energies of α and β phases at various temperatures, calculated based on the quasiharmonic approximation (QHA). The results using (a) the present 2NN MEAM and (b) the present DFT calculation are illustrated. Each energy value was calculated with respect to the reference state of the β phase at 0 K.

As a final confirmation of the transferability of the developed potential, the self-diffusion of liquid Sn was examined using the MD simulation. In the atomic scale, the MD simulation provides a time dependency of the mean square displacement (MSD), and this is related to the bulk diffusivity based on the following equation (Einstein relation),

$$D = \frac{\langle R^2(t) \rangle}{6t},\tag{3}$$

where D is the diffusivity, $\langle R^2(t) \rangle$ is the MSD of atoms, and t is the time.

The self-diffusivity of liquid Sn was calculated using *NPT* ensemble MD runs starting with the β phase of 16,224 atoms. First, the initial structure was melted by maintaining the cell at a high temperature (2000 K) for a sufficient time (40 ps). The temperature was then changed to each target temperature within a range of 600 and 2000 K, and the relaxation was performed for 40 ps. The simulation cell was further maintained at the target temperature for a total simulation time of 2000 ps, and the time evolution of the MSD was counted at every 4 ps. We confirmed that the MSD of atoms accumulated through these conditions shows a clear linear relationship with the time. Figure 7 shows resultant self-diffusivities of the liquid phase at various temperatures (600–2000 K) compared with experimental data. Despite a deviation between experimental results, the present potential successfully reproduces the general trend of an experimental result by Bruson et al. [49]. The calculated self-diffusivities using the present potential are similar to those using the previous potential by Vella et al. [12] even though the values are slightly underestimated at low temperatures. Compared to the previous potentials by Ravelo and Baskes [5] and by Etesami et al. [13], the present potential indicates significantly improved results, especially at high temperatures. It is interesting to note that the self-diffusivities of the liquid phase are sufficiently reproduced despite the discrepancy of properties related to the liquid structure. The result seems to emphasize the effectiveness of the force-matching method to obtain a robust potential that demonstrates a good transferability to kinetic properties such as the migration energy and the attempt frequency of diffusing atoms. However, it should be noted that the transferability of the present potential to kinetic properties of low-coordinated atoms on surfaces (surface diffusion) is not guaranteed because the present fitting was mostly performed by using atomic configurations in the bulk state.

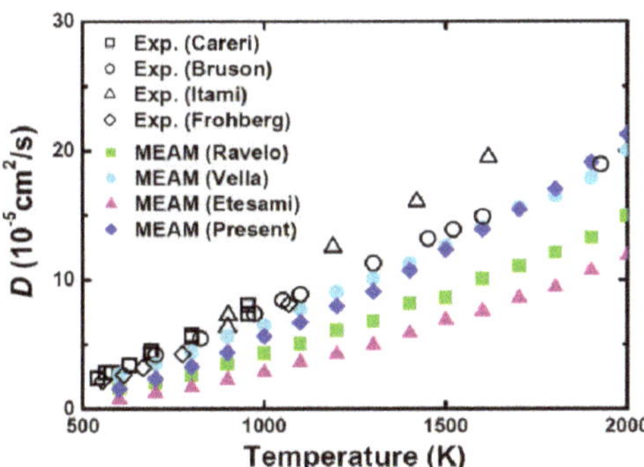

Figure 7. Calculated self-diffusivity of liquid Sn using the present 2NN MEAM potential, in comparison with experimental data (Refs. [49–52]) and the calculation results using previous MEAM potentials by Ravelo and Baskes [5], Vella et al. [12], and Etesami et al. [13].

To summarize, we have shown that the present potential generally reproduces various physical properties of pure Sn, even though there is a difference in the performance for each property. The developed potential performs very well in describing structural, elastic and thermal properties of the β phase, phonon properties of solid phases, and diffusion properties of solid and liquid phases. It is less suited to describing the melting behavior and the liquid structure compared to previous MEAM potentials [5,12,13], which were developed to target specific physical properties of the liquid phase. This fact should be kept in mind in future applications of the present potential. The LAMMPS files implementing the developed potential can be obtained from an online repository [53].

4. Conclusions

We provide a new robust interatomic potential for pure Sn on the basis of the 2NN MEAM model. The potential is developed based on the force-matching method utilizing the DFT database of energies and forces of atomic configurations under various conditions. The developed potential significantly improves the reproducibility of various physical properties compared to previously reported MEAM potentials, especially for properties of the β phase. The present potential exhibits superior transferability to the phonon properties and the properties related to diffusion phenomena. As a possible example, the allotropic phase transformation between α and β phases is investigated based on the QHA. The results indicate that the present potential can be successfully used for the expectation of the transformation temperature at least at the level of the DFT expectation. The present potential also successfully reproduces the self-diffusivity of liquid Sn. The developed potential can be a suitable basis for implementing atomistic simulations of technically important Sn-based alloy systems.

Author Contributions: Conceptualization, W.-S.K. and D.-H.K.; Methodology, W.-S.K. and D.-H.K.; Validation, W.-S.K., D.-H.K, Y.-J. K, and M.H.L.; Formal analysis, W.-S.K. and D.-H.K.; Investigation, W.-S.K. and D.-H.K.; Resources, Y.-J.K.; Data curation, W.-S.K.; Writing—original draft preparation, W.-S.K.; Writing—review and editing, D.-H.K, Y.-J.K, and M.H.L; Visualization, W.-S.K.; Project administration, D.-H.K.; Funding acquisition, M.H.L.

Funding: This work is supported by KITECH research fund (Grant No. UR180012), and National Research Foundation of Korea (NRF) grant funded by the Korea government (MSIP) (Grant No. NRF-2017R1C1B5015038).

Conflicts of Interest: The authors declare no conflict of interest.

References

1. Roshanghias, A.; Vrestal, J.; Yakymovych, A.; Richter, K.W.; Ipser, H. Sn-Ag-Cu nanosolders: Melting behavior and phase diagram prediction in the Sn-rich corner of the ternary system. *Calphad* **2015**, *49*, 101–109. [CrossRef] [PubMed]
2. Scrosati, B.; Garche, J. Lithium batteries: Status, prospects and future. *J. Power Sources* **2010**, *195*, 2419–2430. [CrossRef]
3. Allain, J.P.; Ruzic, D.N.; Hendricks, M.R.D. He and Li sputtering of liquid eutectic Sn-Li. *J. Nucl. Mater.* **2001**, *290–293*, 33–37. [CrossRef]
4. Coenen, J.W.; Temmerman, G.D.; Federici, G.; Philipps, V.; Sergienko, G.; Strohmayer, G.; Terra, A.; Unterberg, B.; Wegener, T.; Bekerom, D.C.M.V.D. Liquid metals as alternative solution for the power exhaust of future fusion devices: Status and perspective. *Phys. Scr.* **2014**, *2014*, 014037. [CrossRef]
5. Ravelo, R.; Baskes, M. Equilibrium and Thermodynamic Properties of Grey, White, and Liquid Tin. *Phys. Rev. Lett.* **1997**, *79*, 2482–2485. [CrossRef]
6. Masaki, T.; Aoki, H.; Munejiri, S.; Ishii, Y.; Itami, T. Effective pair interatomic potential and self-diffusion of molten tin. *J. Non-Cryst. Solids* **2002**, *312–314*, 191–195. [CrossRef]
7. Itami, T.; Munejiri, S.; Masaki, T.; Aoki, H.; Ishii, Y.; Kamiyama, T.; Senda, Y.; Shimojo, F.; Hoshino, K. Structure of liquid Sn over a wide temperature range from neutron scattering experiments and first-principles molecular dynamics simulation: A comparison to liquid Pb. *Phys. Rev. B* **2003**, *67*, 064201. [CrossRef]
8. Mouas, M.; Gasser, J.-G.; Hellal, S.; Grosdidier, B.; Makradi, A.; Belouettar, S. Diffusion and viscosity of liquid tin: Green-Kubo relationship-based calculations from molecular dynamics simulations. *J. Chem. Phys.* **2012**, *136*, 094501. [CrossRef] [PubMed]
9. Sapozhnikov, F.A.; Ionov, G.V.; Dremov, V.V.; Soulard, L.; Durand, O. The Embedded Atom Model and large-scale MD simulation of tin under shock loading. *J. Phys. Conf. Ser.* **2014**, *500*, 032017. [CrossRef]
10. Daw, M.S.; Baskes, M.I. Embedded-atom method: Derivation and application to impurities, surfaces, and other defects in metals. *Phys. Rev. B* **1984**, *29*, 6443–6453. [CrossRef]

11. Baskes, M.I. Modified embedded-atom potentials for cubic materials and impurities. *Phys. Rev. B* **1992**, *46*, 2727–2742. [CrossRef]

12. Vella, J.R.; Chen, M.; Stillinger, F.H.; Carter, E.A.; Debenedetti, P.G.; Panagiotopoulos, A.Z. Structural and dynamic properties of liquid tin from a new modified embedded-atom method force field. *Phys. Rev. B* **2017**, *95*, 064202. [CrossRef]

13. Etesami, S.A.; Baskes, M.I.; Laradji, M.; Asadi, E. Thermodynamics of solid Sn and PbSn liquid mixtures using molecular dynamics simulations. *Acta Mater.* **2018**, *161*, 320–330. [CrossRef]

14. Lee, B.-J.; Baskes, M.I. Second nearest-neighbor modified embedded-atom-method potential. *Phys. Rev. B* **2000**, *62*, 8564–8567. [CrossRef]

15. Lee, B.-J.; Baskes, M.I.; Kim, H.; Cho, Y.K. Second nearest-neighbor modified embedded atom method potentials for bcc transition metals. *Phys. Rev. B* **2001**, *64*, 184102. [CrossRef]

16. Lee, B.-J.; Ko, W.-S.; Kim, H.-K.; Kim, E.-H. The modified embedded-atom method interatomic potentials and recent progress in atomistic simulations. *Calphad* **2010**, *34*, 510–522. [CrossRef]

17. Ercolessi, F.; Adams, J.B. Interatomic Potentials from First-Principles Calculations: The Force-Matching Method. *Eur. Lett.* **1994**, *26*, 583. [CrossRef]

18. Mendelev, M.I.; Han, S.; Srolovitz, D.J.; Ackland, G.J.; Sun, D.Y.; Asta, M. Development of new interatomic potentials appropriate for crystalline and liquid iron. *Philos. Mag.* **2003**, *83*, 3977–3994. [CrossRef]

19. Brommer, P.; Gähler, F. Potfit: Effective potentials from ab initio data. *Model. Simul. Mater. Sci. Eng.* **2007**, *15*, 295. [CrossRef]

20. Ko, W.-S.; Grabowski, B.; Neugebauer, J. Development and application of a Ni-Ti interatomic potential with high predictive accuracy of the martensitic phase transition. *Phys. Rev. B* **2015**. [CrossRef]

21. Ko, W.-S.; Jeon, J.B. Interatomic potential that describes martensitic phase transformations in pure lithium. *Comput. Mater. Sci.* **2017**, *129*, 202–210. [CrossRef]

22. Kresse, G.; Hafner, J. Ab initiomolecular-dynamics simulation of the liquid-metal–amorphous-semiconductor transition in germanium. *Phys. Rev. B* **1994**, *49*, 14251–14269. [CrossRef]

23. Kresse, G.; Furthmüller, J. Efficiency of ab-initio total energy calculations for metals and semiconductors using a plane-wave basis set. *Comput. Mater. Sci.* **1996**, *6*, 15–50. [CrossRef]

24. Kresse, G.; Furthmüller, J. Efficient iterative schemes for ab initio total-energy calculations using a plane-wave basis set. *Phys. Rev. B* **1996**, *54*, 11169–11186. [CrossRef]

25. Blöchl, P.E. Projector augmented-wave method. *Phys. Rev. B* **1994**, *50*, 17953–17979. [CrossRef]

26. Perdew, J.P.; Burke, K.; Ernzerhof, M. Generalized Gradient Approximation Made Simple. *Phys. Rev. Lett.* **1996**, *77*, 3865–3868. [CrossRef] [PubMed]

27. Birch, F. Finite strain isotherm and velocities for single-crystal and polycrystalline NaCl at high pressures and 300°K. *J. Geophys. Res. Solid Earth* **1978**, *83*, 1257–1268. [CrossRef]

28. Murnaghan, F.D. The Compressibility of Media under Extreme Pressures. *Proc. Natl. Acad. Sci. USA* **1944**, *30*, 244–247. [CrossRef] [PubMed]

29. Henkelman, G.; Uberuaga, B.P.; Jónsson, H. A climbing image nudged elastic band method for finding saddle points and minimum energy paths. *J. Chem. Phys.* **2000**, *113*, 9901–9904. [CrossRef]

30. Henkelman, G.; Jónsson, H. Improved tangent estimate in the nudged elastic band method for finding minimum energy paths and saddle points. *J. Chem. Phys.* **2000**, *113*, 9978–9985. [CrossRef]

31. Togo, A.; Oba, F.; Tanaka, I. First-principles calculations of the ferroelastic transition between rutile-type and $CaCl_2$-type SiO_2 at high pressures. *Phys. Rev. B* **2008**, *78*, 134106. [CrossRef]

32. Togo, A.; Tanaka, I. Evolution of crystal structures in metallic elements. *Phys. Rev. B* **2013**, *87*, 184104. [CrossRef]

33. Wei, S.; Chou, M.Y. Ab initio calculation of force constants and full phonon dispersions. *Phys. Rev. Lett.* **1992**, *69*, 2799–2802. [CrossRef] [PubMed]

34. Plimpton, S. Fast Parallel Algorithms for Short-Range Molecular Dynamics. *J. Comput. Phys.* **1995**. [CrossRef]

35. Nosé, S. A unified formulation of the constant temperature molecular dynamics methods. *J. Chem. Phys.* **1984**, *81*, 511–519. [CrossRef]

36. Hoover, W.G. Canonical dynamics: Equilibrium phase-space distributions. *Phys. Rev. A* **1985**, *31*, 1695–1697. [CrossRef]

37. Liu, P.; Wang, S.; Li, D.; Li, Y.; Chen, X.-Q. Fast and Huge Anisotropic Diffusion of Cu (Ag) and Its Resistance on the Sn Self-diffusivity in Solid β-Sn. *J. Mater. Sci. Technol.* **2016**, *32*, 121–128. [CrossRef]

38. Coston, C.; Nachtrieb, N.H. Self-Diffusion in Tin at High Pressure. *J. Phys. Chem.* **1964**, *68*, 2219–2229. [CrossRef]
39. Kittel, C. *Introduction to Solid State Physics*, 8th ed.; Wiley: New York, NY, USA, 2005.
40. Barrett, C.S.; Massalski, T.B. *Structure of Metals*; McGraw-Hill: New York, NY, USA, 1966.
41. Brandes, E.A.; Brook, G.B. *Smithells Metals Reference Book*, 7th ed.; Butterworth-Heinemann: Oxford, UK, 1992.
42. Ihm, J.; Cohen, M.L. Equilibrium properties and the phase transition of grey and white tin. *Phys. Rev. B* **1981**, *23*, 1576–1579. [CrossRef]
43. Harrison, P.G. *Chemistry of Tin*; Blackie: Glasgow, Scotland, 1989.
44. Barin, I.; Knacke, O.; Kubaschewski, O. *Thermochemical Properties of Inorganic Substances*; Springer-Verlag: Berlin, Germany, 1973.
45. Morris, J.R.; Wang, C.Z.; Ho, K.M.; Chan, C.T. Melting line of aluminum from simulations of coexisting phases. *Phys. Rev. B* **1994**, *49*, 3109–3115. [CrossRef]
46. Zhu, L.-F.; Grabowski, B.; Neugebauer, J. Efficient approach to compute melting properties fully from ab initio with application to Cu. *Phys. Rev. B* **2017**, *96*, 224202. [CrossRef]
47. Togo, A.; Tanaka, I. First principles phonon calculations in materials science. *Scr. Mater.* **2015**. [CrossRef]
48. Dean, J.A. *Lange's Handbook of Chemistry*; McGraw-Hill: New York, NY, USA, 1985.
49. Bruson, A.; Gerl, M. Diffusion coefficient of ^{113}Sn, ^{124}Sb, ^{110m}Ag, and ^{195}Au in liquid Sn. *Phys. Rev. B* **1980**, *21*, 5447–5454. [CrossRef]
50. Careri, G.; Paoletti, A.; Vicentini, M. Further experiments on liquid Indium and Tin self-diffusion. *Il Nuovo Cimento* **2008**, *10*, 1088. [CrossRef]
51. Itami, T.; Aoki, H.; Kaneko, M.; Uchida, M.; Shisa, A.; Amano, S.; Odawara, O.; Masaki, T.; Oda, H.; Ooida, T.; et al. Diffusion of Liquid Metals and Alloys—The study of self-diffusion under microgravity in liquid Sn in the wide temperature range. *Jpn. Soc. Micrograv. Appl.* **1998**, *15*, 225.
52. Frohberg, G.; Kraatz, K.H.; Weber, H. Diffusion and transport phenomena in liquids under microgravity. In Proceedings of the 6th European Symposium on Materials Sciences under Microgravity Conditions, Bordeaux, France, 2–5 December 1986; ESA: Paris, France.
53. Available online: http://www.ctcms.nist.gov/potentials (accessed on 1 October 2018).

Article

Identification of the Flow Properties of a 0.54% Carbon Steel during Continuous Cooling

Christoph Rößler [1,2], David Schmicker [1,2], Oleksii Sherepenko [3], Thorsten Halle [3], Markus Körner [2,3], Sven Jüttner [3,*] and Elmar Woschke [1]

[1] Institute of Mechanics, Otto von Guericke University Magdeburg, 39106 Magdeburg, Germany;
christoph.roessler@ovgu.de (C.R.); david.schmicker@ovgu.de (D.S.); elmar.woschke@ovgu.de (E.W.)
[2] Sampro GmbH, 39110 Magdeburg, Germany
[3] Institute of Materials and Joining Technology, Otto von Guericke University Magdeburg,
39106 Magdeburg, Germany; oleksii.sherepenko@ovgu.de (O.S.); thorsten.halle@ovgu.de (T.H.);
markus.koerner@ovgu.de (M.K.)
* Correspondence: sven.juettner@ovgu.de

Received: 16 December 2019; Accepted: 4 January 2020; Published: 9 January 2020

Abstract: The determinination of material properties is an essential step in the simulation of manufacturing processes. For hot deformation processes, consistently assessed Carreau fluid constitutive model derived in prior works by Schmicker et al. might be used, in which the flow stress is described as a function of the current temperature and the current strain rate. The following paper aims to extend the prior mentioned model by making a distinction, whether the material is being heated or cooled, enhancing the model capabilities to predict deformations within the cooling process. The experimental identifaction of the material parameters is demonstrated for a structural carbon steel with 0.54% carbon content. An approach to derive the flow properties during cooling from the same samples used at heating is presented, which massively reduces the experimental effort in future applications.

Keywords: flow stress, hot deformation, carbon steel, continuous cooling, phase transformations

---

1. Introduction

The input of manufacturing process simulations has to include information about the actual process as well as the geometries and the materials being processed. Geometry and process information tend to be easier to obtain, because those are constantly monitored for quality assurance, for instance. The assessment of the material information on the other hand is more challenging, because several factors have to be considered.

A manufacturing process involving a broad range of temperatures and strain rates is rotary friction welding (RFW). In this process, the energy to form a permanent bond is directly introduced in the joining zone in form of frictional heat, which is generated by pressing the parts together while a relative motion is performed. Therefore, through this process temperatures close to the liquidus temperature of the softer material are achieved [1,2] and the strain rates reach one-digit values [3,4]. To simulate RFW processes, Schmicker et al. [4–7] elaborated a non-linear fluid model to describe the material flow. Although there are material models more capable in representing the dependencies of the yield stress on the deformation temperature, the strain rate, the degree of deformation, the phase composition and more, this model is suitable for RFW and for other hot deformation processes, too. A major advantage are the fairly expedient material parameters that can be gathered with only a few samples in a tensile testing routine, in which the test specimen is continuously heated. Besides conducting own experiments, the model can also be parametrized using data from encyclopedias such as [8,9], which significantly lowers the necessary efforts for the application of the process simulation.

As Schmicker et al. define the flow stress solely by steady-state values, the material point history is not taken into account. Similarly, aforementioned encyclopedias may also contain only the first three factors listed previously.

During heating, the microstructure of carbon steels starts to change to austenite by diffusion processes, if the temperature is greater than Ac_1 and which finish at Ac_3. Both, Ac_1 and Ac_3, depend on the rate of heating, which is documented in time-temperature-austenitization (TTA) diagrams as found in [10]. During cooling, the reverse transformation deviates from this and primarly depends on the cooling rate, which is documented in continuous cooling transformation (CCT) diagrams as depicted in Figure 1. Further information in the CCT include the microstructural composition as well as the hardness H, typically in Vickers hardness (HV),which occur at different cooling paths.

The microstructural changes have an impact on the flow properties of the material and the properties during cooling therefore differ from the properties during heating [11]. This should be taken into account in the process simulation if the material is still being deformed during cooling and independently of this in the residual stresses analysis as the yield strength limits the stress formation.

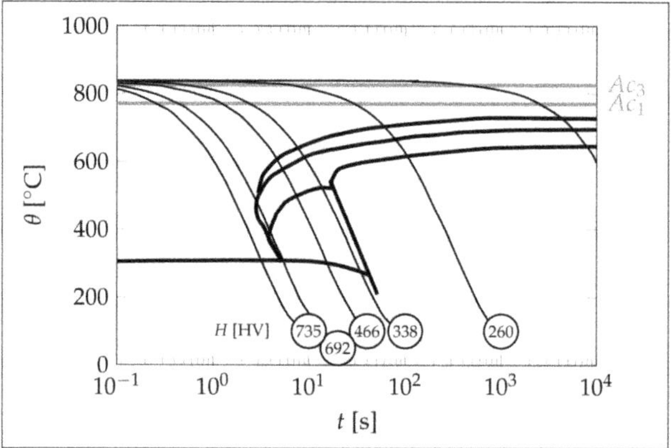

Figure 1. CCT diagram of a 0.55% carbon steel austenitized at 840 °C [12].

The current paper aims to quantify the flow properties of a 0.54% carbon steel during continuous cooling and to link the gathered results to the properties at heating, which are assumed to match the steady-state reference values, because the investigated temperature range is limited around and below Ac_1 temperature, at which the diffusional transformations just start.

To experimentally determine the material properties, a prior developed testing routine by Schmicker is extended by tests with continuous cooling. Since there exist an infinite number of cooling rates, a practical approach is proposed, which involves the hardnesses and transformation temperatures found in CCT diagrams as shown in Figure 1. This approach based on reference flow properties and hardness values aims to ensure that all model parameters can also be determined using the data available in literature collections. Concerning hot forming processes, data found in deformation CCT diagrams [13] indicate that compared to the cooling rate, the effect of deformations on the hardness is significantly smaller.

2. Materials and Methods

2.1. Mathematical Approach

The description and evaluation of the plastic material properties presented below is based on the following assumptions:

- The material behaves isotropic regarding all mechanical properties.

- Strain hardening effects are negligible compared to the phase transformation effects.
- Compared to the plastic deformations, the elastic, thermal and transformation strains are negligible small.

In the Norton–Bailey constitutive model [14], in the double logarithmic depiction the stress–strain rate relation is linear, which can be formulated as

$$\dot{\varepsilon} = A\sigma^n \tag{1}$$

or as

$$\frac{\dot{\varepsilon}}{\dot{\varepsilon}_0} = \left(\frac{\sigma}{\sigma_0}\right)^n \tag{2}$$

in which $\sigma_0(\theta)$ and $n(\theta)$ are material parameters specified for a reference strain rate $\dot{\varepsilon}_0$ and as a function of temperature θ. Equation (2) is valid for a wide range of strain rates and materials and documented in several data collections [8,9]. Under the premise that all isothermal flow curves $\sigma_{iso}(\theta_{iso}, \dot{\varepsilon})$ intersect in one characteristic point $C(\sigma_C, \varepsilon_C)$, Schmicker et al. [7] bypass the determination of $n(\theta)$ for every temperature. Therefore, knowing the point C, the Norton–Bailey exponent can be expressed as

$$n = \frac{\log\left(\frac{\dot{\varepsilon}_C}{\dot{\varepsilon}_0}\right)}{\log\left(\frac{\sigma_C}{\sigma_0}\right)} \tag{3}$$

as a feature in the so-called consistently assessed Carreau fluid model. The strain rate sensitivity typically increases with increasing temperatures and to produce higher strain rates, higher stresses are required, which is ensured by

$$\sigma_C > \max(\sigma_0(\theta, \dot{\varepsilon}_0))$$
$$\dot{\varepsilon}_C > \dot{\varepsilon}_0 \tag{4}$$

To distinguish heating and cooling, the flow properties during continuous cooling are subsequently denoted by an apostrophe. Two effects are taken into account for.

Firstly, a heat treatment effect, in which the material either hardens or softens due to microstructural changes. Rapid quenching causes the formation of martensite, which achieves more than twice the hardnesses than ferrite and pearlite. In annealing processes on the other hand, the cooling is typically slow to avoid this transformation.

Secondly, a transformation inertness that causes the transformation to shift to other temperature ranges depending on the cooling rate. The quicker the cooling process, the less is the time for the carbon diffusion processes. If the diffusion can not take place at all, the carbon become trapped in a body-centered tetragonal lattice configuration below martensite start temperature. In Figure 1 it is also seen that even for very slow cooling, the transformation starts well below Ac_3.

Assuming that σ_0' during cooling is not necessarily identical to σ_0 at heating, but similarly shaped, the two curves are linked by adding offset parameters κ and θ_κ

$$\sigma_0'(\theta, \dot{\theta}) = \kappa\,\sigma_0(\theta + \theta_\kappa) \tag{5}$$

for prior discussed transformation effects, depending on $\theta, \dot{\theta}$.

The hardening factor κ can be interpreted as a vertical scaling of $\sigma_0(\theta)$ to account for the heat treating effect as prior applied by Rößler et al. [15]. For its evaluation the linear relation in-between the yield strength σ_y and the Vickers hardness H [16,17]

$$\sigma_y = aH + b \tag{6}$$

is utilized, in which it is physically reasonable that b is zero. It should be mentioned that for other hardness scales the correlation can be non-linear. At room temperature, κ can be expressed as the proportion

$$\kappa(\theta_0, \dot{\theta}) = \frac{\sigma_0'(\theta_0, \dot{\theta})}{\sigma_0(\theta_0 + \theta_\kappa)} = \frac{H_0'(\dot{\theta})}{H_0} \tag{7}$$

in which H_0 is the initial hardness corresponding to $\sigma_0(\theta_0)$ and H_0' the hardness after cooling with a specific rate. To couple the hardening to the actual transformation, sigmoid function

$$\kappa(\theta, \dot{\theta}) = \kappa(\theta_0, \dot{\theta}) - \frac{\kappa(\theta_0, \dot{\theta}) - 1}{2} \tanh\left(\frac{2}{\theta_r(\dot{\theta}) - Ac_1}(\theta - Ac_1)\right) \tag{8}$$

limits hardening to the lower temperature range. The parameters H_0' and θ_r are either found in the CCT diagram (Figure 2) or can be experimentally determined by indentation and dilatometric testing. The use of the Ac_1 temperature is a recommendation for a free value of this equation, because it guarantees $(\theta(\dot{\theta}) - Ac_1) < 0$, which must always be fulfilled for mathematical reasons.

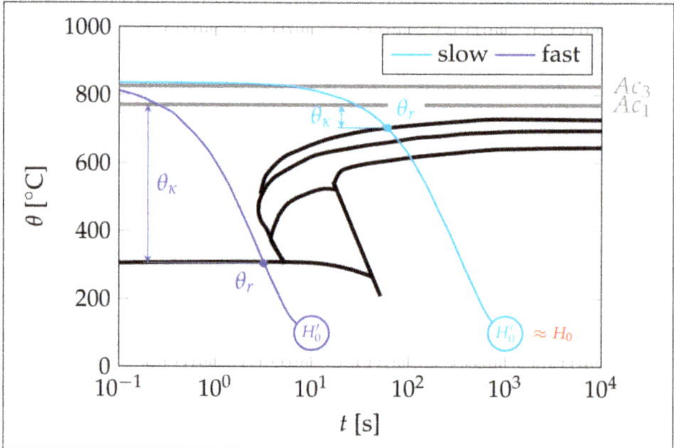

Figure 2. Extraction of κ, θ_κ for fast and slow cooling rates.

To shift $\sigma_0(\theta)$ horizontally due to a transformation inertness, the delay temperature

$$\theta_\kappa(\theta, \dot{\theta}) = \frac{Ac_1 - \theta_r(\dot{\theta})}{2} \tanh\left(\frac{2}{\theta_r(\dot{\theta}) - Ac_1}(\theta - Ac_1)\right) \tag{9}$$

is similar defined as κ. It is worth mentioning that κ and θ_κ actually start to raise above θ_r to compensate discontinuities of σ_0' introduced by large θ_κ for martensitic transformations, for instance.

Concerning Equation (3), the characteristic intersection point C' has to be re-evaluated, too, to satisfy (4) for $\kappa \gg 1$. To maintain the same high strain rate senstivity at high temperatures around the melting point θ_M and assuming that the low sensitivity for low temperatures will not change, point C' is identified using

$$\begin{aligned} n'(\theta_M) &= n(\theta_M) \\ n'(\theta_0) &= n(\theta_0) \end{aligned} \tag{10}$$

To estimate hardnesses not documented in the CCT diagram, law of mixture

$$H = \sum_i \xi_i H_i \tag{11}$$

as presented by Ion et al. [18] can be applied, in which ξ are the phase fractions of ferrite, pearlite, bainite, martensite, and austenite. The calculation of the phase fraction might be done numerically using evolution equation

$$\dot{\xi}(t) = \frac{\xi_\infty(\theta) - \xi(t)}{\tau(\theta)} \tag{12}$$

by Leblond and Devaux [19].

2.2. Materials

A detailed summary of the chemical composition of the investigated steel is provided in Table 1. The round tensile specimens used in tensile testing are machined from rods measuring 10 mm in diameter.

Table 1. Chemical composition of the investigated 0.54% carbon steel.

C	Si	Mn	P	S
0.54	0.21	0.63	0.008	0.006

2.3. Experimental Setup

The uniaxial tensile tests were carried out using Gleeble 3500 testing machine at Otto von Guericke University Magdeburg. Part of its capabilities was the execution of displacement controlled, uniaxial tensile tests, while the temperature of the test specimen could be varied using conductive heating at the same time. Figure 3 provides an overview of the setup. Within the machine, the force $F(t)$ required to produce the stroke $s(t)$ was captured and an additionally installed extensometer monitored the diameter $d(t)$ in the axial center of the test specimen. At the same location, the temperature $\theta(t)$ was measured by welded-on K-type thermo couples, which were used to control the amperage for the conductive heating. The initial diameter in the monitored section was 6 mm based on the permissible force of the testing machine. For cooling upto 30 K/s, two air cooling jets were installed, which aimed at the center of the test specimen. The rate of 30 K/s could be maintained down to 250 °C, which was well below the martensite start temperature.

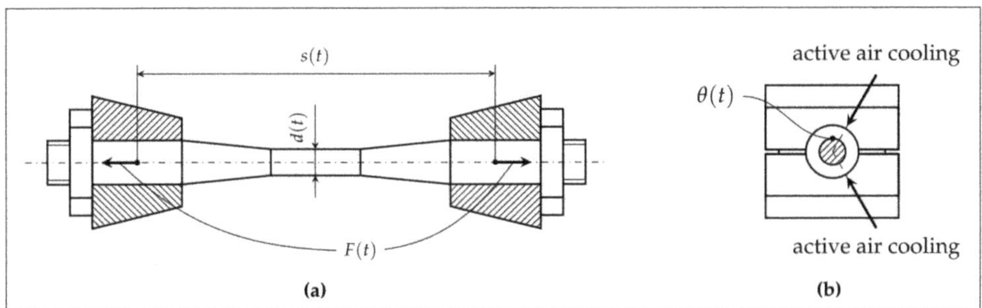

Figure 3. Gleeble 3500 machine setup (**a**) installed test specimen, (**b**) positions of the air cooling jets.

In both routines, the specimens were exposed to linear temperature and displacement profiles in the evaluation range (Figure 4). This allowed us to gather information about the flow properties for a number of temperatures in a single test. In order to examine different ranges of θ and $\dot{\varepsilon}$, the start temperature θ_{in} was varied as well as the stroke rate $\dot{s}$. A drawback of altering these parameters simultaneously was that factors such as crystal recovery, recrystallization, and grain growth remained inseparable from the thermal dependence and were therefore averaged in the evaluation of the tensile tests. To overcome elastic deformations, pre-stroke s_{in} was set to 0.3 mm, before the actual evaluation range begins. A comparitive summary of the test conditions during heating and cooling is provided in Table 2.

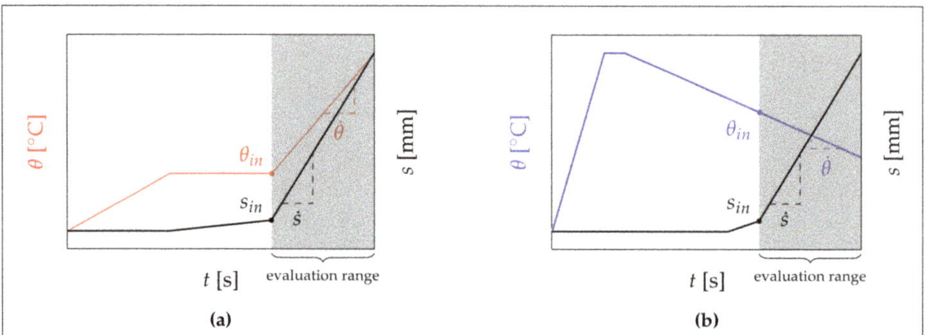

Figure 4. Schematic thermal and axial stroke profiles in the tests during (**a**) heating and (**b**) cooling.

Table 2. Summary of test conditions.

	Heating	Cooling
θ_{in}	$\geq$300 °C	$\leq$800 °C
$\dot{\theta}$	1–2000 K/s	5 K/s, 30 K/s
s_{in}	0.3 mm	
$\dot{s}$	0.003–30 mm/s	0.03–0.3 mm/s

In the cooling tests, all test specimens were heated up to 1200 °C to trigger the phase transformation to austenite. The controlled cooling process was then started by turning on the air jets, where the central thermocouple was used to ensure constant cooling rates. The cooling rates $\dot{\theta}$ were exemplarly set to 5 K/s, at which it was assumed that the microstructure was similar to the initial one, and to 30 K/s, at which martensite is formed. For heating, test data for about 40 specimens were available by Jüttner and Körner [20] and additional 20 tests were performed for both examined $\dot{\theta}$, because of the narrowed temperature evaluation range.

To evaluate Equation (7), Vickers hardness measurements were carried out across the center of the lateral cross-sections of test specimens, tested with the same cooling routine without applying any axial forces. The indentation locations were arranged in three lines, each of which got 10 points spaced over the diameter, with 2 mm axial spacing.

3. Results

Based on the measured data for $F(t)$ and $d(t)$, the longitudinal true stress

$$\sigma = \frac{4F}{\pi d^2} \tag{13}$$

and the longitudinal true strain rate

$$\dot{\varepsilon} = -2\frac{\dot{d}}{d} \tag{14}$$

were derived.

In order to fit the model according to Equation (3), the generated σ-$\dot{\varepsilon}$-θ triples were classified by θ and least squares method was applied to identify $C(\sigma_C, \varepsilon_C)$ and a monotonous $\sigma_0(\theta, \varepsilon_0)$ in a one-step optimization. All measured data and the fitted models are depiced in Figure 5a for the heating experiments respectively Figure 5b for cooling at 5 K/s and Figure 5c at 30 K/s. Tabular summaries of the fitted models are given in Tables 3 and 4. The coefficients of determination

$$R^2 = \frac{\sum_i (\log \sigma(\theta_i, \dot{\varepsilon}_i) - \log \bar{\sigma})^2}{\sum_i (\log \sigma_i - \log \bar{\sigma})^2} \tag{15}$$

are listed in Table 3, confirming the suitablity of the consistently assessed Carreau fluid model for the investigated steel.

Table 5 lists the averaged hardness measurements, which show that at 30 K/s more than twice as high hardnesses arises, which is linked to the formation of martensite. Cooling by 5 K/s produces hardnesses about 10 HV below the initial hardness.

Table 3. Correlation of the experimental data and fitted model.

	Heating	Cooling	
		5 K/s	30 K/s
σ_C	$1.53 \cdot 10^3$	$4.03 \cdot 10^3$	$9.25 \cdot 10^4$
$\dot{\varepsilon}_C$	$5.51 \cdot 10^6$	$7.06 \cdot 10^6$	$3.83 \cdot 10^6$
R^2	0.9084	0.9421	0.8830

Table 4. Determined flow properties for $\dot{\varepsilon}_0 = 0.001$ s^{-1}.

	Heating		5 K/s		30 K/s	
θ [°C]	σ_0 [MPa]	n [-]	σ_0 [MPa]	n [-]	σ_0 [MPa]	n [-]
200	–	–	–	–	440.4	4.13
250	–	–	728.3	13.26	315.7	3.88
300	774.3	32.93	706.8	13.03	232.8	3.69
350	774.3	32.93	670.3	12.64	215.1	3.64
400	759.3	32.01	648.4	12.41	196.4	3.59
450	669.9	27.16	530.6	11.19	165.9	3.49
500	523.3	20.90	399.8	9.82	129.9	3.36
550	381.1	16.14	274.0	8.44	102.8	3.24
600	285.0	13.35	178.9	7.28	84.6	3.15
650	207.0	11.21	118.6	6.43	68.8	3.06
700	147.1	9.58	93.1	6.02	56.0	2.98
750	97.2	8.14	74.6	5.68	46.0	2.90
800	73.0	7.37	56.9	5.32		
850	59.6	6.91				
900	48.6	6.50				
950	39.4	6.13				
1000	32.7	5.83				
1050	26.5	5.53				
1100	22.0	5.29				
1150	18.4	5.08				
1200	17.0	4.98				
1250	13.5	4.74				
1300	10.1	4.47				

Table 5. Vickers hardness measurements (averaged) and transformation temperatures taken from Figure 1.

		Heating	Cooling	
			5 K/s	30 K/s
H_0, H_0'	[HV]	322	311	733
θ_r	[°C]	–	690	305
Ac_1	[°C]	720		
Ac_3	[°C]	850		

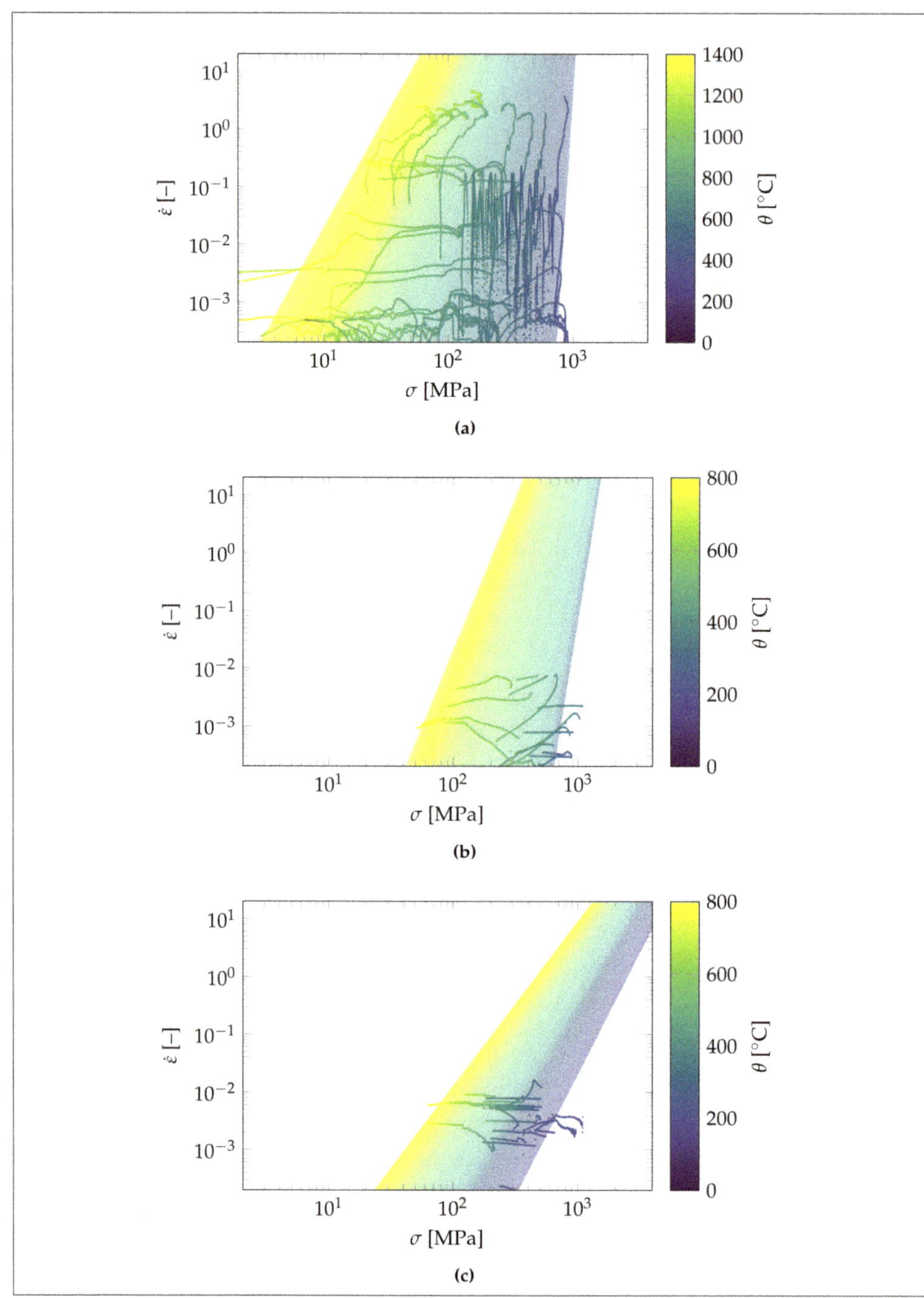

Figure 5. Flow properties examined (**a**) during heating, (**b**) at 5 K/s cooling, (**c**) at 30 K/s cooling; The surface plots represent the fitted consistently assessed Carreau fluid models for each routine.

4. Discussion

Using the presented testing routine during cooling, it becomes difficult to evaluate data for $\theta < 200\,°C$, because the samples start to fracture, due to the increasing brittleness, especially when martensite is formed. Another impediment is the strain hardening at such temperatures. Narrowing down the temperature evaluation ranges in the tests fixes both problems, but involves an increased number of tests, which contradicts the original idea of this routine.

A visualization of $\sigma_0(\theta)$ of Table 4 is presented in Figure 6. The results at heating show that the slope of $\sigma_0(\theta)$ and $n(\theta)$ is small at around $300\,°C$, justifying to extrapolate both values to θ_0.

Comparing σ_0 in the range of $200\,°C < \theta < 800\,°C$, at the same temperature, differences up to 500 MPa occur, which can be explained by the amount of austenite that is still present at $300\,°C$ at sufficient quick cooling. Below $300\,°C$, the slope starts to increase as well, though. The values during cooling at 5 K/s on the other hand almost correspond to the heating values shifted by about $-50\,K$.

Besides the optimzation results from the measurements, Figure 6 also contains the estimated σ_0' using Equation (5) for both cooling rates, which reproduces the experimentally determined curves very well. The parameters in use are listed in Table 5.

The hardness measurements greater than 730 HV confirm the martensite formation in the experiments. Therefore, θ_r for cooling at 30 K/s is assumed to match the martensite start temperature. In this context it is necessary to mention, that CCT diagrams are sensitivite to fluctuations in the chemical composition as well as the specific test conditions for the identification of the CCT diagram. Compared to the tensile tests, the used CCT diagram in Figure 1 has been determined for a lower austenitizing temperature of $840\,°C$. Analogous to deformation CCT diagrams, special weld CCT diagrams exist, which are characterized by high austenitisation temperatures and short holding phases. The described conditions increase the transformation inertness during cooling, which leads to the formation of microstructures of higher hardness at lower cooling rates. Relative to conventional CCT diagrams for heat treatment, the transformation points in the weld CCT diagrams are shifted to the bottom right [21].

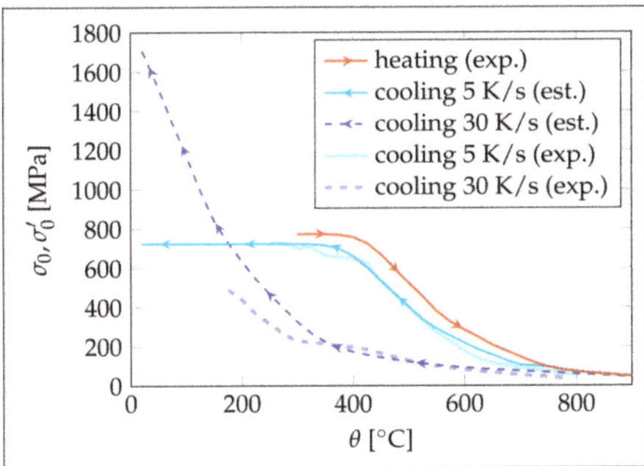

Figure 6. Estimated σ_0' based on σ_0 and parameters from Table 5.

5. Conclusions

The flow properties of a 0.54% carbon steel are identified during continuous cooling and compared to results at heating. The findings are summarized as follows.

- The model by Schmicker et al. is successfully applied to continuous cooling, without violating its consistency.
- The higher the cooling rate, the greater the differences in the flow properties at the same temperature during heating and cooling.
- Using the data available in CCT diagrams, the properties during cooling can be approximated based on the properties determined at heating, which allows to increase the accuracy compared to the model without the presented adaptions.

Author Contributions: Conceptualization, C.R., D.S.; methodology, C.R., D.S., T.H. and E.W.; software, C.R., D.S.; validation, O.S. and M.K.; formal analysis, C.R. and M.K.; investigation, C.R. and O.S.; resources, E.W. and S.J.; data curation, O.S.; writing—original draft preparation, C.R.; writing—review and editing, D.S., O.S. and M.K.; visualization, C.R.; supervision, T.H., S.J. and E.W.; project administration, C.R. and S.J.; funding acquisition, D.S. and S.J. All authors have read and agreed to the published version of the manuscript.

Funding: This research was funded by AIF as part of IGF 18.966 B.

Conflicts of Interest: The authors declare no conflict of interest.

Abbreviations

The following abbreviations are used in this manuscript:

CCT   Continuous cooling transformation
RFW   Rotary friction welding
TTA   Time-temperature-austenitization

References

1. Mousavi, A.; Asghar, S.A.; Rahbar, A. Experimental and Numerical Analysis of the Friction Welding Process for the 4340 Steel and Mild Steel Combinations. *Weld. J.* **2008**, *87*, 178–186.
2. Seli, H.; Ismail, A.I.M.; Rachman, E.; Ahmad, Z.A. Mechanical evaluation and thermal modelling of friction welding of mild steel and aluminium. *J. Mater. Process. Technol.* **2010**, *210*, 1209–1216. [CrossRef]
3. Grant, B.; Preuss, M.; Withers, P.J.; Baxter, G.; Rowlson, M. Finite element process modelling of inertia friction welding advanced nickel-based superalloy. *Mater. Sci. Eng.* **2009**, *513–514*, 366–375. [CrossRef]
4. Schmicker, D. *A Holistic Approach on the Simulation of Rotary Friction Welding*; ePubly GmbH: Berlin, Germany, 2015.
5. Schmicker, D.; Naumenko, K.; Strackeljan, J. A robust simulation of Direct Drive Friction Welding with a modified Carreau fluid constitutive model. *Comput. Methods Appl. Mech. Eng.* **2013**, *265*, 186–194. [CrossRef]
6. Schmicker, D.; Persson, P.O.; Strackeljan, J. Implicit Geometry Meshing for the simulation of Rotary Friction Welding. *J. Comput. Phys.* **2014**, *270*, 478–489. [CrossRef]
7. Schmicker, D.; Paczulla, S.; Nitzschke, S.; Groschopp, S.; Naumenko, K.; Jüttner, S.; Strackeljan, J. Experimental identification of flow properties of a S355 structural steel for hot deformation processes. *J. Strain Anal. Eng. Des.* **2015**, *50*, 75–83. [CrossRef]
8. Spittel, M.; Spittel, T.; Warlimont, H.; Landolt, H.; Börnstein, R.; Martienssen, W. (Eds.) *Numerical Data and Functional Relationships in Science and Technology: New Series, Group VIII, Volume 2, Subvolume C, Part 1: Ferrous Alloys*; Springer: Berlin, Germany, 2009.
9. Spittel, M.; Spittel, T.; Warlimont, H.; Landolt, H.; Börnstein, R.; Martienssen, W. (Eds.) *Numerical Data and Functional Relationships in Science and Technology: New Series, Group VIII, Volume 2, Subvolume C, Part 2: Non-ferrous Alloys–Light Metals*; Springer: Berlin, Germany, 2011.
10. Rech, J.; Hamdi, H.; Valette, S. Workpiece Surface Integrity. In *Machining*; Springer: London, UK, 2008; pp. 59–96. [CrossRef]
11. Radaj, D. *Heat Effects of Welding: Temperature Field, Residual Stress, Distortion*; Springer: Berlin/Heidelberg, Germany, 1992. [CrossRef]

12. Werkstoff-Datenblatt Saarstahl—C55R (Cm55). Available online: http://www.saarstahl.com/sag/downloads/download/11282 (accessed on 20 February 2019).
13. Nürnberger, F.; Grydin, O.; Schaper, M.; Bach, F.W.; Koczurkiewicz, B.; Milenin, A. Microstructure Transformations in Tempering Steels during Continuous Cooling from Hot Forging Temperatures. *Steel Res. Int.* **2010**, *81*, 224–233. [CrossRef]
14. Norton, F.H. *The Creep of Steel at High Temperatures*, 1st ed.; McGraw-Hill Book Company, Inc.: New York, NY, USA, 1929.
15. Rößler, C.; Schmicker, D.; Naumenko, K.; Woschke, E. Adaption of a Carreau fluid law formulation for residual stress determination in rotary friction welds. *J. Mater. Process. Technol.* **2017**. [CrossRef]
16. Prandtl, L. Über die Härte plastischer Körper. *Nachrich. Ges. Der Wiss. GÖTtingen-Math.-Phys. Kl.* **1920**, *1920*, 74–85.
17. Saeed, I. Untersuchungen über die Streuung und Anwendung von Fließkurven. In *Fortschrittberichte der VDI-Zeitschriften*; Grund- und Werkstoffe, VDI-Verlag: Düsseldorf, Germany, 1984; pp. 204–215.
18. Ion, J.C.; Easterling, K.E.; Ashby, M.F. A second report on diagrams of microstructure and hardness for heat-affected zones in welds. *Acta Metall.* **1984**, *32*, 1949–1962. [CrossRef]
19. Leblond, J.B.; Devaux, J. A new kinetic model for anisothermal metallurgical transformations in steels including effect of austenite grain size. *Acta Metall.* **1984**, *32*, 137–146. [CrossRef]
20. Jüttner, S.; Körner, M. *Entwicklung eines Reibgesetzes zur Erfassung des Drehzahleinflusses bei der Reibschweißprozesssimulation*; Otto von Guericke University Library: Magdeburg, Germany, 2019. [CrossRef]
21. Seyffarth, P.; Meyer, B.; Scharff, A. Großer Atlas Schweiss-ZTU-Schaubilder. In *Fachbuchreihe Schweisstechnik*; Dt. Verl. für Schweisstechnik DVS-Verl.: Düsseldorf, Germany, 1992; Volume 110.

Article

A Phenomenological Mechanical Material Model for Precipitation Hardening Aluminium Alloys

Hannes Fröck [1], Lukas Vincent Kappis [1], Michael Reich [1,*] and Olaf Kessler [1,2]

[1]   Chair of Materials Science, Faculty of Mechanical Engineering and Marine Technology, University of Rostock, 18059 Rostock, Germany; hannes.froeck@uni-rostock.de (H.F.); Lukas.kappis@uni-rostock.de (L.V.K.); olaf.kessler@uni-rostock.de (O.K.)

[2]   Competence Centre CALOR, Department Life, Light & Matter, Faculty of Interdisciplinary Research, University of Rostock, 18059 Rostock, Germany

*   Correspondence: michael.reich@uni-rostock.de; Tel.: +49-381-498-9490

Received: 25 September 2019; Accepted: 21 October 2019; Published: 29 October 2019

Abstract: Age hardening aluminium alloys obtain their strength by forming precipitates. This precipitation-hardened state is often the initial condition for short-term heat treatments, like welding processes or local laser heat treatment to produce tailored heat-treated profiles (THTP). During these heat treatments, the strength-increasing precipitates are dissolved depending on the maximum temperature and the material is softened in these areas. Depending on the temperature path, the mechanical properties differ between heating and cooling at the same temperature. To model this behavior, a phenomenological material model was developed based on the dissolution characteristics and experimental flow curves were developed depending on the current temperature and the maximum temperature. The dissolution characteristics were analyzed by calorimetry. The mechanical properties at different temperatures and peak temperatures were recorded by thermomechanical analysis. The usual phase transformation equations in the Finite Element Method (FEM) code, which were developed for phase transformation in steels, were used to develop a phenomenological model for the mechanical properties as a function of the relevant heat treatment parameters. This material model was implemented for aluminium alloy 6060 T4 in the finite element software LS-DYNA (Livermore Software Technology Corporation).

Keywords: aluminium alloy; EN AW-6060; precipitation hardening aluminium alloys; material model; heating; cooling; flow cures; LS-DYNA

---

1. Introduction

Heat treatments are used to influence the mechanical properties of metallic materials. The material properties during and after heat treatment are dependent on different heat treatment parameters. Numerical heat treatment simulation offers the possibility of predicting process results, such as temperature gradients, mechanical properties, residual stress or distortion, in the heat-treated component. By using numerical simulation, process understanding can be further improved by simulating intermediate states, which are very difficult or impossible to measure experimentally. At the same time, the experimental effort for optimizing process parameters can be reduced by numerical simulation.

Precipitation-hardening aluminium alloys achieve their strengths through fine particles in the aluminium matrix. Different stable and metastable secondary phases can be formed or dissolved as a function of the temperature during the short-term heat treatment of precipitation-hardened states. Thus, the precipitation state, and therefore, the mechanical properties, change during short-term heat treatment. In heat treatment simulation, modelling the quenching of steels [1–4] and aluminium alloys [5–8] has been the subject of intensive research and is now used in many industrial applications [9].

In addition, numerical descriptions of the heat treatment processes are also used to simulate various manufacturing processes such as welding steels [10], aluminium alloys [11] or induction hardening [12]. Heat treatment simulation of the short-time heat treatments used for tailored heat-treating blanks (THTB) [13,14] and profiles (THTP) for aluminium has not yet been performed. These short-term heat treatments, which locally heat material areas using a laser, are characterized by very high heating rates of several 100 Ks^{-1}, almost no soaking at peak temperature and subsequent cooling at a few 10 Ks^{-1}. These short-term heat treatments can be used to intentionally soften local areas of semi-finished components and, thus, create a tailored strength layout. By tailoring the strength layout, the forming limits can be extended for subsequent forming. To further advance the development of THTP, a linked heat treatment and bending simulation can be used.

The mechanical properties of an aluminium alloy during a short-time heat treatment depend on the current temperature and the maximum heat treatment temperature. As a result, the mechanical properties during heating differ from those during cooling [15]. Currently, there is no material model for the mechanical properties of aluminium alloys depending on temperature and temperature path. For steels, this problem is solved using the mixture rules of different coarse phases like ferrite, pearlite, austenite, martensite, bainite, etc. [16]. For precipitation-hardened aluminium alloys with fine precipitates, this approach does not work at first glance. For aluminium alloys, some material models have also been developed. However, these are designed for special areas such as quench simulation [5–7] or ageing simulation [17,18]. However, these quench simulation material models can provide only temperature-dependent flow curves, while the ageing models only describe the change in mechanical properties during ageing. The mechanical properties as a function of the relevant parameters during short-time heat treatment cannot provide both types of models. Material models were also developed for producing tailored heat-treated semi-finished products [13,14]. However, these material models were developed for forming simulations and provide the mechanical properties at room temperature depending on the maximum temperature during heat treatment. The mechanical properties at elevated temperatures, as required in a heat treatment simulation, cannot be provided.

In this work, an efficient phenomenological material model is developed, which is based on the mixture rules of imaginary phases. Empirical-phenomenological model approaches provide simple descriptions of the material behavior as a function of the process parameters.

2. Materials and Methods

2.1. Examined Aluminium Alloy

The material model was developed using the precipitation hardening aluminium alloy EN AW-6060 in the natural aged state (T4). As a base material, a 20 mm × 20 mm × 2 mm hollow quadratic extrusion profile was used. The chemical composition of the investigated alloy is given in Table 1.

Table 1. Mass fraction of the alloying elements in the investigated alloy.

Alloy	Mass Fraction in %							
	Si	Fe	Cu	Mn	Mg	Cr	Zn	Al
OES EN AW-6060 T4	0.40	0.22	0.07	0.14	0.56	0.02	0.02	balance
DIN EN 573-3 (6060)	0.3–0.6	0.1–0.3	≤0.1	≤0.1	0.35–0.6	≤0.05	≤0.15	balance

2.2. Database Used

The precipitation and dissolution behavior of the alloy was recorded in previous work by direct [19] and indirect [20] differential scanning calorimetry (DSC) over a wide range of heating rates. The continuous time-temperature dissolution diagram of the alloy EN AW-6060 T4 can be derived from the DSC results. This diagram shows the temperature ranges at which a certain precipitation or dissolution reaction dominates depending on the heating rate. Figure 1 shows the continuous time-temperature dissolution diagram of the investigated alloy, EN AW-6060 T4.

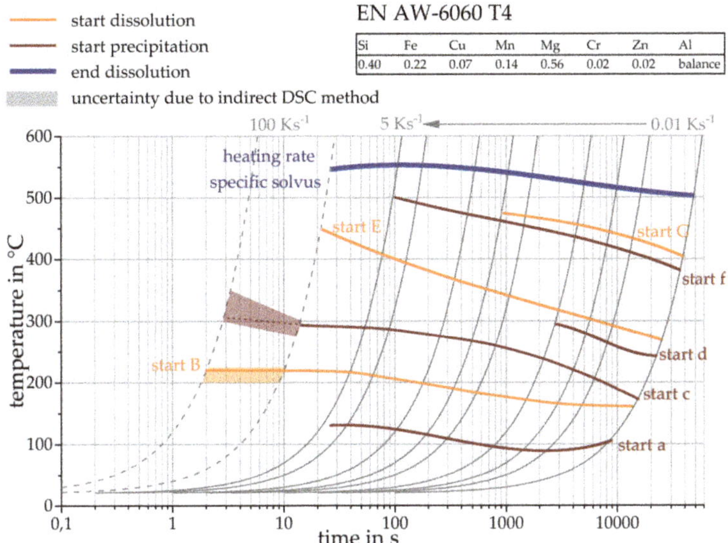

Si	Fe	Cu	Mn	Mg	Cr	Zn	Al
0.40	0.22	0.07	0.14	0.56	0.02	0.02	balance

Figure 1. Continuous heating dissolution diagram of EN AW-6060 T4 [20].

Figure 2 shows a DSC heating curve at 1 Ks^{-1}, together with 0.2% yield strength and tensile strength after heating tensile specimens to different maximum temperatures followed by overcritical quenching [21]. The results indicate that the endothermic peak B, which is interpreted as the dissolution of clusters and GP-Zones [22], is accompanied by a significant decrease in strength. The exothermic peak c + d, which is considered to be the precipitation of the β″ and β′ phases [23], leads to a renewed increase in strength. The subsequent endothermic peak E, is considered to be the dissolution of the β″ and β′ precipitates [24], with further material softening.

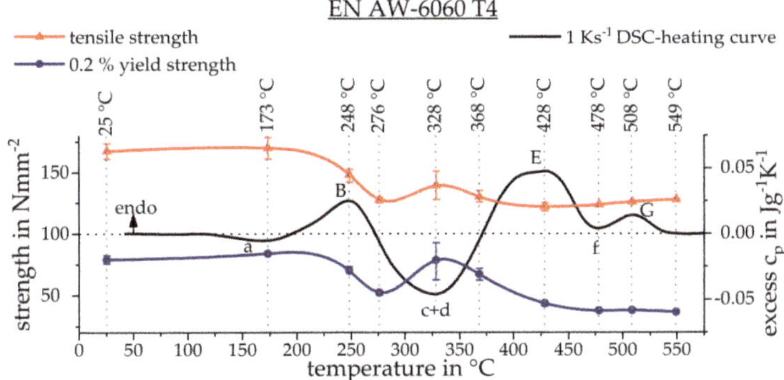

Figure 2. Continuous heating DSC curve of EN AW-6060 T4 at 1 Ks^{-1} correlated with the 0.2% yield and tensile strength after heating tensile specimens to different maximum temperatures followed by overcritical quenching. Experiments described in [21].

Short-term laser heat treatment, as well as recording time-temperature profiles, was carried out at the Institute of Manufacturing Technology of the University of Erlangen-Nürnberg [25]. Laser heat treatment is characterized by a high heating rate of up to several 100 Ks^{-1}, with no soaking at the maximum temperature and a relatively slow, non-linear, cooling with a few 10 Ks^{-1}. The investigated time-temperature courses are shown in Figure 3.

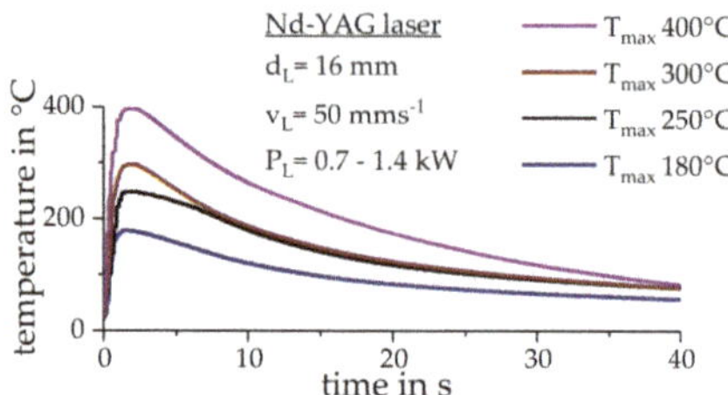

Figure 3. Recorded temperature courses of different laser heat treatments, described in [25].

These time-temperature profiles were imitated in a quenching and deformation dilatometer, interrupted at defined temperatures and tensile tests were carried out immediately at these temperatures in the same device [15]. The schematic measurement plan of the thermo-mechanical analysis, with the major parameters of the tensile test, is shown in Figure 4.

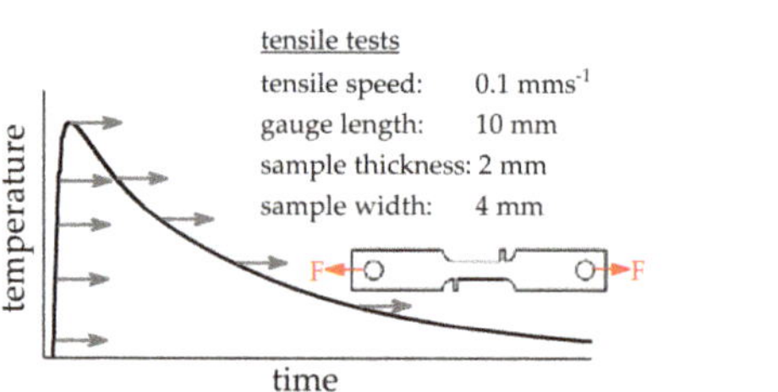

Figure 4. Schematic measurement plan of the thermo-mechanical analysis, with the major parameters of the tensile test. Described in more detail in [15].

It can be seen in Figure 5 that the strength of the alloy decreases during heating with increasing temperature. During subsequent cooling, the strength increases again with decreasing temperature. The mechanical properties during heat treatment with a 180 °C maximum temperature are identical during heating and cooling and only dependent on temperature, as shown in Figure 5A. Up to this temperature, no permanent softening of the material has taken place. In short-term heat treatment with a 250 °C maximum temperature, the strength during cooling does not increase to the same extent as it decreased during heating. The mechanical properties at a certain temperature are, therefore, not identical during heating and cooling, as shown in Figure 5B. The same behavior is evident in the short-term heat treatments at 300 °C and 400 °C maximum temperatures, as seen in Figure 5C,D. For a purposeful simulation of short-term heat treatment, these mechanical properties dependencies must be implemented in a material model.

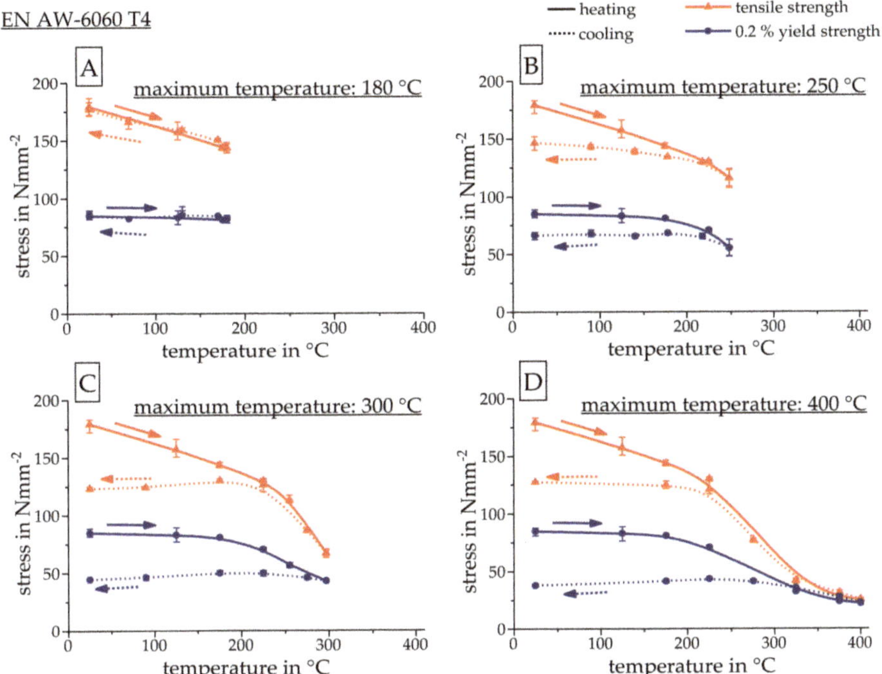

Figure 5. Mechanical properties (0.2% yield and tensile strength) during various short-term heat treatments. Experiments described in [15].

3. Material Model Development

3.1. The Basic Idea

We developed a phenomenological mechanical material model for precipitation hardening aluminium alloys, which can describe flow curves as a function of the actual temperature and the peak temperature of the heat treatment. The basic idea was to define the precipitation-hardened initial state as an imaginary hardened phase (A) and the state in which the strength-increasing precipitates were dissolved as an imaginary softened phase (B). These individual phases can be assigned temperature-dependent flow curves.

From a defined starting temperature (T_{start}), the imaginary hardened phase (A) is transformed into the imaginary softened phase (B) as a function of the temperature during heating. This transformation is completed at a defined finish temperature (T_{finish}). At temperatures between the start and finish temperature, a phase mixture of an imaginary hardened and an imaginary softened phase exists, as shown in Figure 6. At temperatures above the final temperature, the material consists of the softened phase (B). Through developing the material model, it has been found that the accuracy of the model can be increased by introducing an intermediate phase (Z) between the hardened phase (A) and the softened phase (B), see Figure 7. During cooling, no phase transformations are permitted. The resulting mechanical properties of the material state are defined by linearly mixing the mechanical properties of the occurring phases. The schematic of the implemented phase transformation during heat treatment is shown in Figure 7.

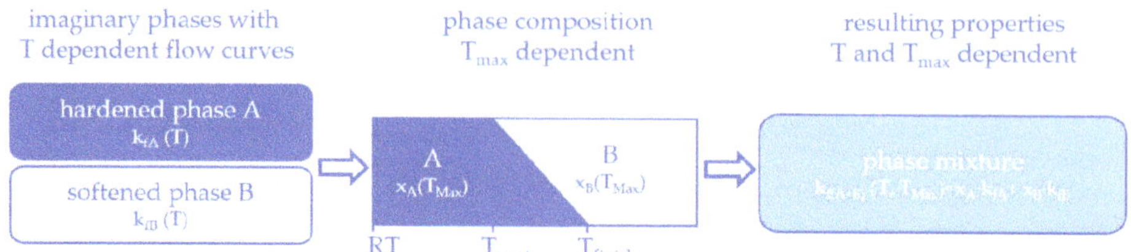

Figure 6. Schematic approach for an empirical-phenomenological material model of precipitation hardening aluminium alloys.

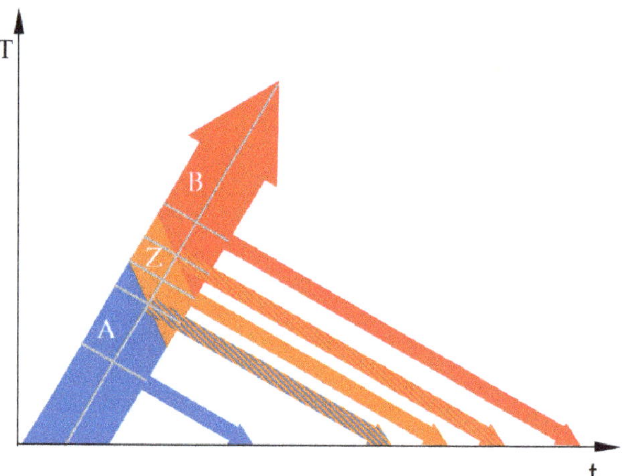

Figure 7. Schematic of the implemented phase transformation during heat treatment.

The following simplifications are assumed in the model:

- The influence of the heating and cooling rate was not considered. Typical heating and cooling rates for short-term laser heat treatment were chosen.
- During cooling, no precipitation reaction took place. This assumption is realistic, as the critical cooling rate of alloy 6060 is far below the cooling rates of short-term laser heat treatment.
- Quenching followed directly after reaching the maximum temperature. There was no isothermal soaking.
- The influence of the strain rate on the mechanical properties was not considered.
- The maximum plastic strains were 10%.
- The influence of ageing after short-time heat treatment was not considered. The model was valid for the as-quenched state.

3.2. Determination of the Phase Transformation Temperatures

To adapt the material model to the investigated alloy, the phase transformation temperature ranges must be determined. This was performed based on the continuous heating dissolution diagram (Figure 1) and mechanical results (Figure 5). Figure 5A shows that the mechanical properties of the heating and cooling step did not differ for short-term heat treatment up to a peak temperature of 180 °C. Figure 1 shows that the clusters and GP-zones of the initial state start to dissolve between 200 °C and 225 °C at a high heating rate of 100 Ks^{-1}. Thus, it was determined that the imaginary hardened phase (A) was completely present up to a maximum temperature of 212.5 °C. At higher temperatures, phase (A) continuously transformed into the intermediate phase (Z). At a maximum temperature of 250 °C, the material consisted only of the imaginary intermediate phase (Z). It can be seen in Figure 1 that at a heating rate of 100 Ks^{-1}, the dissolution of the cluster and GP-zones was completed at 300 °C. This is also reflected in the mechanical properties, as shown in Figure 5C,D. For this reason, it is assumed that the simulated phase transformation Z to B is completed at 300 °C. The properties above this peak temperature are described by phase (B). Figure 8 shows the implemented temperature transformation ranges of the imaginary phases.

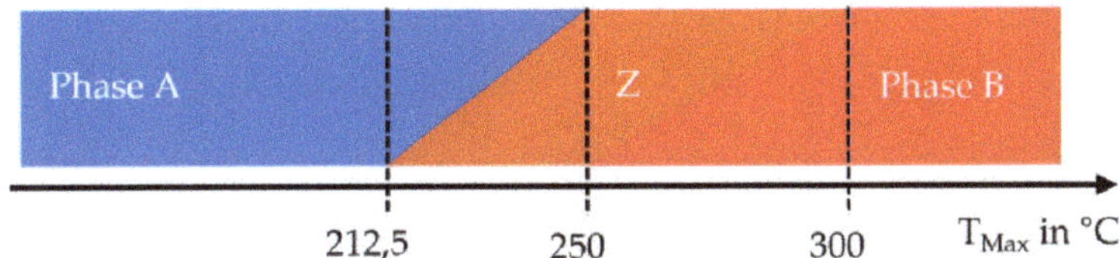

Figure 8. Schematic temperature transformation ranges of the imaginary phases for alloy 6060 T4.

3.3. Calculating the Resulting Flow Curves

The basis for the mechanical properties of individual phases is experimental stress–strain curves after different short-term heat treatments (Figure 5). For the thermo-mechanical simulation, flow curves are required as input data (true stress; k_f vs. logarithmic plastic strain; φ). Experimentally obtained flow curves were converted into a temperature-dependent mathematical description. The flow curves of face-centered cubic metals, like aluminium alloys, can be described well by the Hockett–Sherby [26] relationship, see Equation (1).

$$k_f(\varphi) = k_s - e^{-m\varphi^P}(k_s - k_0) \tag{1}$$

This equation contains four parameters, the initial flow stress (k_0), the saturation stress (k_s) and the hardening exponents, m and P. These parameters were adapted to the experimentally determined flow curves, using OriginPro 2018. By adapting the parameters to the flow curves at different temperatures, the Hockett–Sherby parameters of a phase can be derived over the existence range of this phase.

The Hockett–Sherby parameters of the imaginary phase (B) can be obtained over the entire temperature range directly from the experimental flow curves. The experimental basis for the temperature-dependent flow curves of the imaginary softened phase (B) is the experimentally obtained flow curves during cooling after the 400 °C maximum temperature, see Figure 9, and an additional tensile test at 550 °C. The flow curves are shown in the following figures as true stress k_f vs. logarithmic plastic strain φ. The points in Figure 10 show the Hockett–Sherby parameters for phase (B), which were derived from the experimental data. These parameters from the experimental data can be fit via linear or nonlinear mathematical functions.

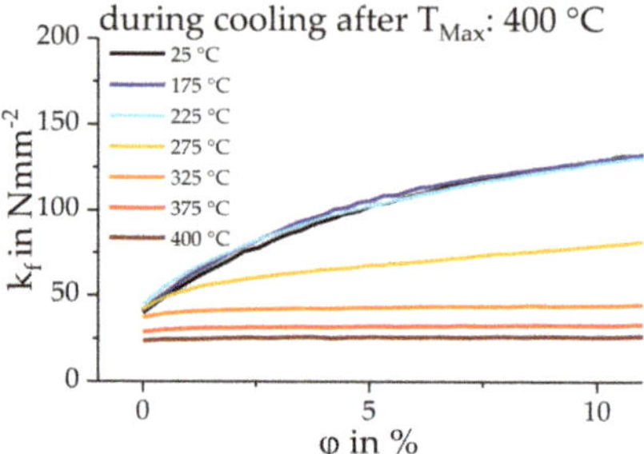

Figure 9. Experimentally obtained flow curves of EN AW-6060 T4 alloy at different temperatures during cooling after the 400 °C maximum temperature, phase (B).

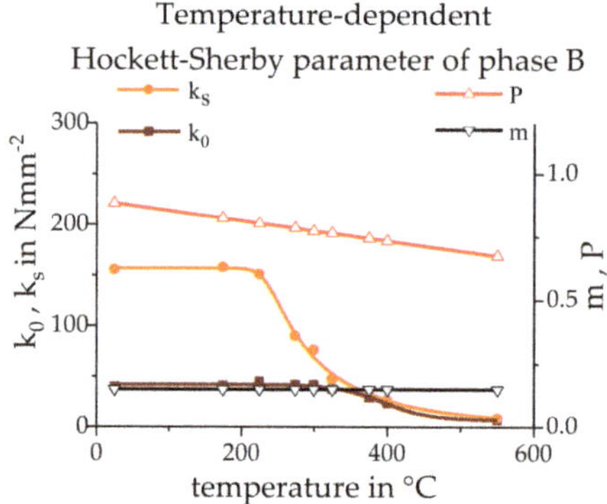

Figure 10. Temperature-dependent Hockett–Sherby parameters of phase (B).

In the temperature range between 212.5 °C and 300 °C, the mechanical properties were represented by an imaginary phase mixture, A + Z respectively Z + B, as shown in Figure 8. The resulting flow curve (k_f) can be described using Equation (2), taking into account the phase proportions of the existing phases (x_i) and the flow curves of the individual phases (k_{fi}). The phase proportions were calculated via a phase transformation model as a function of temperature (Section 3.3), while the individual phases were assigned to temperature-dependent flow curves.

$$k_f = \sum_{i=1}^{n} x_i k_{fi}$$

(2)

The resulting flow curves during the heating of the imaginary phase mixture A + Z respectively Z + B can be determined experimentally by the tensile tests performed during heating. Figure 11 shows the experimentally obtained flow curves of the imaginary phase mixture A + Z respectively Z + B at various temperatures during heating. The points in Figure 12 show the Hockett–Sherby parameters while heating alloy EN AW-6060, which were derived from the experimental data. The points were fit by mathematical formulas, which allows the Hockett–Sherby parameters to be determined for all temperatures during heating.

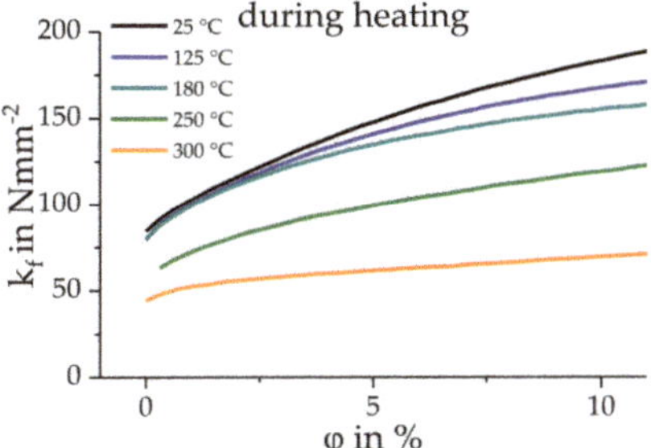

Figure 11. Experimentally obtained flow curves of imaginary phase mixture A + Z respectively Z + B for alloy EN AW-6060 T4 at different temperatures during heating at 100 Ks^{-1}.

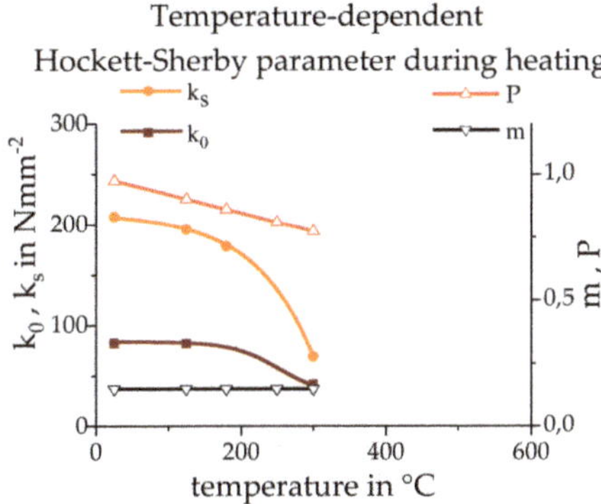

Figure 12. Temperature-dependent Hockett–Sherby parameters of imaginary phase mixture A + Z respectively Z + B during heating.

The simulated microstructure after a 250 °C maximum temperature consists entirely of phase (Z). The temperature-dependent flow curves and the Hockett–Sherby parameter of phase (Z) can thus be determined from tensile tests during cooling after a short-time heat treatment with a 250 °C maximum temperature, as seen in Figures 13 and 14.

As can be seen in Figure 8, phase (Z) is transformed into phase (B) up to a 300 °C maximum temperature. Thus, temperature-dependent flow curves of phase (Z) up to 300 °C are necessary. By definition, the experimentally determined flow curves during heating between 250 °C and 300 °C represent the resulting mechanical properties of phase mixture (Z) and (B). As already described, the temperature-dependent flow curves of phase (B) can be calculated over the entire temperature range. The flow curves of phase (Z) in the interval between 250 °C and 300 °C can thus be determined by the experimentally determined resulting flow curves (k_{fres}) from the phase mixture of (Z) and (B) during heating, as well as the experimentally determined flow curves of phase (B) (k_{fB}) using Equation (3).

$$k_{fZ}(T) = \frac{k_{f_{res}}(T) - x_B(T) \cdot k_{fB}(T)}{x_Z(T)} \tag{3}$$

The phase proportions of the phases are given by the temperatures and can be calculated from the phase transformation (Section 3.3). Figure 14 shows the Hockett–Sherby parameters determined for phase (Z) in its entire existence temperature range from 300 °C to room temperature.

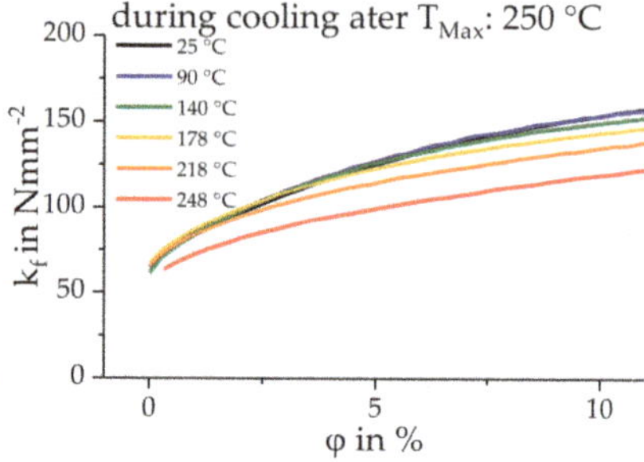

Figure 13. Experimentally obtained flow curves of alloy EN AW-6060 T4 at different temperatures during cooling after a 250 °C maximum temperature (phase Z).

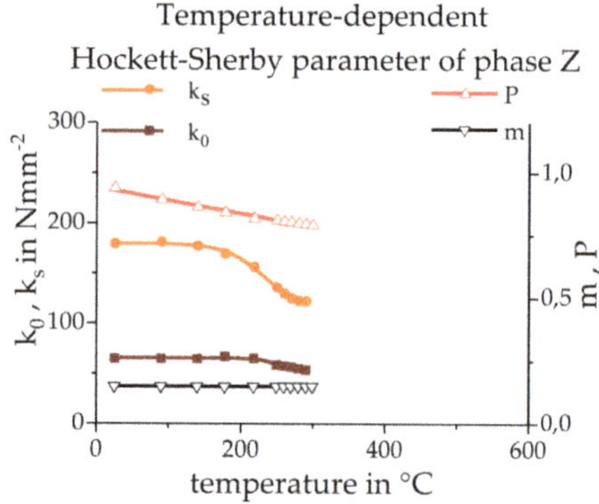

Figure 14. Temperature-dependent Hockett–Sherby parameters of phase (Z).

The experimental database for phase (A) provided the tensile tests performed during the short-time heat treatment with a 180 °C maximum temperature, see Figure 15. The same procedure for determining temperature-dependent Hockett–Sherby parameters (Figure 16) was performed for phase (A) from the imaginary phase mixture A + Z.

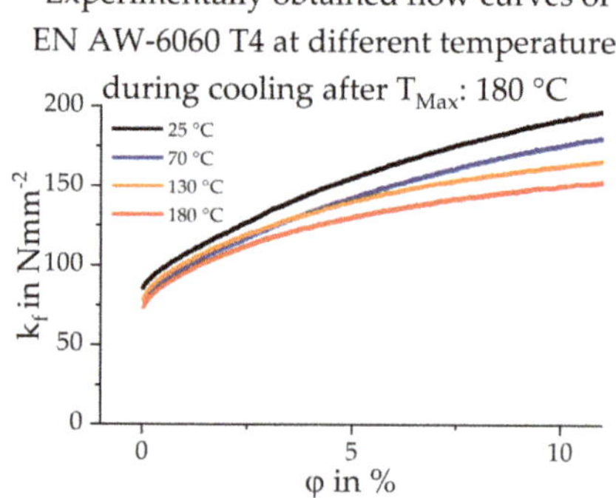

Figure 15. Experimentally obtained flow curves of alloy EN AW-6060 T4 at different temperatures during cooling after a 180 °C maximum temperature (phase A).

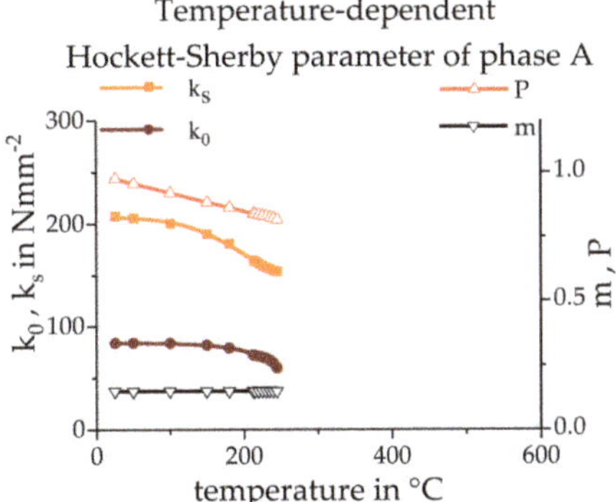

Figure 16. Temperature-dependent Hockett–Sherby parameters of phase (A).

The presented procedure makes it possible to adjust the Hockett–Sherby parameters for each phase. As a result, flow curves can be calculated for each phase at any temperature in the temperature range of phase existence.

3.4. Phase Transformation

According to the model in Figure 6, we have chosen simplified linear transformations of imaginary phases A → Z and Z → B in the specified temperature ranges. Corresponding to the continuous heating dissolution diagram in Figure 1, we have further assumed no dependence of phase transformations on heating rate in the relevant range of approximately 10 Ks^{-1} to 100 Ks^{-1}. Figure 17 shows the phase fractions during a simulated heating up to 400 °C. It becomes clear that an entire linear transformation of the individual phases was achieved as a function of temperature, considering the defined start and end temperatures of the transformations.

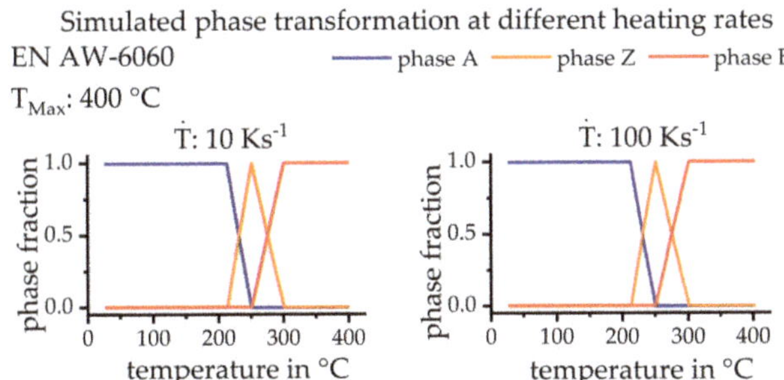

Figure 17. Simulated phase transformations at different heating rates up to 400 °C.

3.5. Simulation Model

The simulation model was implemented using the finite element software LS DYNA (Livermore Software Technology Corporation). The material model MAT_GENERALIZED_PHASECHANGE was released in software version R9.0.1 of LS DYNA (August 2016). It is used to model phase transformations in metallic materials and the associated changes in material properties. Up to 24 individual phases that transform into each other can be defined. The transformation can be defined for heating or cooling [27]. The phase transformation laws according to Koistinen–Marburger, Johnson–Mehl–Avrami–Kolmogorov (JMAK), and Kirkaldy and Oddy are predefined in the keywords.

A simple geometric model was used to verify the phenomenological mechanical material model for precipitation hardening aluminium alloys. The simulation model consisted of a single cube-shaped solid element with a 1 mm edge length. This element was subjected to a defined temperature profile to simulate heat treatment. The element was firmly clamped on one side by boundary conditions. The nodes on the other side were loaded with a predetermined displacement. Thermal expansion was not considered during this simulation to prevent thermally induced deformation.

Figure 18 illustrates an example time-temperature profile and the time-displacement course used. The time-temperature curve was varied in both the maximum temperature and in the temperature during deformation, in different individual simulations. The time-displacement curve remained constant during all individual simulations. The simulated flow curves were then compared with experimental data.

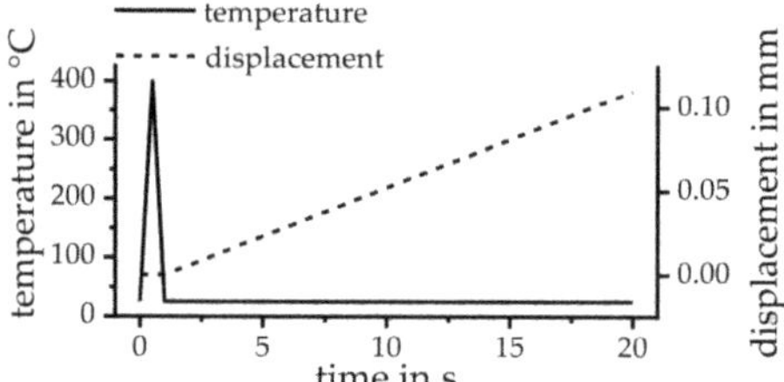

Figure 18. Exemplary time-temperature and time-displacement graph for checking the simulated transformation behavior and the resulting properties.

In the keywords in LS-DYNA, the Hockett–Sherby parameters were not entered directly. The representation of the flow curves via the Hockett–Sherby parameters serves only to calculate an associated flow curve for each phase at any temperature. In LS-DYNA, the flow curves were assigned to a temperature in tabular form for each phase. In the presented material model, a flow curve was tabulated every 25 K for each phase.

4. Results

The experimental and the simulated flow curves during different short-time heat treatments of alloy EN AW-6060 are shown in Figure 19. The experimentally determined flow curves shown represent the average of at least three individual measurements. The true stress over the plastic strain was plotted. The flow curves clearly show that our phenomenological mechanical material model for precipitation hardening aluminium alloys agrees very well with the experiments.

Figure 20 highlights some applications of the model. Figure 20A shows that the material model can provide temperature-dependent flow curves during heating. Strength decreases with increasing temperature, comparable to Figure 5. Figure 20B shows that the simulated flow curves during heating and cooling run differently. By heating to 275 °C, the material is softened. The flow curves after cooling to 200 °C and 25 °C are therefore significantly below the flow curves from heating to the same temperatures. Figure 20C shows that the flow curves at the same current temperature (here 25 °C) depend on the maximum temperature of the previous short-time heat treatment. A short heating to 200 °C does not cause softening and conforms to the flow curve of the initial state. Up to 300 °C, the material is softened, and the strength decreases with the increasing maximum temperature. Above the 300 °C maximum temperature, the material is maximally softened, and the flow curves do not change any further.

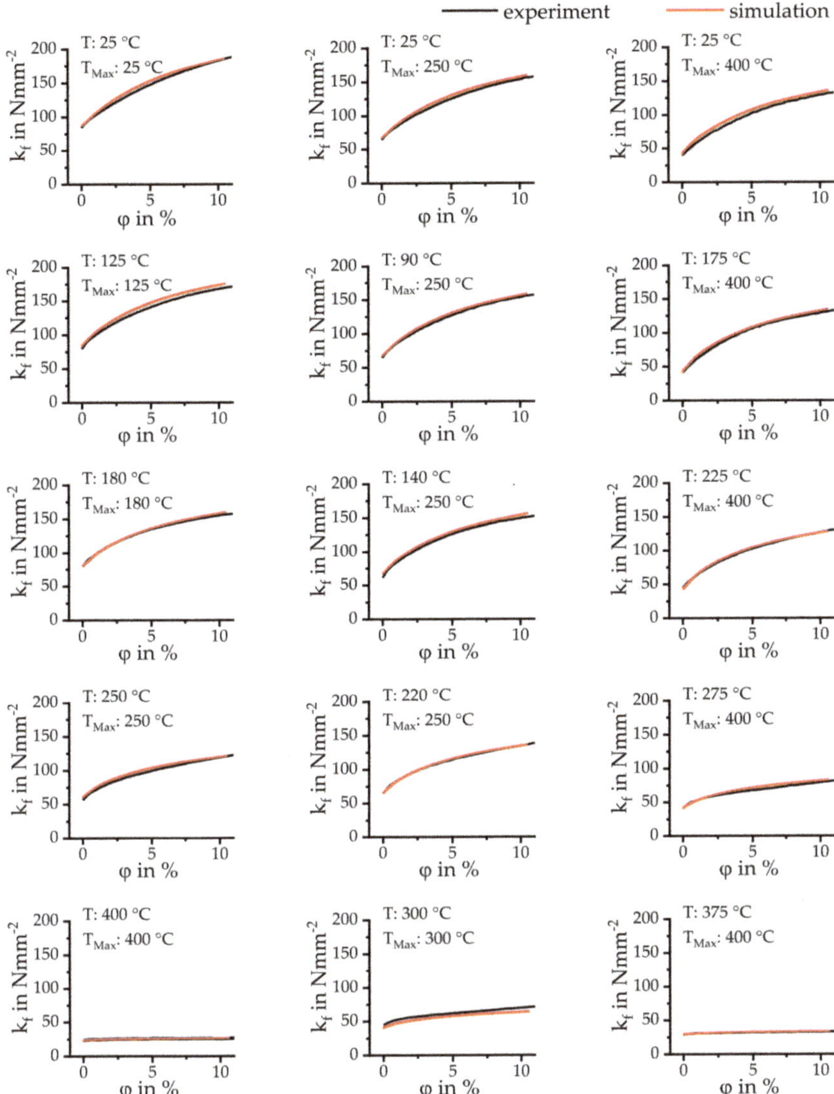

Figure 19. Measured and simulated flow curves of alloy EN AW-6060 T4 during different heat treatments.

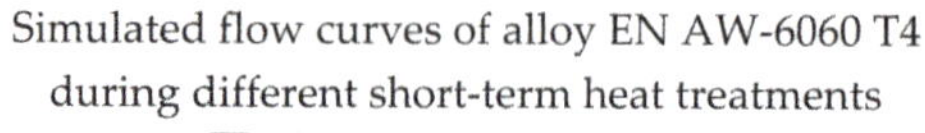

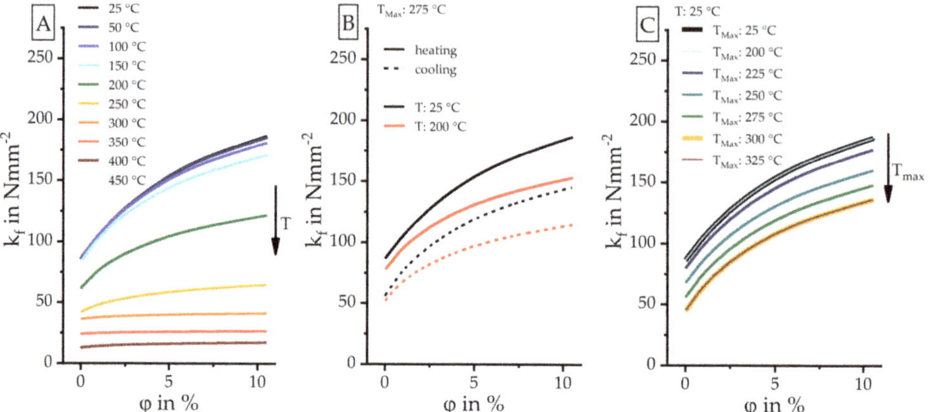

Figure 20. Simulated flow curves of alloy EN AW-6060 T4 during various short-time heat treatments. (**A**) During heating to different temperatures; (**B**) Distinction heating/cooling to 25 °C and 200 °C after T_{Max}: 275 °C; (**C**) At 25 °C according to different maximum temperatures.

The developed material model can provide flow curves that depend on both the current temperature and the previous maximum temperature, and distinguish between heating and cooling. The developed material model thus takes into account the essential influencing factors of a short-term heat treatment on the mechanical properties and is therefore, suitable for the coupled thermal, metallurgical, and mechanical simulation of such a heat treatment. In this case, metallurgical does not mean the real nm-sized precipitation processes in Al-alloys, but a feasible macroscopic approach based on imaginary phases.

5. Conclusions

The two main influencing factors for softening aluminium alloys during short-term heat treatment are the current temperature and the previous maximum temperature [21]. Until now, no material model can describe the mechanical properties as a function of the current temperature and the previous maximum temperature during the short-term heat treatment and can also differentiate between heating and cooling. In this work, a phenomenological mechanical material model was developed for precipitation hardening aluminium alloys. In this model, the real hardening behavior of aluminium alloys, which depends on nm-sized precipitates, was not considered. Instead, the mechanical properties were defined by a mixture of different imaginary phases comparable to coarse phase mixtures in steels. The initial precipitation-hardened state was defined as an imaginary hardened phase (A). During short-term heat treatment, precipitates were dissolved, and the material was softened. This state was defined as an imaginary softened phase (B). Between the hardened phase (A) and the softened phase (B), an imaginary intermediate phase (Z) was defined. These phases transformed into each other during heating depending on the temperature. During cooling, no further phase transformation was assumed. The individual phases were assigned to temperature-dependent flow curves. Due to the phase composition as a function of the maximum temperature and the temperature-dependent flow curves of the individual phases, the developed material model can calculate a flow curve for each current temperature and each previous maximum temperature.

The material model was developed for calculating the mechanical properties during typical short-term heat treatments. For this reason, the following factors deviating from short-term heat treatment conditions were not considered in the material model—heating and cooling rate, strain rate and subsequent natural ageing. In principle, these factors can be considered in the model, but need an even broader experimental database.

The presented material model with the imaginary phases can be adapted to different age hardening aluminum alloys. For this purpose, the flow curves of the corresponding phases, as well as the transformation temperatures of the individual phases must be adapted to the alloy.

The model was implemented in the finite element software LS-DYNA using the keyword MAT_GENERALIZED_PHASECHANGE. The laser short-time heat treatment of Al-alloys can be simulated with this material model. Residual stresses and distortions can be calculated and handed over to a subsequent forming simulation, to realize a through process simulation.

Author Contributions: H.F., L.V.K., M.R. and O.K. conceived and designed the experiments; H.F. performed the experiments and analysed the data, L.V.K. implemented the model; H.F., L.V.K., M.R., and O.K. discussed and interpreted the results together; H.F. wrote the paper.

Funding: This research was funded by the German Research Foundation (DFG), within the scope of the research project Improvement of formability of extruded aluminium profiles by a local short-term heat treatment (DFG KE616/22-2).

Conflicts of Interest: The authors declare no conflict of interest.

References

1. Mackerle, J. Finite element analysis and simulation of quenching and other heat treatment processes. *Comput. Mater. Sci.* **2003**, *27*, 313–332. [CrossRef]
2. Oliveira, W.P.; de Savi, M.A.; Pacheco, P.M.C.L.; de Souza, L.F.G. Thermomechanical analysis of steel cylinders quenching using a constitutive model with diffusional and non-diffusional phase transformations. *Mech. Mat.* **2010**, *42*, 31–43. [CrossRef]
3. Deng, X.; Ju, D. Modeling and simulation of quenching and tempering process in steels. *Phys. Procedia* **2013**, *50*, 368–374. [CrossRef]
4. Carlone, P.; Palazzo, G.S.; Pasquino, R. Finite element analysis of the steel quenching process: Temperature field and solid-solid phase change. *Comput. Math. Appl.* **2010**, *59*, 585–594. [CrossRef]
5. Reich, M.; Kessler, O. Numerical and experimental analysis of residual stresses and distortion in different quenching processes of aluminum alloy profiles. In *ASM International*; Quenching Control and Distortion; ASM International: Chicago, OH, USA, 2012; pp. 563–574.
6. Reich, M.; Schöne, S.; Kessler, O.; Nowak, M.; Grydin, O.; Nürnberger, F.; Schaper, M. Simulation of gas and spray quenching during extrusion of aluminium alloys. *Key Eng. Mater.* **2010**, *424*, 57–64. [CrossRef]
7. Yang, X.; Zhu, J.; Nong, Z.; Lai, Z.; He, D. FEM simulation of quenching process in A357 aluminum alloy cylindrical bars and reduction of quench residual stress through cold stretching process. *Comput. Mater. Sci.* **2013**, *69*, 396–413. [CrossRef]
8. Wang, M.J.; Gang, Y.A.N.G.; Huang, C.Q.; Bin, C.H.E.N. Simulation of temperature and stress in 6061 aluminum alloy during online quenching process. *Trans. Nonferrous Met. Soc. China* **2014**, *24*, 2168–2173. [CrossRef]
9. Felde, I.; Simsir, C. Simulation trends in quenching technology for automotive components. *Int. Heat Treat. Surf. Eng.* **2014**, *8*, 42–48. [CrossRef]
10. Deng, D.; Kiyoshima, S. FEM prediction of welding residual stresses in a SUS304 girth-welded pipe with emphasis on stress distribution near weld start/end location. *Comput. Mat. Sci.* **2010**, *50*, 612–621. [CrossRef]
11. Donati, L.; Troiani, E.; Proli, P.; Tomesani, L. FEM analysis and experimental validation of friction welding process of 6xxx alloys for the prediction of welding quality. *Mater. Today Proc.* **2015**, *2*, 5045–5054. [CrossRef]
12. Hömberg, D.; Liu, Q.; Montalvo-Urquizo, J.; Nadolski, D.; Petzold, T.; Schmidt, A.; Schulz, A. Simulation of multi-frequency-induction-hardening including phase transitions and mechanical effects. *Finite Elem. Anal. Des.* **2016**, *121*, 86–100. [CrossRef]
13. Geiger, M.; Merklein, M.; Vogt, U. Aluminum tailored heat treated blanks. *Prod. Eng.* **2009**, *3*, 401–410. [CrossRef]
14. Kahrimanidis, A.; Lechner, M.; Degner, J.; Wortberg, D.; Merklein, M. Process design of aluminum tailor heat treated blanks. *Materials* **2015**, *8*, 8524–8538. [CrossRef] [PubMed]

15. Fröck, H.; Graser, M.; Reich, M.; Lechner, M.; Merklein, M.; Kessler, O. Linked Heat Treatment and Bending Simulation of Aluminium Tailored Heat Treated Profiles. In *Proceedings of the 4th World Congress on Integrated Computational Materials Engineering*; Mason, P., Fisher, C.R., Glamm, R., Manuel, M.V., Schmitz, G.J., Singh, A.K., Strachan, A., Eds.; Springer: Cham, Germany, 2017; pp. 237–248.

16. Chen, X.; Xiao, N.; Li, D.; Li, G.; Sun, G. The finite element analysis of austenite decomposition during continuous cooling in 22MnB5 steel. *Modell. Simul. Mater. Sci. Eng.* **2014**, *22*, 65005. [CrossRef]

17. Esmaeili, S.; Lloyd, D.J.; Poole, W.J. Modeling of precipitation hardening for the naturally aged Al-Mg-Si-Cu alloy AA6111. *Acta Mater.* **2003**, *51*, 3467–3481. [CrossRef]

18. Esmaeili, S.; Lloyd, D.J.; Poole, W.J. A yield strength model for the Al-Mg-Si-Cu alloy AA6111. *Acta Mater.* **2003**, *51*, 2243–2257. [CrossRef]

19. Fröck, H.; Graser, M.; Reich, M.; Lechner, M.; Merklein, M.; Kessler, O. Influence of short-term heat treatment on the microstructure and mechanical properties of EN AW-6060 T4 extrusion profiles: Part, A. *Prod. Eng.* **2016**, *10*, 383–389. [CrossRef]

20. Fröck, H.; Reich, M.; Milkereit, B.; Kessler, O. Scanning rate extension of conventional DSCs through indirect measurements. *Materials* **2019**, *12*, 1085. [CrossRef]

21. Fröck, H.; Graser, M.; Milkereit, B.; Reich, M.; Lechner, M.; Merklein, M.; Kessler, O. Precipitation behaviour and mechanical properties during short-term heat treatment for tailor heat treated profiles (THTP) of aluminium alloy 6060 T4. *Mat. Sci. Forum* **2016**, *877*, 400–406. [CrossRef]

22. Murayama, M.; Hono, K. Pre-precipitate clusters and precipitation processes in Al-Mg-Si alloys. *Acta Mater.* **1999**, *47*, 1537–1548. [CrossRef]

23. Tsao, C.S.; Chen, C.Y.; Jeng, U.S.; Kuo, T.Y. Precipitation kinetics and transformation of metastable phases in Al-Mg-Si alloys. *Acta Mater.* **2006**, *54*, 4621–4631. [CrossRef]

24. Edwards, G.A.; Stiller, K.; Dunlop, G.L.; Couper, M.J. The precipitation sequence in Al-Mg-Si alloys. *Acta Mater.* **1998**, *46*, 893–3904. [CrossRef]

25. Fröck, H.; Graser, M.; Reich, M.; Lechner, M.; Merklein, M.; Kessler, O. Influence of short-term heat treatment on the microstructure and mechanical properties of EN AW-6060 T4 extrusion profiles—Part, B. *Prod. Eng.* **2016**, *10*, 391–398. [CrossRef]

26. Hockett, J.E.; Sherby, O.D. Large strain deformation of polycrystalline metals at low homologous temperatures. *J. Mech. Phys. Solids* **1975**, *23*, 87–98. [CrossRef]

27. Loose, T.; Klöppel, T. An LS-DYNA material model for the consistent simulation of welding forming and heat treatment. In Proceedings of the 11th Int Seminar Numerical Analysis of Weldability, Seggau, Austria, 28 September 2015.

Article

Influence of Niobium or Molybdenum Addition on Microstructure and Tensile Properties of Nickel-Chromium Alloys

Marisa Aparecida de Souza [1], Bárbara de Oliveira Fiorin [2], Tomaz Manabu Hashimoto [1], Ana Paula Rosifini [1], Carlos Angelo Nunes [3], Carlos Antônio Reis Pereira Baptista [3] and Alfeu Saraiva Ramos [1,2,*]

[1] Department of Materials and Technology, Campus de Guaratinguetá, São Paulo State University, 12516-410 Guaratinguetá, Brazil; mari_sou@hotmail.com (M.A.d.S.); tmanabu@feg.unesp.br (T.M.H.); rosifini@feg.unesp.br (A.P.R.)

[2] Institute of Science and Technology, Federal University of Alfenas, 11999, 37715-400 Poços de Caldas-MG, Brazil; barbarafiorin@hotmail.com

[3] Engineering School of Lorena, University of São Paulo, 12602-810 Lorena, SP, Brazil; cnunes@demar.eel.usp.br (C.A.N.); baptista@demar.eel.usp.br (C.A.R.P.B.)

* Correspondence: alfeu.ramos@unifal-mg.edu.br; Tel.: +55-35-36974600

Received: 26 March 2019; Accepted: 9 May 2019; Published: 22 May 2019

Abstract: This work discusses on influence of niobium or molybdenum addition on microstructure and tensile properties of NiCr-based dental alloys. In this regard, the Ni-24Cr-8Nb, Ni-22Cr-10Nb and Ni-20Cr-12Nb (wt. %) alloys produced by arc melting process. To compare the typical Ni-22Cr-10Mo dental alloy was also produced. These ternary alloys were analyzed by chemical analyses, X-ray diffraction (XRD), scanning electron microscopy (SEM), electron dispersive spectrometry (EDS), thermogravimetric analysis (TG), Vickers hardness and tensile tests. Although the mass losses of the samples during arc melting, the optical emission spectrometry showed that the initial compositions were kept. The Ni-22Cr-10Mo alloy produced a matrix of Ni_{ss} (ss—solid solution), whereas Ni_3Nb disperse in a Ni_{ss} matrix was also identified in Ni-Cr-Nb alloys. Excepting for the Ni-22Cr-10Nb alloy with mass gain of 0.23%, the as-cast Ni-Cr alloys presented mass gains close to 0.4% after heating up to 1000 °C under synthetic airflow. The hardness values, the modulus of elasticity, yield strength and ultimate tensile strength have enhanced, whereas the ductility was reduced with increasing niobium addition of up to 12 wt.-%.The Ni-22Cr-10Mo alloy presented an intergranular fracture mechanism containing deep dimples and quasi-cleavage planes, whereas the shallow dimples were identified on fracture surface of the as-cast Nb-richer Ni-Cr alloys due to the presence of higher Ni_3Nb amounts.

Keywords: microstructure; tensile properties; intermetallics; casting; phase transformation

1. Introduction

As-cast nickel-chromium (Ni-Cr) and nickel-chromium-molybdenum (Ni-Cr-Mo) based alloys are applied in dentistry to partial removable prosthesis since 1930 because of their excellent characteristics, such as corrosion resistance and mechanical properties, compared with the more expensive gold-based alloys. Thus, the less expensive and lighter Ni-Cr alloys with high elastic module (170–190 GPa) have been developed for producing metal-ceramic restorations, intra-radicular pins, crowns and removable partial prosthesis [1–3]. In these alloys, the chromium is responsible for the formation of an outer protective oxide layer [4]. Further, the control on the quantities released from dental materials has also been considered for these alloy designs.

Typical chemical compositions of different commercial Ni-Cr alloys used for metal-ceramic restorations presented the Cr and Mo contents varying between 18.5–20.4 wt. % and 4.2–9.5 wt. %, respectively [5]. Other alloys and traces such as cobalt, iron, zinc and manganese containing up to 2.5, 2.5, 0.65 and 0.1 wt. %, respectively, can be found in these alloys. Studies with Be-free Ni-Cr alloys have been also considered for dental applications [6]. In accordance with the phase diagrams of the nickel-chromium [7] and nickel-molybdenum [8] systems, the adopted chromium and molybdenum amounts in these commercial Ni alloys are located in regions of Ni solid solutions, respectively. For comparison, the commercial Ni-20Cr-5Mo (wt. %) alloy with a microstructure based on Ni_{ss} (ss—solid solution) was adopted, and its mechanical properties are based mainly on the strengthening solid solution and grain refining mechanisms. However, the use of metal matrix-intermetallic composites can also improve the need flexural strength for dentistry devices [9].

It is well known on the use of niobium as alloying in the new Ti alloys for orthopedic and dentistry devices, since their superior biocompatibility and osseointegration features compared to commercial Ti-6Al-4V alloy [10–12]. In accordance with the phase diagram of the nickel-niobium system [13], the nickel can dissolve up to 4 wt. % (11 at. %) Nb whereas the eutectic reaction L $\leftrightarrow$ Ni + Ni_3Nb at 1216 °C can happen during solidification of these richer-Ni alloys. The liquidus projection of the Ni-Cr-Nb system indicates the presence of a ternary eutectic reaction [L $\leftrightarrow$ $NbCr_2$ (high-temperature) + $NbNi_3$ + Nb_6Ni_7] at 1170 °C in the rich-Ni region [14].

In this sense, this work reports on the influence of niobium or molybdenum addition in microstructure and tensile properties of nickel-chromium alloys.

2. Materials and Methods

In this work, high-purity starting powders of Ni, Cr and Nb were used to prepare the Ni-22Cr-10Mo, Ni-24Cr-8Nb, Ni-22Cr-10Nb and Ni-20Cr-12Nb alloys (wt. %) by arc melting process. For comparison, the typical Ni-20Cr-4Mo dental alloy (wt. %) was also evaluated. In this way, the elemental Ni-Cr-Nb and Ni-Cr-Mo powder mixtures were mixed and then compacted square-section blocks with a length of 55 mm and a 5 mm thickness were obtained by axial compaction using 7 ton (approximately 25 MPa). The Ni-Cr-Mo and Ni-Cr-Nb ingots weighing close to 50 g were prepared in an arc melting furnace under argon atmosphere using a water-cooled copper hearth (own USP project, Lorena, Brazil), non-consumable tungsten electrode, and gettered by titanium. Five melting steps were carried out for each alloy in order to obtain homogeneous ingots. The mass losses occurred during the melting process were lower than 0.6%, and the nominal compositions were then adopted.

The as-cast Ni-24Cr-8Nb, Ni-22Cr-10Nb, Ni-20Cr-12Nb and Ni-22Cr-10Mo (wt. %) alloys were characterized by chemical analysis, scanning electron microscopy (SEM, Carl Zeiss Microscopy GmbH, Jena, Germany), X-ray diffraction (XRD, Shimadzu Corporation, Kyoto, Japan), electron dispersive spectrometry (EDS, INCA Energy Oxford, Oxfordshire, UK), Vickers hardness and tensile tests (Buehler Global Head Quarters, Lake Bluff, IL, USA) at room temperature.

Chemical analysis of as-cast Ni-Cr-Nb and Ni-Cr-Mo alloys was conducted in an ARL 3460 Optical Emission Spectrometer (OES, Thermo Cientific, Waltham, MA, USA). The total C, S, N and O contents of as-cast Ni-Cr-Nb and Ni-Cr-Mo alloys were measured by combustion analyses in CS-125 and TC-436 LECO analyzers (Leco Corporation, St. Joseph, MI, USA). Conventional metallographic techniques, including the use of silica colloidal suspension for polishing and chemical attack with aqueous acid solution (20 vol. % $HF{:}HNO_3$ (2:1)), were adopted in this work.

XRD experiments at room temperature of the monolithic Ni-Cr-Nb and Ni-Cr-Mo alloys were performed in XRD-6000 Shimadzu equipment (Shimadzu Corporation, Kyoto, Japan) using Ni-filtered Cu-Kα radiation, voltage of 40 kV, current of 30 mA, angular range (2θ) from 10° to 80°, angular pass of 0.05° and counting time per pass of 1 s. Based on the JCPDS database [15] and Powdercell computer program [16], the phases formed in microstructure of as-cast Ni-Cr-Nb and Ni-Cr-Mo alloys were indexed.

Electron images of as-cast Ni-Cr-Nb and Ni-Cr-Mo alloys were obtained in a Zeiss (LEO) 1450-VP SEM (Carl Zeiss Microscopy GmbH, Jena, Germany) using the secondary electron and back-scattered electron (BSE) detectors. EDS analysis was adopted to measure the amounts of Ni, Cr, N and Mo presents in phases of as-cast Ni-Cr-Nb and Ni-Cr-Mo alloys. At least three measurements per phase were obtained for each sample. The amounts of precipitates in as-cast Ni-Cr-Nb alloys were determined using the ImageJ computer program [17] from their SEM images.

To determine the oxidation resistance of the Ni-Cr alloys evaluated in this work, the thermogravimetric analysis of these as-cast samples was conducted after heating up to 1000 °C in alumina crucible under synthetic air flow with heating rate of 10 °C per minute.

Vickers hardness tests of as-cast Ni-Cr-Nb and Ni-Cr-Mo samples were performed in a Wolpert durometer using 9.81 N load and loading time of 15 s, according to ASTM E-92 standard [18]. At least eight measurements were obtained for each specimen.

Tensile tests of as-cast Ni-Cr-Nb and Ni-Cr-Mo samples were conducted according to the ASTM E-8M [19] standard in an INSTRON model 8801 servo hydraulic testing machine (Instron Company, Norwood, MA, USA) with load cell of 100 kN capacity. Cylinder specimens with 20 mm gage length and diameter of 4 mm were carefully machined. At least three specimens per alloy were tested. Details on their fractured surfaces were obtained by SEM analysis.

3. Results

Table 1 shows the Ni, Cr, Nb and Mo contents (wt. %) of as-cast Ni-Cr-Nb and Ni-Cr-Mo alloys measured by ICP-AES analysis.

Table 1. Elemental contents (wt. %) measured by an ICP-AES analysis of the Ni-Cr-Nb and Ni-Cr-Mo alloys evaluated in this work.

Element	Ni-20Cr-10Mo	Ni-24Cr-8Nb	Ni-22Cr-10Nb	Ni-20Cr-12Nb	Commercial Alloy
C	0.014	0. 032	0.036	0.027	ND
S	0.011	0. 012	0.015	0.012	ND
N	0.016	0.019	0.021	0.021	ND
Al	0.002	0.000	0.001	0.001	ND
Si	0.170	0.220	0.210	0.170	ND
P	0.004	0.003	0.003	0.003	ND
Ti	0.011	0.113	0.136	0.162	ND
V	0.003	0.006	0.006	0.007	ND
Cr	21.92	24.29	22.04	20.08	19,1
Mn	0.010	0.030	0.010	0.010	0.65
Ni	69.00	66.30	66.20	65.30	70.0
Cu	0.003	0.018	0.018	0.023	ND
Nb	0.010	8.090	9.840	11.52	ND
Mo	9.110	0.045	0.011	0.006	4.2

ND—not determined.

Table 2 shows the C, S, O and N contents (wt. %) of the as-cast NiCr alloys evaluated in this work, which were determined utilizing either the gas fusion method (nitrogen and oxygen) or the combustion method (carbon and sulfur).

Table 2. Interstitial contents (wt. %) measured by LECO combustion analyses of the Ni-Cr-Nb and Ni-Cr-Mo alloys.

Alloy	C (wt. %)	S (wt. %)	N (wt. %)	O (wt. %)
Ni-22Cr-10Mo	0.020	0.020	0.016	0.099
Ni-20Cr-12Nb	0.010	0.007	0.022	0.039
Ni-22Cr-10Nb	0.010	0.010	0.025	0.059
Ni-24Cr-8Nb	0.020	0.099	0.019	0.054
Commercial	0.260	0.002	0.032	0.027

XRD patterns of the Ni-22Cr-10Mo and Ni-20Cr-12Nb alloys as well as the commercial Ni-Cr alloy are presented in Figure 1.

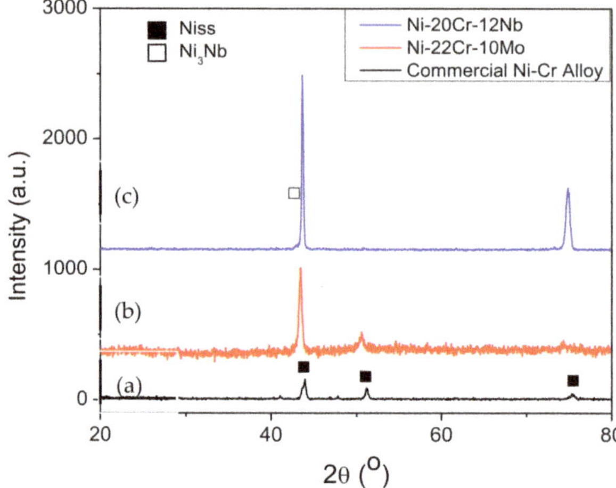

Figure 1. XRD patterns of Ni-Cr alloys evaluated in this work: (**a**) Commercial Ni-Cr alloy, (**b**) Ni-22Cr-10Mo alloy and (**c**) Ni-20Cr-12Nb alloy.

Figure 2 displays the SEM images of the Ni-22Cr-10Mo alloy as well as the commercial Ni-Cr alloy evaluated in this work.

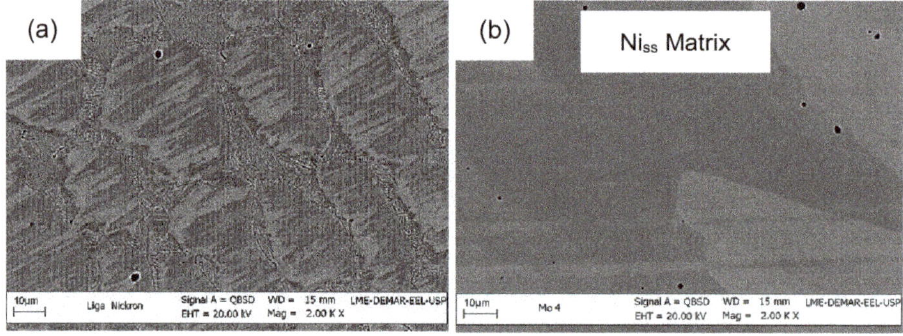

Figure 2. SEM images of the (**a**) commercial Ni-Cr alloy and (**b**) as-cast Ni-22Cr-10Mo alloy without a chemical attack.

SEM images of as-cast Ni-Cr-Nb alloys evaluated in this work are presented in Figure 3.

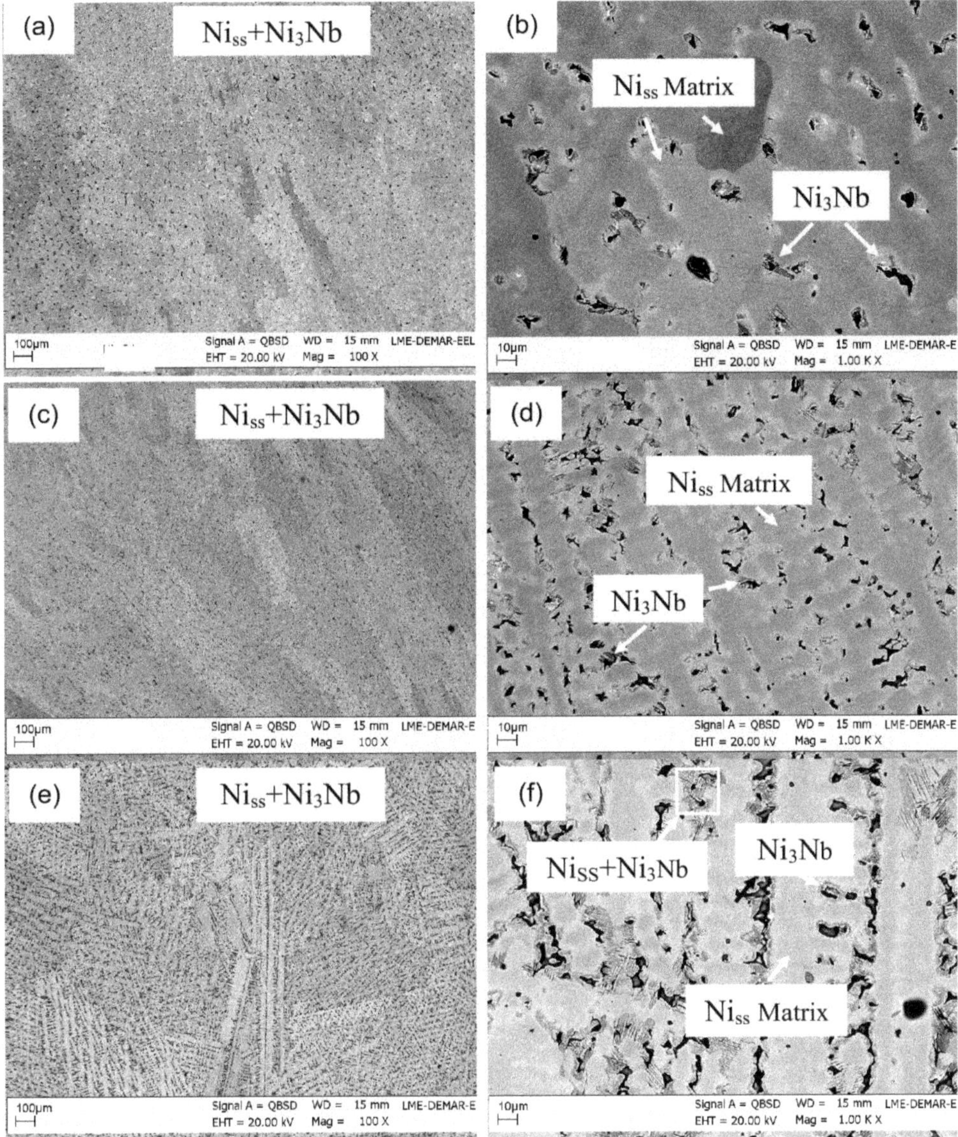

Figure 3. SEM images of the as-cast (**a,d**) Ni-24Cr-8Nb, (**b,e**) Ni-22Cr-10Nb and (**c,f**) Ni-20Cr-12Nb alloys: (**a,c**) Without chemical attack and (**b,d,f**) after chemical attack.

The contents (at. %) of Ni, Cr, Mo, Nb and other elements measured by EDS analysis of phases formed in the as-cast Ni-Cr-Mo and Ni-Cr-Nb alloys evaluated in this work are presented in Table 3.

Table 3. The elemental contents (at. %) measured by electron dispersive spectrometry (EDS) analysis of phases formed in the as-cast Ni-Cr-Nb and Ni-Cr-Mo alloys.

Alloy	Phase	Ni	Cr	Mo	Nb	Others
Ni-22Cr-10Mo	Ni_{ss}	68.5–70.5	24.0–25.5	5.1–6.0	-	-
Ni-24Cr-8Nb	Ni_{ss}	68.9–69.4	25.4–27.8	-	2.8–5.5	-
Ni-20Cr-12Nb	Ni_{ss}	70.2–71.1	22.5–25.0	-	3.9–7.1	-
	Ni_3Nb	72.3	9.6–10.5	-	17.2–18.1	-
Commercial Ni-Cr Alloy	Ni_{ss}	59.5–70.7	22.0–26.7	2.0–4.5	-	Si (3.6–8.6); Mn (0.6–0.7)

Representative thermogravimetric curves of the as-cast Ni-Cr-Mo and Ni-Cr-Nb alloys evaluated in this work are illustrated in Figure 4.

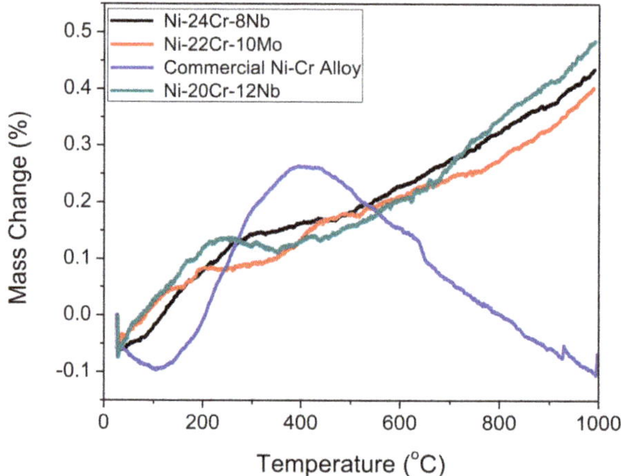

Figure 4. Representative thermogravimetric curves of the commercial Ni-Cr alloy, and as-cast Ni-Cr-Nb and Ni-Cr-Mo alloys evaluated in this work.

The Vickers hardness results of as-cast Ni-Cr-Nb and Ni-Cr-Mo alloys investigated in this work are indicated in Table 4.

Table 4. Vickers hardness of the as-cast Ni-Cr-Mo and Ni-Cr-Nb alloys investigated in this work.

Alloy	Vickers Hardness (HV)
Ni-22Cr-10Mo	150.0 ± 5.3
Ni-24Cr-8Nb	243.0 ± 19.8
Ni-22Cr-10Nb	309.0 ± 6.7
Ni-20Cr-12Nb	374.0 ± 21.1
Commercial Ni-Cr Alloy	231.0 ± 14.6

The representative tensile $\sigma-\varepsilon$ curves of the as-cast Ni-Cr-Mo and Ni-Cr-Nb alloys are illustrated in Figure 5 whereas their compression mechanical properties are presented in Table 5.

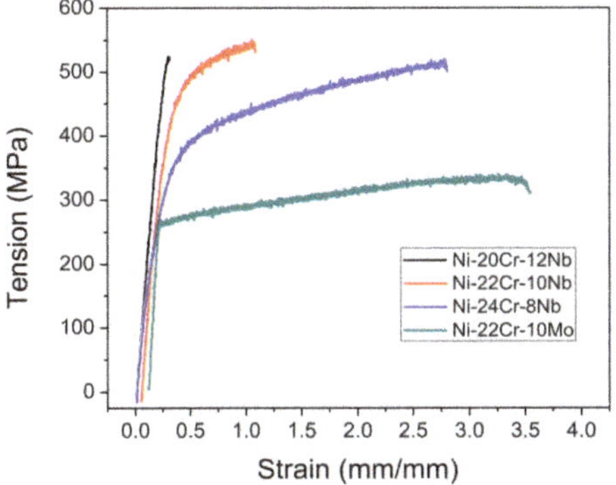

Figure 5. Representative tensile σ-ε curves of the as-cast Ni-Cr-Mo and Ni-Cr-Nb alloys investigated in this work.

Table 5. Tensile mechanical properties of the as-cast Ni-Cr-Mo and Ni-Cr-Nb alloys evaluated in this work.

Alloy	Yield Strength (MPa)	Ultimate Tensile Strength (MPa)	Total Strain (%)	Elastic Module (GPa)
Ni-22Cr-10Mo	212.4 ± 46.2	380.6 ± 36.5	4.2 ± 0.2	28.9 ± 16.4
Ni-24Cr-8Nb	334.2 ± 22.3	505.3 ± 22.2	6.5 ± 0.5	41.4 ± 0.6
Ni-22Cr-10Nb	448.4 ± 48.4	551.3 ± 51.7	4.5 ± 0.7	61.0 ± 5.8
Ni-20Cr-12Nb	510.5 ± 81.5	576.5 ± 133.4	1.8 ± 0.5	69.5 ± 18.3

Details on the fracture surface of the Ni-22Cr-10Mo and Ni-20Cr-12Nb alloys after tensile tests are illustrated in Figure 6a,b.

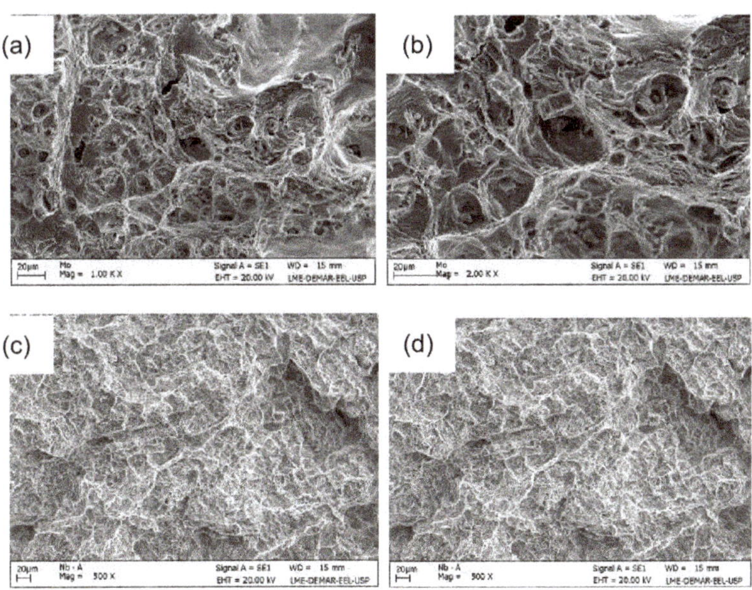

Figure 6. SEM fractographies of the as-cast (**a,b**) Ni-22Cr-10Mo and (**c,d**) Ni-20Cr-12Nb alloys after tensile tests.

4. Discussion

Chemical analysis by ICP method of the as-cast Ni-Cr-Mo and Ni-Cr-Nb alloys has indicated that the Cr, Nb and Mo amounts were kept close to their nominal compositions (see Table 1). The commercial Ni-Cr alloy presented the Ni, Cr and Mo contents of 70, 19.1 and 4.2 wt. %, respectively. Trace of manganese (0.65 wt. %) was also detected in this commercial alloy.

Similar values of C, S and N were measured from the combustion gases within the as-cast Ni-Cr-Mo and Ni-Cr-Nb alloys (see Table 2). Despite the route adopted in this work the amount of total oxygen of up to 0.099 wt. % were picked up during arc melting from the compacted elemental powder mixtures only. On contrary, this route has produced alloys with lower C and N interstitial amounts than the commercial alloy. Nitrogen contents higher than 0.1 wt. % can contribute to a reduction on the ductility of Ni-based alloys [20]. Similar results were found for a silver-palladium alloy [21]. The commercial alloy had also amounts of Co, Fe, Mn and Zn of 0.25, 0.68, 0.65 and 0.1 wt. %, respectively.

Both the commercial alloy (with 19 wt. % Cr and 4 wt. % Mo) and the experimental Ni-22Cr-10Mo alloy presented similar XRD results, as are shown in Figure 1. In these alloys, the intense Ni_{ss} peaks can be observed only, which could be related for X-ray diffraction experiments with monolithic samples. However, these Ni_{ss} peaks have shifted toward the direction of smaller diffraction angles for the Mo-richer alloy, suggesting that the larger molybdenum amount in Ni-22Cr-10Mo alloy is preferentially dissolved into the Ni lattice. Traces of others peaks besides the intense Ni peaks were also identified in XRD patterns of the Ni-Cr-Nb alloys evaluated in this work. Nonetheless, it is very important to note that there are two overlapping peaks between the major diffraction lines of Ni, Ni_8Nb, Ni_3Nb and Ni_6Nb_7 [22].

The commercial Ni-Cr alloy evaluated in this work presents primary Ni_{ss} grains and a eutectic region located in their grain boundary as is indicated in the SEM micrograph of Figure 2a. It can be also observed the presence of regions with different contrasts into the Ni_{ss} grains, which had the contents of Ni, Cr, Mo, Mn and Si measured by EDS analysis varying between 59.5–70.0, 26.7–22.0, 4.5–2.0, 0.7–0.6 and 8.6–3.8 at. %, respectively (see Table 3). In this sense, the brighter regions presented the higher Nb contents. Excepting to the manganese content with 0.9 at. %, similar elemental values were found in eutectic regions.

Agreeing with the phase diagram of the Ni-Cr-Mo system [23], the microstructure of the as-cast Ni-22Cr-10Mo alloy presented in Figure 2b has revealed the presence of a homogeneous structure formed by coarse Ni_{ss} (ss—solid solution) grains, and the different contrasts observed are relative to their different crystallographic orientations only. No segregation tendency was detected by EDS analysis in these regions in which the Cr and Mo contents varied between 24.0–25.5 and 5.1–6.0 at. %, respectively. Some diffusion pores can be also noted in microstructure of this ternary alloy.

The microstructures of the as-cast Ni-Cr-Nb alloys presented in Figure 3 have indicated the same solidification path. Agreeing with the phase diagram of the Ni-Cr-Nb system [14], the solidification is started with the formation of primary Ni_{ss} grains followed by simultaneous precipitation of Ni_{ss} and Ni_3Nb for Ni-rich alloys. Comparing with the Ni-24Cr-8Nb alloy, it can be noted the presence of larger Ni_3Nb amounts in microstructure of the Ni-20Cr-12Nb alloy. Coherently with the liquidus projection of this ternary system, this alloy containing more eutectic regions is located closer to the L + Ni + Ni_3Nb monovariant line [14]. For the as-cast Ni-24Cr-8Nb alloy, the Ni_{ss} grains containing different contrasts have presented 25.4–27.8 at. % Cr and 2.8–5.5 at. % Nb, whereas the Nb-rich precipitates containing up to 15 at. % Nb were also measured by EDS analysis from very fine regions. In addition, it was observed that the brighter Ni_3Nb precipitates were darker after chemical attack with aqueous acid solution. EDS analysis also indicated that the Nb contents dissolved into the Ni_{ss} lattice were increased in the as-cast Ni-20Cr-12Nb alloy, which have varied from 3.9 at. % to 7.1 at. % Nb. In this ternary alloy, the coarser Nb-rich precipitates presented 17.2–18.2 at. % Nb, indicating to be the Ni_3Nb phase with highest enthalpy of formation (31 kJ·mol^{-1}) instead to the Ni_8Nb phase (with 15 kJ·mol^{-1}) containing contents near the 11.1 at. % Nb [21]. In this way, these EDS results have confirmed that the traces identified in XRD patterns of as-cast Ni-20Cr-12Nb alloy could be associated with the Ni_3Nb

phase. According with the ImageJ computer program [17], the SEM images of the as-cast Ni-24Cr-8Nb, Ni-22Cr-10Nb and Ni-20Cr-12Nb alloys have indicated the presence of precipitates for 6.3%, 6.9% and 14.4% in area, respectively.

Remelting operations and thermal cycles are commons for Ni-Cr alloys used in dental restorations and high-temperature applications. Similar behavior was noted during heating up to 1000 °C of Ni-based ternary alloys investigated in this work (see Figure 4). Conversely, the commercial Ni-Cr alloy presented a mass gain of 0.3% after heating up to 400 °C followed by a continuous mass loss under heating in synthetic air between 400 °C and 1000 °C. These facts can be associated with initial oxidation and subsequent reactive evaporation of chromium oxide at high temperatures in the presence of oxygen and/or water vapor, respectively [24]. Thermogravimetric curves of the as-cast Ni-22Cr-10Mo, Ni-24Cr-8Nb, Ni-22Cr-10Nb and Ni-20Cr-12Nb alloys are illustrated in Figure 4, which have indicated mass gains of 0.4, 0.43, 0.23 and 0.49%, respectively. Excepting to the Ni-22Cr-10Nb alloy, similar oxidation resistance was noted for experimental Ni-Cr alloys evaluated in this work.

The Ni-22Cr-10Mo presented lower hardness values than the commercial alloy containing lesser Mo contents, which could be related with its lower interstitial amounts (see Table 4). Similar hardness values were noted between the Ni-24Cr-8Nb alloy and commercial Ni-Cr alloy. In contrast, the as-cast Ni-Cr alloys with 10 wt. % and 12 wt. % Nb presented higher hardness values than the commercial Ni-Cr alloy probably due to the higher Ni_3Nb amounts in microstructure of these as-cast ternary alloys.

Depending on the composition and processing route, the Ni-Cr alloys can present the yield strength and ultimate tensile strength of 260–830 MPa and 440–1200 MPa, respectively [25–28]. The commercial Ni-Cr alloy evaluated in this work presented the yield strength (offset 0.2%) and ultimate tensile strength of 417 MPa and 550 MPa, respectively. Table 5 displays the tensile mechanical properties at room temperature of the as-cast Ni-Cr-Mo and Ni-Cr-Nb alloys investigated in this work. Coherently to the hardness results, the Ni-22Cr-10Mo alloy exhibited the lowest values of yield stress (offset 0.2%) and ultimate tensile strength besides the elastic module of 29 GPa close to that related to bone tissues [25–28]. Similarly, the as-cast Ni-24Cr-8Nb alloy containing small amounts of Ni_3Nb also exhibited lower yield strength and ultimate tensile strength. Nevertheless, these tensile properties of as-cast Ni-20Cr-12Nb alloy were slightly increased and kept close to those of the commercial Ni-Cr alloy. Consequently, the total strain of the Ni-24Cr-8Nb alloy was reduced from 6.5% to 1.8% in samples of as-cast Ni-20Cr-12Nb alloy whereas their elastic module was increased between 41 GPa and 70 GPa, respectively. It is well known that the Ni_3Nb platelets has a detrimental effect on ductility and other mechanical properties that are related to precipitate within the inter-dendritic regions of the as-cast microstructures [29]. In this regard, the development of multicomponent alloys containing the Mo and Nb additions could produce both the high ductility and mechanical strength for structural applications.

The fracture surface of the as-cast Ni-22Cr-10Mo alloy presented an intergranular fracture mechanism and two different regions constituted with variable dimple sizes, which were associated mainly to crystallography orientation of Ni_{ss} grains formed during solidification. Moreover, it was also noted the existence of a small amount of quasi-cleavage planes. Shallow dimples were identified on the fracture surfaces of the as-cast Ni-Cr-Nb alloys, which were less deep with increasing Ni_3Nb amount as shown in Figure 3c,d, denoting that the Ni_3Nb precipitates favored the occurrence of brittle fracture.

5. Conclusions

Arc melting successfully produced the low-interstitial Ni-Cr-Nb and Ni-Cr-Mo alloys from elemental powder mixtures.

The values of Vickers hardness, elastic module, yield strength and ultimate tensile strength were enhanced from 243 HV, 41 GPa, 334 MPa and 505 MPa to 374 HV, 70 GPa, 510 MPa and 576 MPa, respectively, with increasing Nb addition up to 12% in the alloy composition owing the Ni_3Nb formation during solidification of Ni-Cr-Nb alloys.

The Ni-22Cr-10Mo alloy with deep dimples and quasi-cleavage planes on its fracture surface has exhibited intergranular fracture mechanism whereas the shallow dimples were identified on fracture surface of the as-cast Nb-richer Ni-Cr alloys due to the presence of higher Ni_3Nb amounts.

Author Contributions: M.A.d.S. produced the Ni-Nb alloys and participated in all stages; B.d.O.F. performed the XRD experiments; T.M.H. participated in all stages; A.P.R. conducted the thermal analysis; C.A.N. obtained SEM images and EDS analysis; C.A.R.P.B. conducted the compression mechanical tests; A.S.R. conceived, participated and coordinate the present work in all stages as well as wrote the paper.

Funding: This research received no external funding.

Acknowledgments: Authors thank to FINEP-, CNPq-, CAPES-, FAPEMIG-, and FAPESP-Brazil by technical supports provided for this work.

Conflicts of Interest: The authors declare no conflict of interest.

References

1. Bauer, J.R.D.O.; Loguercio, A.D.; Reis, A.; Rodrigues Filho, L.E. Microhardness of Ni-Cr alloys under different casting conditions. *Braz. Oral Rev.* **2006**, *20*, 40–46. [CrossRef]
2. Bezzon, O.L.; Mattos, M.G.; Rollo, J.M.; Panzeri, H. Desenvolvimento de uma liga experimental de Ni-Cr para restaurações metalocerâmicas: Ensaios de dureza e resistência mecânica. *Rev. Odontol. Univ. São Paulo* **1995**, *9*, 145–149.
3. Chen, W.C.; Teng, F.Y.; Hung, C.C. Characterization of Ni–Cr alloys using different casting techniques and molds. *Mater. Sci. Eng. C* **2014**, *35*, 231–238. [CrossRef] [PubMed]
4. Ho, K.H.; Kim, J.H.; Chang, K.; Kwon, J. The role of Cr on oxide formation in Ni-Cr alloys: A theoretical study. *Comput. Mater. Sci.* **2018**, *142*, 185–191.
5. Huang, H.H. Surface characterization of passive film on Ni-Cr based dental casting alloys. *Biomaterials* **2003**, *24*, 1575–1582. [CrossRef]
6. Bezzon, O.L.; de Mattos, M.D.G.; Ribeiro, R.F.; de Almeida Rollo, J.M. Effect of beryllium on the castability and resistance of ceramometal bonds in nickel-chromium alloys. *J. Prosthet. Dent.* **1998**, *80*, 5–570. [CrossRef]
7. Turchi, P.; Kaufman, L.; Liu, Z.K. Modeling of Ni-Cr-Mo based alloys: Part I—Phase Stability. *Calphad* **2006**, *30*, 70–87. [CrossRef]
8. Franke, P.; Neuschütz, D. Scientific Group Thermodata Europe (SGTE) Mo-Ni. In *Binary Systems. Part 4: Binary Systems from Mn-Mo to Y-Zr*; Franke, P., Neuschütz, D., Eds.; Landolt-Börnstein—Group IV Physical Chemistry (Numerical Data and Functional Relationships in Science and Technology); Springer: Berlin/Heidelberg, Germany, 2006; Volume 19B4.
9. Wataha, J.C. Biocompatibility of dental casting alloys. *Crit. Rev. Oral Biol. Med.* **2002**, *87*, 351–363.
10. Bania, P.J. Beta titanium alloys and their role in the titanium industry. *J. Met.* **1994**, *46*, 16–19. [CrossRef]
11. Rosenstiel, S.F.; Land, M.F.; Fujimoto, J. *Prótese Fixa Contemporânea*, 3rd ed.; Editora Santos: São Paulo, Brazil, 2002; p. 868.
12. Silva, H.M.; Schneider, S.G.; Neto, C.M. Estudo das propriedades mecânicas das ligas Ti-8Nb-13Zr e Ti-18Nb-13Zr para aplicação como biomaterial. *Revista Brasileira de Aplicações de Vácuo* **2003**, *22*, 5–7.
13. Chen, H.; Du, Y.; Xu, H.; Liu, Y.; Schuster, J.C. Experimental investigation of the Nb-Ni phase diagram. *J. Mater. Sci.* **2005**, *40*, 6019–6062. [CrossRef]
14. Du, Y.; Liu, S.; Chang, Y.A.; Yang, Y. A thermodynamic modeling of the Cr–Nb–Ni system. *Calphad* **2005**, *29*, 140–148. [CrossRef]
15. JCPDS: International Centre for Diffraction Date. *Powder Diffraction File (Inorganic Phases)*, 1st ed.; JCPDS, International Centre for Diffraction Date: Swarthmore, PA, USA, 1988; Volume 2.
16. Kraus, W.; Nolze, G. Powdercell—A program for the representation and manipulation of crystal structures and calculation of the resulting X-ray powder patterns. *J. Appl. Crystallogr.* **1996**, *29*, 301–303. [CrossRef]
17. Rasband, W.S. *ImageJ*; U. S. National Institutes of Health: Bethesda, MD, USA. Available online: https://imagej.nih.gov/ij/ (accessed on 29 April 2019).
18. *ASTM E92-97: Standard Test Method for Vickers Hardness of Metallic Materials*; ASTM International: West Conshohocken, PA, USA, 1982; pp. 1–9.

19. *ASTM E8M-00a: Standard Test Methods for Tension Testing of Metallic Materials [Metric]*; ASTM International: West Conshohocken, PA, USA, 2001; pp. 1–22.
20. Craig, R.G.; Powers, J.M. *Restorative Dentals Materials*, 11th ed.; Mosby: St. Louis, MO, USA, 2002.
21. Horasawa, N.; Marek, M. The effect of remolding on corrosion of a silver-palladium alloy. *Dent. Mater.* **2004**, *20*, 352–357. [CrossRef]
22. Stachurski, Z.H. Section 3.4.3—X-ray scattering from a-NiNb alloy. In *Fundamentals of Amorphous Solids: Structure and Properties*; Higher Education Press, Wiley-VCH Verlag GmbH & Co. KGaA: Weinheim, Germany, 2015; p. 136.
23. American Society for Metals. *Alloy Phase Diagrams*; ASM Handbook, Metals Park, Ohio: Russell Township, OH, USA, 2016; Volume 3, pp. 2314–2321.
24. Holcomb, G.R.; Alman, D.E. The effect of manganese additions on the reactive evaporation of chromium in Ni–Cr alloys. *Scr. Mater.* **2006**, *54*, 1821–1825. [CrossRef]
25. Bauer, J.R.; Reis, A.; Loguercio, A.D.; Filho, L.E. Resistência à tração e alongamento de ligas de Ni-Cr fundidas sob diferentes condições. *RPG Rev. Pós. Grad.* **2006**, *13*, 83–88.
26. Pimenta, A.R.; Diniz, M.G.; Paciornik, S.; Sampaio, C.A.F.; Miranda, M.S.; Paolucci-Pimenta, J.M. Mechanical and Microstructural properties of a nickel-chromium alloy after casting process. *RSBO* **2012**, *9*, 17–24.
27. Olivieri, K.A.N.; Neisser, M.P.; Souza, P.C.; Bottino, M.A. Mechanical properties and micro structural analyses of a Ni-Cr alloy cast under different temperatures. *Braz. J. Oral Sci.* **2004**, *3*, 414–419.
28. Anusavice, K.J. *Phillips: Materiais Dentários*, 10th ed.; Guanabara Koogan: Rio de Janeiro, Brazil, 1998.
29. Lass, E.A.; Stoudt, M.R.; Williams, M.E.; Katz, M.B.; Levine, L.E.; Phan, T.Q.; Gnaeupel-Herold, T.H.; Ng, D.S. Formation of the Ni_3Nb δ-Phase in Stress-Relieved Inconel 625 Produced via Laser Powder-Bed Fusion Additive Manufacturing. *Metall. Trans. A* **2017**, *48*, 5547–5558. [CrossRef]

Article

Numerical Simulation of Three-Dimensional Mesoscopic Grain Evolution: Model Development, Validation, and Application to Nickel-Based Superalloys

Zhao Guo, Jianxin Zhou *, Yajun Yin, Xu Shen and Xiaoyuan Ji

State Key Laboratory of Materials Processing and Die & Mould Technology, Huazhong University of Science and Technology, Wuhan 430074, China; zg2014@hust.edu.cn (Z.G.); Yinyajun436@hust.edu.cn (Y.Y.); shenxu@hust.edu.cn (X.S.); jixiaoyuan@hust.edu.cn (X.J.)
* Correspondence: zhoujianxin@hust.edu.cn; Tel./Fax: +86-027-87541922

Received: 27 November 2018; Accepted: 5 January 2019; Published: 9 January 2019

Abstract: The mesoscopic grain model is a multiscale model which takes into account both the dendrite growth mechanism and the vast numerical computation of the actual castings. Due to the pursuit of efficient computation, the mesoscopic grain calculation accuracy is lower than that of dendrite growth model. Improving the accuracy of mesoscopic grain model is a problem to be solved urgently. In this study, referring to the calculation method of solid fraction in microscopic dendrite growth model, a cellular automata model of 3D mesoscopic grain evolution for solid fraction calculated quantitatively at the scale of cell is developed. The developed model and algorithm validation for grain growth simulation is made by comparing the numerical results with the benchmark experimental data and the analytical predictions. The results show that the 3D grain envelopes simulated by the developed model and algorithm are coincident with the shape predicted by the analytical model to a certain extent. Then, the developed model is applied to the numerical simulation of solidification process of nickel-based superalloys, including equiaxed and columnar dendritic grain growth. Our results show good agreement with the related literature.

Keywords: numerical simulation; cellular automaton; dendritic grain growth; quantitative prediction

1. Introduction

Grain evolutions including nucleation and growth during solidification process have a profound effect on the material mechanical properties of alloys. In particular, the competitive growth of equiaxed and columnar dendritic grains determine the microstructure of alloys. Over the past few decades, many computer models and methods have been developed to study theoretically and experimentally the grain growth in alloy solidification, such as the volume-averaged, front tracking, cellular automata (CA), and phase field models [1–5]. Among them, due to the advantages of clearly describing the physical phenomena and high computational efficiency of solidification transition, the CA model has been widely used and developed for prediction of grain growth in additive manufacturing [6,7], welding [8,9], and casting [10,11].

In 1993, the CA model was applied for the first time in alloy solidification by Rappaz and Gandin [12]. It mainly dealt with nucleation and growth of grains in a uniform temperature field. At this point, the basic framework of the CA model for grain growth simulation was constructed, although it still had some drawbacks. Subsequently, in order to satisfy the adaptability of non-uniform temperature field at the same time, the "2D rectangular algorithm" [13] and "decentered square algorithm" [14,15] coupled with finite element (FE) calculations were proposed. Similarly, the CA model could also be coupled with the finite difference (FD) method to form the CA-FD model [16], or coupled with the finite volume (FV) method to form the CA-FV model [17]. As one of the effective methods to simulate the growth of grain, the CA-FE model integrated into the casting commercial software of ProCAST have been widely used [18,19].

However, the growing octahedron in these models is no longer considered if the corresponding cell is fully surrounded by the mushy cells [15]. In this way, this CA-FE model is still a weak coupling method and overestimates the heat release of cell transformation in the mushy zone [20]. In the CA model, the solid fraction of growing cell or interface cell in the mushy zone cannot be calculated quantitatively at the scale of CA grid. And it is calculated approximately at the scale of FE grid by the Scheil's equation and the Scheil microsegregation model. Recently, Guillemot et al. [20] have employed a front tracking method proposed by Gandin [21] to deal with the coupling problem. Similar works have also been conducted by Carozzani et al. [22] and Liu et al. [23,24].

Additionally, the CA model of grain growth, ignoring the dendritic morphology of the grain itself, is generally referred to as mesoscopic model. Corresponding CA models of dendritic growth called microscopic model have been proposed and developed, typically works like Nastac's [25], Zhu's [26] and Pan's [27]. In this regard, it is a full coupling model that accounts for the time evolution of the fraction of solid predicted at the scale of CA grid in the CA model of dendritic growth. It is further observed in the CA models of two scales that both are faced with the same problems of grid anisotropy [13,28], transformation rules [14,29], and coupling methods [30]. From the point of view of the simulation method, the method for dealing with the same problem in the CA models of two scales are interrelated [16,31]. Nevertheless, there is a problem that the CA model of dendritic growth requires sufficiently higher mesh resolution than the CA model of grain growth [32].

The purpose of this paper is to achieve the full coupling simulation of grain growth so as to improve the simulation accuracy. A 3D modified cellular automaton coupling model of grain growth for solid fraction calculated quantitatively at the scale of CA gird is proposed. The relevant models and algorithms are presented and validated. Then, the developed model is applied to simulate the formation of equiaxed and columnar dendritic grains during solidification process of nickel-based superalloys.

2. Model Description and Numerical Algorithm

A 3D computation domain is firstly divided into a series of regular hexahedral cells. The cell spacing is equal to the corresponding mesh size. Each cell is characterized with such variables as solid fraction, grain growth rate, grain orientation, etc. All the cell states are defined as three types: liquid ($f_s = 0$), solid ($f_s = 1$), or interface ($0 < f_s < 1$). And the cell neighbors are characterized by the first-order neighbors (6 neighborhoods), second-order neighbors (12 neighborhoods), and third-order neighbors (8 neighborhoods). At the beginning of the numerical simulation, the cells in computation domain are initialized with given initial and boundary conditions. When the undercooling reaches or exceeds the critical nucleation undercooling, the grain begins to nucleate and grow. As the dendritic grain grows, the solid fraction of growing cell or interface cell is calculated quantitatively at the scale of CA grid. The governing equations for calculating the nucleation distribution and growth rate of dendritic grains are presented. The numerical algorithm about capture rules for new interface cells and the calculation of solid fraction at the scale of CA grid are described in detail below.

2.1. Grain Nucleation

Only heterogeneous nucleation is considered in this study, since high undercooling is not achieved due to the presence of foreign phases. For the sake of practicability, the continuous nucleation model proposed by Rappaz and Gandin et al. [12] is used to describe the grain nucleation. The total grain density $n(\Delta T)$ is calculated by:

$$n(\Delta T) = \int_0^{\Delta T} \frac{dn}{d\Delta T'} d(\Delta T') \tag{1}$$

where ΔT is the undercooling.

The continuous nucleation distribution $dn/d\Delta T'$ is given by:

$$\frac{dn}{d\Delta T'} = \frac{n_{\max}}{\sqrt{2\pi}\Delta T_\sigma} \exp\left[-\frac{1}{2}\left(\frac{\Delta T - \Delta T_N}{\Delta T_\sigma}\right)\right] \tag{2}$$

where $n_{\max}$ is the total density of grains, ΔT_σ is the standard deviation of undercooling, and ΔT_N is the mean undercooling in nucleation, respectively. Additionally, the total density of grains on the two-dimensional section $(n_{max})_{2D}$ can be obtained by the stereological relationships [12], $(n_{max})_{2D} = (6 \cdot n_{max}^2/\pi)^{1/3}$.

Unlike the numerical realization of nucleation process determined by random number method, in this present study, the nucleation position and critical undercooling are randomly assigned to the nucleation region in advance, and then the actual undercooling and the critical undercooling of each cell are compared in each time step to determine the nucleation.

2.2. Grain Growth

Grain growth rate is mainly determined by the growth behavior of dendrite tips. According to the Lipton-Glicksman-Kurz (LGK) growth model [33,34], the dendrite tips growth of alloys are affected by thermal undercooling (ΔT_t), constitutional undercooling (ΔT_c) and curvature undercooling (ΔT_R). With the growth of dendrite tips, latent heat and solute are released from tips. At the interface of dendrite tips, heat and solute transfer lead to changes in thermal and constitutional undercooling. In addition, the growth of dendrite tips also lead to the change of curvature undercooling. At a given total undercooling, the grains will grow under the interaction of various factors until they grow up completely or encounter obstacles to stop growth.

The thermal undercooling, constitutional undercooling and curvature undercooling are defined as:

$$\Delta T_t = Iv(P_t)\frac{\Delta H}{c_p} \tag{3}$$

$$\Delta T_c = m_L C_0 \left(1 - \frac{1}{1 - (1 - k_0)Iv(P_c)}\right) \tag{4}$$

$$\Delta T_R = \frac{2\Gamma}{R} \tag{5}$$

where $Iv(P)$ is the Ivantsov function of the Peclet number; $P_t = RV/(2\alpha)$ and $P_c = RV/(2D)$ are the heat Peclet number and solutal Peclet number, respectively; ΔH is the latent heat; c_p is the specific heat; m_L is the liquidus slope; C_0 is the initial components concentration of alloys; k_0 is the solute partition coefficient; Γ is Gibbs-Thompson coefficient; R is the tip radius; V is the interface growth rate; α is thermal diffusion coefficient; and D is solute diffusion coefficient.

The 3D Ivantsov function is expressed as:

$$Iv(P) = P\exp(P)E_1(P) \tag{6}$$

where $E_1(P)$ is the exponential integral function.

The relationship between the tip radius and the interface growth rate can be established by combining the above equations, but it cannot be solved, and other more definite conditions are needed.

Based on the stability criterion of a dendrite tip, the tip radius is equal to the shortest wavelength λ_i, which can be evaluated by:

$$\lambda_i = \left(\frac{\Gamma}{\sigma^*(m_L G_c - G)}\right)^{0.5} \tag{7}$$

where G_c is the concentration gradient; G is the temperature gradient; σ^* is a stability parameter, it should be noted here that the stability parameter according to the microsolvability theory varies with the degree of anisotropy of the interface energy, and its value can be obtained by the linearized solvability theory [35].

For ease of numerical calculations, the above equations are realized and solved using Matlab. The growth rate is simplified as a polynomial function of local undercooling.

2.3. Capture Rules for New Interface Cells and Calculation of Solid Fraction at the Scale of the CA Grid

A single regular octahedron is defined as a constituent element of a grain. Each cell corresponds to a regular octahedron. The sizes of regular octahedrons are characterized by the distance (S_L) between any two adjacent vertices. Its three orthogonal axes ([100]/[010]/[001]) represent the directions of dendrite tips. Meanwhile, in order to avoid the infinite growth problem of the single regular octahedron, the distance (S_L) is limited to $2 \cdot \Delta x$ (Δx is the cell size). The grain envelope, which consists of a number of regular octahedrons, is also shown as a larger regular octahedron in the uniform temperature field. A single grain whose crystallographic orientation is assumed to be the Euler angles (φ, θ, ψ) with respect to the axis of CA, as shown in Figure 1.

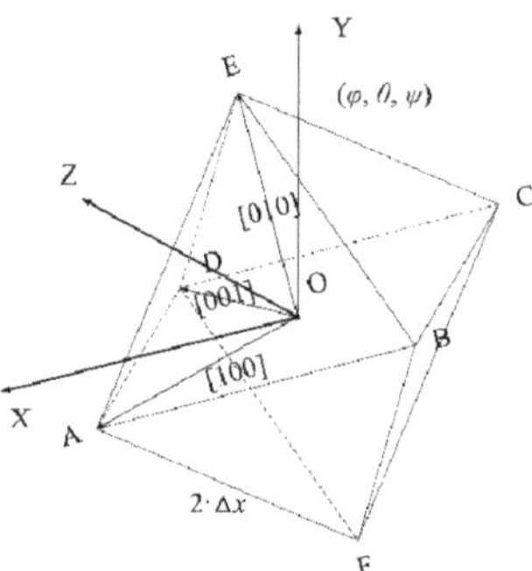

Figure 1. Schematic diagram of a single regular octahedron with arbitrary orientation (φ, θ, ψ).

In this present work, the nucleated cell is defined as a solid cell. And all its neighbor cells are defined as interface cells with a solid fraction of ε ($\varepsilon \rightarrow 0$). When the solid fraction of the interface cells reaches 1 or greater than 1, all the first-order, second-order, and third-order neighbors' liquid cells of the former interface cell transform to interface cells. At this point, it is guaranteed that at least one of the 26 neighbor cells of the interface cells evolves to a solid cell. The Schematic diagram of cell state is shown in Figure 2. For the octahedron corresponding to the nucleated cell, the center of octahedron coincides with the center of nucleated cell. While the center of octahedron corresponding to the interface cell also coincides with the center of nucleated cell. Similar to 3D decentered square algorithm, the tips length of octahedron (L) from its center to any of its six vertices are given by:

$$L(t_2) = L(t_1) + \int_{t_1}^{t_2} V(\Delta T(\tau))d\tau \tag{8}$$

where t_2 represents the next time, t_1 represents the current time. The initial tip length of the grain corresponding to the nucleated cell is taken as the maximum value ($\sqrt{2} \cdot \Delta x$). The initial tip length of the grain corresponding to the interface cell is taken as ε ($\varepsilon \rightarrow 0$).

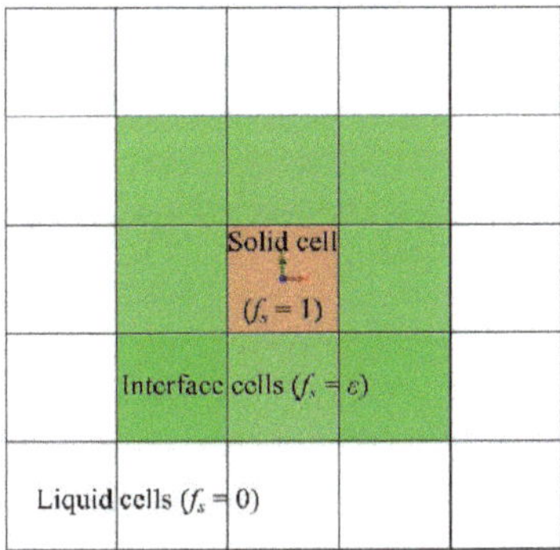

Figure 2. Schematic diagram of the cell state.

During the evolution of grains, due to the displacement between the center of the interface cell and the center of the growing octahedron, the growing octahedron corresponding to the interface cell will capture the cell center of the neighbors of solid cell. When the interface cell is captured, it is then switched to the solid cell. At this moment, all the remaining liquid cells being neighbors of this new solid cell will be automatically transferred from liquid to interface cells. The new octahedron corresponding to this new interface cell is initialized. Once the tip length of the growing octahedron is equal to or greater than the maximum value ($\sqrt{2}{\cdot}\Delta x$), the growing octahedron will cease to grow. The corresponding interface cell will be switched to the solid cell regardless of the solid fraction of the interface cell.

In order to realize the self-consistency between the tip length of the growing octahedron and the solid fraction of interface cell, a quantitative method for calculating the two is described below.

Figures 3 and 4 sketch the capture rule for solid cell in a time step calculation. When the center of growing octahedron corresponding to the interface cell (for example, I_J) coincides with the center of nucleated cell, the fraction increment of the interface cell I_J is calculated by:

$$\Delta f_s = \frac{\Delta L}{L_1} with \qquad L_1 \leq \sqrt{2}\Delta x \tag{9}$$

where ΔL represent the increment of the tip length of the growing octahedron, L_1 represent the final tip length (OA_1) of the growing octahedron.

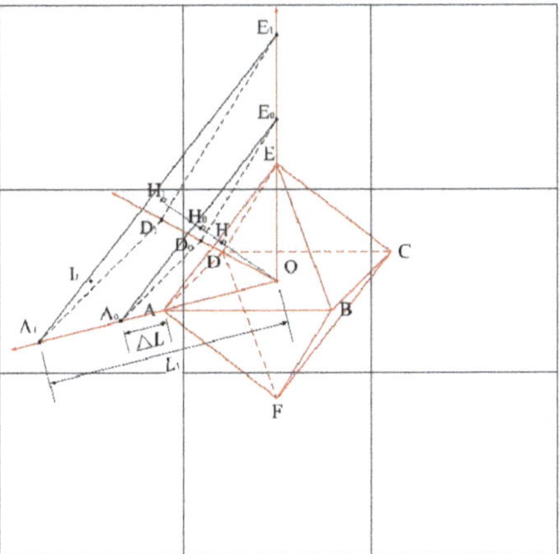

Figure 3. Schematic diagram of the capture rule for solid cell in a time step calculation, the center of I_J is in the final plane $A_1D_1E_1$ of the growing octahedron with $L_1 \leq \sqrt{2}{\cdot}\Delta x$.

Subsequently, as soon as the center of the interface cell I_J is captured by the growing octahedron, this interface cell is immediately switched to solid cell and the remaining liquid cells adjacent to this new solid cell will be transformed into new interface cells (for example, I_{JK}), as shown in Figure 4. As mentioned above, the new growing octahedron corresponding to the new interface cell I_{JK} is initialized. Especially, the center (O_0) of the new growing octahedron is also located at the center of the octahedron corresponding to the original interface cell I_J. And the tip length of the new growing octahedron is initialized with the maximum value L_m ($L_m = \sqrt{2} \cdot \Delta x - \varepsilon$). As a result, before the solid fraction of the new interface cell I_{JK} reaches 1, the center (O_2) of the new growing octahedron has moved gradually along the direction of growth so as to maintain its continuous growth. In this case, the fraction increment of the new interface cell I_{JK} is calculated by:

$$\Delta f_s = \frac{\Delta L}{L_2 - L_m} \tag{10}$$

where L_2 represent the final tip length ($O_0 A_3$) of the growing octahedron.

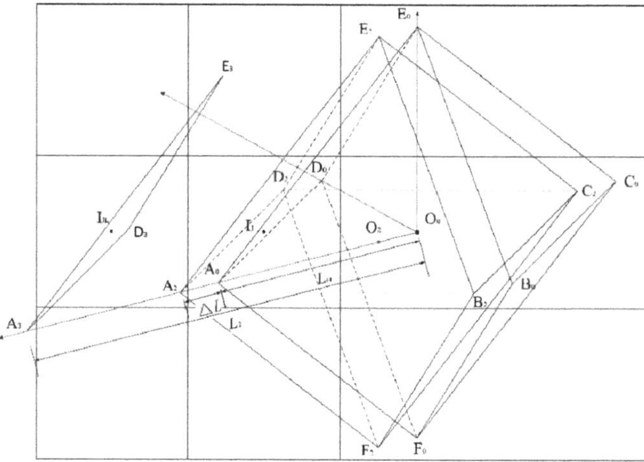

Figure 4. Schematic diagram of the capture rule for solid cells in a time step calculation, the center of I_{JK} is in the final plane $A_3 D_3 E_3$ of the growing octahedron with $L_m = \sqrt{2} \cdot \Delta x - \varepsilon$.

3. Results and Discussion

3.1. Comparison with the Benchmark Experimental Data

Firstly, to validate the numerical calculation program for mesoscopic grain growth in this study, the dendritic tip growth rate of a succinonitrile-acetone mixtures are calculated and compared with benchmark experimental data [34]. According to the microsolvability theory, the stability parameter remains unchanged within a certain range of Peclet number. Here, two stability parameters are considered. Figure 5 presents the dendritic tip growth rate as a function of initial concentration in succinonitrile-acetone mixtures.

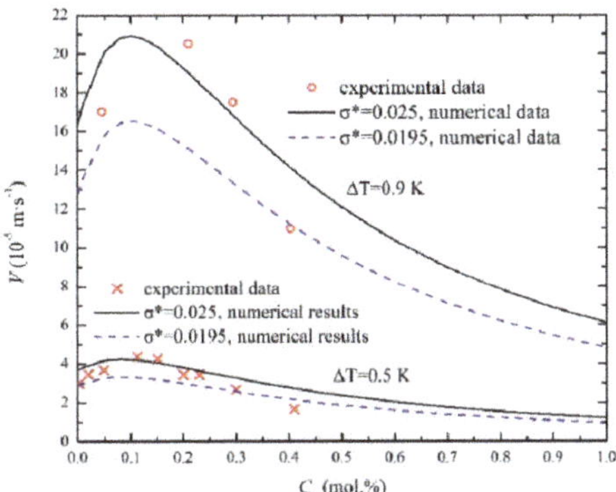

Figure 5. Dendritic tip growth rate as a function of initial concentration in succinonitrile-acetone mixtures.

As shown in Figure 5, under both undercooling conditions of 0.5 K and 0.9 K, the dendrite tip growth rate calculated by numerical methods increases first and then decreases with the initial concentration, which is consistent with the experimental results. For the initial concentration range of 0–0.25 (mol. %), the curve of the dendritic tip growth rate with a stability parameter of 0.025 is approximately matched with the experimental results. For the initial concentration range of 0.25–0.45 (mol%), the curve of the dendritic tip growth rate with a stability parameter of 0.0195 is approximately matched with the experimental results. It is confirmed that the stability parameters match the experimental results only in a certain range of Peclet numbers.

3.2. Comparison with the Analytical Model of Grain Growth

The model and algorithm validation for grain growth simulation is made by comparing the numerical results with the analytical model predictions [14,36]. In the analytical growth model, it was assumed that the grain growth was controlled by the solute diffusion and only applicable to the low Peclet number regime. This is basically similar to the assumption of the LGK model. There are two kinds of grain branches predicted by the analytical model, here the external envelope is assumed to be the grain envelope. Figure 6 presents the schematic diagram of two-dimensional dendrite growth envelope trajectory under directional temperature. In the process of two-dimensional dendritic growth under directional solidification, the coordinates of the external envelope trajectory can be calculated by Equation (11) [36]:

$$\begin{cases} x = x_0 + (y - y_0) \cdot f(\theta) \\[2mm] \Delta T = \dfrac{\Delta T_L}{\sqrt{g(\theta)}} \cdot \dfrac{h\left(G \cdot A \cdot \Delta T_L \cdot \sqrt{g(\theta)} \cdot (t - t_0)\right) + \frac{\Delta T_0 \cdot \sqrt{g(\theta)}}{\Delta T_L}}{1 + S_{\langle 10 \rangle} \cdot \frac{\Delta T_0 \cdot \sqrt{g(\theta)}}{\Delta T_L} \cdot h\left(G \cdot A \cdot \Delta T_L \cdot \sqrt{g(\theta)} \cdot (t - t_0)\right)} \\[2mm] y = v_L \cdot t - \dfrac{\Delta T}{G} \end{cases} \tag{11}$$

where V_L is the moving rate of liquid isotherm, ΔT_L is the dendrite tip undercooling, ΔT_0 is the undercooling at initial time t_0, θ is the dendrite orientation angle, A is a constant coefficient in LGK model, $f(\theta)$, $g(\theta)$ and $S_{\langle 10 \rangle}$ are directional functions related to the orientation angle respectively, which are given by Equations (12)–(14).

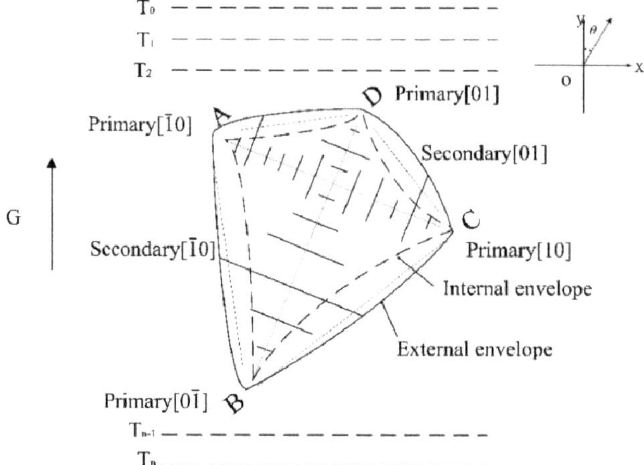

Figure 6. Schematic diagram of two-dimensional dendrite growth envelope trajectory under directional temperature.

$$\begin{cases} f(\theta) = \frac{1}{\tan(\theta)} & \text{for the } [10] \text{ and } [\bar{1}0] \text{ branches} \\ f(\theta) = -\tan(\theta) & \text{for the } [01] \text{ and } [0\bar{1}] \text{ branches} \end{cases} \tag{12}$$

$$\begin{cases} g(\theta) = \sin(\theta) & \text{for the } [10] \text{ and } [\bar{1}0] \text{ branches} \\ g(\theta) = \cos(\theta) & \text{for the } [01] \text{ and } [0\bar{1}] \text{ branches} \end{cases} \tag{13}$$

$$\begin{cases} S_{\langle 10 \rangle} = 1 & \text{for the } [10] \text{ and } [01] \text{ branches} \\ S_{\langle 10 \rangle} = -1 & \text{for the } [\bar{1}0] \text{ and } [0\bar{1}] \text{ branches} \end{cases} \tag{14}$$

Since the above analytical model is based on two-dimensional dendrite growth, it is necessary to extend this method to three-dimensional grain growth. In this present study, the projection method and the three cosine theorem are used to calculate the outermost envelope trajectory of a three-dimensional grain. Besides, since the analytical model was restricted to a quadratic approximation of the velocity-undercooling relationship, the growth rate was temporarily simplified as the same quadratic approximation relationship of local undercooling in the numerical simulation. Figure 7 presents the 3D envelopes and tips length of a single grain growing in a uniform temperature field and in a Bridgman-like situation. In Figure 7a, the cooling rate is set to 0 K/s and the thermal gradient is set to 0 K/m, the tips length (L_1, L_2, L_3, L_4, L_5, L_6) of the grain corresponding to Figure 7a are shown in Figure 7b. In Figure 7c, the cooling rate is set to -0.1 K/s and the thermal gradient along the Z-axis is set to 250 K/m, the tips length (L_1, L_2, L_3, L_4, L_5, L_6) of the grain corresponding to Figure 7c are shown in Figure 7d. The cell size Δx is set to 0.0001 m.

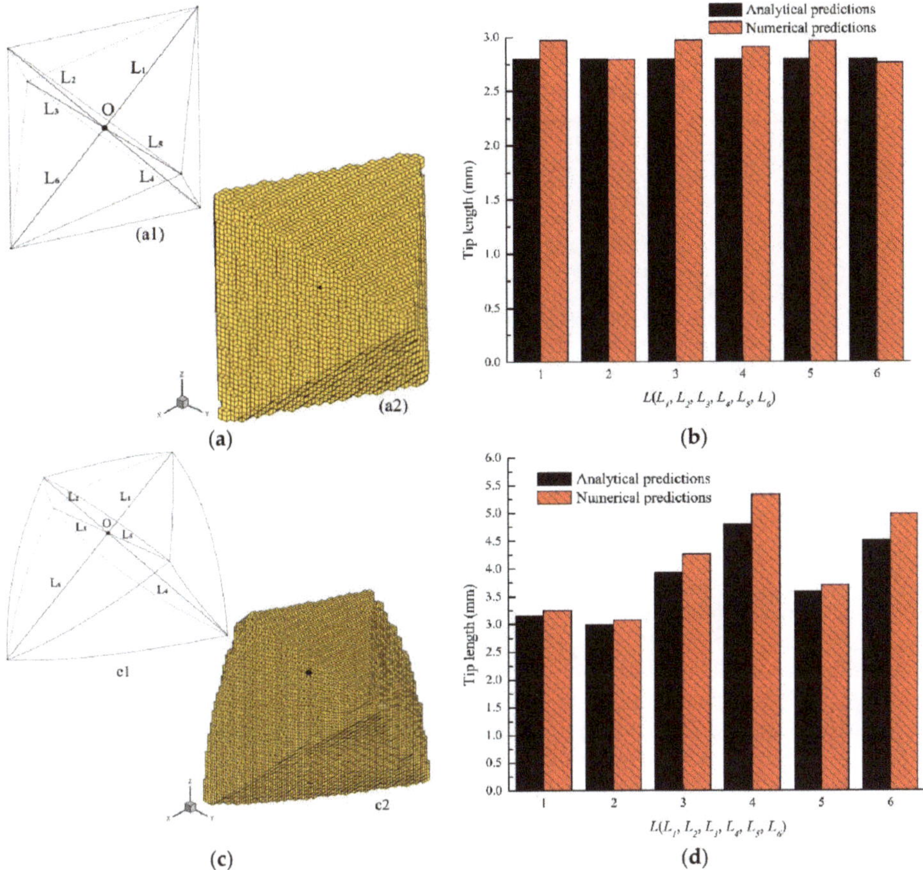

Figure 7. 3D envelopes and tips length of a single growing grain at time t = 7 s after nucleation, where the grain is assumed to nucleate and grow with an initial undercooling ΔT_0 = 2 K, the crystallographic orientation (20°, 40°, 60°) and the growth constant A = 0.0001 m/(s·K²). (**a**) 3D envelopes in a uniform temperature field; (**b**) tip length in a uniform temperature field; (**c**) 3D envelopes in a Bridgman-like situation; and (**d**) tip length in a Bridgman-like situation.

As shown in Figure 7a,c the 3D views of analytical and numerical predictions, for a given crystallographic orientation, are approximately regular octahedrons in a uniform temperature field. In Bridgman-like situations, due to the effects of deep undercooling, the tips length (L_4, L_6) of the grain closest to the negative thermal gradient direction are the longer distances. Altogether, the 3D grain envelopes (a2, c2) simulated by the developed model and algorithm in both cases are visually reproduced and coincident with the shape (a1, c1) predicted by the analytical model. Furthermore, from the Figure 7b,d, it can be noted that there are only a few mesh size gaps among the tips length of the grain in both predictions, which also approximately present the same distribution.

3.3. Model Application to Nickel-Based Superalloys

The present model is used to simulate the formation of equiaxed and columnar dendritic grains during nickel-based superalloys solidification process. The elemental composition (mass fraction, wt.%) of the material is Ni–7.82Cr–5.34Co–2.25Mo–4.88W–6.02Al–1.94Ti–3.49Ta. For multicomponent alloys, a large number of simplifications and calculations are required due to the complex multicomponent coupling solidification process. Here, a multicomponent pseudo-binary alloy method [37] is used. The material properties and model parameters used in these simulations are listed in Table 1.

Table 1. Material properties and model parameters.

Parameters	Value
Density (kg·m^3)	7223.48
Specific heat (J·(kg·K)$^{-1}$)	1074.60
Thermal conductivity (W·(m·K)$^{-1}$)	37.43
Latent heat (J·kg^{-1})	23,3371.59
Initial equivalent concentration of solute (wt.%)	31.74
Equivalent coefficient of partition (wt.%)	0.57
Equivalent slope of liquidus	-1.96
Gibbs Thomson Coefficient (K·m)	1.0×10^{-7}
Solute diffusion coefficient in the liquid (m^2·s^{-1})	3.0×10^{-9}
Solute diffusion coefficient in the solid (m^2·s^{-1})	3.0×10^{-12}

3.3.1. Equiaxed Dendritic Grains

The calculation domain is uniformly divided into 200 × 200 × 200 meshes with mesh size of 0.0001 m. It is assumed that the temperature field is homogeneous and cooled down from liquidus temperature at different cooling rates leading to nucleation and growth process. The cooling rates are set to 1 K·s^{-1}, 10 K·s^{-1}, and 100 K·s^{-1}, respectively. Here, the nucleation process only takes into account the nucleation inside the melt. It must be stressed that the nucleation parameters are affected by many factors, such as random nucleation, solidification conditions, etc. It cannot be uniquely determined or calculated in advance. For the ease of numerical calculations, the nucleation parameter is assumed to be known in advance. In the bulk of the melt, the total density of grains is 1×10^{10} m^{-3}, the median undercooling is 3 K, and the standard deviation of undercooling is 0.5 K. Figure 8 presents the size distribution of equiaxed dendritic grains at different cooling rates.

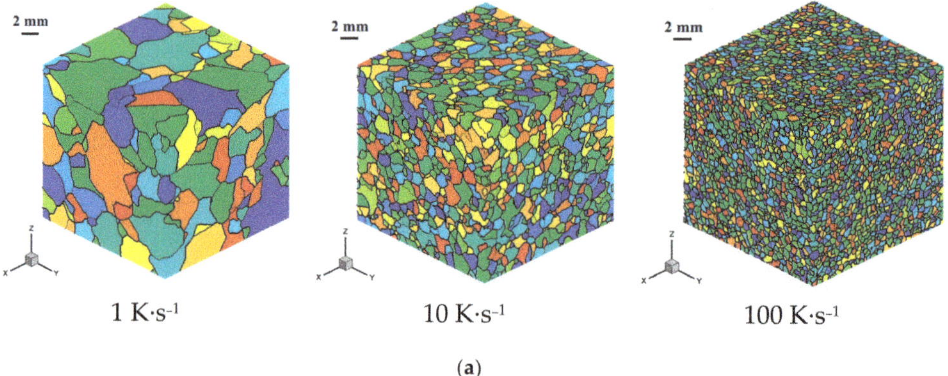

(**a**)

Figure 8. *Cont.*

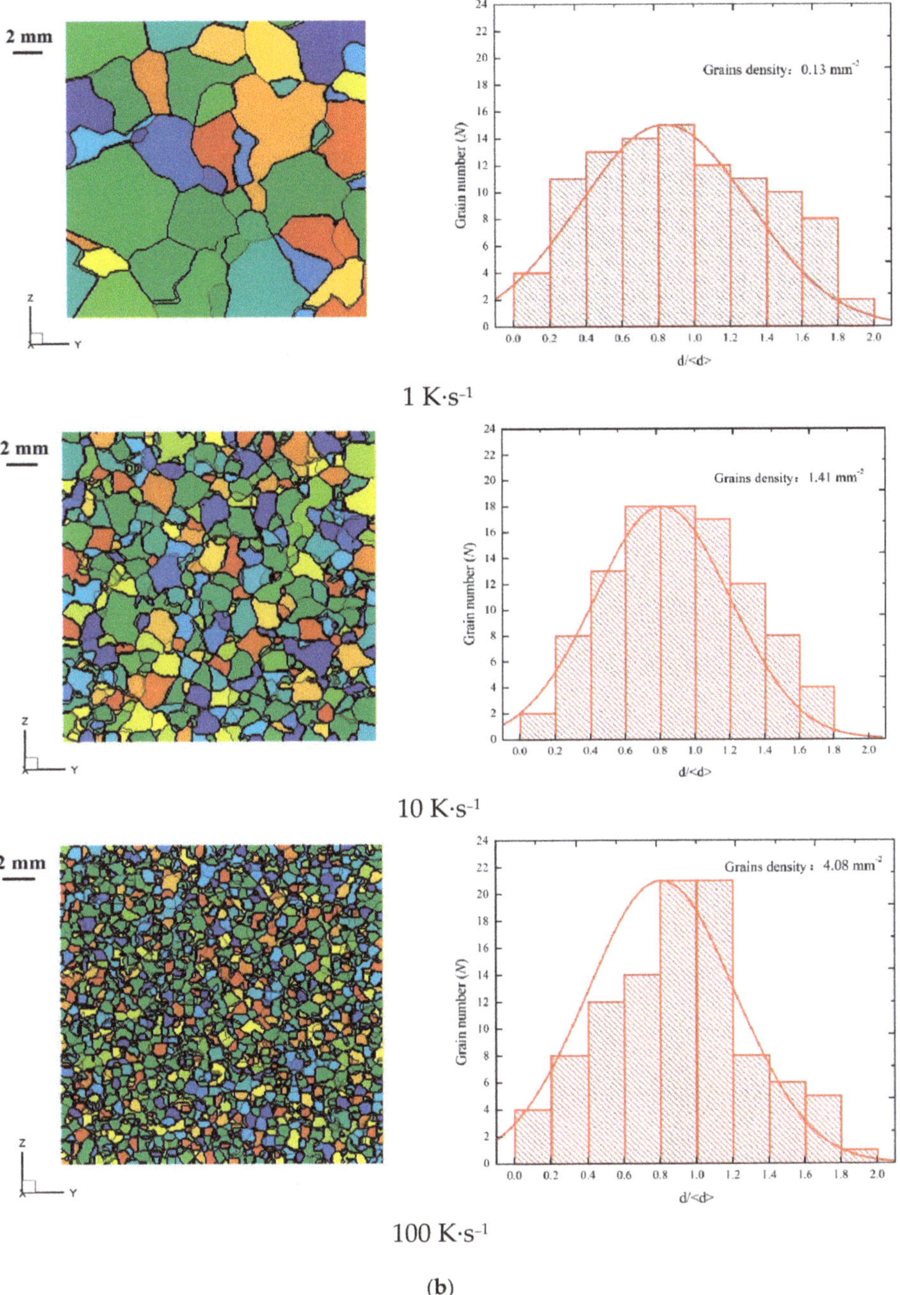

Figure 8. Size distribution of equiaxed dendritic grains at different cooling rates. (**a**) Simulation results of three-dimensional equiaxed dendritic grains; and (**b**) size distribution of equiaxed dendritic grains in a two-dimensional section.

Figure 8 shows that, with the increase of cooling rate, the grain size of equiaxed grains decreases obviously, and the grain density increases from 0.13 mm^{-2} to 4.08 mm^{-2}. From the data analysis of the two-dimensional section, it can be seen that the number of grains approximately presents a normal distribution curve at d/<d> from 0 to 2.0. With higher cooling rates the maximum of the grain number as function of d/<d> grows, while its dispersion in grain-size lowers. The highest grain number-density is always given close to the median grain-size. It is indicated that the increase of cooling rate can significantly promote the grain refinement. These results are consistent with the experimental results of solidification microstructures [38], which preliminarily validate the effectiveness of the proposed model and algorithm.

3.3.2. Columnar Dendritic Grains

Cooling rate and temperature gradient are two important parameters for directional solidification microstructure. Since the thermal diffusion rate is more than several orders of magnitude larger than the solute diffusion rate, it can be assumed that the temperature is uniform at the scale of a grain. At a fixed temperature gradient (G) and cooling rate (C) the temperature isotherms of the whole calculation domain will change at the same drawing speed ($R = C/G$). The calculation domain is uniformly divided into 60 × 180 × 600 cubic cells with size of 0.0001 m. Two types of nucleation parameters are considered. At the bottom surface, the total density of grains is 5.0 × 10^5 m^{-3}, the mean undercooling is 0 K, and the standard deviation is 0 K. In the bulk of the melt, the total density of grains is 1 × 10^{10} m^{-3}, the mean undercooling is 3 K, and the standard deviation is 0.5 K.

Equiaxed-to-Columnar Transformation (ECT) for Different Cooling Rates

The temperature gradient is set to 1000 K·m^{-1}. And the cooling rates are set to 1 K·s^{-1}, 10 K·s^{-1}, and 100 K·s^{-1}, respectively. Figure 9 presents the grain size distributions for different cooling rates.

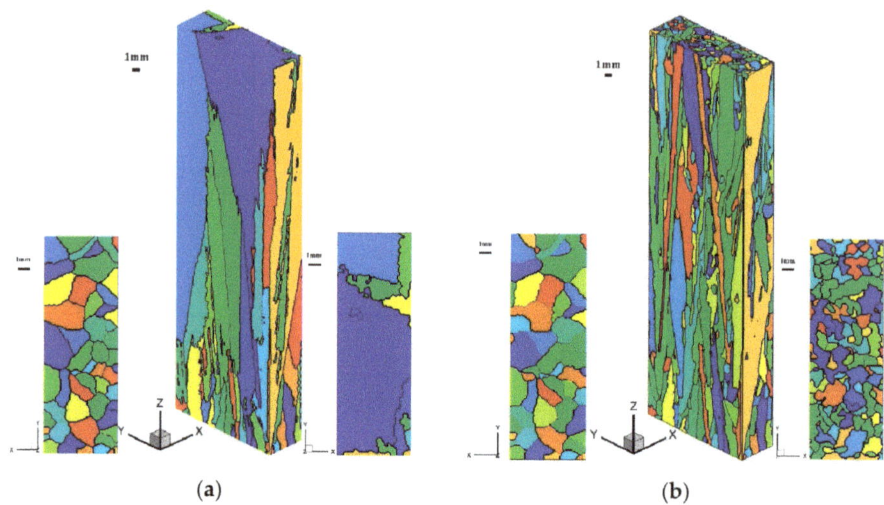

(a)   (b)

Figure 9. Cont.

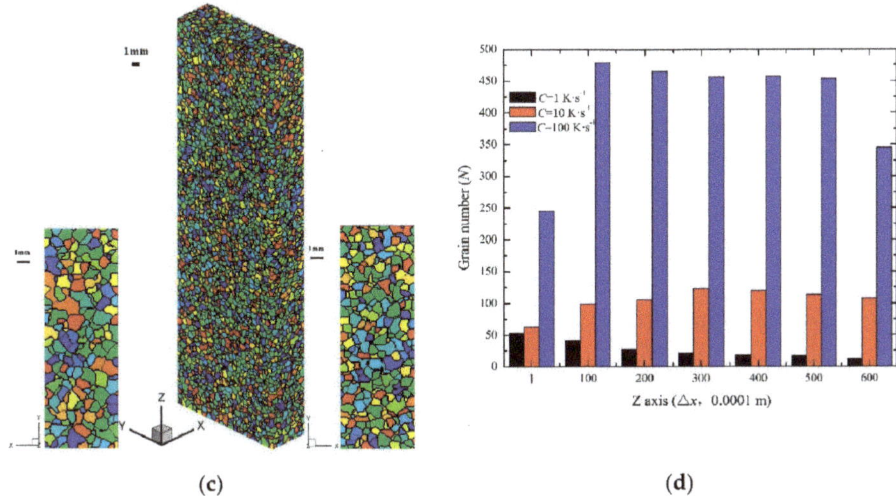

(c)   (d)

Figure 9. Grain size distributions for different cooling rates. (**a**) C = 1 K·s^{-1}; (**b**) C = 10 K·s^{-1}; (**c**) C = 100 K·s^{-1}; and (**d**) grain number along the Z axis.

Figure 9 shows, that for different cooling rates, the developed solidification microstructures of the calculated domain gradually changes (with respect to the size distribution of grains). At the cooling rate of 1 K·s^{-1}, some thick columnar dendritic grains are shown in the calculation domain, and the number of grains decreases gradually along the Z axis. At the cooling rate of 10 K·s^{-1}, the grain structure in the calculation domain is composed of both columnar and fine equiaxed dendritic grains, and the number of grains increases first and then decreases. At the cooling rate of 100 K·s^{-1}, almost all grains in the calculation domain are small equiaxed dendritic grains, and the number of grains increases suddenly and decreases slowly with increasing Z. From the above results, it can be inferred that at low cooling rate the grain growth rate of directional solidification is faster than the grain nucleation rate in the supercooled region of melt. In addition, due to the boundary limitation of the calculation domain, some grains deviating from the direction of temperature gradient will be eliminated at the boundary, resulting in the gradual decrease of the number of grains along the Z axis.

Equiaxed-to-Columnar Transformation (ECT) for Different Temperature Gradients

The temperature gradient is set to 500 K·m^{-1}, 1000 K·m^{-1}, and 1500 K·m^{-1}, respectively. The cooling rates are set to 10 K·s^{-1}. Figure 10 presents the grain size distributions for different temperature gradients.

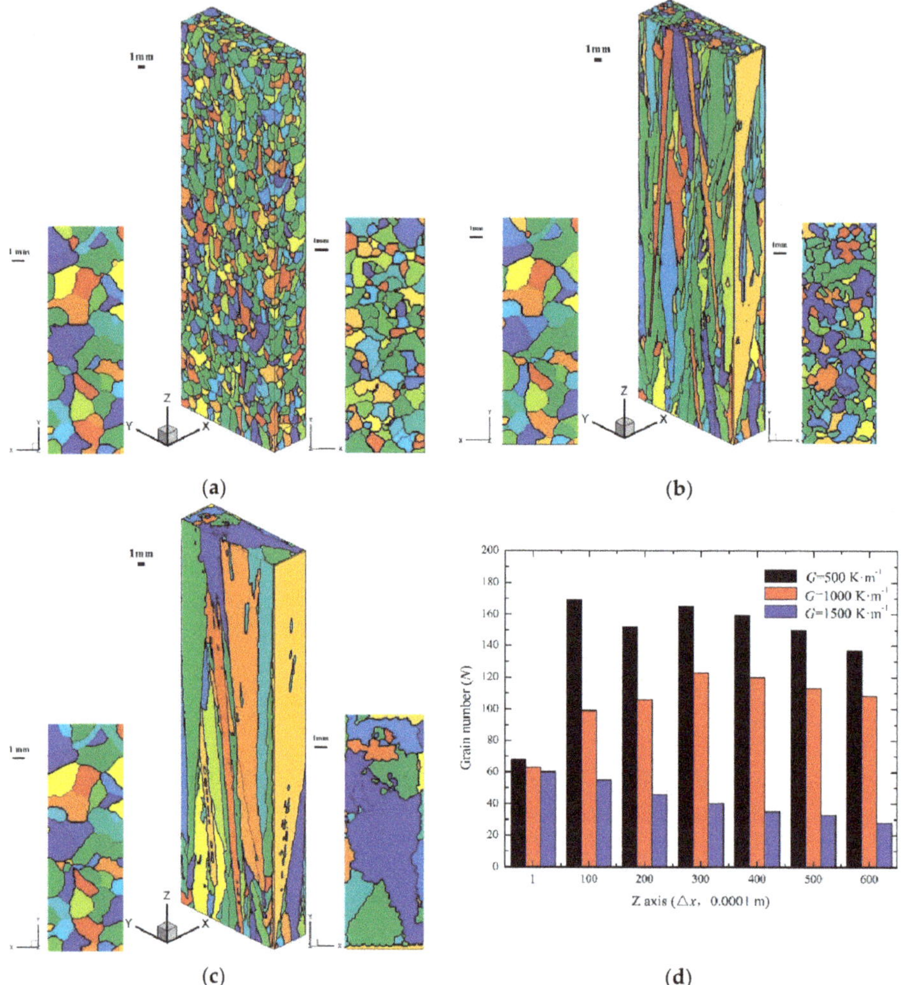

Figure 10. Grain size distributions for different temperature gradients. (**a**) $G = 500$ K·m^{-1}; (**b**) $G = 1000$ K·m^{-1}; (**c**) $G = 1500$ K·m^{-1}; and (**d**) grain number along the Z axis.

Figure 10 shows for the calculated domain, that with increasing temperature gradient, the solidification microstructure develops to the opposite of the results presented in Figure 8. In other words, at the temperature gradient of 500 K·m^{-1}, almost all grains in the calculation domain are small equiaxed dendritic grains, and the number of grains increases suddenly and decreases slowly. At the temperature gradient of 1500 K·m^{-1}, some thick columnar dendritic grains are shown in the calculation domain, and the number of grains decreases gradually along the Z axis. It can also be deduced that a rapid growth of columnar dendritic grains is promoted more by an increasing temperature gradient than by an increasing nucleation rate.

3.3.3. Columnar-to-Equiaxed Transformation (CET)

The ECT and CET are common phenomena of the grain evolution in solidification process. Under different solidification conditions, the transformations rule of the two influences the grain size distribution in the casting. During directional solidification, the ECT transformation is beneficial to the growth of columnar dendritic grains, and it is also affected by the nucleation parameters and cooling rate. Since the cooling rate and the temperature gradient are difficult to control, the casting of the metallic alloy may exhibit a non-uniform distribution of the temperature field. Especially, when the temperature gradient gradually decreases to a negative temperature gradient, this will lead to the CET transformation. To characterize the positive-to-negative transition in the temperature gradient, the temperature gradient is set to 1000 K·m^{-1} from mesh point 0 to 300 in the direction of the Z axis, and the temperature gradient is set to −300 K·m^{-1} from mesh point 300 to 600 in the direction of the Z axis. The cooling rate is set to 10 K·s^{-1}. The other grain nucleation and growth parameters set here are the same as those above Section 3.3.2. Figure 11 presents the simulation results of the CET transformation.

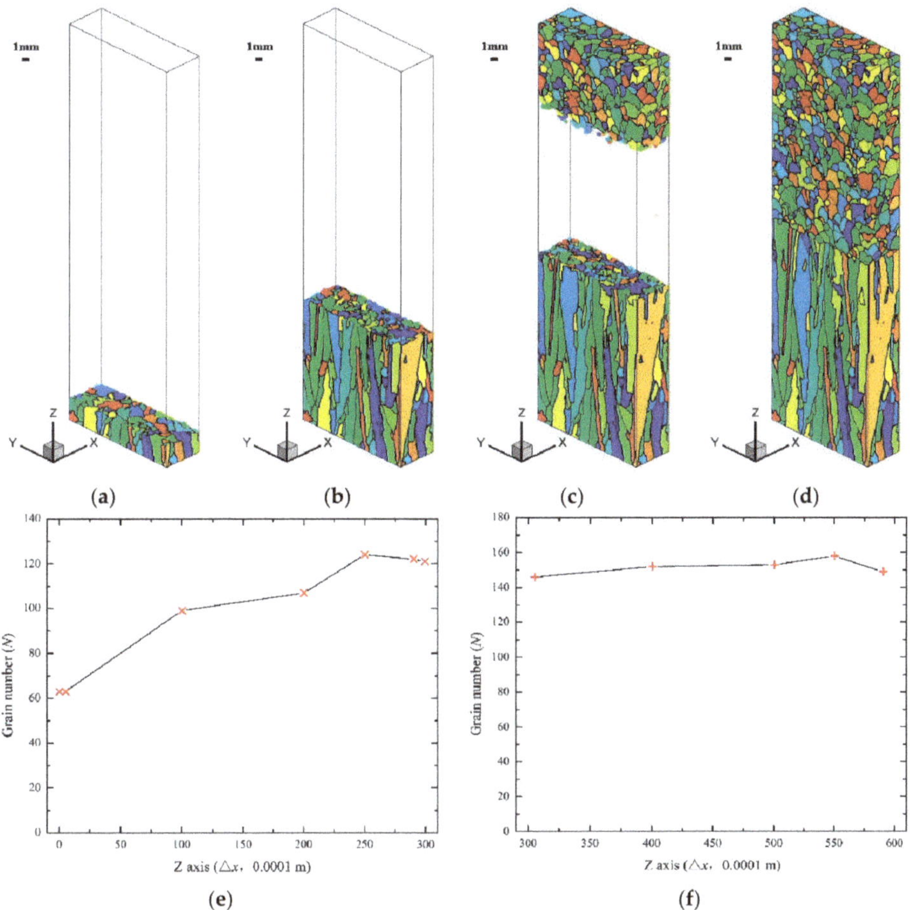

Figure 11. CET transformation. (**a**) t = 0.5 s; (**b**) t = 2.0 s; (**c**) t = 2.7 s; (**d**) t = 3.3 s; (**e**) grain number (0 ≤ Z ≤ 300); and (**f**) grain number (300 ≤ Z ≤ 600).

As shown in Figure 11, in the calculation domain, the grains first nucleate and grow at the bottom surface. From mesh point 0 to 300 in the direction of the Z axis, under the positive temperature gradient, grains begin to compete with each other along the Z axis, which is similar to the ECT transformation process mentioned above. From mesh point 300 to 600 in the direction of the Z axis, the grains nucleate and grow at the top surface under the negative temperature gradient. Compared with the ECT transformation process, the grains grow in the form of a large number of equiaxed crystals. Finally, the two parts of the grain stop growing at intermediate position, thus forming the CET transformation process. The above results show a good agreement with the results of related literature [23,39,40], which further validates the applicability of the proposed model and algorithm.

4. Conclusions

A cellular automata model of 3D mesoscopic grain evolution is developed. In this model, the 3D grain envelope trajectory is coupled with the solid fraction in a cell. The solid fraction in grain growth is calculated quantitatively at the scale of a CA grid. The detailed model and algorithm validations are performed by comparing the numerical results with the analytical model predictions. The simulated 3D grain envelopes are found to be in agreement with the analytical predictions. The developed model was then applied to simulate the formation of equiaxed and columnar dendritic grains during the solidification process of nickel-based superalloys. The results of this simulation could well explain the evolution law of grain growth in solidification microstructures, including equiaxed dendritic grain growth, columnar dendritic grain growth, and microstructure transformation.

Although the mesoscopic grain growth in a simple temperature field is studied in this present work, the proposed grain envelope trajectory coupled with the cell solid fraction method can be easily extended to complex multiscale and multiphysical field calculations. Such a 3D mesoscopic grain model coupled to a macroscopic temperature field and solute field is in progress and can be used to simulate the solidification process of actual castings.

Author Contributions: Z.G. and J.Z. wrote the paper; Z.G. and Y.Y. did the simulations; Z.G. and X.S. analyzed the data; Z.G. and X.J. revised the paper.

Funding: This work was supported by the National Natural Science Foundation of China (No. 51605174, No. 51775205) and AECC Beijing Institute of Aeronautical Materials.

References

1. Wu, M.; Ludwig, A. A three-phase model for mixed columnar-equiaxed solidification. *Metall. Mater. Trans. A* **2006**, *37*, 1613–1631. [CrossRef]
2. Satbhai, O.; Roy, S.; Ghosh, S. A parametric multi-scale, multiphysics numerical investigation in a casting process for Al-Si alloy and a macroscopic approach for prediction of ECT and CET events. *Appl. Therm. Eng.* **2017**, *113*, 386–412. [CrossRef]
3. Mcfadden, S.; Browne, D.J.; Gandin, C.A. A comparison of columnar-to-equiaxed transition prediction methods using simulation of the growing columnar front. *Metall. Mater. Trans. A* **2009**, *40*, 662–672. [CrossRef]
4. Martorano, M.A.; Biscuola, V.B. Predicting the columnar-to-equiaxed transition for a distribution of nucleation undercoolings. *Acta Mater.* **2009**, *57*, 607–615. [CrossRef]
5. Souhar, Y.; Felice, V.F.D.; Beckermann, C.; Combeau, H.; Založnik, M. Three-dimensional mesoscopic modeling of equiaxed dendritic solidification of a binary alloy. *Comput. Mater. Sci.* **2016**, *112*, 304–317. [CrossRef]
6. Lopez-Botello, O.; Martinez-Hernandez, U.; Ramírez, J.; Pinna, C.; Mumtaz, K. Two-dimensional simulation of grain structure growth within selective laser melted AA-2024. *Mater. Des.* **2017**, *113*, 369–376. [CrossRef]
7. Tian, F.; Li, Z.; Song, J. Solidification of laser deposition shaping for TC4 alloy based on cellular automation. *J. Alloys Compd.* **2016**, *676*, 542–550. [CrossRef]
8. Chen, S.; Guillemot, G.; Gandin, C.A. Three-dimensional cellular automaton-finite element modeling of solidification grain structures for arc-welding processes. *Acta Mater.* **2016**, *115*, 448–467. [CrossRef]
9. Valvi, S.R.; Krishnan, A.; Das, S.; Narayanan, R.G. Prediction of microstructural features and forming of friction stir welded sheets using cellular automata finite element (CAFE) approach. *Int. J. Mater. Form.* **2016**, *9*, 115–129. [CrossRef]

10. Zhang, Q.; Xue, H.; Tang, Q.; Pan, S.Y.; Rettenmayr, M.; Zhu, M.F. Microstructural evolution during temperature gradient zone melting: Cellular automaton simulation and experiment. *Comput. Mater. Sci.* **2018**, *146*, 204–212. [CrossRef]

11. Hu, Y.; Xie, J.; Liu, Z.; Ding, Q.; Zhu, W.; Zhang, J.; Zhang, W. CA method with machine learning for simulating the grain and pore growth of aluminum alloys. *Comput. Mater. Sci.* **2018**, *142*, 244–254. [CrossRef]

12. Rappaz, M.; Gandin, C.A. Probabilistic modelling of microstructure formation in solidification processes. *Acta Metall. Mater.* **1993**, *41*, 345–360. [CrossRef]

13. Gandin, C.A.; Rappaz, M. A coupled finite element-cellular automaton model for the prediction of dendritic grain structures in solidification processes. *Acta Metall. Mater.* **1994**, *42*, 2233–2246. [CrossRef]

14. Gandin, C.A.; Rappaz, M. A 3D cellular automaton algorithm for the prediction of dendritic grain growth. *Acta Mater.* **1997**, *45*, 2187–2195. [CrossRef]

15. Gandin, C.A.; Desbiolles, J.L.; Rappaz, M.; Thevoz, P. A three-dimensional cellular automation-finite element model for the prediction of solidification grain structures. *Metall. Mater. Trans. A* **1999**, *30*, 3153–3165. [CrossRef]

16. Wang, W.; Lee, P.D.; Mclean, M. A model of solidification microstructures in nickel-based superalloys: Predicting primary dendrite spacing selection. *Acta Mater.* **2003**, *51*, 2971–2987. [CrossRef]

17. Zhu, M.F.; Hong, C.P. A Three Dimensional Modified Cellular Automaton Model for the Prediction of Solidification Microstructures. *Trans. Iron Steel Inst. Jpn.* **2002**, *42*, 520–526. [CrossRef]

18. Pozdniakov, A.N.; Monastyrskiy, V.P.; Ershov, M.Y.; Monastyrskiy, A.V. Simulation of competitive grain growth upon the directional solidification of a Ni-base superalloy. *Phys. Met. Metallogr.* **2015**, *116*, 67–75. [CrossRef]

19. Carter, P.; Cox, D.C.; Gandin, C.A.; Reed, R.C. Process modelling of grain selection during the solidification of single crystal superalloy castings. *Mater. Sci. Eng. A* **2000**, *280*, 233–246. [CrossRef]

20. Guillemot, G.; Gandin, C.A.; Combeau, H.; Heringer, R. A new cellular automaton—finite element coupling scheme for alloy solidification. *Model. Simul. Mater. Sci. Eng.* **2004**, *12*, 545–556. [CrossRef]

21. Gandin, C.A. From constrained to unconstrained growth during directional solidification. *Acta Mater.* **2000**, *48*, 2483–2501. [CrossRef]

22. Carozzani, T.; Digonnet, H.; Gandin, C.A. 3D CAFE modeling of grain structures: Application to primary dendritic and secondary eutectic solidification. *Model. Simul. Mater. Sci. Eng.* **2012**, *20*, 15010–15028. [CrossRef]

23. Liu, D.R.; Mangelinck-noël, N.; Gandin, C.A.; Zimmermann, G.; Sturz, L.; Nguyen-Thi, H.; Billia, B. Structures in directionally solidified Al–7 wt.% Si alloys: Benchmark experiments under microgravity. *Acta Mater.* **2014**, *64*, 253–265. [CrossRef]

24. Liu, D.R.; Mangelinck-noël, N.; Gandin, C.A.; Zimmermann, G.; Sturz, L.; Nguyen-Thi, H.; Billia, B. Simulation of directional solidification of refined Al–7 wt.% Si alloys–Comparison with benchmark microgravity experiments. *Acta Mater.* **2015**, *93*, 24–37. [CrossRef]

25. Nastac, L. Numerical modeling of solidification morphologies and segregation patterns in cast dendritic alloys. *Acta Mater.* **1999**, *47*, 4253–4262. [CrossRef]

26. Zhu, M.F.; Stefanescu, D.M. Virtual front tracking model for the quantitative modeling of dendritic growth in solidification of alloys. *Acta Mater.* **2007**, *55*, 1741–1755. [CrossRef]

27. Pan, S.Y.; Zhu, M.F. A three-dimensional sharp interface model for the quantitative simulation of solutal dendritic growth. *Acta Mater.* **2010**, *58*, 340–352. [CrossRef]

28. Beltran-Sanchez, L.; Stefanescu, D.M. A quantitative dendrite growth model and analysis of stability concepts. *Metall. Mater. Trans. A* **2004**, *35*, 2471–2485. [CrossRef]

29. Reuther, K.; Rettenmayr, M. Perspectives for cellular automata for the simulation of dendritic solidification–A review. *Comput. Mater. Sci.* **2014**, *95*, 213–220. [CrossRef]

30. Zhu, M.F.; Pan, S.Y.; Sun, D.K.; Zhao, H.L. Numerical simulation of microstructure evolution during alloy solidification by using cellular automaton method. *ISIJ Int.* **2010**, *50*, 1851–1858. [CrossRef]

31. Chen, R.; Xu, Q.; Liu, B. Cellular automaton simulation of three-dimensional dendrite growth in Al–7Si–Mg ternary aluminum alloys. *Comput. Mater. Sci.* **2015**, *105*, 90–100. [CrossRef]

32. Zinovieva, O.; Zinoviev, A.; Ploshikhin, V.; Romanova, V.; Balokhonov, R. A solution to the problem of the mesh anisotropy in cellular automata simulations of grain growth. *Comput. Mater. Sci.* **2015**, *108*, 168–176. [CrossRef]

33. Kurz, W.; Giovanola, B.; Trivedi, R. Theory of microstructural development during rapid solidification. *Acta Metall.* **1986**, *34*, 823–830. [CrossRef]

34. Lipton, J.; Glicksman, M.E.; Kurz, W. Dendritic growth into undercooled alloy metals. *Mater. Sci. Eng.* **1984**, *65*, 57–63. [CrossRef]

35. Barbieri, A.; Langer, J.S. Predictions of dendritic growth rates in the linearized solvability theory. *Phys. Rev. A* **1989**, *39*, 5314–5325. [CrossRef]

36. Gandin, C.A.; Schaefer, R.J.; Rappax, M. Analytical and numerical predictions of dendritic grain envelopes. *Acta Mater.* **1996**, *44*, 3339–3347. [CrossRef]

37. Stefanescu, D.M. *Science and Engineering of Casting Solidification*, 3nd ed.; Springer: Cham, Switzerland, 2015; pp. 145–196.

38. Kavoosi, V.; Abbasi, S.M.; Ghazi Mirsaed, S.M.; Mostafaei, M. Influence of cooling rate on the solidification behavior and microstructure of IN738LC superalloy. *J. Alloys Compd.* **2016**, *680*, 291–300. [CrossRef]

39. Mishra, S.; DebRoy, T. Non-isothermal grain growth in metals and alloys. *Mater. Sci. Technol.* **2006**, *22*, 253–278. [CrossRef]

40. Spittle, J.A. Columnar to equiaxed grain transition in as solidified alloys. *Int. Mater. Rev.* **2006**, *51*, 247–269. [CrossRef]

Article

Analysis of Forming Parameters Involved in Plastic Deformation of 441 Ferritic Stainless Steel Tubes

Orlando Di Pietro [1], Giuseppe Napoli [1], Matteo Gaggiotti [1], Roberto Marini [2] and Andrea Di Schino [1,*]

[1] Engineering Department, University of Perugia, Via G. Duranti 93, 06125 Perugia, Italy; orlando.dipietro@studenti.unipg.it (O.D.P.); giuseppe.napoli@studenti.unipg.it (G.N.); matteo.gaggiotti@studenti.unipg.it (M.G.)

[2] Acciai Speciali Terni S.p.A., Viale B. Brin, 051100 Terni, Italy; roberto.marini@acciaiterni.it

* Correspondence: andrea.dischino@unipg.it

Received: 19 June 2020; Accepted: 23 July 2020; Published: 28 July 2020

Abstract: A welded stainless steel tube is a component used in several industrial applications. Its manufacturing process needs to follow specific requirements based on reference standards. This calls for a predictive analysis able to face some critical issues affecting the forming process. In this paper, a model was adopted taking into account the tube geometrical parameters that was able to describe the deformation process and define the best industrial practices. In this paper, the effect of different process parameters and geometric constraints on ferritic stainless steel pipe deformation is studied by finite element method (FEM) simulations. The model sensitivity to the input parameters is reported in terms of stress and tube thinning. The feasibility of the simulated process is assessed through the comparison of Forming Limit Diagrams. The comparison between the calculated and experimental results proved this approach to be a useful tool in order to predict and properly design industrial deformation processes.

Keywords: stainless steels; plastic deformation; mechanical properties

1. Introduction

Stainless steels are nowadays used in many applications targeting corrosion resistance coupled to mechanical resistance [1,2]. In particular, they are employed in automotive [3], construction and building [4–6], energy [7–11], aeronautics [12], food [13–15] and three-dimensional (3D) printing [16] applications. Stainless steels find application in automotive field, thanks to their ability to be worked into complex geometries [17–19]. Compared to the most common techniques, the metal forming is used to manufacture complex geometries in efficient way. In order to demonstrate compliance with the most relevant standards, many quality tests are commonly carried out in the industrial sector: this implies a strong increase in terms of cost and time-consumption.

The modelling and simulation of manufacturing processes is a well-established prerequisite for cost-effective manufacturing and quality-controlled manufacturing plant operation. Therefore, the management and control of manufacturing, which have the potential to be game changers, are a key issue. A game-changing technology is one which either directly provides a new and advantageous method of manufacture or else provides an important missing link, thereby enabling a method of manufacturing which would not otherwise be viable or so significantly advantageous. This should be based on the development and application of models able to support manufacturing process stability and product quality.

In the case of ferritic steels, for example, the plastic processing of welded pipes is characterized by a poor behavior homogeneity [20]; because of the nature of this material, characterized by an intrinsic inhomogeneity, a certain percentage of tests will not be reliable. The quality control, performed after the manufacturing process, is generally carried out by means of tensile tests according to specifications. In many cases, the measured tensile properties are found to be not sufficient to guarantee compliance with the required standards. The numerical simulation of sheet metal forming processes has become an indispensable tool for the design of components and their forming processes [21,22]. This role was attained due to the strong impact in reducing the time to market and the cost of developing new components in industries ranging from automotive to packing, as well as enabling an improved understanding of the deformation mechanisms and their interaction with process parameters. Despite being a consolidated tool, its potential for application continues to be discovered with the continuous need to simulate more complex processes, including the integration of the various processes involved in the production of a sheet metal component and the analysis of the in-service behavior. The request for more robust and sustainable processes has also changed its deterministic character into stochastic to be able to consider the scatter in mechanical properties induced by previous manufacturing processes.

In this framework, many tube bending approaches have been developed in response to the different demands of tube specifications, shapes, materials and forming tolerances [23–25]. Many mathematical models able to describe the steel plastic deformation behavior have been developed, considering the type of material and application field [26]. All of these approaches require the analysis of the mechanical properties, considering the steel in its macroscopic and microscopic structure. Another relevant characteristic concerns the anisotropic behavior caused by the plastic deformation due to the pipe manufacturing. He et al. [27] reported about the occurring mechanisms, accurate prediction, and efficient controlling of multiple defects/instabilities during bending forming. Wu et al. [28] investigated the effects of the temperature, bending velocity, and grain size on the wall thickness change, cross-section distortion, and spring-back of a tube in rotary draw bending. Liu et al. [29,30] analyzed the effect of dies and process parameters on the wall thickness distribution and cross-section distortion of a thin-walled alloy tube. Tang [31] deduced several bending-related equations to predict the stress distributions of bent tubes, wall thickness variation, bending moment, and flattening based on the plastic-deformation theory. Al-Qureshi et al. [32] derived the approximate equations of tube bending to predict spring-back and residual stress quantitatively with assumptions of the ideal elastic-plastic material, plane strain condition, absence of defects, and "Bauschinger effect". E et al. [33] derived an analytical relation to predict stress distributions, neutral layer shift, wall thickness change, and cross-section distortion based on in-plane strain assumption in tube bending with the exponential hardening law. Li et al. [34], based on the plastic-deformation theory, established the analytical prediction model for spring-back, taking into account the tube specifications and material properties. Jeong et al. [35] proposed equations able to calculate the bending moment and spring-back of bent tubes, considering the work hardening in the plastic-deformation region. The cross-section distortion formula was deduced based on the virtual force principle by Liu et al. [36]. It is noted that, though many factors cannot be considered, such as contact conditions and unequal stress/strain distributions, the analytical models are still very useful to estimate and predict the forming quality for tube bending.

Considering the complexity of the tube bending forming process, the finite element numerical simulation method has been widely used to explore the bending deformation under various forming conditions [37–39]. The main goal of simulations is to predict the behavior of different tubes' geometries during the bending process or the cold metal forming of steel sheets. In fact, a proper procedure of tube bending and a correct simulation of pipe yielding, after bending, turns out to be critical. Several approaches devoted to ferritic and austenitic stainless steel grades are based on the Von Mises and Johnson–Cook criteria [40,41]. Such criteria describe the elastic-plastic behavior of isotropic materials, defining the stress induced as a function of the deformation, strain rate, and temperature. A step forward is based on Hill's criterion, which introduces different equations for orthotropic and anisotropic materials [42,43].

In this paper, the deformation process of a ferritic stainless steel tube is simulated by means of a commercial software package adopting Hill's criterion. The results of the simulations are compared to those coming from experimental tests, with the aim to develop a tool able to predict the bending behavior of stainless steel tubes easily applicable in industrial applications. As a matter of fact, even if many researchers are active in this field and fundamental papers have been published, stainless steel tube bending is commonly performed in industries just based on an empirical approach (trial and error). The main novelty of the paper is to present an approach able to cover the gap between fundamental theoretical research and industrial application, putting in evidence the effect of bending process parameters. A proper forming process can therefore be designed by the comprehension of the effect of different pipe geometric parameters on the plastic deformation operation, thus allowing an optimum process fitting to the component under examination.

2. Materials and Methods

2.1. Materials

The steel chemical composition of the selected material (EN 1.4509 441) and the geometrical parameters of the tubes considered in this paper are reported in Tables 1 and 2, respectively.

Table 1. Chemical analysis of 441 (main elements, mass. %).

Steel Grade	C	Cr	Ni	Mo	Ti+Nb	Fe
441	0.02–0.04	17.5–18.5	–	–	0.55%	Balance

Table 2. Materials selected for the simulations with their geometric characteristics.

Steel Grade	Tube Diameter [mm]	Tube Thickness [mm]
441	40; 45; 50; 55; 60	1.0; 1.2; 1.5; 1.8

2.2. Methods

The steel, for each combination of diameter and thickness, has been tested according to the UNI EN 69892−1 and UNI EN 6892−2 standard. A Formability Limit Diagram (FLD) determination has been carried out to describe the sample deformation paths. This type of diagram contains the Formability Limit Curve (FLC), which shows the maximum capacity of a material to be deformed, calculated by carrying out repeated Nakazima tests and measuring the deformation along the two perpendicular directions [44]. Based on this test, the strains are measured with the conventional grid method, plotting a pattern of circles on the specimen. Such circles are deformed into ellipses during the deformation process. On these ellipses, strains are measured on minor and major sizes, thus identifying on the FLC diagram the deformation state points of the material. The materials properties needed as input parameters for the subsequent modelling are reported in Table 3. They include the steel density, Young modulus, Poisson ratio, Lankford value, and strain hardening coefficient.

Table 3. Steel properties [2].

Density [$\frac{g}{cm^3}$]	Young Modulus [$\frac{N}{mm^2}$]	Poisson Ratio	Lankford Value	Strain Hardening
7.8×10^{-9}	210,000.0	0.30	1.30–1.40	0.20–0.25

2.3. Model

The FEM model used in such an investigation is based on the rigid-plastic variational principle following:

$$\pi = \int_V \overline{\sigma}\dot{\overline{\varepsilon}}dV - \int_{S_F} F_i u_i dS, \tag{1}$$

which defines the allowed deformation rates. In Equation (1), the incompressibility condition can be driven by adding a penalty constant α, as reported in Equation (2):

$$\pi = \int_V \bar{\sigma}\dot{\bar{\varepsilon}}dV + \int_V \frac{\alpha}{2}\dot{\varepsilon}_V^2 dV - \int_{S_F} F_i u_i dS,\tag{2}$$

thus:

$$\delta\pi = \int_V \bar{\sigma}\delta\dot{\bar{\varepsilon}}dV + \alpha\int_V \dot{\varepsilon}_V\delta\dot{\varepsilon}_V dV - \int_{S_F} F_i\delta u_i dS,\tag{3}$$

where $\bar{\sigma}$, $\dot{\bar{\varepsilon}}$, ε_V, F_i, u_i are the average stress, deformation coefficient, volume deformation coefficient, surface stress, and speed of the pipe, respectively. The 8-nodes element is adopted to simulate the pipe bending process. The position (x,y,z) and speed vectors (u_x, u_y, u_z) of node-i are therefore defined as follows:

$$\begin{aligned}
x &= \sum_{i=1}^{8} N_i(\xi,\eta,\zeta)x_i,\\
y &= \sum_{i=1}^{8} N_i(\xi,\eta,\zeta)y_i,\\
z &= \sum_{i=1}^{8} N_i(\xi,\eta,\zeta)z_i,\\
u_x &= \sum_{i=1}^{8} N_i(\xi,\eta,\zeta)u_x,\\
u_y &= \sum_{i=1}^{8} N_i(\xi,\eta,\zeta)u_y,\\
u_z &= \sum_{i=1}^{8} N_i(\xi,\eta,\zeta)u_z,
\end{aligned}\tag{4}$$

where ξ,η,ζ are the co-ordinates in the cell space.

The model takes into account of the friction subjected to the pipe during bending according to the Tresca model [20]:

$$f = -\frac{2}{\pi}mk\,\tan^{-1}\left(\frac{u_s}{u_0}\right)t,\tag{5}$$

where m is the friction factor, k is the effectve shear stress, u_s is the speed of the pipe, and t is the unit vector in the u_s direction. Therefore, a FEM model has been implemented, considering the tube as discretized with a 4×4 mm $\times$ mm mesh. The pressure die, wiper die, and bending die tools are discretized, according to the theory of plastic deformation, such as rigid body.

The input related to the shape of tools are described in Figure 1.

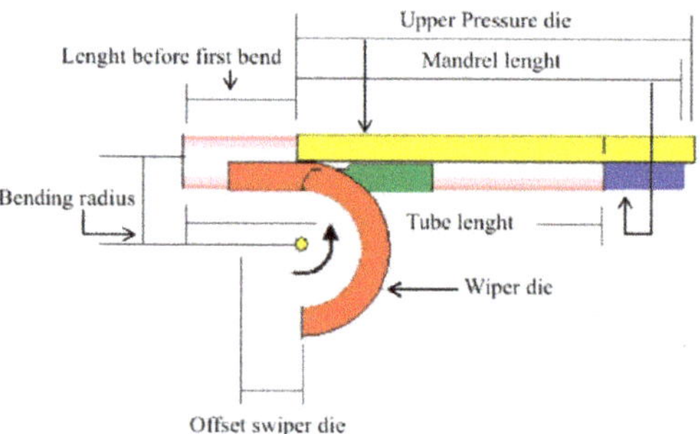

Figure 1. Tools for the bending simulation.

The input tools sizes are reported in Table 4.

Table 4. Adopted bending tool sizes.

Mandrel [mm]	Upper Pressure Die [mm]	Offset Swiper Die [mm]	Bending Radius [mm]
750	750	100	100

The other related tools' geometric parameters are calculated by simulation. Internal spherical elements supporting the bending process (so called balls) are present even if they are not reported in Figure 1. In this paper, four balls with 3.3 mm decreasing diameter sizes are considered. The starting diameter of the first ball is calculated by decreasing the tube diameter by the same measure. It is worth mentioning that an additional support machinery element (called a booster) is not taken into account in the calculations.

The numerical calculation was performed using the Altair HyperWorks™ (2017 version, Troy, MI, USA, commercial software). Such software is able to take into account the material properties reported in Table 3 together with the true stress–true strain steel curves. As far as concerns the tensile material properties, five true stress–true strain curves have been considered, aiming to obtain an average curve. The above materials properties, in particular, are collected in a file readable by the commercial package solver (namely, RADIOSS™). The solver is able to take into account the strain rate by reading the tensile curves performed at different deformation rate conditions. This software allows the adoption of the Hill 48' yield function. Such a function is known to be ideal for small-sized tubular geometries [45] as a constitutive equation for stainless steel behavior, taking into account the following parameters in order to simulate the bending process:

- bending radius;
- bending angle;
- rotational speed;
- bending temperature.

The analysis of the simulation outputs is carried out through mappings of the values calculated by the solver (e.g., internal stress, thinning, and deformation). An example of the obtained maps is reported in Figure 2.

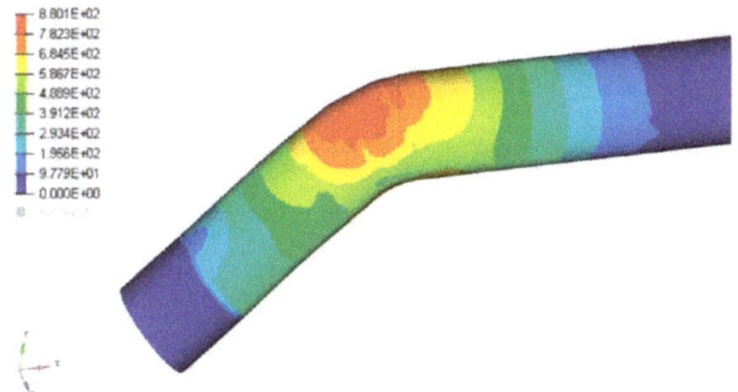

Figure 2. Stress mapping on bent tube.

In order to analyze the parameters' effects on the final process and its feasibility, the maximum stress values obtained on the mappings will be considered in order to consider the critical points of the geometry.

3. Results and Discussion

The effects of the geometric and operational inputs parameters described for the tubes' plastic deformation are reported below.

3.1. Effect of Tube Diameter

In this section, the effect of the tube's diameter on the bending process is analyzed, keeping the *R/D* ratio as a fixed value. *R/D* = 1 is considered being such a value significative in order to simulate what is commonly performed in the industrial bending process. The tube's stress behavior as a function of the diameters is reported in Figure 3.

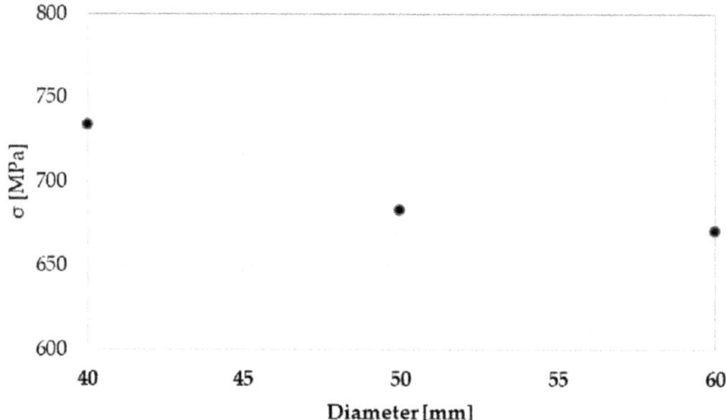

Figure 3. Mean maximum stress behavior as a function of diameter size for 441 steel with a 1.5 mm thickness.

The results allow us to evaluate an effect of about 5% of the tube's diameter on the internal stress. The same effect is also found for the tube's thinning (Figure 4). Such results well fit with what is expected from common industrial experience. In addition, they allow us to estimate the stress and thinning loss as a function of the tube's diameter, thus allowing a process optimization for different tube geometries.

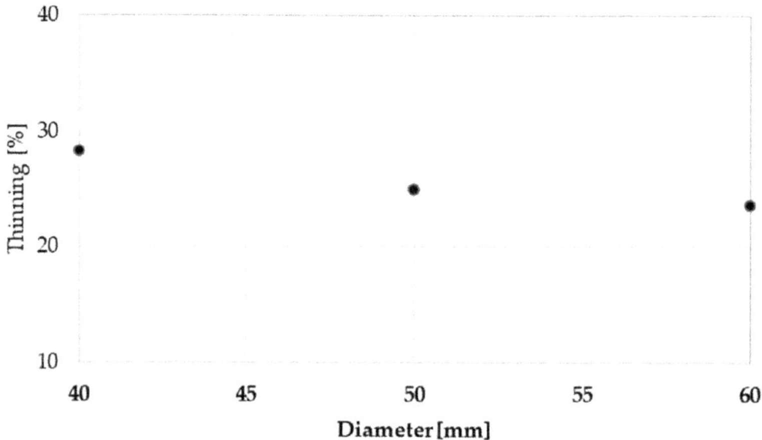

Figure 4. Maximum thinning as a function of the diameter size for 441 steel with a 1.5 mm thickness.

In order to evaluate the deformation capacity and support the above calculation outcomes, the results from FLD diagrams were considered. Figure 5 shows the formability limit (%) as reached during deformation for 441 stainless steel (diameter sizes ranging from 40 to 60 mm). The dashed red line represents the sample breakage. The FLD diagrams confirms that the deformation of the various geometric elements is affected by the diameter size. The results show that an increase in the diameter size will allow a decrease in the breakage risk. Such a result allows us to design the forming process for different pipes sizes, avoiding pipe breakage.

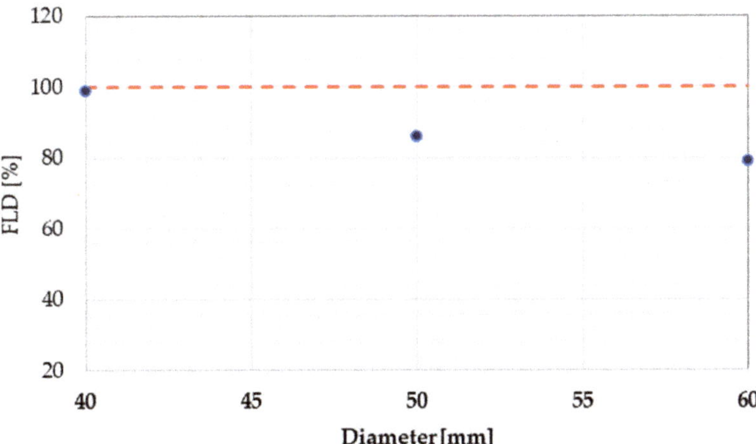

Figure 5. Formability limit (%) as a function of the tube's diameter for 441 with a 1.5 mm thickness.

3.2. Tubes Thickness Effect

The stress and the tube's thinning dependence on the thickness are reported in Figures 6 and 7, respectively. The results show that a not-significative stress distribution was found. In particular, the total variation is lower than 2%. On the other hand, the thinning (%) increases as the tube's thickness increases (Figure 7). Although there have been not relevant variations in terms of stress and thinning, the trend reported in Figure 7 is significant. As a matter of fact, while in the case reported in Section 3.1 the thickness reduction (which is kept constant) has the possibility of being distributed over ever larger diameters, in this case the diameter was fixed as a constant parameter; following that, an increase in thickness led to an increase in the tube's thinning.

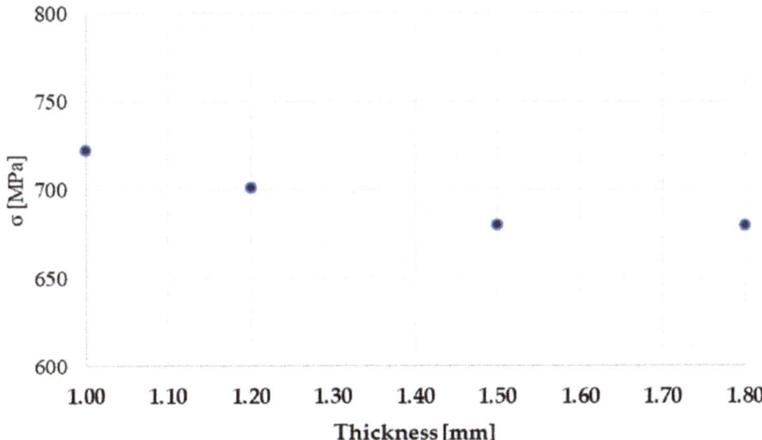

Figure 6. Mean maximum stress behavior as a function of thickness for 441 steel with a 50 mm diameter.

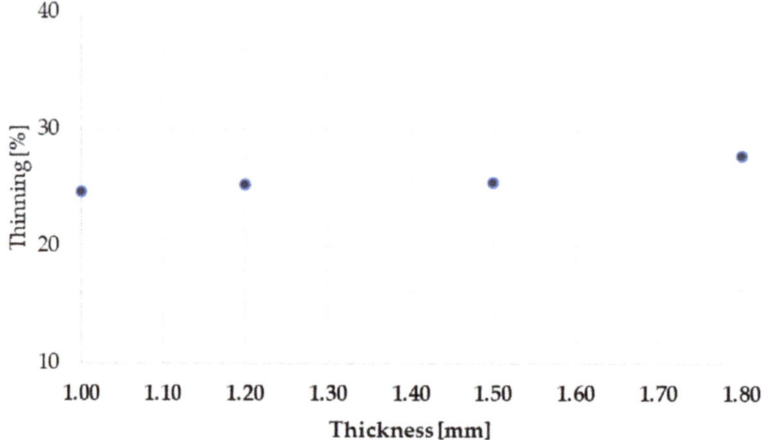

Figure 7. Maximum thinning as a function of thickness size for 441 steel with a 50 mm diameter.

An FLD plot as a function of the tube's thickness is reported in Figure 8. The results show that the success of the manufacturing process depends on the tube's thickness. Figure 8 clearly shows the strong effect exercised by the initial thickness on the bending process success.

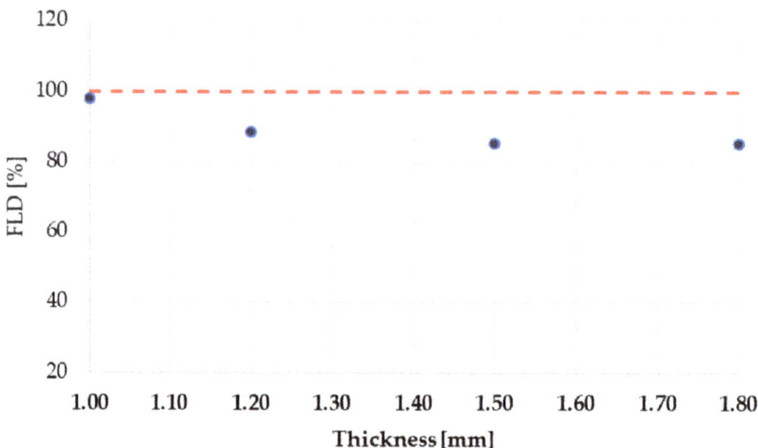

Figure 8. Formability limit (%) as a function of the thickness for 441 with a 50 mm diameter.

3.3. Speed and Bend Angle Effects

The effect of the speed and bending angle are reported in the following Figure 9.

As far as concerns the speed, the influence of its variation for every bending angle (in the range of 30° and 90° for a common manufacturing process) has been considered. The formability limit (%) for the combination of angle and thickness is shown in Figure 9. In particular, the geometric parameters were fixed, together with the relationship between the diameter and the thickness.

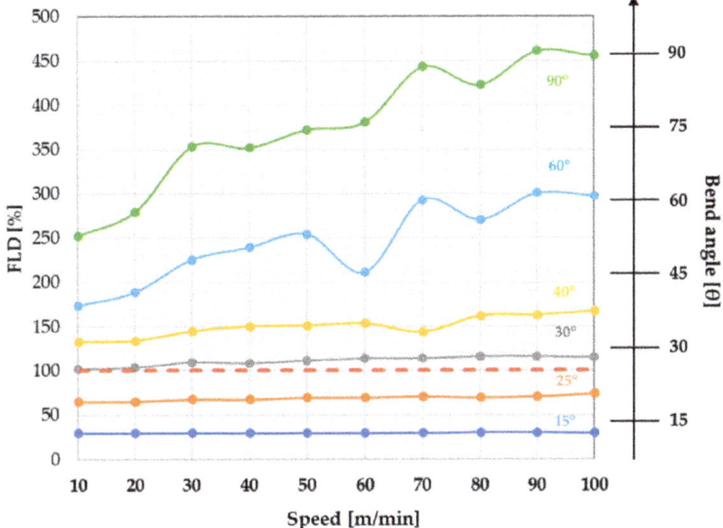

Figure 9. A 2D plot of the formability limit (%) for different speed and angle combinations.

In order to analyze the data, the lines shown in Figure 9 have been interpolated and reported in Figure 10 to better evaluate the influence of the speed. For this reason, the percentage variations between the percentages of formability limits obtained at minimum and maximum feed speeds for each angle, calculated according to Equation (6), are reported in Figure 11.

$$\Delta FLD = FLD_{v_{max}} - FLD_{v_{min}}. \tag{6}$$

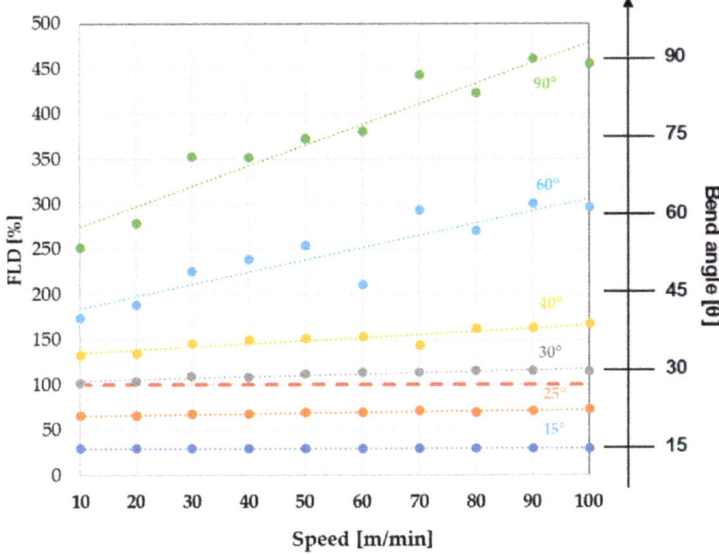

Figure 10. Formability Limit Diagram (FLD) dependence on speed for different bending angles (linear interpolation).

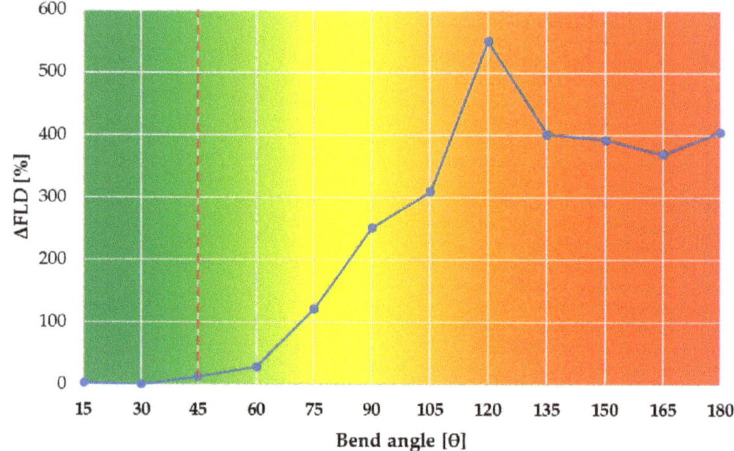

Figure 11. Goodness of the simulation output beyond the breaking of the worked piece (red dotted line).

ΔFLD is closely related to the slope of the interpolated line. Figure 11 shows that in the angle region ranging between 30° and 90° degrees (the most interesting for common industrial processes), the ΔFLD varies almost linearly with respect to the bending angle. If higher angle values are considered the results show that such a parameter reaches a maximum at about 120 degrees of bending, then decreases. The motivations leading to this particular behavior need to be deeply analyzed, but currently we can hypothesize that this is due to the stress concentration being more localized in the first 90° of bending.

In order to obtain more consistent results, the analysis has been repeated using conditions better reproducing the industrial process. Higher curvature radius values (and the R/D ratio, consequently) were considered. After that, the data were newly interpolated (Figure 12) and the FLD deltas were calculated for the new data set (Figure 13).

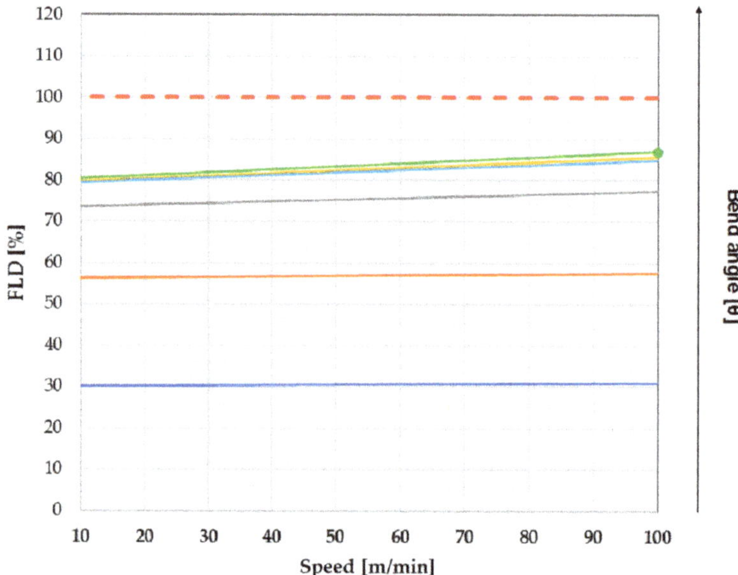

Figure 12. Linear interpolation of the formability limit (%) for each combination of speed and angle.

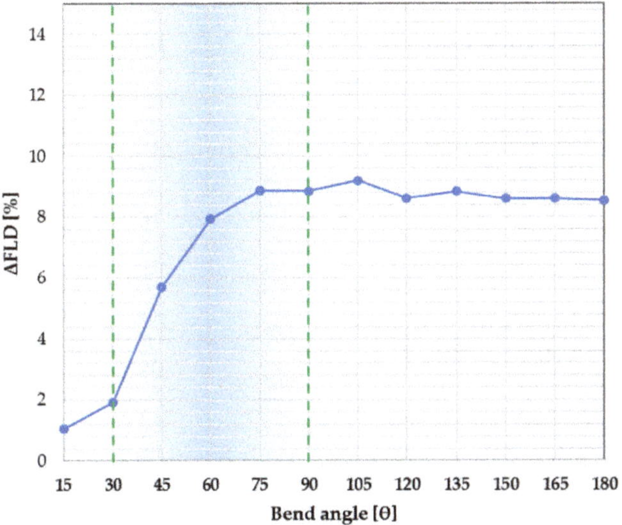

Figure 13. FLD dependence on the bending angle.

Figure 13 shows that the better choice of the *R/D* ratio leads to an improvement in the sample formability. ΔFLD increases in the bending angle range 30°–90°; it then tends to be stable, keeping away from breaking conditions. Such results suggests that a focus should be put on the *R/D* ratio on the process. As a matter of fact, this is one of the key issues in the setup of an industrial bending process.

3.4. R/D Effect

The *R/D* ratio is commonly adopted as a feasibility index for the bending industrial process. In common industrial practice, this value ranges between 1.0 and 1.5. In fact, *R/D* < 1 values increase the breakage risk; on the other hand, an *R/D* > 1.5 is not commonly adopted in the automotive field. The results reported in Figure 14 show a negligible effect of *R/D* on stresses. On the other hand, a marked *R/D* ratio effect is found on the tube's thinning (Figure 15). The results show that the tube's thinning decreases as the *R/D* increases. Additionally, in this case the results well fit with what is expected from common industrial experience. In addition, they allow us to estimate the stress and thinning loss as a function of the tubes' *R/D* parameter.

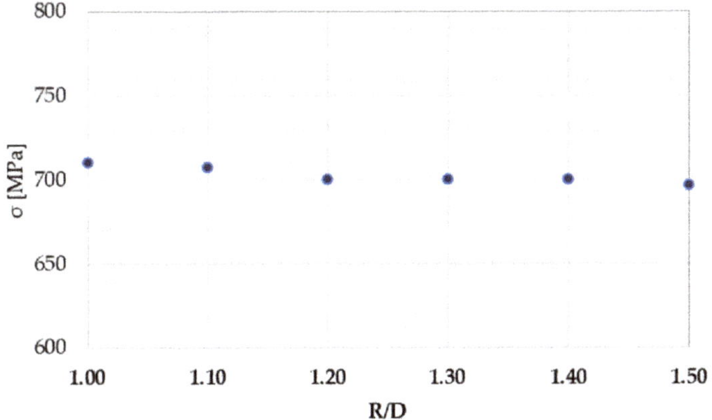

Figure 14. Maximum equivalent stress dependence on the *R/D* ratio.

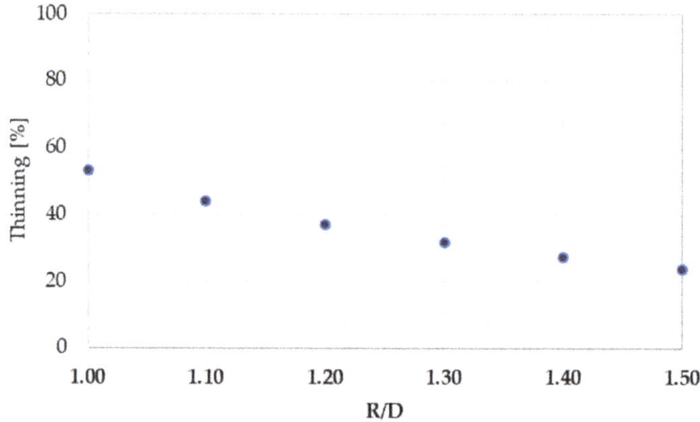

Figure 15. Maximum thinning dependence on the *R/D* ratio.

3.5. Experimental Validation

Six samples of 441 steel tubes (50 mm diameter and 1.2 mm thickness size) were considered for validation. All of the considered samples have been chosen with sizes corresponding to the above reported simulations. One of the six samples (for each group) was subjected to tensile tests in order to measure the related stress–strain curve and reduce the uncertainty due to the use of a mean curve. Experimental tests have been carried out according to operational parameters (e.g., rotational velocity and bending radius), as reported in Table 5. The chosen values are those commonly adopted by industries operating in pipe forming for the typical finishing process. Both the thicknesses reached during the bend along the backbone at specific angles, as shown in Figure 16, were measured.

Table 5. Testing conditions.

Rotational Velocity [rad/sec]	Bending Radius [mm]	Temperature [°C]	Bend Angle [ϑ]
1.6235	100.0	25.0	90

Figure 16. Thickness measuring grid on the backbone.

The thickness values were measured for each marked angle, and the average value was considered. The obtained values are shown in Table 6.

Table 6. Thickness for the 441 steel samples as measured at different considered angles (mm).

Angle	Thickness (mm)						
	Specimen n° 1	Specimen n° 2	Specimen n° 3	Specimen n° 4	Specimen n° 5	Specimen n° 6	Mean Value
0°	1.169	1.200	1.180	1.124	1.250	1.235	1.193
22.5°	1.003	1.019	1.026	1.023	1.123	1.101	1.049
45°	0.982	0.993	1.002	1.058	1.157	1.016	1.035
67.5°	1.086	1.016	1.050	1.029	1.166	1.052	1.067
90°	1.200	1.180	1.180	1.152	1.149	1.161	1.166

The measured thickness variation with the angle and the percentage discrepancy between the measured and calculated thickness as a function of the angle are reported in Table 7. In Table 7, ΔThickness is defined as the difference between the simulation thickness and the sample mean thickness, as conventionally adopted by companies operating in the tube forming sector.

Table 7. Comparison between the calculated and measured thickness at different angles.

Measurement Angle	Simulation Thickness [mm]	Sample Mean Thickness [mm]	ΔThickness [mm]	Percentage Variation [%]
0°	1.168	1.193	−0.025	−2.10
22.5°	1.014	1.049	−0.035	−3.35
45°	0.876	1.035	−0.159	−15.34
67.5°	0.892	1.067	−0.175	−16.36
90°	1.171	1.170	0.001	0.06

The deviation of the model from the experimental results (maximum underestimation of 16.36%) corresponds to 67.5 degrees. Such a deviation is probably related to the presence in the experimental tests of an additional support machinery element (the so-called booster), which was not considered in the simulation model. The booster effect in the industrial deformation process is that of pushing the tube during bending in order to avoid the deformations or failure caused by the friction between the element and the machine or by stress concentration. Its action also affects the distribution of the thinning caused by the deformation. As a matter of fact, the tube being pushed by the booster will have more evenly distributed stresses. As a consequence, the deformations and the thinning will take place on a wider area and will not lead to the failure of the piece. Anyway, the accuracy between the modelling and experimental tests is considered good and quite promising, since the discrepancy between the calculated and experimental values is lower than 15% for an angle position lower than 45°. Moreover, the fact that the model appears to underestimate the thickness values allows us to consider its results conservative with respect to the real behavior.

4. Conclusions

This paper analyzes the influence of geometric and operational parameters on the bending process of 441 ferritic stainless steel pipes. Experimental investigations combined with simulations highlighted the effect of each parameter, both operational and geometric, on the final results. The main novelty of the paper is to present an approach able to bridge the gap between fundamental theoretical research and industrial application, putting in evidence the effect of bending process parameters.

Based on the reported results, the following conclusions can be drawn:

- Pipe diameter effect on forming: If the same parameters are considered and the R/D ratio between the radius and diameter is kept constant, an increase in the diameter size (in the typical automotive range) will result in a −9% variation in terms of internal stresses, evaluated according to the von Mises criterion. The thinning of the tube will decrease by −4%, and the feasibility characteristics of the process will improve; in fact, the percentage reaching the formability limit will drop by −20%. The diameter size choice appears therefore to be a key issue in terms of process feasibility. In this sense, the reported approach is a useful tool aimed to properly design, also in quantitative terms, the forming process for different pipe sizes.

- Pipe thickness effect on forming: It is reported that a pipe thickness increase implies more safety for the bending process. In fact, keeping the operating conditions constant, the increase from a thickness of 1.0 mm to a thickness of 1.8 mm will lead to a reduction in the internal stresses of the −6% f, −3% for thinning and −13% in FLD.
- The combined study of bending angle and speed has shown how the simulation has incorrect results when the tube collapses. In non-breaking conditions, the speed has relevant importance in terms of feasibility in the range of the bending angle 30°–90°.
- The simulations show the *R/D* ratio as the most important parameter in the bending process. An increase in it from 1.0 to 1.5 entails a 30% reduction in thinning and a 60% increase in the bending process success.

The comparison between the calculation and experimental results proved the reported approach to be a useful tool in order to predict and properly design industrial deformation processes. The experimental validation showed a deviation of the model from the experimental case, with a maximum underestimation of 16.36%. Such discrepancy is probably related to the presence in the experimental tests of a booster, which was not included in the simulation model.

Author Contributions: Conceptualization, O.D.P. and R.M.; methodology, O.D.P. and R.M.; formal analysis, O.D.P., G.N., and M.G.; paper preparation, O.D.P. and A.D.S.; supervision, R.M. and A.D.S. All authors have read and agreed to the published version of the manuscript.

Funding: This research received no external funding.

References

1.　Marshall, P. *Austenitic Stainless Steels: Microstructure and Mechanical Properties*; Elsevier Applied Science Publisher: Amsterdam, The Netherlands, 1984.
2.　Di Schino, A. Manufacturing and application of stainless steels. *Metals* **2020**, *10*, 327. [CrossRef]
3.　Liu, W.; Lian, J.; Munstermann, S.; Zeng, C.; Fang, X. Prediction of crack formation in the progressive folding of square tubes during dynamic axial crushing. *Int. J. Mech. Sci.* **2020**, *176*, 105534. [CrossRef]
4.　Corradi, M.; Osofero, A.I.; Borri, A. Repair and Reinforcement of Historic Timber Structures with Stainless Steel—A Review. *Metals* **2019**, *9*, 106. [CrossRef]
5.　Gedge, G. Structural uses of stainless steel—Buildings and civil engineering. *J. Constr. Steel Res.* **2008**, *64*, 1194–1198. [CrossRef]
6.　Di Schino, A. Analysis of heat treatment effect on microstructural features evolution in a micro-alloyed martensitic steel. *Acta Met. Slovaca* **2016**, *22*, 266–270. [CrossRef]
7.　Sharma, D.K.; Filipponi, M.; Di Schino, A.; Rossi, F.; Castaldi, J. Corrosion behavior of high temperature fuel cells: Issues for materials selection. *Metalurgija* **2019**, *58*, 347–351.
8.　Gennari, C.; Lago, M.; Bogre, B.; Meszaros, I.; Calliari, I.; Pezzato, L. Microstructural and corrosion properties of cold rolled laser welded UNS S32750 duplex stainless steel. *Metals* **2018**, *8*, 1074. [CrossRef]
9.　Fava, A.; Montanari, R.; Varone, A. Mechanical spectroscopy investigation of point defect-driven phenomena in a Chromium martensitic steel. *Metals* **2018**, *8*, 870. [CrossRef]
10.　Di Schino, A.; Kenny, J.M.; Abbruzzese, G. Analysis of the recrystallization and grain growth processes in AISI 316 stainless steel. *J. Mat. Sci.* **2002**, *37*, 5291–5298. [CrossRef]
11.　Di Schino, A.; Testani, C. Corrosion behavior and mechanical properties of AISI 316 stainless steel clad Q235 plate. *Metals* **2020**, *10*, 552. [CrossRef]
12.　Rufini, R.; Di Pietro, O.; Di Schino, A. Predictive simulation of plastic processing of welded stainless steel pipes. *Metals* **2018**, *8*, 519. [CrossRef]

13. Di Schino, A.; Valentini, L.; Kenny, J.M.; Gerbig, Y.; Ahmed, I.; Hefke, H. Wear resistance of high-nitrogen austenitic stainless steel coated with nitrogenated amorphous carbon films. *Surf. Coat. Technol.* **2002**, *161*, 224–231. [CrossRef]

14. Di Schino, A.; Di Nunzio, P.E.; Turconi, G.L. Microstructure evolution during tempering of martensite in medium carbon steel. *Mater. Sci. Forum* **2007**, *558*, 1435–1441.

15. Di Schino, A.; Alleva, L.; Guagnelli, M. Microstructure evolution during quenching and tempering of martensite in a medium C steel. *Mater. Sci. Forum* **2012**, *715–716*, 860–865.

16. Zitelli, C.; Folgarait, P.; Di Schino, A. Laser powder bed fusion of stainless steel grades: A review. *Metals* **2019**, *9*, 731. [CrossRef]

17. Mancini, S.; Langellotto, L.; Di Nunzio, P.E.; Zitelli, C.; Di Schino, A. Defect reduction and quality optimisation by modelling plastic deformation and metallurgical evolution in ferritic stainless steels. *Metals* **2020**, *10*, 186. [CrossRef]

18. Gardner, L. The use of stainless steel in structures. *Prog. Struct. Eng. Mat.* **2005**, *7*, 45–55. [CrossRef]

19. Mulidran, P.; Siser, M.; Slota, J.; Spisak, E.; Sleziak, T. Numerical Prediction of Forming Car Body Parts with Emphasis on Springback. *Metals* **2018**, *8*, 60435. [CrossRef]

20. Mei, Z.; Khun, G.; He, Y. Advances and trends in plastic forming technologies for welded tubes. *Chin. J. Aeron.* **2016**, *29*, 305–315.

21. Oliveira, M.C.; Fernandes, J.B. Modelling and simulation of sheet metals forming processes. *Metals* **2019**, *9*, 1356. [CrossRef]

22. Cherouat, A.; Borouchaki, H.; Zhang, J. Simulation of Sheet Metal Forming Processes Using a Fully Rheological-Damage Constitutive Model Coupling and a Specific 3D Remeshing Method. *Metals* **2018**, *8*, 991. [CrossRef]

23. Bong, H.J.; Barlat, F.; Lee, M.; Ahn, D.C. The forming limit diagram of ferritic stainless steel sheets: Experiments and modeling. *Int. J. Mech. Sci.* **2012**, *64*, 1–10. [CrossRef]

24. Zhang, H.; Liu, Y. The inverse parameter identification of Hill'48 yield function for small-sized tube combining response surface methodology and three-point bending. *J. Mat. Res.* **2017**, *32*, 2343–2351. [CrossRef]

25. Lehmberg, A.; Suderkoetter, C.; Glaesner, T.; Brokmeier, H.G. Forming behavior of stainless steel sheets at different material thicknesses. *Mater. Sci. Eng.* **2019**, *651*, 012082.

26. Scott, T.; Kotadia, H. Microstructural evolution of 316L austenitic stainless steel during in-situ biaxial deformation and annealing. *Mat. Chat.* **2020**, *163*, 110288.

27. He, Y.; Heng, L.; Zhiyoing, Z.; Mei, Z.; Jing, L.; Guangjun, L. Advances and Trends on Tube Bending Forming Technologies. *Chin. J. Aeron.* **2012**, *25*, 1–12.

28. Wu, W.Y.; Zhang, P.; Zeng, X.Q.; Li, J.; Yao, S.S.; Luo, A.A. Bendability tubes using a rotary draw bender. *Mater. Sci. Eng. A* **2008**, *486*, 596–601. [CrossRef]

29. Liu, K.X.; Liu, Y.L.; Yang, H.; Zhao, G.Y. Experimental study on cross-section distortion of thin-walled rectangular tube by rotary draw bending. *Int. J. Mater. Prod. Technol.* **2011**, *42*, 110–120. [CrossRef]

30. Liu, K.X.; Liu, Y.L.; Yang, H. Experimental study on the effect of dies on wall thickness distribution in NC bending of thin-walled rectangular tube. *Int. J. Mater. Prod. Technol.* **2013**, *68*, 1867–1874.

31. Tang, N. Plastic-deformation analysis in tube bending. *Int. J. Pres. Vessels. Pip.* **2000**, *77*, 751–759. [CrossRef]

32. Al-Qureshi, H.; Russo, A. Spring-back and residual stresses in bending of thin-walled tubes. *Mater. Des.* **2002**, *23*, 217–222. [CrossRef]

33. Chen, J.; Ding, J.; Bai, X. In-plane strain solution of stress and defects of tube bending with exponential hardening law. *Mech. Based Des. Struct. Mach. Int. J.* **2012**, *40*, 257–276.

34. Li, H.; Yang, H.; Song, F.F.; Zhan, M.; Li, G.J. Spring-back characterization and behaviors of high-strength tube in cold rotary draw bending. *J. Mater. Process. Technol.* **2012**, *212*, 1973–1987. [CrossRef]

35. Jeong, H.S.; Ha, M.Y.; Cho, J.R. Theoretical and FE analysis for fine tube bending to predict spring-back. *Int. J. Precis. Eng. Manuf.* **2012**, *13*, 2143–2148. [CrossRef]

36. Liu, J.; Tang, C.; Ning, R. Deformation calculation of cross section based on virtual force in thin-walled tube bending process. *Chin. J. Mech. Eng.* **2009**, *22*, 696–701. [CrossRef]

37. Li, H.; Yang, H.; Yan, J.; Zhan, M. Numerical study on deformation behaviors of thin-walled tube NC bending with large diameter and small bending radius. *Comput. Mat. Sci.* **2009**, *45*, 921–934. [CrossRef]

38. Napoli, G.; Di Schino, A.; Paura, M.; Vela, T. Colouring titanium alloys by anodic oxidation. *Metalurgija* **2018**, *57*, 111–113.

39. Di Schino, A.; Di Nunzio, P. Metallurgical aspects related to contact fatigue phenomena in steels for back up rolling. *Acta Metall. Slovaca* **2017**, *23*, 62–71. [CrossRef]

40. Di Schino, A. Analysis of phase transformation in high strength low alloyed steels. *Metalurgija* **2017**, *56*, 349–352.

41. Wang, S.; Wei, K.; Li, J.; Liu, Y.; Huang, Z.; Mao, Q.; Li, Y. Enhanced tensile properties of 316L stainless steel processed by a novel ultrasonic resonance plastic deformation technique. *Mater. Lett.* **2019**, *236*, 342–345. [CrossRef]

42. Kaushal, M.; Joshi, Y.M. Three-dimensional yielding in anisotropic materials: Validation of Hill's criterion. *Soft Matter* **2019**, *15*, 4915–4920. [CrossRef] [PubMed]

43. Cazacu, O.; Revil-Baudard, B.; Chandola, N. Yield criteria for anisotropic polycrystals. In *Plasticity-Damage Couplings: From Single Crystal to Polycrystalline Materials. Solid Mechanics and Its Applications*; Springer International Publishing: Cham, Switzerland, 2019; volume 253.

44. Lumelskyj, D.; Rojek, J.; Tkocz, M. Numerical simulations of Nakazima formability tests with prediction of failure. *Appl. Mech.* **2015**, *60*, 184–194.

45. Yang, T.B.; Yu, Z.Q.; Xu, C.B.; Li, S.H. Numerical analysis for forming limit of welded tube in hydroforming. *J. Shanghai Jiaotong Univ.* **2011**, *45*, 6–10.

Article

Effect of Cutting Crystal Directions on Micro-Defect Evolution of Single Crystal γ-TiAl Alloy with Molecular Dynamics Simulation

Jianhua Li [1,2], Ruicheng Feng [1,2,3,*], Haiyang Qiao [1,2,*], Haiyan Li [1,2,3], Maomao Wang [1], Yongnian Qi [1] and Chunli Lei [1,2]

[1] School of Mechanical and Electrical Engineering, Lanzhou University of Technology, Lanzhou 730050, China; li_jh@vip.sina.com (J.L.); y5217@163.com (H.L.); 15620864891@163.com (M.W.); 13893591492@163.com (Y.Q.); lclyq2004@163.com (C.L.)

[2] Key Laboratory of Digital Manufacturing Technology and Application, Ministry of Education, Lanzhou University of Technology, Lanzhou 730050, China

[3] Centre for Efficiency and Performance Engineering, University of Huddersfield, Huddersfield HD1 3DH, UK

* Correspondence: postfeng@lut.edu.cn (R.F.); honey@lut.edu.cn (H.Q.); Tel.: +86-1500-260-6830 (R.F.)

Received: 30 October 2019; Accepted: 25 November 2019; Published: 28 November 2019

Abstract: In this work, the distribution and evolution of micro-defect in single crystal γ-TiAl alloy during nanometer cutting is studied by means of molecular dynamics simulation. Nanometer cutting is performed along two typical crystal directions: $[\bar{1}00]$ and $[\bar{1}01]$. A machined surface, system potential energy, amorphous layer, lattice deformation and the formation mechanism of chip are discussed. The results indicate that the intrinsic stacking fault, dislocation loop and atomic cluster are generated below the machined surface along the cutting crystal directions. In particular, the Stacking Fault Tetrahedron (SFT) is generated inside the workpiece when the cutting crystal direction is along $[\bar{1}00]$. However, a "V"-shape dislocation loop is formed in the workpiece along $[\bar{1}01]$. Furthermore, atomic distribution of the machined surface indicates that the surface quality along $[\bar{1}00]$ is better than that along $[\bar{1}01]$. In a certain range, the thickness of the amorphous layer increases gradually with the rise of cutting force during nanometric cutting process.

Keywords: molecular dynamics; nano-cutting; crystal direction; γ-TiAl alloy; stacking fault

---

1. Introduction

With the development of modern industry, nano-processing technology has been widely used in aerospace, biomedical, national defense and military and other high-tech fields [1]. In the fabrication of nanodevices, the basic understanding of the material removal mechanism and the evolution of defects is becoming more and more important. Nanofabrication involves material deformation of only a few atomic layers of about 2–10 nm, and the removal of atoms is discrete and discontinuous. Obviously, experimental and continuous theory is inaccurate. Since the 1990s, Molecular Dynamics (MD) simulation technology has been successfully applied to study the removal mechanism of various materials [2]. The MD can reveal the phenomena that cannot be observed by the traditional method, and capture the structure of material, position and velocity of atoms.

Over past decades, many scholars have done a lot of work by MD method. These studies were mainly focused on monocrystalline silicon, single crystal copper, silicon carbide and monocrystalline nickel that are described by Embedded Atom Method (EAM), Morse and Tersoff potentials. Wang et al. [3] studied the formation mechanism of stress-induced SFT and results show that the complex SFT is nucleated due to tensile and compressive stress. Chuvashia et al. [4] analyzed the mechanism of chip formation under high temperature conditions, and the key conclusion was that less heat is released and larger chips are formed when cutting at higher temperatures or on (111) crystal faces. Besides, the cutting force is reduced by 24% when the temperature at 1173 K compared with room temperature. Xie et al. [5] found that, as cutting-edge radius increases, both the dislocation slip zone and chip formation zone are suppressed while the elastic deformation zone tends to continually grow in monocrystalline copper nano-cutting process by MD simulations. Dai et al. [6] discussed the effects of different tool structures on nanometer cutting of single crystal silicon and found that some structural tools can cause a lower temperature. Fung et al. [7] researched the effect of surface defects of tool on tool wear during nano-machining and the statistical results showed that atoms loss in the defective region of tool is one order of magnitude greater than that in the non-defective region. Ren et al. [8] analyzed the grinding process of monocrystalline nickel at various speeds and depths with the molecular dynamics simulation, and revealed that, owing to the effect of the elastic deformation, a part of the defects can be eliminated after grinding and the dislocations and the variation in the number of HCP atoms cause the change of grinding force. Xu et al. [9] investigated the effect of hard particles on surface generation in nanometer cutting. Liu et al. [10] expounded the evolution of SFT and the work hardening effect in monocrystalline copper. Xu et al. [11] explained the effects of recovery on surface generation in nanometric cutting. Zhu et al. [12] simulated monocrystalline Nickel nano-cutting and found that a stacking fault is main cause of work hardening. Wang et al. [13] found that the tangential and normal forces of the cutting are different in different lattice orientations, and the stacking fault induced by cutting propagates in the basal plane and depends on the angle between the cutting direction and the c-axis through the MD simulations of four off-axis 4H-SiC nano-cutting processes.

The above literature focused on revealing the mechanism of material removal, microdefect evolution, the correlation between chip removal and cutting speed or temperature and cutting force change. However, few studies have focused on the effect of cutting crystal orientation on the evolution of micro-defect in the materials during nano-cutting, and, in particular, the evolution of a micro-defect along different cutting crystal orientation of γ-TiAl alloy has not been found. Additionally, γ-TiAl alloy has the advantage of low density, good high temperature strength and oxidation resistance, so it is widely used in aeroengine and automobile industries [14,15]. γ-TiAl alloy has face-centered tetragonal (FCT) with an $L1_0$ structure [16], as shown in Figure 1. Ti atoms and Al atoms are alternately arranged along the [001] direction. Furthermore, the crystal axis ratio of γ-TiAl is $c/a = 1.02$, which reduces the crystal symmetry and reduces the slip system. Therefore, the mechanical properties are different when cutting in different crystal directions. The purpose of this paper is to research the evolution of micro-defect by cutting along different crystal orientations of γ-TiAl alloy using MD methods. Section 2 will introduce the cutting model and methods. Section 3 will discuss the results of simulation and analyze the evolution of the micro-defect. Finally, the results will be summarized in Section 4.

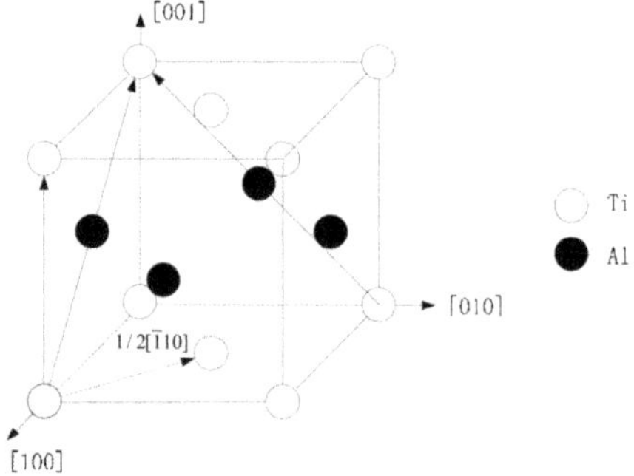

Figure 1. L10 structure of γ-TiAl.

2. Model and Methods

2.1. The MD Simulation Model

As shown in Figure 1, the x, y, and z coordinate axes correspond to the orientations [100], [010] and [001], respectively. The lattice constants are $a_0 = b_0 = 0.4001$ nm and $c_0 = 0.4181$ nm, which are consistent with the experimental values $a_0 = b_0 = 0.4005$ nm and $c_0 = 0.40707$ nm [17]. The nano-cutting model is shown in Figure 2, consisting of a monocrystalline γ-TiAl workpiece and a diamond tool. The workpiece contains 131,175 atoms and has a size of $20 \times 10 \times 11$ nm^3 along the [100], [010] and [001], respectively, and the diamond tool contains 7793 atoms. A periodic boundary condition along the z direction is adopted to eliminate the influence of size effect. In addition, five layers of the right and the bottom edges are fixed, and five layers next to them are a thermostat region which surrounds the Newton atoms. The thermostat region serves as a heat reservoir at a constant temperature 293 K for Newton atoms [18]. Newton atoms are used to study all aspects of the cutting process, the motion of Newton atoms obeys Newton's second law. Moreover, the selection of cutting speed is based on references [19,20], which combines with the characteristics of the γ-TiAl alloy.

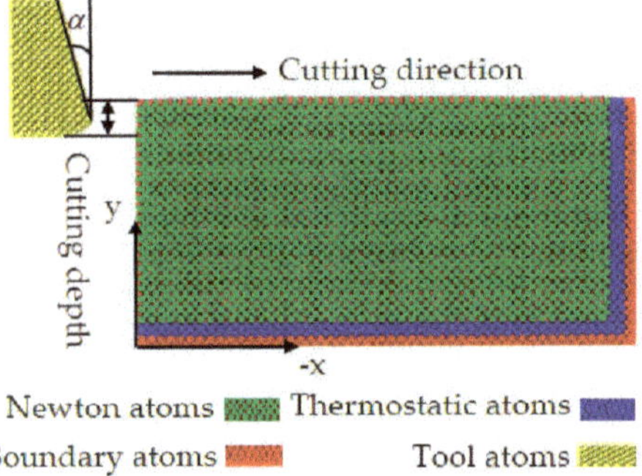

Figure 2. The MD simulation model of nano-cutting.

2.2. Interatomic Potential

The reliability of any MD simulation greatly depends on the quality of the interatomic potential employed. For metallic systems, the EAM is a widely used technique [21], which is fitted from a large database of both experiment and abinitio data. The total potential is expressed by

$$E = \sum_i F_i(\rho_i) + \frac{1}{2} \sum_{ij(i \neq j)} \varphi_{ij}(r_{ij}) \tag{1}$$

$$\rho_i = \sum_{j(j \neq i)} f_j(r_{ij}) \tag{2}$$

where F_i is the embedding energy of atom i, whose electron density is ρ_i, ϕ_{ij} is the relative potential energy of atom i and atom j, r_{ij} represents the distance between atom i and atom j, $F_{i(\rho_i)}$ is the sum of electron cloud density produced by all other atoms, extranuclear electron to atom i, and $f_j(r_{ij})$ is the electron density produced by atom j.

The interaction between Al-C and Ti-C is described by Morse potential, and the expression is given by

$$V(r_{ij}) = D_e\left[e^{2\alpha(r_e-r_{ij})} - 2e^{\alpha(r_e-r_{ij})}\right] \tag{3}$$

where D_e is the cohesive energy between atoms i and j, α is the elastic modulus of the material, r_{ij} represents the instantaneous distance between atoms i and j, and r_e is the equilibrium distance between atoms i and j. The parameter of Morse potential can be obtained from Zhu et al. [22]. As $D_{Ti\text{-}C} = 0.982$ eV, $\alpha_{Ti\text{-}C} = 22.83$ nm^{-1}, $r_{Ti\text{-}C} = 0.1892$ nm, $D_{Al\text{-}C} = 0.28$ eV, $\alpha_{Al\text{-}C} = 27.8$ nm^{-1}, $r_{Al\text{-}C} = 0.22$ nm.

As the rigidity of diamond is much higher than that of γ-TiAl, the tool is modeled as a rigid body. Table 1 shows the MD simulation parameters.

Table 1. The MD simulation parameters.

Machining Parameters	Value
Potential function	EAM, Morse
Workpiece	Single crystal γ-TiAl
Tool	Diamond
Workpiece size	$20 \times 10 \times 11$ nm^3
Lattice structure	L_{10}
Tool rake angle α	15°
Tool clearance angle β	0°
Tool edge radius R	1.5 nm
Cutting speed	200 m/s
Cutting depth	1.5 nm
Cutting direction	$< 100 > \left[\bar{1}00\right], < 110 > \left[\bar{1}01\right]$
Timestep	1fs

In this paper, Common Neighbor Analysis (CNA) [23] is utilized to identify the defect type, such as atomic cluster, dislocation, stacking fault etc.

3. Results and Discussion

3.1. Material Removal Mechanism in Nanometric Cutting Process

In the nano-cutting process, substrate atoms adjacent to the tool tip are subjected to the high compressive energy, resulting in highly displaced and disordered atoms. Figure 3 shows the diagram of material deformation; (a) and (b) represent the cutting crystal direction along $\left[\bar{1}01\right]$ and $\left[\bar{1}00\right]$, respectively. As shown in Figure 3, in the first place, the lattice orientation is 45° to the tool when the cutting crystal direction is along $\left[\bar{1}01\right]$. However, the lattice orientation is 0° or 180° to the tool along $\left[\bar{1}00\right]$. In the second instance, the shape of the chip is also different: The chip along $\left[\bar{1}00\right]$ is higher than that along $\left[\bar{1}01\right]$, and the atomic packing is more compact. During the machining process, the elastic and plastic deformation of γ-TiAl occurs with the increase of cutting distance, which is consistent with the literature [24]. The atoms of workpieces are distributed symmetrically on both sides of the tool to form the side-flow, and the others pile up in front of the tool to form chips. The chips are formed by extrusion force from the tool, and is removed in the form of atomic group. Then, the machined surface is generated with a certain elastic recovery.

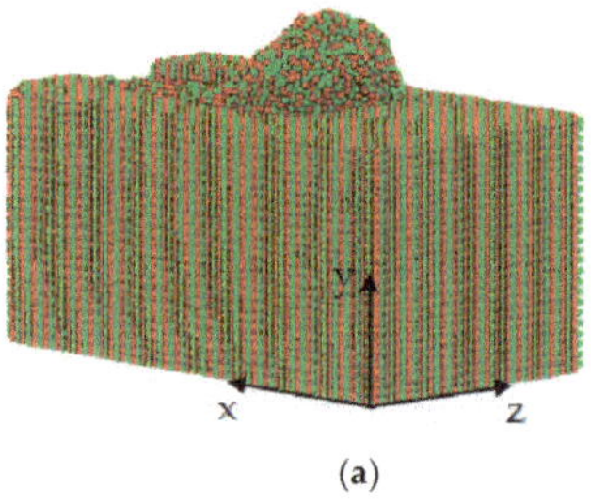

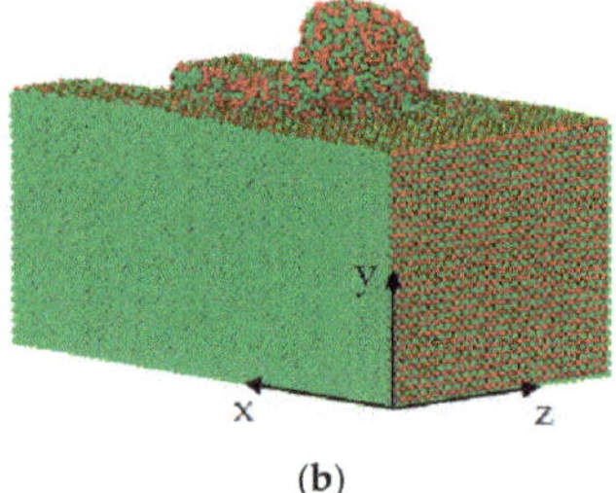

Figure 3. The diagram of cutting distances at 12 nm and the cutting crystal directions along $\left[\bar{1}01\right]$, $\left[\bar{1}00\right]$ respectively, (green represents Ti and red represents Al). (**a**) Cutting crystal direction along $\left[\bar{1}01\right]$. (**b**) Cutting crystal direction along $\left[\bar{1}00\right]$.

3.2. System Potential Energy and Lattice Deformation in Nanometric Cutting Process

The system potential energy and lattice deformation can be used as indicators of material deformation during the cutting process. Figure 4 shows the effect of different cutting crystal directions on the potential energy of the system, and the two potential energy curves rise with the increase of the cutting distance and the trend of the grows are the same. However, the potential energy of the system is higher than that along $[\bar{1}00]$ when the cutting direction is along $[\bar{1}01]$. On one hand, the different lattice arrangements result in different surface energy of the workpiece. On the other hand, the lattice bond breaks more and the internal energy releases more when the cutting direction is along $[\bar{1}01]$. However, the lattice bond breaks less and the internal energy releases less during nanometric cutting along $[\bar{1}00]$.

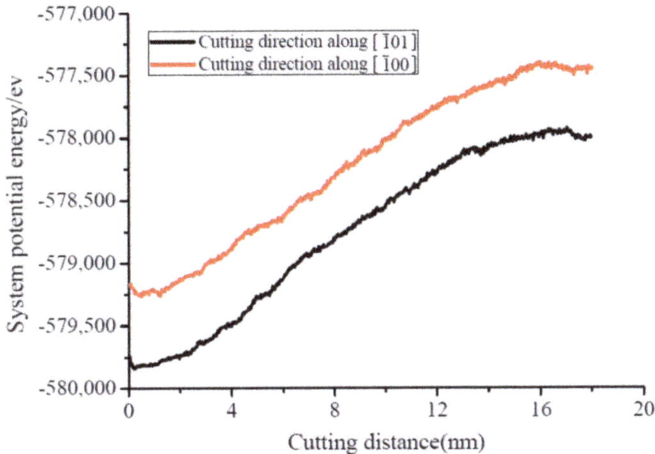

Figure 4. Change curve of system potential energy in the nanometric cutting process.

The CNA method is used to identify the crystal structure transformation during nano-cutting process. Figure 5 shows the effect of cutting directions on atomic structure transition. Figure 5a,b represents the number of structural transitions of HCP (Hexagonal Close-Pack) and BCC (Body-Centered Cubic), respectively. Two bar graphs are compared between the number of HCP transition crystal structure and the number of BCC transition along $[\bar{1}01]$. In the initial stage of cutting, it increases with the increase of cutting distance. The value fluctuates greatly and shows the opposite trend when the cutting distance is 8–16 nm. This is due to the complex stacking faults nucleation, expansion, annihilation. There is a large area of stacking fault at 8–16 nm, which increases the number of HCP transformations and inhibits the number of BCC. The HCP structure plays a dominant role in the whole cutting process. Comparing the number of HCP and BCC transitions along $[\bar{1}00]$, we can see that the BCC crystal structure predominates when the cutting distance is less than 6 nm and more than 14 nm, while the HCP crystal structure dominates and the value fluctuates sharply between 8–14 nm. As the cutting force increases with the increase of cutting distance, which leads to the formation, expansion and annihilation of SFT, thus increasing the number of HCP and decreasing the number of BCC. This indicates that there is a competitive relationship between the transformation of HCP and BCC crystal structures in nano-cutting. It can be seen from Figure 5a that the number of transitions along $[\bar{1}01]$ is larger than that along $[\bar{1}00]$, while Figure 5b is just the opposite. As the crystal orientation of the workpiece is 45° with the tool along $[\bar{1}01]$, it is easier to form intrinsic stacking faults; therefore, the number of HCP is greater along $[\bar{1}01]$.

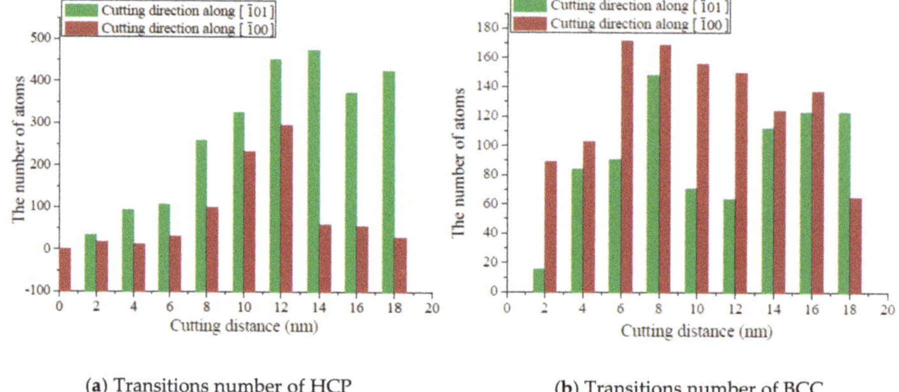

(a) Transitions number of HCP

(b) Transitions number of BCC

Figure 5. Graph of atomic structure variations with the cutting distance on the surface.

3.3. Evolution of Dislocation in Different Cutting Crystal Direction

In order to clearly understand the evolution process of surface defects during nano-cutting process, Figure 6 shows the effect of different cutting crystal directions on the cutting surface. Figure 6a,b represents dislocation nucleation and dislocation slip along $[\bar{1}00]$ and the corresponding time is 124.5 ps and 125.5 ps, respectively. Figure 6c,d represents dislocation nucleation and dislocation slip along $[\bar{1}01]$ and the corresponding time is 98 ps and 100 ps, respectively. Among them, white, red and green represent surface atoms (dislocation atoms), HCP atoms and FCC atoms, respectively. The nucleation and slip of dislocations are marked in Figure 6. In the early stage of nano-machining, it can be seen clearly the dislocation nucleated, slide and emitted at 45° to the cutting tool, which is explained by the Schmid rule; the resolved shear stress reaches the maximum value at ±45° in front of the tool and slips in these directions. The dislocation nucleation along $[\bar{1}00]$ is late; however, its dislocation slip region is larger than that along $[\bar{1}01]$, as a result of the formation of SFT along $[\bar{1}00]$, which is a reaction between dislocations. Finally, the SFT annihilates to the boundary of the workpiece. Due to the certain image force to dislocation from free surface, the slip area of dislocation increases gradually with the increase of cutting distance, as shown in Figure 6. This is owing to the rise of cutting distance, cutting force and temperature, which contributes to the slip of dislocations during the nanometric cutting process.

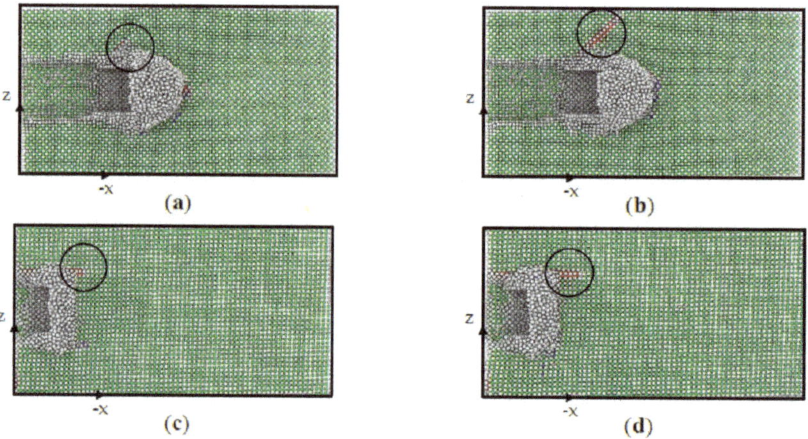

Figure 6. Variation of cutting surface in nanometric cutting process. (**a**) Dislocation nucleation at 124.5 ps along $[\bar{1}00]$. (**b**) Dislocation slip at 125.5 ps along $[\bar{1}00]$. (**c**) Dislocation nucleation at 98 ps along $[\bar{1}01]$. (**d**) Dislocation slip at 100 ps along $[\bar{1}01]$.

The evolution of internal defects in the nanometric cutting process is a dynamic process. Figure 7 shows the internal defects evolution in the workpiece along $[\bar{1}01]$; atoms are colored according to the CNA. For a better observation of the evolution of defects, ordinary FCC atoms are removed. The white, red and blue represents Other, HCP and BCC atoms, respectively. Two layers of red atom represent an intrinsic stacking fault [25]. In the early stage of the nano-machining, the stacking fault is generated along $(0\bar{1}1)$ crystal planes for the instability of the cutting, as shown in Figure 7a. The stacking fault nucleates and extends at shear-slip zone below the cutting tool, and finally annihilates at free surface of workpiece, as depicted in Figure 7b. Two Shockley partial dislocations extend along (011) and $(0\bar{1}1)$ planes and form a Lomer-Cottrell lock [26], which finally forms a "V"-shaped dislocation loop, as illustrated in Figure 7b. In the middle stage of nanomachining, it can be seen clearly that the complex dislocation loop and the stacking fault are generated beneath the tool attributing to the rise of cutting force and temperature, as shown in Figure 7c. With the increase of cutting distance, the stacking fault and the complex dislocation loop disappear gradually and are removed in the form of chips, as depicted in Figure 7d. However, the defects continue to evolve in the workpiece, and a small number of the atomic clusters are generated inside the workpiece, as illustrated in Figure 7d. The atomic clusters are formed as a result of the combined effects of temperature and plastic deformation during nano-cutting.

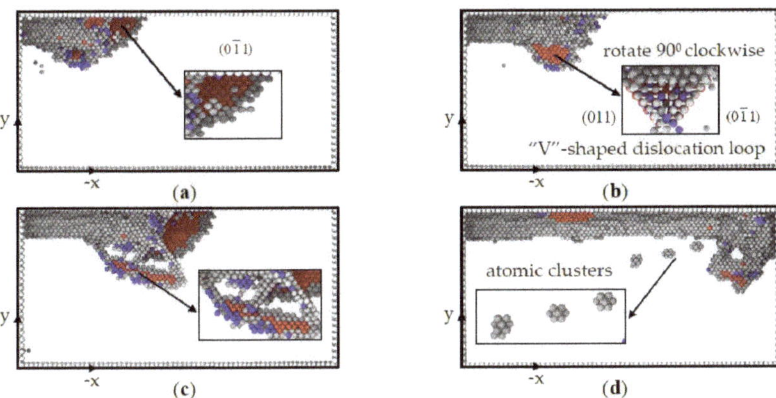

Figure 7. Internal defect evolution of workpiece along $[\bar{1}01]$. (**a**) Stacking fault along $(0\bar{1}1)$. (**b**) "V"-shaped dislocation loop. (**c**) Dislocation loop and stacking fault. (**d**) Atomic clusters.

Figure 8 shows the internal defect evolution inside the workpiece along $[\bar{1}00]$. It can be seen that the stacking fault, the SFT, the dislocation loop and the atomic cluster are formed in the subsurface. The stacking fault in the shear-slip zone extends along the $(\bar{1}\,\bar{1}1)$ and $(1\bar{1}1)$ planes, as shown in Figure 8a,b. After processing, there are two stacking faults slipping along each slip direction, which finally annihilate at free surface of the workpiece. As can be seen from Figure 8c,d, the complex SFT consists of three edges, three stacking fault planes and one workpiece surface, which are formed in the subsurface of the workpiece. The order of the nucleation of the intrinsic stacking fault is marked "S1", "S2", "S3", where 1/3<001> Hirth dislocation is formed between "S1" and "S2", 1/6<112> Shockley partial dislocation is produced between "S2" and "S3", and 1/6<110> Stair-rod dislocation is generated between "S1" and "S3" along the respective slip surfaces, in which Stair-rod and Hirth are immobile dislocations to preserve the shape of the SFT. As a result of the instability of external forces during the cutting process, the SFT is eventually removed in the form of chips, and a small number of atomic clusters and dislocation loops are formed, as illustrated in Figure 8e.

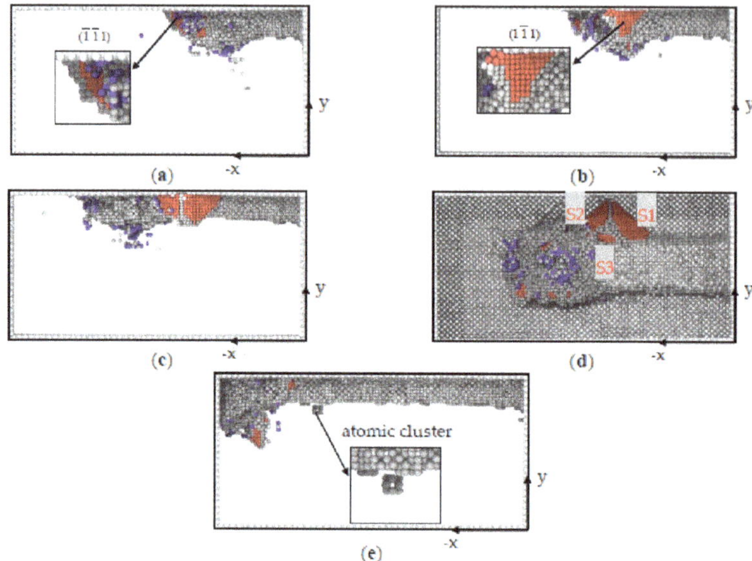

Figure 8. Internal defect evolution of the workpiece along $\left[\bar{1}00\right]$. (**a**) Stacking fault along $\left(\bar{1}\,\bar{1}1\right)$. (**b**) Stacking fault along $\left(1\bar{1}1\right)$. (**c**) SFT. (**d**) Nucleation of the intrinsic stacking fault. (**e**) Atomic clusters.

3.4. Effect of Cutting Crystal Direction on Cutting Force, Temperature and Cutting Surface Quality

The cutting force is an indispensable factor in the analysis of the nanometer cutting process. Figure 9 shows tangential force, F_x and F_y, in the cases with different cutting crystal directions. It is noticeable that when the tool and the workpiece are close to a certain distance, the workpiece atoms give a certain repulsive force to the tool, where the cutting force has a negative value. Subsequently, the F_x increases rapidly with cutting distance and becomes gradually stable. This conforms with reference [27]. The tangential force is basically stable and fluctuates within a certain range when the cutting distance exceeds 6 nm, and the fluctuation of the curve is caused by crystal structure transformation, stacking fault and lattice reconstruction. In addition, there is a consistent trend in the curve of F_x for the cutting crystal directions. In Figure 10, the temperature of the cutting zone grows along with increase of cutting distance before the distance reaches 15 nm. Comparing derivation of the temperature and the cutting force, the temperature of the workpiece grows along with the increase of cutting force at the early stage of cutting. During the final of the cutting process, the temperature goes down, which happens as the cutting force drops. The temperature of workpiece with cutting direction of $\left[\bar{1}00\right]$ is higher when the cutting distance is between 12 nm and 16 nm than that of the one along the $\left[\bar{1}01\right]$ direction. This phenomenon results in the resistance of materials varying along different directions and more heat being generated during the cutting of a workpiece along the $\left[\bar{1}00\right]$ direction.

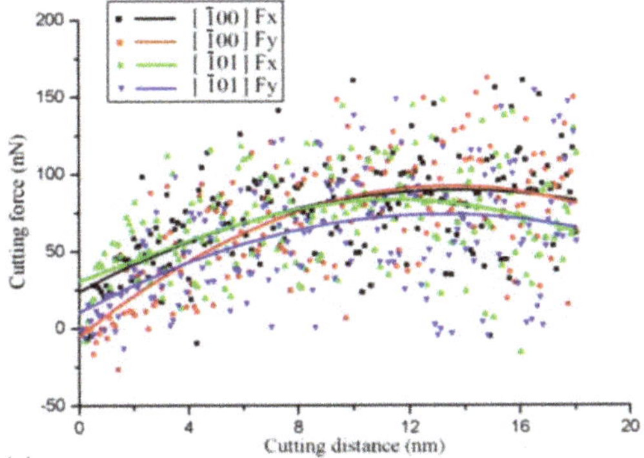

Figure 9. Tangential force.

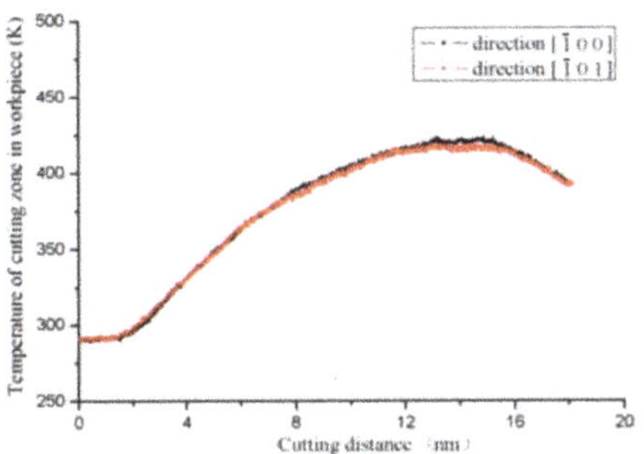

Figure 10. Temperature of cutting zone.

In order to clearly distinguish the size of the cutting force, this article details statistics of the average cutting force. Besides, the cutting distance is selected between 6–18 nm, which aims to reduce error of the cutting force. The average cutting force is shown in Table 2. It can be clearly seen that the average cutting force along $[\overline{1}00]$ is greater than that along $[\overline{1}01]$. This results from the interaction of dislocations along $[\overline{1}00]$, which activates more dislocation slip systems and intersects in the cutting direction, and, in consequence, the greater cutting force is required for material removal than that along $[\overline{1}01]$. This is in line with reference [28]. When the cutting crystal direction changes from $[\overline{1}01]$ to $[\overline{1}00]$, the tangential force and normal force is increased by 13.32%, 10.25%, respectively.

Table 2. Cutting force for the two simulation cases.

Average Stress Type	Cutting Crystal Directions	
	$[\overline{1}01]$	$[\overline{1}00]$
ave F_x (nN)	81.763	92.655
ave F_y (nN)	79.391	87.531

The quality of the cutting surface is crucial during nanometer cutting process. Figure 11 shows the atomic distribution of machined surface under different cutting crystals. Figure 11a,b represents the cutting direction along $[\overline{1}01]$ and along $[\overline{1}00]$, respectively. The blank area in the figure means the atom lost in cutting, and there is a large atom loss area. The number of atoms remaining on the machined surfaces in Figure 11a,b is 706 and 697, respectively, and these two numbers are close. In order to analyze the surface quality more accurately, the y-direction distribution of atoms on the top layer of the machined surface is calculated. As shown in Figure 11, the black and red curves represent the y-direction distribution of atoms along $[\overline{1}01]$ and $[\overline{1}00]$, respectively. It can be seen that the atom distribution span along $[\overline{1}01]$ is 0.01–0.1 nm, with the largest number of atoms at 0.05 nm, and its peak value is 145. However, the atom distribution span along $[\bar{1}00]$ is 0.01–0.1 nm, and the peak value is 211 at 0.06 nm. From the analysis of data and curve distribution, it can be seen that the atomic spans of the two cutting directions are the same. Furthermore, the *y*-direction atom distribution along $[\bar{1}00]$ is more concentrated and compact, which shows that the machined surface quality along $[\bar{1}00]$ is better than that along $[\bar{1}01]$.

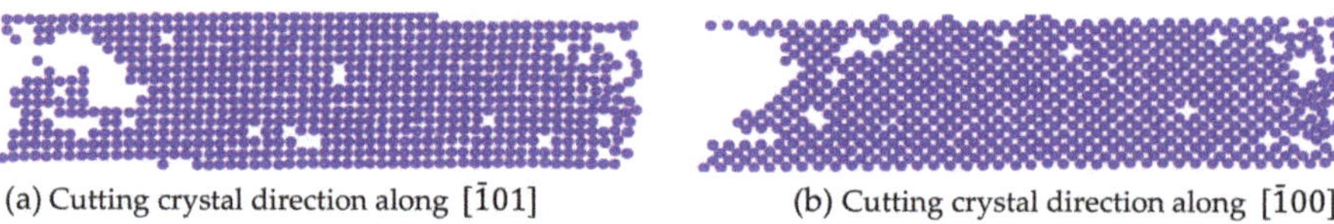

(a) Cutting crystal direction along $[\bar{1}01]$ (b) Cutting crystal direction along $[\bar{1}00]$

Figure 11. Atomic distribution of machined surface under different cutting crystals.

In order to analyze the effect of different cutting directions on the quality of the cutting surface, the atomic distribution in the First layer along the *y*-direction was quantified, that is, the unevenness of the First layer atomic plane. Figure 12 shows the effect of different cutting directions on atomic distribution. It can be seen that both curves have multiple peaks; the maximum number of atoms is 135, 87 at 0.15 nm, 0.145 nm. Secondly, the distribution of atoms along $[\bar{1}00]$ is concentrated, while the distribution of atoms along $[\bar{1}01]$ is dispersed. This indicates that the distance between atoms is smaller and the planeness is better than that along $[\bar{1}01]$. When the cutting direction is along $[\bar{1}01]$, the orientation of the lattice is 45° to the tool, resulting in smaller fluctuations of atoms in the nano-cutting process.

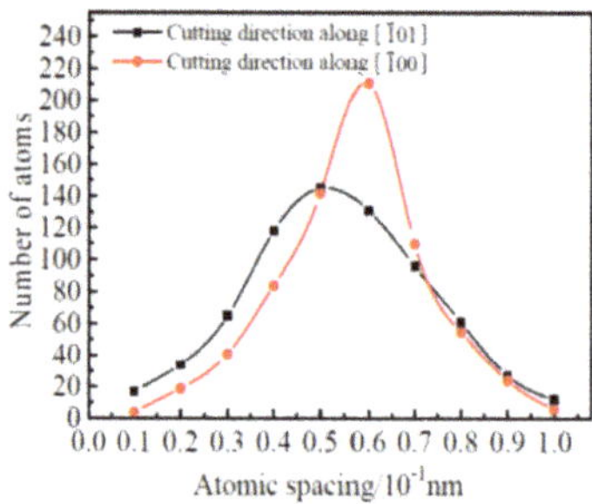

Figure 12. *Y*-direction atom distribution in the top layer of machined surface under different cutting crystals.

As shown in Figure 13, the effect of different cutting directions on the amorphous layer is displayed. The amorphous layer is composed of dislocation and disordered atoms, which is formed below the machined surface. Figure 13a,b represents the cutting direction along $[\bar{1}01]$ and $[\bar{1}00]$, respectively. It is clear that the thickness of amorphous layer along $[\bar{1}00]$ is slightly larger than that along $[\bar{1}01]$. The cutting force along $[\bar{1}00]$ is slightly higher than that along $[\bar{1}01]$. This, in a certain range, the deeper the thickness of the amorphous layer, the greater the cutting force during the nanometric cutting process.

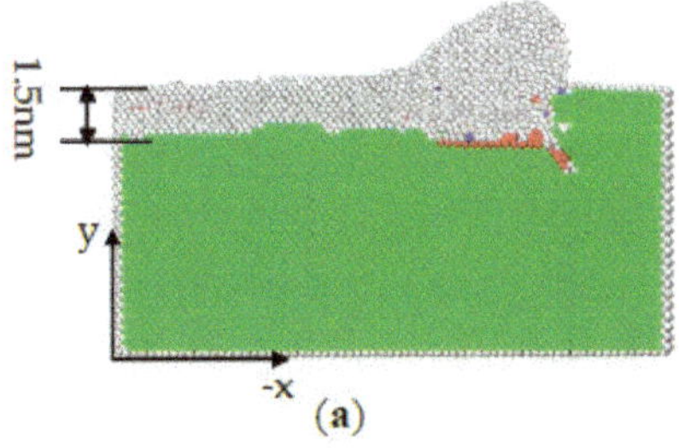
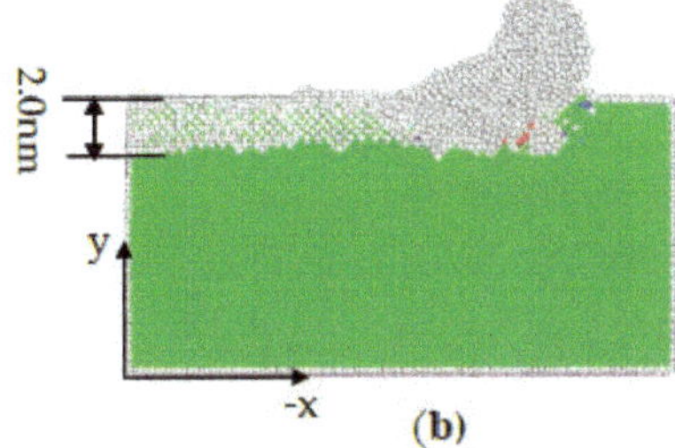

Figure 13. Effect of cutting direction on thickness of amorphous layer. (**a**) Thickness of amorphous layer along $[\bar{1}01]$. (**b**) Thickness of amorphous layer along $[\bar{1}00]$.

4. Conclusions

In this paper, the nanometric cutting process of single crystal γ-TiAl alloy has been studied by molecular dynamics simulation. The effect of cutting directions on the material deformation, system potential energy, lattice deformation, machined surface, cutting force and the evolution of dislocation are thoroughly discussed. The conclusions are as follows:

(1) The orientation of the lattice is 45° to the tool when the cutting crystal direction is along $[\bar{1}01]$. However, the arrangement of the lattice is 0° or 180° to the tool when the cutting direction is along $[\bar{1}00]$, and the chips are removed as atomic groups. The potential energy of cutting orientation along $[\bar{1}01]$ is larger than that along $[\bar{1}00]$.

(2) The evolution of defects is affected by cutting directions in the workpiece. Dislocation slip starts at the cutting surface. Inside the workpiece, the intrinsic stacking faults, dislocation loops and atomic clusters are generated. The SFT is produced inside the workpiece when the cutting direction is along $[\bar{1}00]$. However, a "V"-shape dislocation loop is generated when the cutting direction is along $[\bar{1}01]$.

(3) As the cutting direction changes from $[\bar{1}01]$ to $[\bar{1}00]$, the tangential force and normal force is increased by 13.32% and 10.25%, respectively. Quantitative analysis of atomic distribution along the *y*-direction proves that the machined surface quality along $[\bar{1}00]$ is better than that along $[\bar{1}01]$. The thickness of the amorphous layer along $[\bar{1}00]$ is slightly deeper than that along $[\bar{1}01]$.

Author Contributions: Conceptualization, R.F.; methodology, R.F. and J.L.; software, J.L.; formal analysis, H.Q.; investigation, H.Q. and Y.Q.; resources, H.L.; data curation, H.L. and M.W.; writing—original draft preparation, H.Q.; writing—review and editing, R.F. and J.L.; visualization, H.Q. and J.L.; supervision, R.F. and C.L.

Funding: This research was funded by a grant from the China Scholarship Council, grant number 201808625035, the National Natural Science Foundation of China, grant number 51865027; the Program for Changjiang Scholars and Innovative Research Team in University of Ministry of Education of China, grant number IRT_15R30; and the Hongliu First-class Disciplines Development Program of Lanzhou University of Technology.

Acknowledgments: The technical support of Gansu Province Supercomputer Centre is gratefully acknowledged.

Conflicts of Interest: The authors declare no conflict of interest.

References

1. Gong, Y.D.; Zhu, Z.X.; Zhou, Y.G.; Sun, Y. Research on the nanometric machining of a single crystal nickel via molecular dynamics simulation. *Sci. Chin. Technol. Sci.* **2016**, *59*, 1–10. [CrossRef]
2. Guo, Y.B.; Liang, Y.C. Atomistic simulation of thermal effects and defect structures during nanomachining of copper. *Trans. Nonferrous Met. Soc. China* **2012**, *22*, 2762–2770. [CrossRef]
3. Wang, Q.L.; Bai, Q.S.; Chen, J.X.; Guo, Y.B.; Xie, W.K. Stress-induced formation mechanism of stacking fault tetrahedra in nano-cutting of single crystal copper. *Appl. Surf. Sci.* **2015**, *355*, 1153–1160. [CrossRef]
4. Chuvashia, S.Z.; Luo, X. An atomistic simulation investigation on chip related phenomena in nanometric cutting of single crystal silicon at elevated temperatures. *Comput. Mater. Sci.* **2016**, *113*, 1–10. [CrossRef]

5. Xie, W.K.; Fang, F.Z. Effect of tool edge radius on material removal mechanism in atomic and close-to-atomic scale cutting. *Appl. Surf. Sci.* **2019**, 144451. [CrossRef]

6. Dai, H.F.; Chen, G.Y.; Zhou, C.; Fang, Q.H.; Fei, X.J. A numerical study of ultraprecision machining of monocrystalline silicon with laser nano-structured diamond tools by atomistic simulation. *Appl. Surf. Sci.* **2017**, *393*, 405–416. [CrossRef]

7. Fung, K.Y.; Tang, C.Y.; Cheung, C.F. Molecular dynamics analysis of the effect of surface flaws of diamond tools on tool wear in nanometric cutting. *Comput. Mater. Sci.* **2017**, *133*, 60–70. [CrossRef]

8. Ren, J.; Hao, M.R.; Lv, M.; Wang, S.Y.; Zhu, B.Y. Molecular dynamics research on ultra-high-speed grinding mechanism of monocrystalline nickel. *Appl. Surf. Sci.* **2018**, *455*, 629–634. [CrossRef]

9. Xu, F.; Fang, F.; Zhang, X. Hard particle effect on surface generation in nano-cutting. *Appl. Surf. Sci.* **2017**, *425*, 1020–1027. [CrossRef]

10. Liu, H.T.; Zhu, X.F.; Sun, Y.Z.; Xie, W.K. Evolution of stacking fault tetrahedral and work hardening effect in copper single crystals. *Appl. Surf. Sci.* **2017**, *422*, 413–419. [CrossRef]

11. Xu, F.; Fang, F.; Zhang, X. Effects of recovery and side flow on surface generation in nano-cutting of single crystal silicon. *Comput. Mater. Sci.* **2018**, *143*, 133–142. [CrossRef]

12. Zhu, Z.X.; Gong, Y.D.; Zhou, Y.G.; Gao, Q. Molecular dynamics simulation of single crystal Nickel nanometric machining. *Sci. China Technol. Sci.* **2016**, *59*, 867–875. [CrossRef]

13. Wang, M.; Zhu, F.; Xu, Y.; Liu, S. Investigation of Nanocutting Characteristics of Off-Axis 4H-SiC Substrate by Molecular Dynamics. *Appl. Sci.* **2018**, *8*, 2380. [CrossRef]

14. Appel, F.; Clemens, H.; Fischer, F.D. Modeling concepts for intermetallic titanium aluminides. *Prog. Mater Sci.* **2016**, *81*, 55–124. [CrossRef]

15. Schwaighofer, E.; Rashkova, B.; Clemens, H.; Stark, A.; Mayer, S. Effect of carbon addition on solidification behavior, phase evolution and creep properties of an intermetallic β-stabilized γ-TiAl based alloy. *Intermetallics* **2014**, *46*, 173–184. [CrossRef]

16. Chubb, S.R.; Papaconstantopoulos, D.A.; Klein, B.M. First-principles study of L10 Ti-Al and V-Al alloys. *Phys. Rev. B* **1988**, *38*, 120–124. [CrossRef]

17. Tang, F.L.; Cai, H.M.; Bao, H.W.; Xue, H.T.; Lu, W.J.; Zhu, L.; Rui, Z.Y. Molecular dynamics simulations of void growth in γ-TiAl single crystal. *Comput. Mater. Sci.* **2014**, *84*, 232–237. [CrossRef]

18. Romero, P.A.; Anciaux, G.; Molinari, A.; Molinar, J.F. Insights into the thermo-mechanics of orthogonal nanometric machining. *Comput. Mater. Sci.* **2013**, *72*, 116–126. [CrossRef]

19. Liu, Y.; Li, B.; Kong, L. A molecular dynamics investigation into nanoscale scratching mechanism of polycrystalline silicon carbide. *Comput. Mater. Sci.* **2018**, *148*, 76–86. [CrossRef]

20. Wang, J.; Zhang, X.; Fang, F. Molecular dynamics study on nanometric cutting of ion implanted silicon. *Comput. Mater. Sci.* **2016**, *117*, 240–250. [CrossRef]

21. Daw, M.S.; Baskes, M.I. Embedded-atom method: Derivation and application to impurities, surfaces, and other defects in metals. *Phys. Rev. B* **1984**, *29*, 6443–6453. [CrossRef]

22. Zhu, Y.; Zhang, Y.; Qi, S.; Xiang, Z. Titanium Nanometric Cutting Process Based on Molecular Dynamics. *Rare Met. Mater. Eng.* **2016**, *45*, 897–900. [CrossRef]

23. Faken, D.; Jónsson, H. Systematic analysis of local atomic structure combined with 3D computer graphics. *Comput. Mater. Sci.* **1994**, *2*, 279–286. [CrossRef]

24. Zhu, P.; Fang, F. On the mechanism of material removal in nanometric cutting of metallic glass. *Appl. Phys. A* **2014**, *116*, 605–610. [CrossRef]

25. Swygenhoven, H.V.; Derlet, P.M.; Frøseth, A.G. Stacking fault energies and slip in nanocrystalline metals. *Nat. Mater.* **2004**, *3*, 399–403. [CrossRef]

26. Liang, Y.C.; Wang, Q.L.; Yu, N.; Chen, J.X.; Zha, F.S.; Sun, Y.Z. Study of Dislocation Nucleation Mechanism in Nanoindentation Process. *Nanosci. Nanotechnol. Lett.* **2013**, *5*, 536–541. [CrossRef]

27. Sharma, A.; Datta, D.; Balasubramaniam, R. An investig- ation of tool and hard particle interaction in nanoscale cutting of copper beryllium. *Comput. Mater. Sci.* **2018**, *145*, 208–223. [CrossRef]

28. Dai, H.; Chen, G. A molecular dynamics investigation into the mechanisms of material removal and subsurface damage of nanoscale high speed laser-assisted machining. *Mol. Simul.* **2017**, *43*, 42–51. [CrossRef]

 *metals*

Article

Fracture Toughness Calculation Method Amendment of the Dissimilar Steel Welded Joint Based on 3D XFEM

Yuwen Qian [1,2] and Jianping Zhao [1,2,*]

1 School of Mechanical and Power Engineering, Nanjing Tech University, Nanjing 211816, China; reasonqiean@gmail.com
2 Jiangsu Key Lab of Design and Manufacture of Extreme Pressure Equipment, Nanjing 211816, China
* Correspondence: jpzhao71@njtech.edu.cn

Received: 21 March 2019; Accepted: 27 April 2019; Published: 30 April 2019

Abstract: The dissimilar steel welded joint is divided into three pieces, parent material–weld metal–parent material, by the integrity identification of BS7910-2013. In reality, the undermatched welded joint geometry is different: parent material–heat affected zone (HAZ)–fusion line–weld metal. A combination of the CF62 (parent material) and E316L (welding rod) was the example undermatched welded joint, whose geometry was divided into four pieces to investigate the fracture toughness of the joint by experiments and the extended finite element method (XFEM) calculation. The experimental results were used to change the fracture toughness of the undermatched welded joint, and the XFEM results were used to amend the fracture toughness calculation method with a new definition of the crack length. The research results show that the amendment of the undermatched welded joint geometry expresses more accuracy of the fracture toughness of the joint. The XFEM models were verified as valid by the experiment. The amendment of the fracture toughness calculation method expresses a better fit by the new definition of the crack length, in accordance with the crack route simulated by the XFEM. The results after the amendment coincide with the reality in engineering.

Keywords: undermatched; integrity identification; XFEM; fracture toughness calculation method

---

1. Introduction

1.1. The Undermatched Welded Joint

High-strength low-alloy (HSLA) steel is commonly used for petrochemical equipment. The development of special electrodes for the new HSLA is expensive and time-consuming. Ready-made lower strength electrodes for HSLA are a good choice to fabricate the undermatched welded joint. The undermatched welded joint has a low crack tendency and is economical within a wide range of matching ratios [1].

1.2. The Integrity Identification of the Undermatched Welded Joint

The integrity identification of BS7910-2013 [2] shows the three pieces of the undermatched welded joint, parent material–weld metal–parent material, which is like a sandwich structure. In reality, the undermatched welded joint geometry is different [3], parent material–heat affected zone (HAZ)–fusion line–weld metal, as Figure 1a shows. Figure 1b shows the different grain sizes of the different areas of the undermatched welded joint. The blue line in the figure is the line between the HAZ and the parent material. The combination of CF62 (parent material) and E316L (welding rod) is the example undermatched welded joint in this study. The fracture toughness of the different areas was researched by experiments and extended finite element method (XFEM) simulations in this paper. The XFEM models of the different areas of the undermatched welded joint were built and verified as valid by the experiments.

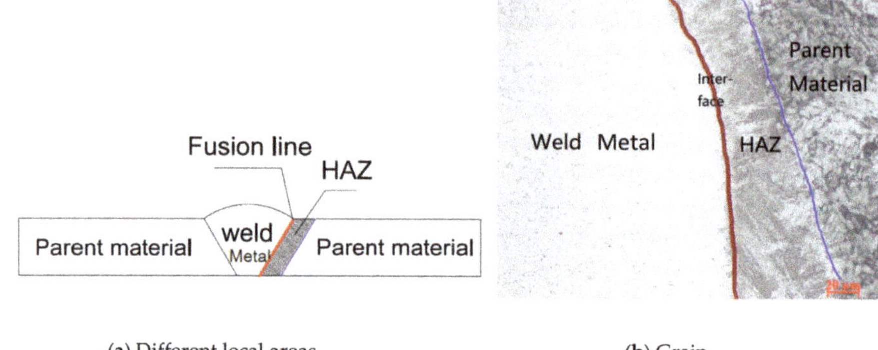

(**a**) Different local areas   (**b**) Grain

Figure 1. Different areas of the investigated welded joint. (**a**) Different local areas; (**b**) Grain.

1.3. The Research Status of XFEM

The extended finite element method (XFEM) was put forward by Belytschko [4], an American professor, and was used to research the crack growth route. Compared to the traditional finite element, the XFEM simulated the crack initiation and growth without the need to refresh the meshing, reducing the workload. Sukumar et al. [5] promoted the 2D XFEM crack model to a 3D XFEM crack model. Legay [6] and Ventura et al. [7] researched the blending element (shown in Figure 2) to improve the convergence and precision of the simulation. Menouillard [8] improved integral stability. Zhuang and Cheng [9] realized the fusion line crack extension between the dissimilar steels by the XFEM.

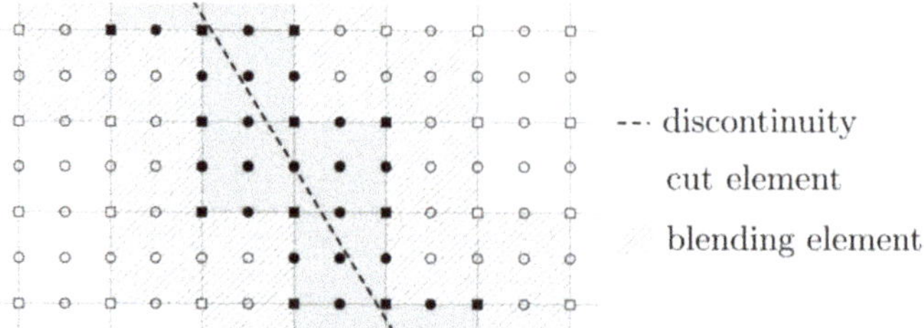

Figure 2. Planar crack, cut element, and blending element.

Research of the XFEM began late in China. Xiujun Fang of Tsinghua University researched the cohesive crack model based on the XFEM in 2007 [10]. Hai Xie of Shanghai Jiaotong University developed the XFEM with the ABAQUS User Subroutine in 2009 [11]. Yehai Li of Nanjing University of Aeronautics and Astronautics researched the interfacial crack growth of the dissimilar material in 2012 [12]. Shaoyun Zhang of Zhejiang University researched the helical crack growth by the XFEM in 2013 [13]. Zhifeng Yang of Nanjing Tech University developed the 2D elastic-plastic crack XFEM model and verified its validity by the *J* integral in 2014 [14]. Yixiu Shu of the Northwestern Polytechnical University analyzed the multiple crack interaction problems by the XFEM in 2015 [15]. Yang Zhang of Harbin Engineering University applied a 3D crack to the pressure vessel by the XFEM in 2015 [16]. Zhen Wang and Tiantang Yu researched the adaptive multiscale XFEM model for 3D crack problems in 2016 [17].

1.4. The Welding of the Undermatched Joint

The size of the weld plate is 600 × 400 × 40 mm (length, width, and thickness, respectively) with an X-shaped groove, shown in Figure 3. The welding rod (E316L) is the special electrode for 316L stainless steel. The welding parameters are listed in Table 1, in accordance with the criteria of ASME-VIII [18]. Tables 2 and 3 list the chemical composition tested results of CF62 and E316L. Tables 4 and 5 list the mechanical properties of the CF62 steel and the E316L steel, respectively.

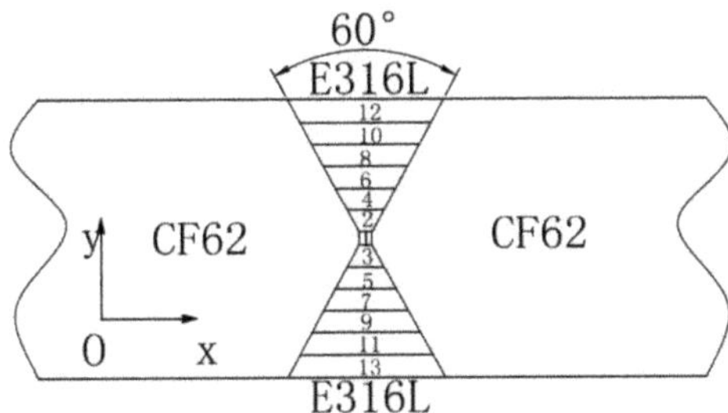

Figure 3. The investigated welded joint.

Table 1. Welding parameters of the undermatched welded joint adapted from [18], with permission from American Society of Mechanical Engineers, 2019.

Layer Name	Layer Numbers	Voltage (V)	Current (A)	Welding Speed (cm/min)
Backing Welding	1	20	75	9
Filler Welding	2–11	22	79.5	10.5
Cover Welding	12–13	24	105	15

Table 2. Chemical composition of CF62 (wt.%).

C	Mn	Si	S	P	Cr	Mo	V	B
0.121	1.260	0.188	0.002	0.011	0.224	0.213	0.035	0.001

Table 3. Chemical composition of E316L(wt.%).

C	Cr	Ni	Mo	Si	Mn	P	S	Co
0.039	18.96	10.73	2.33	0.62	1.06	0.02	0.0064	0.025

Table 4. Mechanical properties of CF62 reproduced from [19], with permission from Taiyuan University of Science and Technology, 2019.

Temperature (°C)	Heat Conductivity Coefficient (W·m⁻¹·°C⁻¹)	Density (kg·m⁻³)	Specific Heat (J·kg⁻¹·°C⁻¹)	Elastic Modulus (GPa)	Poisson's Ratio	Yield Strength (MPa)	Thermal Expansivity (10⁻⁶·mm/mm/K)
25	51.25	7860	450	209	0.29	575	0.01
100	47.95	7840	480	204	0.29	550	0.09
200	44.13	7810	520	200	0.30	520	0.23
400	38.29	7740	620	175	0.31	438	0.51
600	34.65	7670	810	135	0.31	280	0.82
800	29.59	7630	990	78	0.33	80	0.97
1000	29.35	7570	620	15	0.35	30	1.24
1200	31.88	7470	660	3	0.36	10	1.70
1400	34.4	7380	690	1	0.38	5	2.16
1600	34.79	6940	830	1	0.50	5	4.39
1640	335	6940	830	1	0.50	5	4.39

Table 5. Mechanical properties of E316L reproduced from [20], with permission from welding technology, 2019.

Temperature (°C)	Heat Conductivity Coefficient (W·m^{-1}·°C^{-1})	Density (kg·m^{-3})	Specific Heat (J·kg^{-1}·°C^{-1})	Elastic Modulus (GPa)	Poisson's Ratio	Yield Strength (MPa)	Thermal Expansivity (10^{-6}·mm/mm/K)
20	13.31	7966	470	195.1	0.267	325	15.24
200	16.33	7893	508	185.7	0.29	226	16.43
400	19.47	7814	550	172.6	0.322	180	17.44
600	22.38	7724	592	155	0.296	165	18.21
800	25.07	7630	634	131.4	0.262	153	18.83
900	26.33	7583	655	116.8	0.24	100	19.11
1000	27.53	7535	676	100.1	0.229	53	19.38
1100	28.67	7486	698	81.1	0.223	23	19.66
1200	29.76	7436	719	59.5	0.223	15	19.95
1420	31.95	7320	765	2	0.223	3.3	20.7
1460	320	7320	765	2	0.223	3.3	20.7

1.5. The Fracture Toughness Calculation Method

ASTM 1820-18 [21] is the newest standard test method for the measurement of fracture toughness. Fracture toughness is calculated by the relationship between the *J* integral and the crack length, a_i [21]:

$$J = \frac{K^2\left(1 - v^2\right)}{E} + J_{pl} \tag{1}$$

$$J_{pl} = \frac{\eta_{pl} A_{pl}}{B_N b_o} \tag{2}$$

where:

A_{pl} is the area under force versus displacement, as shown in Figure 4;

η_{pl} is either 1.9 if the load-line displacement is used for A_{pl} or $3.667 - 2.199(a_0/w) + 0.437(a_0/w)^2$ if the recorded crack mouth opening displacement is used for A_{pl};

B_N is the net specimen thickness;

b_o is W $- a_0$.

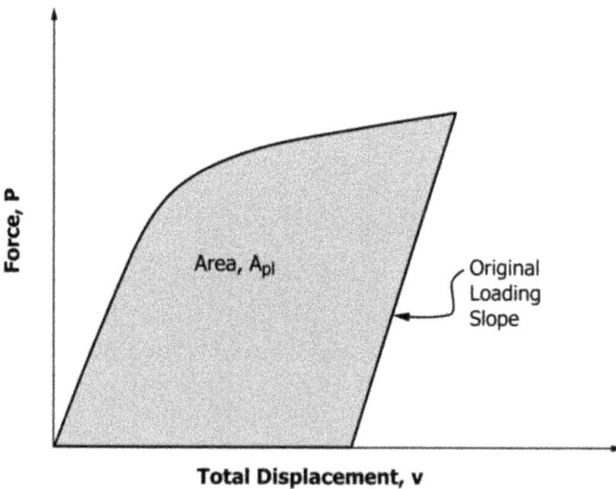

Figure 4. Definition of the area for the *J* calculation using the basic method.

The crack length, a_i, is calculated by the displaced force point with Equation (3):

$$\frac{a_i}{W} = 1.000196 - 4.06319\mu + 11.242\mu^2 - 106.043\mu^3 + 464.335\mu^4 - 650.677\mu^5 \tag{3}$$

where:

$$\mu = \frac{1}{\left(B_e E\, C_{c(i)}\right)^{1/2} + 1}$$
(4)

$$B_e = B - (B - B_N)^2 / B$$
(5)

$$C_{c(i)} = \frac{C_i}{\left(\frac{H^*}{R_i}\sin\theta_i - \cos\theta_i\right)\left(\frac{D}{R_i}\sin\theta_i - \cos\theta_i\right)}$$
(6)

where:

C_i is the measured specimen elastic compliance ($\Delta v_m / \Delta P_m$) (at the load-line);
H^* is the initial half-span of the load points (center of the pin holes);
D is one-half of the initial distance between the displacement measurement points;
R_i is the radius of the rotation of the crack center line, (W+a)/2, where a is the updated crack size.

Figure 5a shows the relationship between the crack rotation angle, θ_i, and the respective crack length, a_i [21]. The crack rotation angle, θ_i, is calculated by Equation (7). Figure 5a shows that the crack length, a_i, was equal to the length parallel to the center line of the compact tension (CT) samples. The crack deflection has not been considered in the crack length calculation in the criteria.

$$\theta_i = \sin^{-1}\left\{\left(D + \frac{V_{m(i)}}{2}\right) / \left[D^2 + R_i^2\right]^{1/2}\right\} - tan^{-1}\left(\frac{D}{R_i}\right)$$
(7)

where $V_{m(i)}$ is the total measured load-line displacement at the beginning of the *i*-th unloading/reloading cycle.

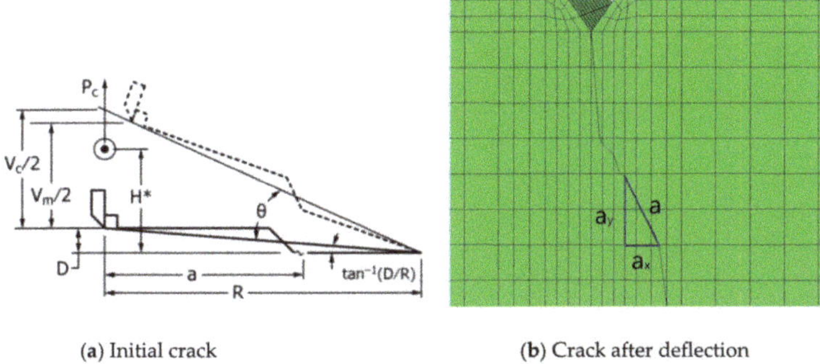

(a) Initial crack   (b) Crack after deflection

Figure 5. Crack rotation angle, θ, of the compact tension (CT) specimens. (a) Initial crack; (b) Crack after deflection.

Figure 5b shows the crack route after the crack deflection. It shows the crack length after the deflection, $a = \sqrt{a_x^2 + a_y^2}$. The crack length in the criteria of ASTM 1820-18 is $a_i = a_y$, which means that $a_i < a$. Therefore, the crack length, a_i, calculated by the criteria of ASTM 1820-18, was smaller than the true crack length after the deflection, a. The fracture toughness calculation method was then amended by the crack length after the deflection was simulated by the XFEM.

The division of the geometry (parent material–HAZ–fusion line–weld metal) of the undermatched welded joint and the experimental results changed the fracture toughness of the dissimilar steel welded joint. The crack length amendment after the deflection defined a new fracture toughness calculation method of the dissimilar welded joint.

1.6. J Integral and the Fracture Toughness Value,K_{mat}

The *J* integral was put forward by Rice [22]. Figure 6a shows the integral loop/area around the crack tip. Equation (8) is calculated with the equivalent integral region to deal with the stress and the strain of Figure 6b [22]. Figure 6c shows the integral area, which contains a heterogeneous material interface. In [23], it was verified that the interface integral area $I^{th}_{A+} = I^{th}_{A-} = I_{interface} = 0$ when the crack tip is under a constant temperature with a pure mechanical load. Equation (8) is applicable to calculate the *J* integral for the investigated welded joint in this paper.

$$J = \int_A \left(\sigma_{ij} \frac{\partial u_j}{\partial x_1} - \omega \delta_{1i} \right) \frac{\partial q}{\partial x_i} dA \tag{8}$$

where u_i is the component of the displacement vector; dA is the tiny increment area on the integral path, Γ; ω is the strain energy density factor; and $q(x,y)$ is a mathematical function, which is $q = 0$ in the outer boundary and $q = 1$ in the other places of the regional; and ω, T_i are listed in Equation (9).

$$\omega = \int_0^{\varepsilon_{ij}} \sigma_{ij} d\varepsilon_{ij}$$
$$T_i = \sigma_{ij} n_j \tag{9}$$

where σ_{ij} is the stress tensor and ε_{ij} is the strain tensor.

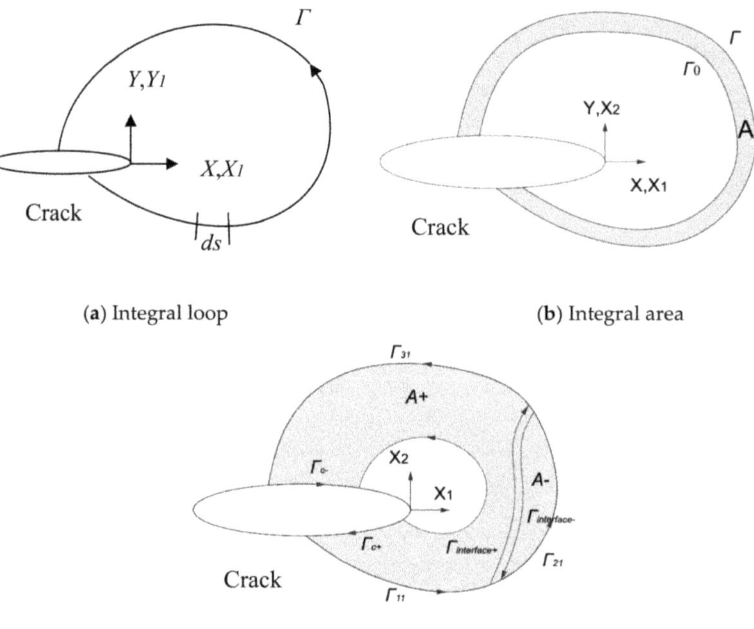

Figure 6. Counterclockwise integral loop around the crack tip. (**a**) Integral loop; (**b**) Integral area; (**c**) Interface integral area.

BS7910-2013 [2] gives Equation (10), which shows the relationship between the J_{mat} value and the fracture toughness value, K_{mat}, during the elastic stage:

$$J_{mat} = \frac{K_{mat}^2 \left(1 - v^2\right)}{E} \tag{10}$$

where E is the elastic modulus and v is the Poisson's ratio.

2. Experiments

2.1. The Fracture Toughness Experiments

The fracture toughness of the different areas of the joint geometry (parent material–HAZ–fusion line–weld metal) was tested with the individual CT specimens that were taken from the undermatched welded joint. The crack initiated and grew along the center line of the CT specimens during the experiments. The center lines in the different areas of the joint present the locations of the CT specimens in the dissimilar joint, as seen in Figure 7. The CT size is shown in Figure 8.

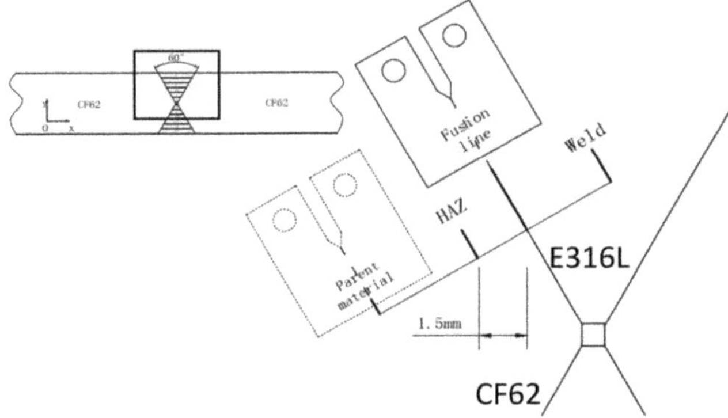

Figure 7. Center lines of the CT specimens in the investigated welded joint.

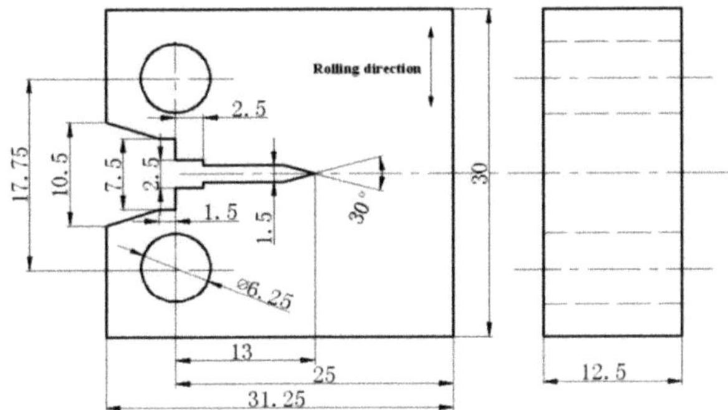

Figure 8. Size of the CT specimens.

The fracture toughness was tested with the MTS-809 (MTS Systems Cor., Eden Prairie, MN, USA). The local fracture surfaces taken by SEM of the different areas of the joint geometry are shown in Figure 9. Figure 9b shows that the crack initiated from the high-strength region (parent material) and deflected to the lower strength region (weld metal) in the HAZ. Figure 10 depicts the HAZ fracture surface macro-image, which indicates the parent material (PM) region and the weld metal region. It means that the crack deflection existed in the dissimilar material during the experiment. The experimental process was in accordance with the criteria of ASTM 1820-18 [21], and the results of the *J* integral and the crack length, Δa, were managed under the regulations.

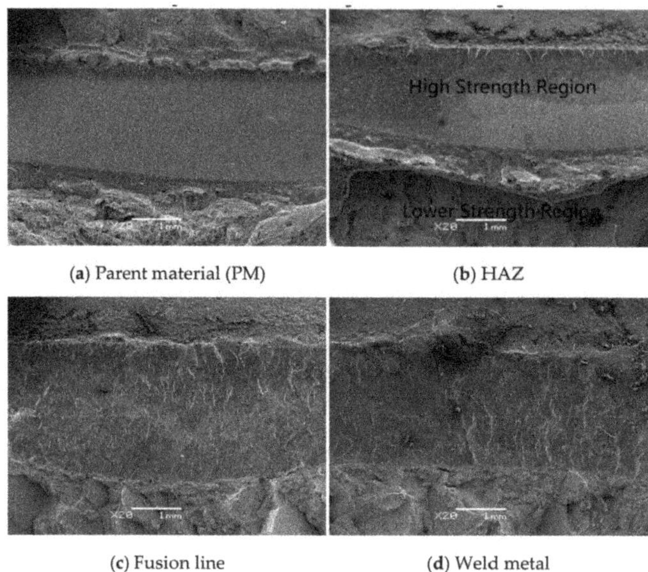

(a) Parent material (PM)   (b) HAZ

(c) Fusion line   (d) Weld metal

Figure 9. Fracture surfaces after the fracture toughness experiment. (**a**) Parent material (PM); (**b**) HAZ; (**c**) Fusion line; (**d**) Weld metal.

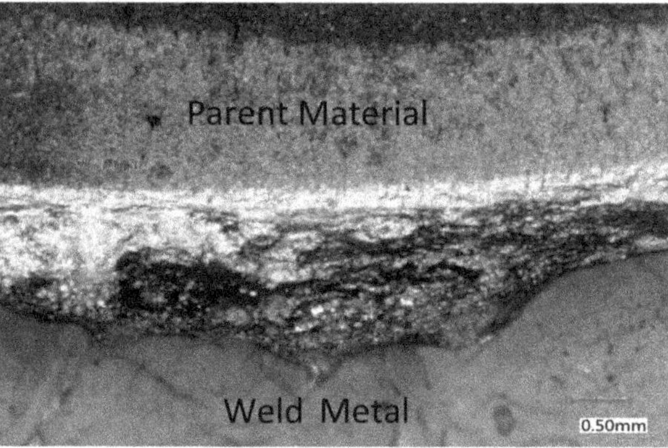

Figure 10. HAZ fracture surface macro-image.

2.2. The Tensile Experiments

The maximum stress damage of the parent material and the weld metal was needed to build the XFEM models. It was tested with a flat tensile specimen, which was 0.8 mm in thickness. Figure 11a shows the location of the tensile specimen, which was taken from the parent material to the weld metal and is parallel to the fusion line. Figure 11b shows the shape and dimensions of the flat tensile specimen.

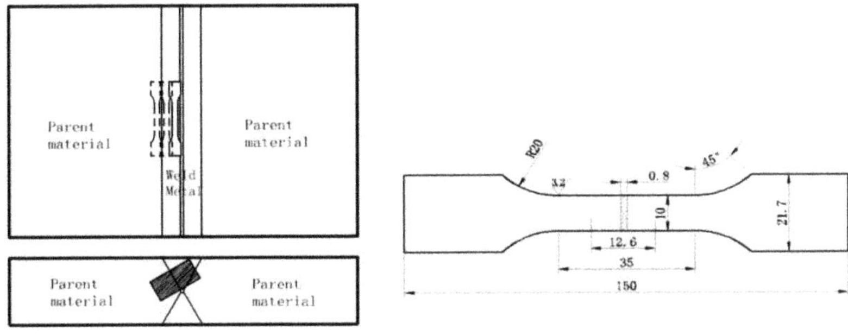

(a) Location   (b) Shape and dimensions

Figure 11. The location and size of the flat tensile specimen. (**a**) Location; (**b**) Shape and dimensions.

3. XFEM Models

A pre-crack was needed when the 1/2 CT specimen was tested and the pre-crack length was 1.5 mm, in accordance with the criteria of the ASTM 1820-18 regulation. The XFEM models were built with a 1.5 mm pre-crack.

The four XFEM models were the parent material, weld metal, fusion line (a combination of PM and weld metal), and HAZ. The mechanical properties were tested by the flat tensile specimen. There were two important parameters—the maximum stress damage, $-\sigma_{max}$, and the equivalent energy release rate, $G_{\theta c}$. In [14,24], the relationship between, σ_{max}, and the tensile strength, σ_b, is given as $\sigma_{max} = \sigma_b$. When the maximum energy release rate (MERR), $G_{\theta max}$, is equal to the threshold, $G_{\theta c}$, the crack begins to initiate. The MERR, $G_{\theta max}$, is calculated in Equation (11), and the threshold is calculated in Equation (12) [13]:

$$G_{\theta max} = \frac{1-v^2}{4E}1 + cos\theta_0\left[K_I^2[1 + cos\theta_0] - 4K_IK_{II}sin\theta_0 + K_{II}^2[5 - 3cos\theta_0]\right] = G_{\theta c} \tag{11}$$

where E is the elastic modulus, v is the Poisson's ratio, K_I is the applied tensile stress intensity factor of Mode I, and K_{II} is the applied tensile stress intensity factor of Mode II.

There are three kinds of crack mode—the opening crack mode (Mode I), the sliding crack mode (Mode II), and the tearing crack mode (Mode III). The experimental and XFEM simulation load is perpendicular to the crack surface. The research crack in this paper is the opening crack mode(Mode I). The MERR threshold, $G_{\theta c}$, is related to the critical stress intensity factor of Mode I, which is $K_{II} = 0$:

$$G_{\theta c} = G_{Ic} = \frac{1-v^2}{E}K_{Ic}^2 \tag{12}$$

Since the crack mode in this paper is Mode I, $G_{Ic} = G_{IIc} = G_{IIIc} = G_{\theta c}$ [13]. The XFEM simulation properties of CF62 and E316L are listed in Table 6.

Table 6. Mechanical properties of the crack propagation.

Material	The Maximum Stress Damage σ_{max}/MPa	MERR $G_{\theta c}$/N/mm
CF62	650	146
E316L	550	160

Figure 12 shows the meshing of the XFEM of the 1/2CT. The sparse meshing elements in the XFEM are applicable [25–27]. Figure 13 shows the dissimilar material XFEM models of the joint geometry with the material printed on their parts. The parent material and weld metal XFEM models are not shown since they are the uniform material model, i.e., they are the material printed on the whole model.

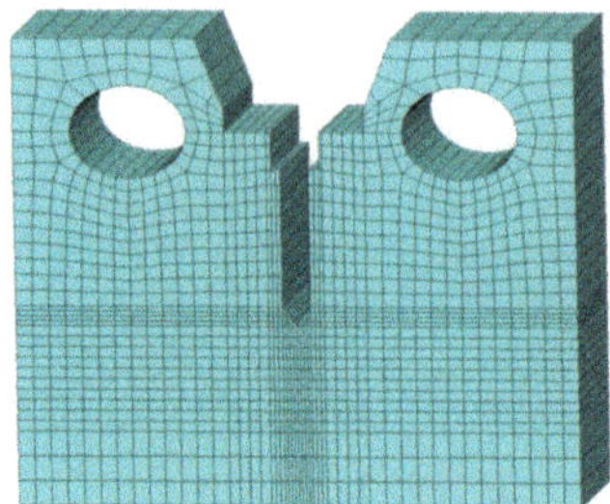

Figure 12. The meshing model of the extended finite element method(XFEM).

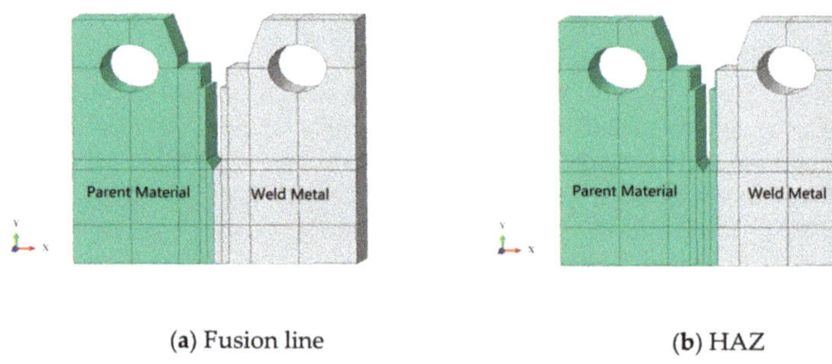

(a) Fusion line   (b) HAZ

Figure 13. The dissimilar material XFEM models. (a) Fusion line; (b) HAZ.

4. The Fracture Toughness of the Undermatched Welded Joint

4.1. Fracture Toughness of the Undermatched Welded Joint

The test results of the *J* integral and crack length, Δa, are fitted by Equation (13) in the criteria of ASTM 1820-18 [21]:

$$J = \alpha + \beta(\Delta a)^r \tag{13}$$

The experimental data and the resistance equations of the different areas of the joint geometry are listed in Tables 7 and 8, respectively. The resistance curves are shown in Figure 14.

Table 7. Experimental data.

Data Points	Parent Material		HAZ		Fusion Line		Weld Metal	
	Δa/mm	J/KJ/m^2	Δa/mm	J/KJ/m^2	Δa/mm	J/KJ/m^2	Δa/mm	J/KJ/m^2
1	0	0	0	0	0	0	0	0
2	0.133	118.03	0.478	276.52	0.439	181.30	0.265	116.20
3	0.318	191.35	0.623	315.80	0.624	230.76	0.509	223.11
4	0.496	309.24	1.277	438.68	0.869	245.48	0.611	278.27
5	1.073	423.93	1.821	562.78	1.101	296.98	0.872	331.57
6	1.234	514.53	3.396	328.69	1.478	332.97	1.144	359.87
7	1.433	606.27	3.851	326.04	1.939	354.85	1.535	373.71

Table 8. The resistance equations and the J_{mat} and K_{mat} value of the undermatched welded joint.

Areas of Joint Geometry	Equations	J_{mat}/KJ/m^2	K_{mat}/MPa*m$^{1/2}$
Parent material	$J = 451.8*(\Delta a)^{0.825}$	146.7	176.7
HAZ	$J = 330.3*(\Delta a)^{0.286}$	232.8	222.6
Fusion Line	$J = 268*(\Delta a)^{0.314}$	181.1	156.1
Weld Metal	$J = 280.5*(\Delta a)^{0.437}$	160.3	146.9

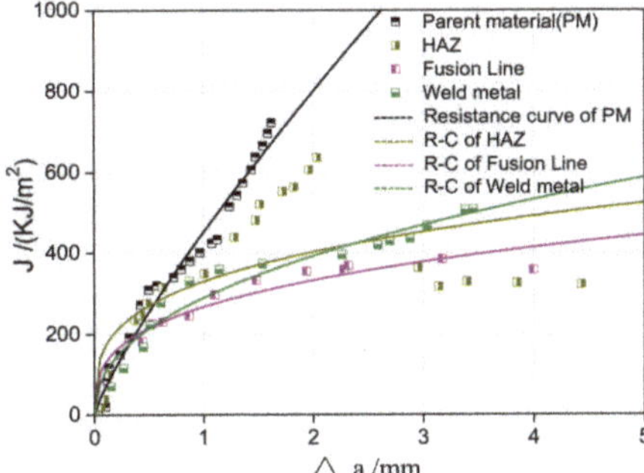

Figure 14. The ensemble of the *J*-Δa resistance curves of the undermatched welded joint.

4.2. Amendment of HAZ Fracture Toughness

Figure 15a shows the fracture surface of the HAZ and shows that there was a 1.04 mm length of the smooth fracture surface and a 1.07 mm length of the parent material fracture surface. The fusion line between the parent material and the weld metal is marked with a red line. The weld metal fracture surface is below the red line.

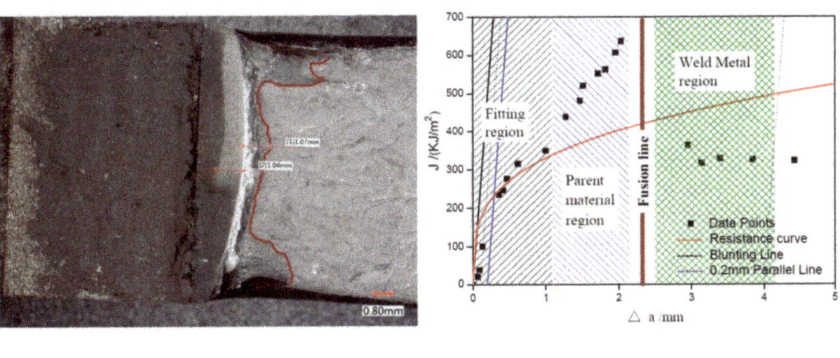

(a) Fracture surface   (b) Resistance curve regions of the HAZ

Figure 15. Fracture toughness results of the HAZ. (**a**) Fracture surface; (**b**) Resistance curve regions of the HAZ.

Figure 15b shows the resistance curve regions of the HAZ. It shows that in the initial crack (Δa < 1.04 mm) the *R*-curve fits the resulting data points well. The parent material region is next to the fitting region: 1.04 mm < Δa < 2.11 mm. The *J* integral is larger than the curve point which shows the parent material's fracture toughness. After the crack grew across to the weld metal region, the *J* integral was smaller than the *R*-curve points.

Haitao Wang tested and simulated the crack growth paths in a dissimilar metal welded joint and found that the crack deflected to the soft area during the experiment [28,29]. Longchao Cao [30] and Pavel Podany [31] researched the microstructure and the local mechanical properties of the dissimilar butt joints. Jianbo Li [32] studied the fracture toughness near the fusion line (HAZ and fusion line) of the dissimilar material and, in addition, found the crack deflection. The HAZ experimental results demonstrated the crack deflection when the CT specimens were tested. The resistance curve of the HAZ (Figure 15b) shows that the fracture toughness of the undermatched welded joint needs to be amended. The crack deflection shows that the crack length test method needs to be amended as well.

As there was a misfit of the data points in the two regions and the lower strength region was the weld metal region, which was below the high-strength region, the HAZ was considered to be the high-strength region. The crack initiated in the HAZ high-strength region and then deflected to the weld metal region. During the crack growth process, the crack crossed the fusion line between the high-strength region and the weld metal region. Figure 9b shows the local fracture surface figure taken by SEM, which shows the location of the high-strength region and the weld metal region (lower strength region).

Table 9 lists the resistance equations after the amendment of the different areas (parent material–HAZ–fusion line–weld metal) of the undermatched welded joint geometry. The fusion line crack deflection initiated in the pre-crack stage and the crack growth of the fusion line (the heterogeneous material XFEM model) were always in the weld metal region.

Figure 16 shows the resistance curves of the high-strength regions of the HAZ after amending. The figure shows that the HAZ is more applicable between the resistance curve and the data points after amending.

Table 9. The resistance equations and the J_{mat} and K_{mat} values of the undermatched welded joint after amending.

Area of the Joint Geometry		Resistance Equations	J_{mat}/KJ/m^2	K_{mat}/MPa*m$^{1/2}$
	Parent material	$J = 451.8*(\Delta a)^{0.825}$	146.7	176.7
HAZ	(High-strength region)	$J = 389.9*(\Delta a)^{0.647}$	167.0	188.5
	Fusion line	$J = 268*(\Delta a)^{0.314}$	181.1	156.1
	Weld Metal	$J = 280.5*(\Delta a)^{0.437}$	160.3	146.9

Note: the crack growth of the fusion line was always in the weld metal region with the deflection of the pre-crack.

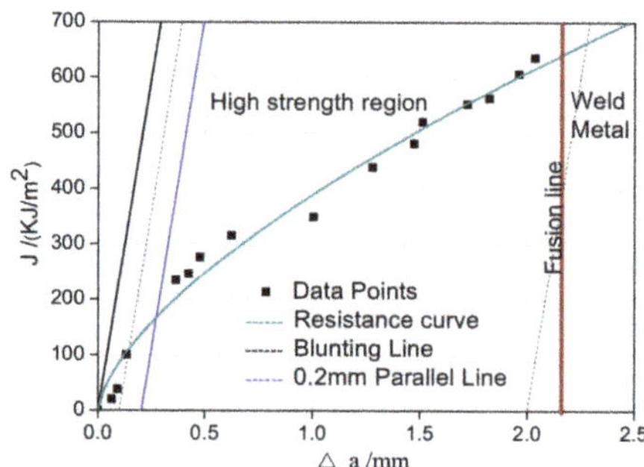

Figure 16. Resistance curves of the HAZ after amending.

4.3. J Integral Verification of XFEM

The XFEM gives the *J* integral of the crack growth, which is used to compare the experimental *J*-Δ*a* resistance curves. The XFEM models of the parent material and the weld metal were verified, respectively, by the experimental results. Figures 17 and 18 are the XFEM results of the parent material and the weld metal. Figures 17a and 18a show the CT stress distribution, and Figures 17b and 18b show the crack route.

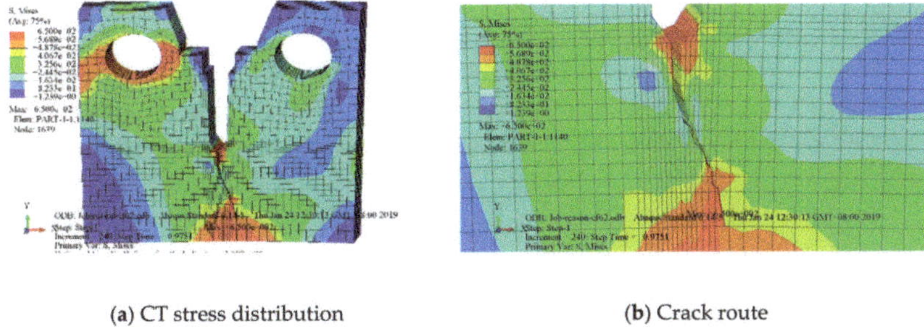

(**a**) CT stress distribution   (**b**) Crack route

Figure 17. XFEM result of the parent material. (**a**) CT stress distribution; (**b**) Crack route.

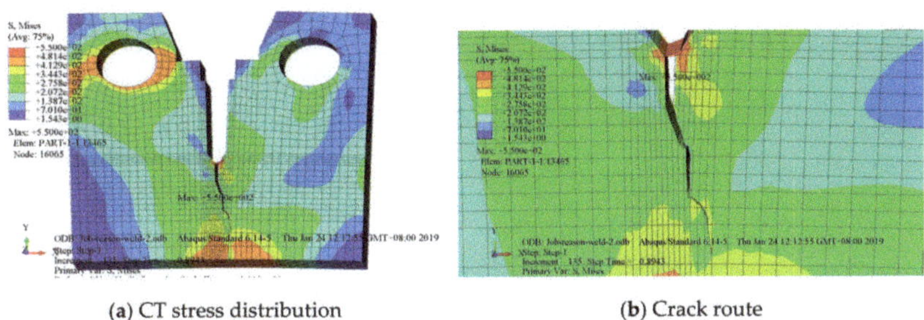

(**a**) CT stress distribution   (**b**) Crack route

Figure 18. XFEM result of the weld metal. (**a**) CT stress distribution; (**b**) Crack route.

There are five J integral routes when the XFEM model outputs the J integral and the results are the same. The J integral result is independent of the route, which has been demonstrated previously [27,33,34].

Figure 19 shows the comparison of the experimental J-Δa resistance curves and the J integral data point output of the XFEM. The figure shows that the data point output of the XFEM coincides well with the experimental J-Δa resistance curves. This means that the XFEM models in this paper are valid for the undermatched welded joint.

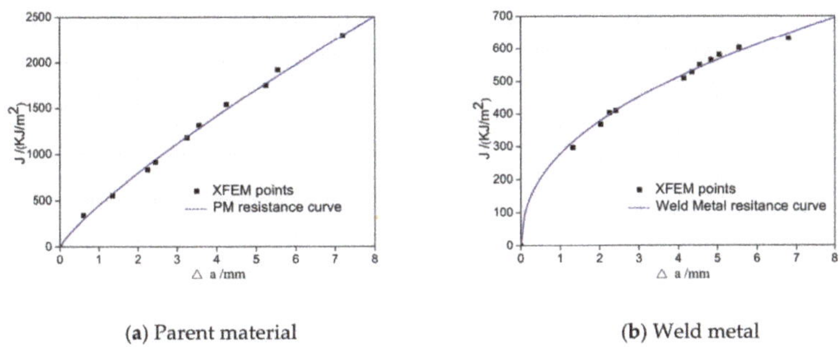

(**a**) Parent material   (**b**) Weld metal

Figure 19. J integral verification of the XFEM. (**a**) Parent material; (**b**) Weld metal.

5. Amendment of the Fracture Toughness Calculation Method

Table 10 lists the resistance equations of the investigated welded joint after the crack length amendment. Figure 20 shows the homogeneous material XFEM models' (PM and weld metal) resistance curves after amendment, while Figure 21 shows the heterogeneous material XFEM models' (fusion line and HAZ) resistance curves after amendment. Figure 21a shows the resistance curve of the fusion line, and Figure 21b shows that of the HAZ high-strength region.

Table 10. The resistance equations and the J_{mat} and K_{mat} values of the undermatched welded joint after the crack amendment based on the XFEM.

Regions of the Joint		Resistance Equations	J_{mat}/KJ/m^2	K_{mat}/MPa*m$^{1/2}$
	Parent material	$J = 432.5*(\Delta a)^{0.823}$	141.7	173.6
HAZ	(High-strength region)	$J = 402.3*(\Delta a)^{0.572}$	194.6	203.5
	Fusion line	$J = 251.6*(\Delta a)^{0.319}$	167.6	150.2
	Weld metal	$J = 267.0*(\Delta a)^{0.461}$	146.4	140.4

Note: the crack growth of the fusion line was always in the weld metal region with the deflection of the pre-crack.

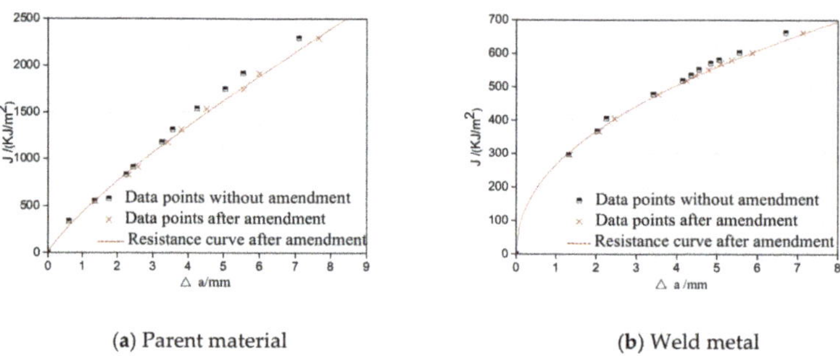

(**a**) Parent material   (**b**) Weld metal

Figure 20. Resistance curves after the crack length amendment of the homogeneous material. (**a**) Parent material; (**b**) Weld metal.

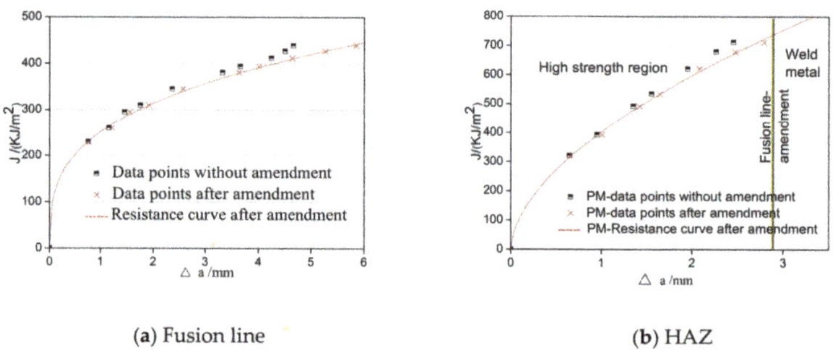

(**a**) Fusion line   (**b**) HAZ

Figure 21. Resistance curves after the crack length amendment of the heterogeneous material. (**a**) Fusion line; (**b**) HAZ.

The crack length amendment results show that the fracture toughness calculation method of the undermatched welded joint needed to be amended, due to the crack growth deflection of the heterogeneous material.

The fracture toughness of the investigated welded joint was given by experimenting with the different areas of the joint (parent material–fusion line–HAZ–weld metal). In addition, the fracture toughness calculation method was amended based on the XFEM models of the heterogeneous material in this paper.

6. Conclusions

The division of the dissimilar steel welded joint geometry of the current integrity identification needed to be amended to four pieces: parent material–HAZ–fusion line–weld metal. In accordance with this new division of the undermatched welded joint, XFEM models of the undermatched welded joint were built to calculate their fracture toughness. The conclusions are as follows:

(1) The different areas (parent material–HAZ–fusion line–weld metal) of the undermatched welded joint geometry were tested to obtain the fracture toughness of the investigated welded joint. During the experiments, the fracture toughness of the joint changed in accordance with the crack deflection.

(2) In this study, XFEM models of the homogeneous material and the heterogeneous material, i.e., the different areas of the undermatched welded joint geometry, were built to calculate the crack route.

(3) The experimental results of the J-Δa resistance curve were compared to the XFEM calculation results to verify the validity of the XFEM to the undermatched welded joint.

(4) The crack length amendment was in accordance with the XFEM simulation result—the crack route. The fracture toughness calculation method was amended by the crack length amendment after the deflection of the heterogeneous material by the XFEM.

Author Contributions: Y.Q. is mainly on the Methodology, Software, Validation, Formal Analysis, Investigation, Data Curation, Writing-Original Draft Preparation, Writing-Review & Editing. J.Z. is mainly on Writing-Review & Editing, Visualization, Supervision, Funding Acquisition.

Funding: This research was funded by the National Key Research and Development Program of China [2016YFC0801905].

Conflicts of Interest: The authors declare no conflict of interest.

References

1. Wen, X.; Wang, P.; Dong, Z.B.; Liu, Y.; Fang, H.Y. Nominal stress-based equal-fatigue bearing-capacity design of under-matched HSLA steel butt-welded joints. *Metals* **2018**, *8*, 880. [CrossRef]

2. *BS7910:2013+A1:2015, Guide to Methods for Assessing the Acceptability of Flaws in Metallic Structures*; BSI Standards Limited: London, UK, 2015.

3. Baris Basyigit, A.; Murat, M.G. The Effects of TIG Welding Rod Compositions on Microstructural and Mechanical Properties of Dissimilar AISI 304L and 420 Stainless Steel Welds. *Metals* **2018**, *8*, 972. [CrossRef]

4. Belytschko, T.; Black, T. Elastic crack growth in finite elements with minimal remeshing. *Int. J. Numer. Methods Eng.* **1999**, *45*, 601–620. [CrossRef]

5. Sukumar, N.; Moes, N.; Moran, B.; Belytschko, T. Extended finite element method for three-dimensional crack modeling. *Int. J. Numer. Methods Eng.* **2000**, *48*, 1549–1570. [CrossRef]

6. Legay, A.; Wang, H.W.; Belytschko, T. Strong and weak arbitrary discontinuities in spectral finite elements. *Int. J. Numer. Methods Eng.* **2005**, *64*, 991–1008. [CrossRef]

7. Ventura, G.; Moran, B.; Belytschko, T. Dislocations by partition of unity. *Int. J. Numer. Methods Eng.* **2005**, *62*, 1463–1487. [CrossRef]

8. Menouillard, T.; Rethore, J.; Combescure, A.; Bung, H. Efficient explicit time stepping for the extended finite element method (XFEM). *Int. J. Numer. Methods Eng.* **2006**, *68*, 911–939. [CrossRef]

9. Zhuang, Z.; Cheng, B.B. Equilibrium state of mode-I sub-interfacial crack growth in bi-materials. *Int. J. Fract.* **2011**, *170*, 27–36. [CrossRef]

10. Fang, X.J.; Jin, F.; Wang, J.T. Cohesive crack model based on extended finite element method. *J. Tsinghua Univ. Sci. Tech.* **2007**, *47*, 344–347. (In Chinese)

11. Xie, H.; Feng, M.L. Implementation of Extended Finite Element Method with ABAQUS User Subroutine. *J. Shanghai Jiaotong Univ.* **2009**, *43*, 1644–1648. (In Chinese)

12. Li, Y. Research on Interfacial Crack Propagation Using ABAQUS: Application to Artificial Ice Shape. Ph.D. Thesis, Nanjing University of Aeronautics and Astronautics, Nanjing, China, 2012. (In Chinese).

13. Zhang, S. Study on Helical Crack Propagation by Extended Finite Element Method. Ph.D. Thesis, Zhejiang University, Hangzhou, China, 2013. (In Chinese).

14. Yang, Z.; Zhou, C.; Dai, Q. Elastic-plastic crack propagation based on extended finite element method. *J. Nanjing Tech Univ. (Nat. Sci. Ed.)* **2014**, *36*, 50–57. (In Chinese)

15. Shu, Y.X.; Li, Y.Z.; Jiang, W.; Jia, Y.X. Analyzing Multiple Crack Propagation Using Extended Finite Element Method(X-FEM). *J. Northwest. Polytech. Univ.* **2015**, *33*, 197–203. (In Chinese)

16. Zhang, Y. Application Research of the Extended Finite Element Method on the Cracks in Pressure Vessel. Ph.D. Thesis, Harbin Engineering University, Harbin, China, 2015. (In Chinese).

17. Wang, Z.; Yu, T.T. Adaptive multiscale extended finite element method for modeling three-dimensional crack problems. *Eng. Mech.* **2016**, *33*, 32–38. (In Chinese)

18. ASME. *ASME Boiler & Pressure Vessel Code, Section VIII, Division 1*; American Society of Mechanical Engineers: New York, NY, USA, 2010.

19. Gan, H. Effect of Cold Treatment Temperature on Microstructure and Mechanical Properties of Forged Low Alloy High Strength Steel. Ph.D. Thesis, Taiyuan University of Science and Technology, Taiyuan, China, 2012. (In Chinese).

20. Zhu, P.; Chen, Y.; Wang, G.G. Effect of high temperature tempering heat treatment on deposited metal properties of nuclear grade E316L stainless steel welding rod. *Weld. Technol.* **2016**, *45*, 51–54. (In Chinese)

21. ASTM E1820-18. Standard test methods for measurement of fracture toughness. In *Annual Book of ASTM Standards*; American Society for Testing and Materials: Philadelphia, PA, USA, 2018; Volume 3.01.

22. Rice, J.R. A Path independent integral and the approximate analysis of strain concentration by notched and cracks. *J. Appl. Mech.* **1968**, *35*, 379–386. [CrossRef]

23. Guo, F. Investigation on the Thermal Fracture Mechanics of Nonhomogeneous Materials with Interfaces. Ph.D. Thesis, Harbin Institute of Technology, Harbin, China, 2014. (In Chinese).

24. Yang, Z. Limit Load Research of Crack Structure Based on Extended Finite Element Method. Ph.D. Thesis, Nanjing Tech University, Nanjing, China, 2014. (In Chinese).

25. Banasal, M.; Singh, I.V.; Mishra, B.K.; Bordas, S.P.A. A parallel and efficient multi-split XFEM for 3-D analysis of heterogeneous materials. *Comput. Methods Appl. Mech. Eng.* **2019**, *347*, 365–401. [CrossRef]

26. Sim, J.M.; Chang, Y.-S. Crack growth evaluation by XFEM for nuclear pipes considering thermal aging embrittlement effect. *Fatigue Fract. Eng. Mater. Struct.* **2019**, *42*, 775–791. [CrossRef]

27. Dai, Q.; Zhou, C.Y.; Peng, J.; He, X.H. Finite Element Analysis and EPRI Solution for J-integral of Commercially Pure Titanium. *Rare Metal. Mater. Eng.* **2014**, *43*, 257–263.

28. Wang, H.T.; Wang, G.Z.; Xuan, F.Z.; Tu, S.T. An experimental investigation of local fracture resistance and crack growth paths in a dissimilar metal welded joint. *Mater. Des.* **2013**, *44*, 179–189. [CrossRef]

29. Wang, H.T.; Wang, G.Z.; Xuan, F.Z.; Tu, S.T. Numerical investigation of ductile crack growth behavior in a dissimilar metal welded joint. *Nucl. Eng. des.* **2011**, *241*, 3234–3243. [CrossRef]

30. Cao, L.C.; Shao, X.Y.; Jiang, P.; Zhou, Q.; Rong, Y.N.; Geng, S.N.; Mi, G.Y. Effects of Welding Speed on Microstructure and Mechanical Property of Fiber Laser Welded Dissimilar Butt Joints between AISI316L and EH36. *Metals* **2013**, *7*, 270. [CrossRef]

31. Podany, P.; Reardon, C.; Koukolikova, M.; Prochazka, R.; Franc, A. Microstructure, Mechanical Properties and Welding of Low Carbon, Medium Manganese TWIP/TRIP Steel. *Metals* **2018**, *8*, 263. [CrossRef]

32. Li, J.B.; Liu, B.; Wang, Y. A Study on the Zener-Holloman Parameter and Fracture Toughness of an Nb-Particles-Toughened TiAl-Nb Alloy. *Metals* **2018**, *8*, 387. [CrossRef]

33. Dai, Q.; Zhou, C.; Peng, J. Non-probabilistic Failure Assessment Methodology about Titanium Pipes with Cracks. In Proceedings of the ASME 2011 Pressure Vessels and Piping Conference, Baltimore, MD, USA, 17–21 July 2011.

34. Dai, Q.; Zhou, C.Y.; Peng, J. Non-probabilistic defect assessment for structures with cracks based on interval model. *Nucl. Eng. Des.* **2013**, *262*, 235–245. [CrossRef]

Article

Using DEFORM Software for Determination of Parameters for Two Fracture Criteria on DIN 34CrNiMo6

Ivana Poláková *, Michal Zemko, Martin Rund and Ján Džugan

COMTES FHT, Průmyslová 995, Dobřany 334 41, Czech Republic; mzemko@comtesfht.cz (M.Z.); mrund@comtesfht.cz (M.R.); jdzugan@comtesfht.cz (J.D.)
* Correspondence: ivana.polakova@comtesfht.cz; Tel.: +420-377-197-321

Received: 27 February 2020; Accepted: 26 March 2020; Published: 28 March 2020

Abstract: The aim of this study was to calibrate a material model with two fracture criteria that is available in the DEFORM software on DIN 34CrNiMo6. The purpose is to propose a type of simple test that will be sufficient for the determination of damage parameters. The influence of the quantity of mechanical tests on the accuracy of the fracture criterion was explored. This approach was validated using several tests and simulations of damage in a tube and a round bar. These tests are used in engineering applications for their ease of manufacturing and their strong ability to fracture. The prediction of the time and location of the failure was based on the parameters of the relevant damage model. Normalized Cockroft-Latham and Oyane criteria were explored. The validation involved comparing the results of numerical simulation against the test data. The accuracy of prediction of fracture for various stress states using the criteria was evaluated. Both fracture criteria showed good agreement in terms of the fracture locus, but the Oyane criterion proved more suitable for cases covering larger triaxiality ranges.

Keywords: flow stress; stress triaxiality; material damage; FEM simulation

---

1. Introduction

The aim of this paper is to develop a procedure for failure prediction through simulations when a limited amount of tests are available. The purpose of the paper is to demonstrate quick and sufficiently accurate predictions based on very few—and if possible simple—experiments. From engineering and practical points of view, the number of tests on specimens should be no higher than absolutely necessary. When it comes to calibrating fracture criteria, the challenge lies in choosing the right tests [1]. DEFORM software (developed by Scientific Forming Technologies Corporation, Columbus, OH, USA) was used for all computational modelling tasks presented in this paper. Two fracture criteria available in DEFORM were chosen for validating the procedure of failure prediction: the Normalized Cockcroft-Latham criterion (NCL) and the Oyane criterion.

An accurate description of material behaviour is a prerequisite for successful modelling of cold metal forming processes. Strain hardening (flow stress) and fracture behaviour depend on basic material properties, which are governed by processing history, typically the temperature and strain rate profiles. Such properties can be determined through experiments and testing. In forming processes, where fracture must be avoided, identification of forming limits is a matter of key importance. However, the ability to determine the conditions that lead to fracture is important for processes in which it is undesirable, as well as in processes where material separation is the desired outcome (such as cutting).

Numerous researchers have attempted to predict damage and fracture [2–4] and a multitude of models have been proposed over the years. These include Bridgman [5], McClintock [6] and Rice and Tracey [7], who demonstrated the effect of hydrostatic pressure on fracture. Wilkins et al. postulated that both hydrostatic pressure and deviatoric stress contribute to damage. Johnson and Cook [8] introduced a fracture model that accounted for strain rate and temperature.

Bao and Wierzbicki [9] performed numerous experiments involving a wide range of triaxialities and suggested that the dependence of fracture strain on stress triaxiality was not a monotonic curve. They claimed that there were different fracture mechanisms operating in different stress triaxiality ranges. Later, these authors proposed a cut-off value for triaxiality at negative 1/3, where fracture strain becomes infinity. Kim et al. [10], Gao et al. [11], and Brünig et al. [12] studied triaxiality and Lode angle. This has been followed up by many authors, including our laboratory, to develop testing procedures and the shape of the samples [13–15].

The failure prediction procedure described in this paper consists of defining material behaviour for numerical simulation on the basis of several different tests and finding correlation between them and the accuracy and usability of the prediction. For this purpose, mechanical tests classified into three groups were performed. The first one involved tensile and compression tests through which flow stress data were acquired. The second group comprised notched tensile test, shear test and plane strain test conducted to determine fracture parameters. The last group of tests were performed to validate the fracture criteria. These included radial compression of a ring specimen and axial compression of a notched cylinder.

2. Material

Fracture modelling for tests was carried out on DIN 34CrNiMo6 steel at room temperature and at quasi-static strain rates. This steel was chosen due to its versatile engineering applications [16]. It is used for heavy-duty components, such as axles, shafts, crankshafts, connecting rods, valves, propeller hubs, gears, couplings, torsion bars, aircraft components, and heavy parts of rock drills.

The advantages of DIN 34CrNiMo6 martensitic steel include high ductility, hardenability and strength [16]. In fact, it provides the highest yield strength and ultimate tensile strength from the entire class of high strength steels. Generally, martensitic steels are characterized by a martensitic matrix produced by quenching with small amounts of ferrite or bainite. This steel is strengthened by the precipitation of a fine dispersion of carbides during tempering. Its nominal chemical composition, in weight percent, is presented in Table 1. The material was heat-treated to 34 HRC. Basic measured mechanical data are shown in Table 2.

Table 1. Nominal chemical composition of DIN 34CrNiMo6.

Element	C	Si	Mn	Cr	Mo	Ni
Weight %	0.34	≤0.40	0.65	1.50	0.22	1.50

Table 2. Measured mechanical data of DIN 34CrNiMo6.

Yield Strength (MPa)	Tensile Strength (MPa)	Elongation (%)
877	990	6.6

3. Uncoupled Fracture Models

Fracture models are divided into coupled and uncoupled categories. Coupled models are complex, require a great many material constants, and are computationally demanding. In uncoupled models, the plasticity model is separate from the failure model. This means a significant reduction in the computation effort. The DEFORM software only offers uncoupled fracture criteria.

Two fracture criteria were selected for this paper. Their applicability to industrial problems was evaluated. The NCL criterion has only one parameter which governs the critical damage. Therefore, the nature of the problem at hand needs to be accounted for. Identifying the applicability range of a criterion is an essential task. Nevertheless, the NCL criterion is based on maximal principal stress, which is dependent on both the triaxiality and the Lode angle. The NCL criterion only averages the effect of both parameters using maximal principal stress. Generally, fracture criteria with multiple parameters are applicable to a wider range of stress states. One of them is the Oyane criterion, the other criterion explored in this work. The Oyane criterion only includes stress triaxiality dependence. Oyane prefers triaxiality, which seems to be more significant for crack initiation and evolution.

The stress state in a material is affected by external loads. Tensile stress leads to damage more rapidly than compression. Physical damage is activated by microvoid nucleation and propagates through their growth and coalescence. The development of damage is studied in correlation to stress triaxiality. Stress triaxiality is defined as a ratio of the mean hydrostatic stress σ_m to the equivalent stress σ_{eq}:

$$\eta = \frac{\sigma_m}{\sigma_{eq}} = \frac{\sigma_1 + \sigma_2 + \sigma_3}{3\sigma_{eq}}, \tag{1}$$

where σ_i ($i = 1, 2, 3$) denotes the principal stress. The equivalent stress σ_{eq} is defined in terms of three principal stresses. Assuming that the material is isotropic, the equivalent stress is given by the von Mises equation:

$$\sigma_{eq} = \frac{1}{\sqrt{2}} \sqrt{\left[(\sigma_1 - \sigma_2)^2 + (\sigma_2 - \sigma_3)^2 + (\sigma_3 - \sigma_1)^2\right]}, \tag{2}$$

The equivalent strain ε_{eq} is expressed as:

$$\varepsilon_{eq} = \frac{\sqrt{2}}{3} \sqrt{\left[(\varepsilon_1 - \varepsilon_2)^2 + (\varepsilon_2 - \varepsilon_3)^2 + (\varepsilon_3 - \varepsilon_1)^2\right]}, \tag{3}$$

Many published experimental reports [17] have shown that fracture strain decreases with increasing triaxiality, although the dependence is not monotonic as in Figure 1. A peak is normally found near the 1/3 value, which corresponds to a smooth tensile test. Another factor in the occurrence of fracture is the Lode parameter. This is related to the third deviatoric invariant and reflects shear deformation. The value of the Lode parameter is 0 for plane strain conditions, +1 for axisymmetric tension, and −1 for axisymmetric compression. Data from various experimental tests plotted in a fracture locus graph provide the basis for fracture prediction, see example in Figure 1, where the dependence of the fracture strain on stress triaxiality and the Lode parameter is plotted. The black line in the graph correspond to plane stress. The critical level of accumulated damage is reached when equivalent strain exceeds a limit value.

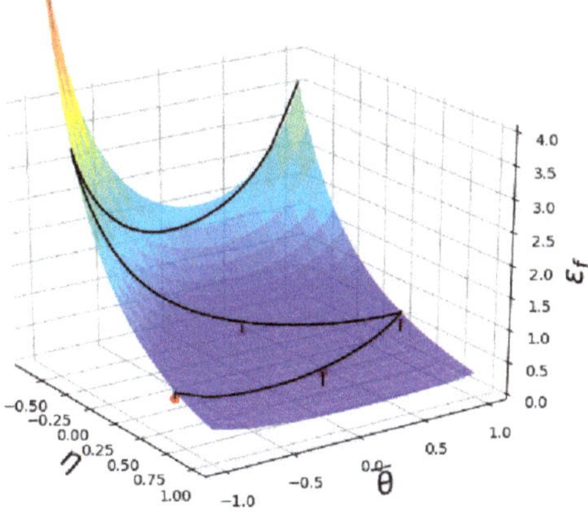

Figure 1. Representation of fracture locus in space.

In computer simulation, the critical value of a damage criterion indicates the initiation of fracture. The actual fracture strain can be determined from a simulation of the real test. In DEFORM code, the growth and coalescence of voids that lead to the evolution of fracture can be modelled by either the softening method or by element deletion. The critical damage value depends on the stress state in the test and on the fracture model used.

In the NCL criterion, the largest principal stress is the most relevant factor for the initiation of fracture:

$$D_C = \int_0^{\varepsilon_f} \frac{\sigma_1}{\sigma_{eq}} d\varepsilon , \tag{4}$$

where D_C denotes the critical damage value, ε_f is the fracture strain, σ_1 stands for the maximal principal stress and σ_{eq} is the equivalent stress. The maximal principal stress can be expressed by stress triaxiality and Lode angle [18,19]:

$$\sigma_1 = \sigma_{eq}\left(\eta + \frac{2}{3}\cos\frac{\pi}{6}(1-\bar\theta)\right), \tag{5}$$

where η is triaxiality and $\bar\theta$ is a normalized Lode angle.

The Oyane fracture criterion is based on stress triaxiality as the main reason for fracture:

$$D_C = \int_0^{\varepsilon_f}\left[1 + \frac{1}{a_0}\frac{\sigma_m}{\sigma_{eq}}\right]d\varepsilon, \tag{6}$$

where a_0 is a material coefficient to be determined experimentally and the ratio σ_m/σ_{eq} denotes the stress triaxiality.

The dimensionless normalized Lode angle [17] is defined in terms of invariants of the stress tensor:

$$I_1 = p = -\sigma_m = -\frac{1}{3}(\sigma_1 + \sigma_2 + \sigma_3), \tag{7}$$

$$I_2 = q = \sigma_{eq} = \sqrt{\frac{1}{2}\left[(\sigma_1 - \sigma_2)^2 + (\sigma_2 - \sigma_3)^2 + (\sigma_3 - \sigma_1)^2\right]}, \tag{8}$$

$$I_3 = r = \frac{27}{2}\left[(\sigma_1 - \sigma_m)(\sigma_2 - \sigma_m)(\sigma_3 - \sigma_m)\right]^{\frac{1}{3}}, \tag{9}$$

$$\bar\theta = 1 - \frac{2}{\pi}\arccos\left(\left(\frac{r}{q}\right)^3\right), \tag{10}$$

where I_i $(i = 1, 2, 3)$ are invariants of the stress tensor and $\bar\theta$ is the normalized Lode angle parameter. Only the NCL criterion has a built-in dependence on the Lode angle. The effect of the Lode angle on Oyane criterion can be done by separating the fracture coefficients for tests with different Lode angle.

4. Experiments

In the following sections, three groups of tests are described. The basic tests include tensile and compression tests. Tests for fracture coefficients determination comprised a tensile test of a notched specimen, plane strain, and shear tests. Validation tests involved notched and unnotched rings and notched cylinder specimens.

4.1. Basic Tests

Tensile testing was carried out at room temperature. The test data were used for flow stress calibration. A drawing of the smooth tensile specimen is shown in Figure 2; the dimensions are in millimetres. The figure also shows the locations and the distance of the landmarks (red crosses) monitored by an optical (laser) extensometer. The rate of loading for smooth tensile specimens was 2 mm/min. The initial strain rate was 0.001 s^{-1}. Elongation was measured by optical and mechanical extensometers but only the optical extensometer data were used for calculations.

The other tests described below were recorded using a DIC (Digital Image Correlation) system, except for the tensile tests of smooth and notched specimens, which were measured by an optical extensometer. A stochastic pattern was applied by airbrush on samples recorded with the DIC system. The actual plastic strain on the specimen surface was calculated using a correlation algorithm implemented in GOM Aramis software [20]. The resolution of cameras was 4248 × 2832 pixels. The frame rate was 5 Hz.

Cylindrical specimens 10 mm in diameter and 15 mm in height were used for compression testing. The locations of and the distance between the landmarks for load-displacement measurement are indicated with red crosses as indicated in Figure 3. The displacement was measured using DIC. The camera was pointed at the specimen and dies, onto which a pattern was applied by spraying to provide contrast for strain measurement and for calculation by means of GOM Aramis software.

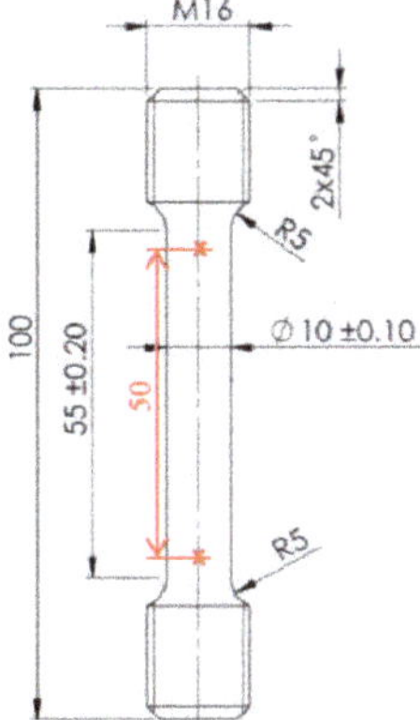

Figure 2. Drawing of tensile specimen.

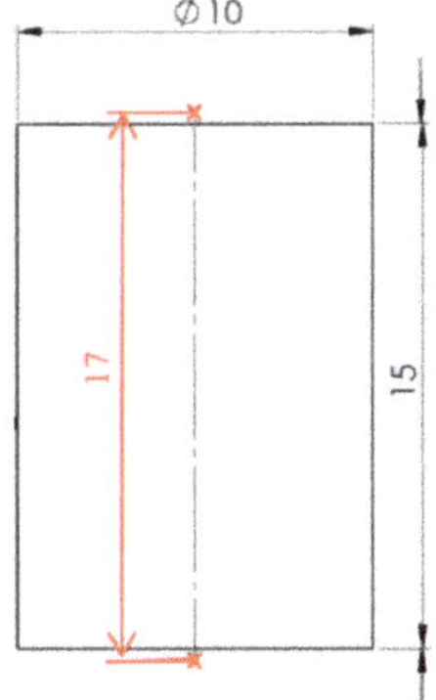

Figure 3. Drawing of compression specimen. The distance shown in red refers to the dies.

4.2. Tests for Fracture Coefficient Determination

The critical value of damage is estimated from plastic strain accumulated prior to fracture, which depends on the stress state. The choice of these tests for calibrating the fracture criteria was intended to cover a wide range of triaxialities and was based on literature survey and previous experience of the authors. They included the notched tensile test (Figure 4), shear test (Figure 5), and plane strain test (Figure 6).

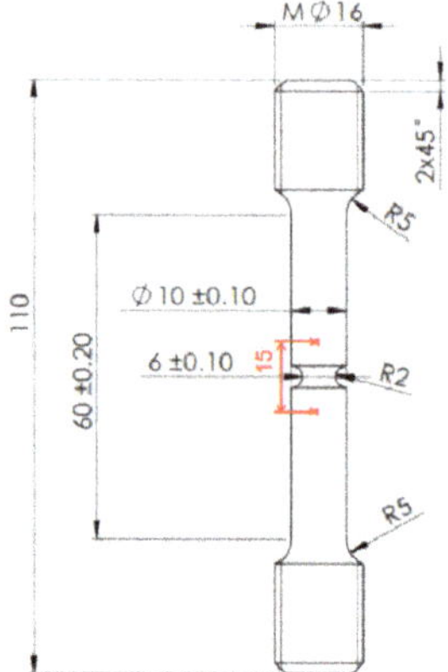

Figure 4. Drawing of notched tensile test specimen.

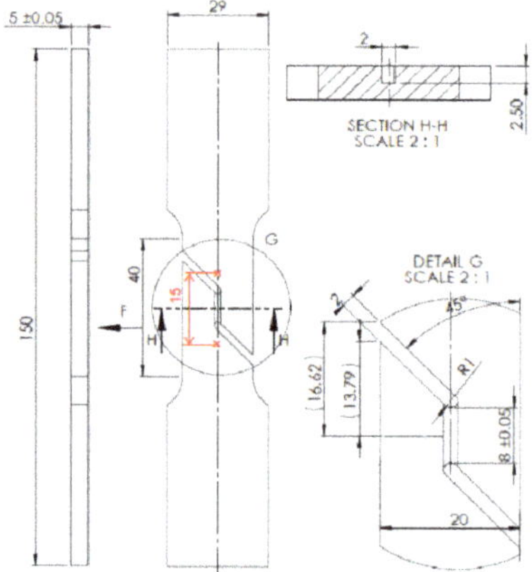

Figure 5. Drawing of shear test specimen.

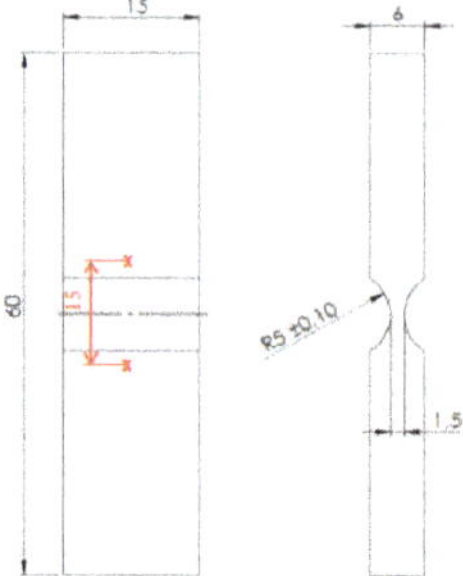

Figure 6. Drawing of plane strain test specimen.

Tensile testing of notched specimens at a loading velocity of 0.25 mm/min produced the same strain rate as in the previous tests. Optical and mechanical extensometers were used for measuring the gauge length by the red crosses as indicated in Figure 4. Fractured tensile specimens after testing are shown in Figure 7.

Figure 7. Notched tensile test specimens.

The purpose of the shear test and plane strain test was to cover a wider range of stress states for this study. The loading velocity was set at 0.5 mm/min in order to obtain the same strain rate in the relevant region as in notched tensile test. In both tests, extension was measured by DIC. Drawings of shear test and plane strain test specimens are presented in Figures 5 and 6, respectively. The gauge lengths are indicated by red crosses. Some fractured specimens with their surface patterns are shown in Figures 8 and 9.

Figure 8. Shear test specimens.

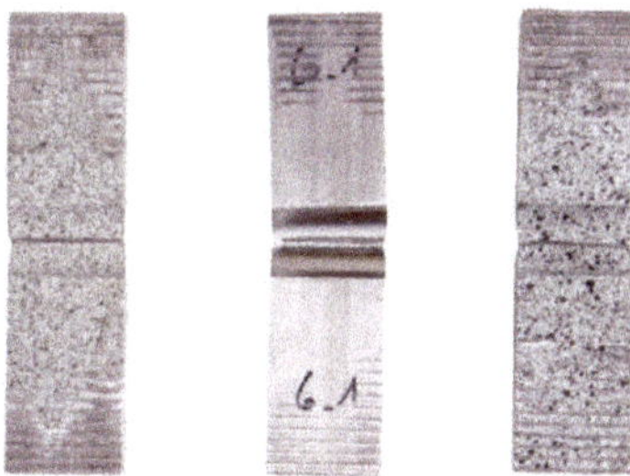

Figure 9. Plane strain test specimens.

4.3. Validation Tests

The proposed fracture criteria parameters were validated using two simple tests. One involved radial compression of notched and unnotched rings, see Figures 10 and 11. The other was axial compression of a notched cylinder, as shown in Figure 12.

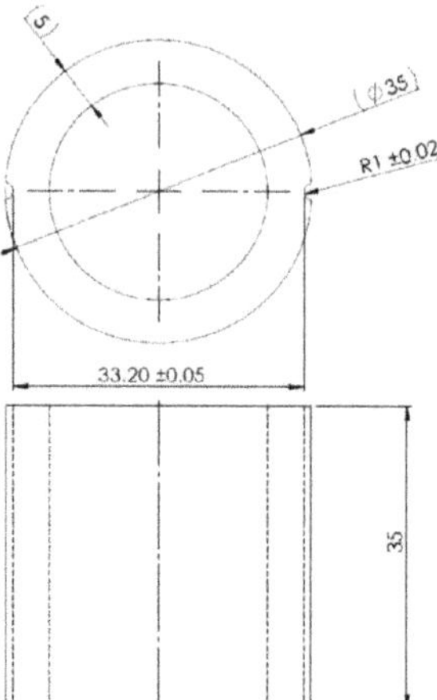

Figure 10. Drawing of notched ring specimen.

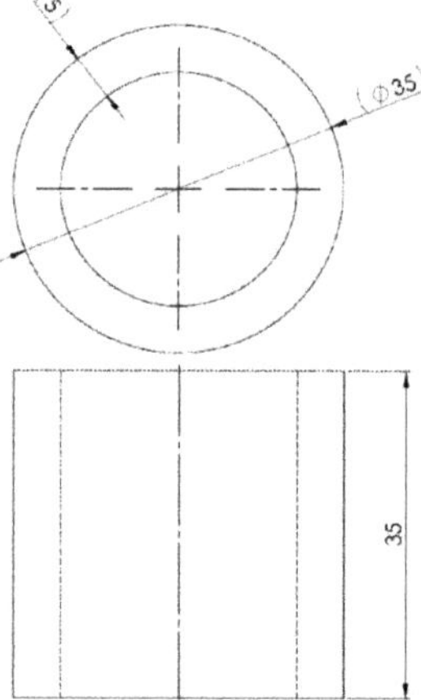

Figure 11. Drawing of unnotched ring specimen.

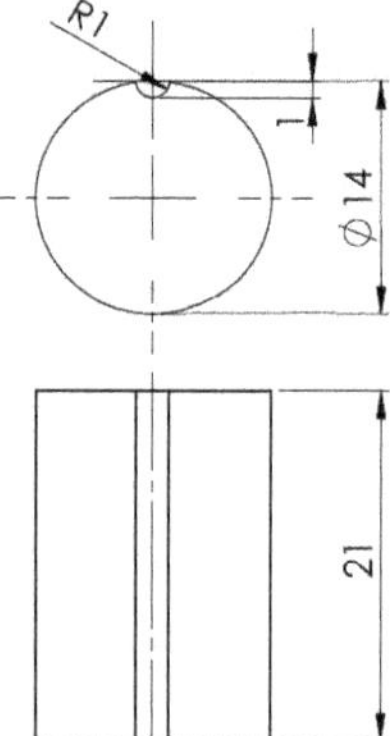

Figure 12. Drawing of the notched cylinder.

Based on an extensive literature survey, it was found that the presented geometry of test specimens is easy to manufacture and has a strong capacity for fracture initiation. These tests are not standardized, and are used for engineering purposes, such as assessing the damage mechanisms in pipes with and without defect, or the load response of a round bar with a defect.

A loading velocity of 2 mm/min was applied at room temperature. A DIC system was used for measuring the specimen displacement. Photographs of notched and unnotched specimens after testing are shown in Figure 13. Notched cylinders after axial compression are displayed in Figure 14. The cracked specimens clearly demonstrate that the failure was a shear-dominated process in this case.

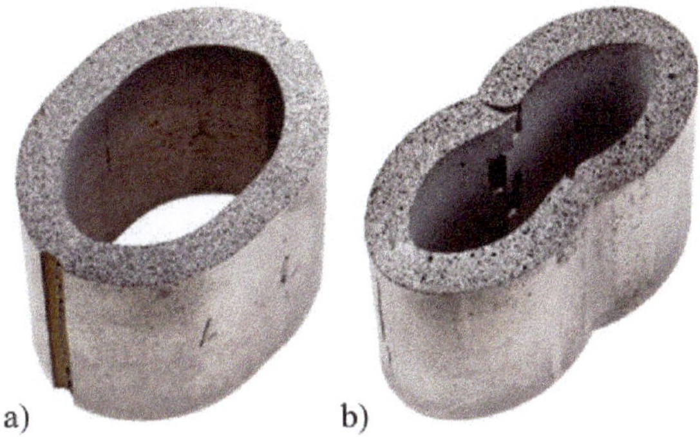

Figure 13. (**a**) Notched and (**b**) unnotched rings after testing.

Figure 14. Notched cylinder specimens after testing with fractures in the notch region.

5. Numerical Simulation Using DEFORM Software

5.1. Construction of Plasticity Models

Stress–strain (flow stress) curves for the material were calibrated against data from the smooth tensile and compression tests. Obtaining flow stress curves from uniaxial tensile test data up to the onset of necking is relatively straightforward. This technique has been reported by several authors [21]. However, a disadvantage of tensile testing is that its flow stress data applies at very small strains only.

Much higher strain values are attained in compression than in tension. In compression tests, friction can be a significant issue. It should be reduced as much as possible to prevent specimen barrelling. Using compression test data to expand the flow stress range obtained from tensile tests is a common technique. The resulting flow stress curve is then more accurate at larger strains. The complete set of flow stress data used for simulations in this study is summarized in Table 3 and plotted in Figure 15.

Table 3. Flow stress data in a tabular form.

Strain (-)	Flow Stress (MPa)	Strain (-)	Flow Stress (MPa)
0	978	0.2	1210
0.005	1004	0.3	1230
0.01	1025	0.4	1242
0.015	1046	0.5	1255
0.02	1068	0.6	1268
0.03	1100	0.7	1282
0.04	1122	0.8	1295
0.05	1135	1	1325
0.06	1150	1.15	1348
0.1	1180	1.2	1356

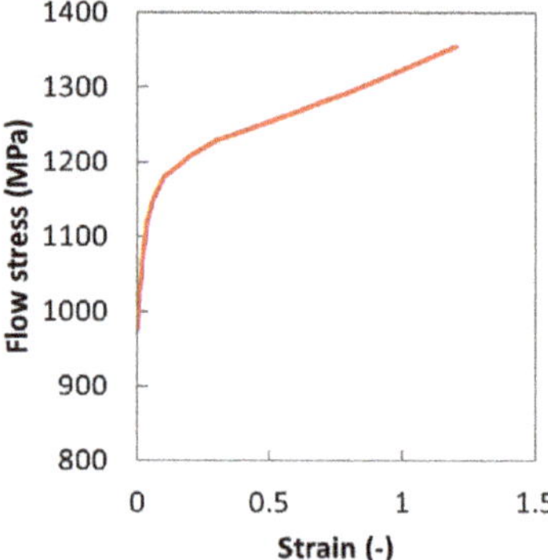

Figure 15. Flow stress plot.

Using the flow stress data calculated from tensile and compressive test data, these tests were simulated, replicating the real test conditions. Force-extension diagrams with the actually measured curves and the curves obtained by simulation using fitted flow stress data are shown in Figures 16 and 17 for the smooth tensile test and the compression test, respectively. Both tests were modelled as axisymmetric problems. A fine mesh with elements of 0.25 mm in size was used in the region where fracture was expected to occur. The reason was that strain gradient becomes steep in this region. Remeshing was applied when the specimen shape changed.

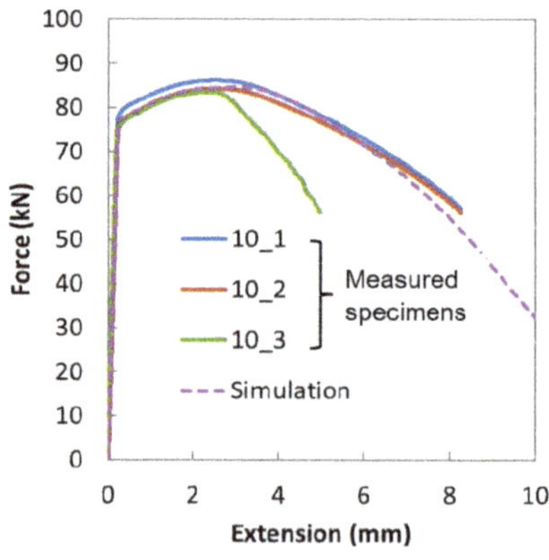

Figure 16. Comparison of tensile test curves.

None of the smooth tensile specimens fractured at mid-length. Specimens 10_1 and 10_2 exhibited almost identical trends, see Figure 16. Specimen 10_3 fractured earlier than the other two. All the measured compression test curves were almost identical. The simulated load response is in very good agreement with the measured one. The aim of these simulations was to verify the plasticity data to be used for modelling. No fracture criteria were considered at this stage.

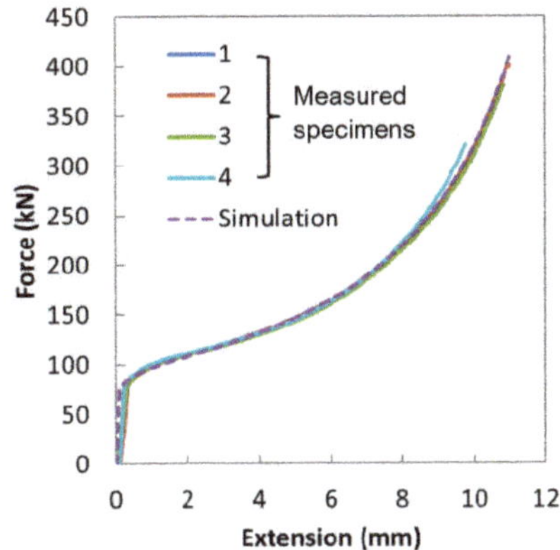

Figure 17. Comparison of compression test curves.

5.2. Identification of Fracture Parameters

To determine the critical damage threshold for simulation, experimental fracture data are needed [22]. Fracture is assumed to occur when the force-displacement curve begins to drop. This indicates fracture initiation and the amount of fracture strain. However, since the specimen deforms rapidly during necking, the exact fracture strain value it is very difficult to identify from measured test plots. An estimate of fracture strain can also be obtained by measuring the cross-sectional area of a specimen before and after testing [23]. Using FE simulation of tests, the onset of fracture and the fracture strain can be determined more accurately.

The measured and simulated forces are compared in Figure 18 for the notched tensile test, in Figure 19 for the shear test, and in Figure 20. for the plane strain test. These simulations were based on flow stress data from the previous tests. The measured and simulated forces are in very good agreement for all the tests, which means that the flow stress data from smooth tensile and compression tests had been determined accurately. The conditions and constraints used in the simulations of notched tensile test, shear and plane strain tests replicated the actual test conditions. A fine mesh with elements 0.15–0.3 mm in size was used at the mid-length region of the specimen.

The amount of fracture strain in the simulated notched tensile test can be found at the notch tip where failure was expected to occur. The point is indicated by an arrow in Figure 21. In the shear test simulation, a point on the specimen surface was chosen for this purpose where strain reaches the highest values, as in Figure 22. The location of the highest strain in the plane strain specimen is not on the surface but inside the specimen, see Figure 23. The moment when the crack is detected by DIC on the surface of the plane strain specimen may be different. The average strain on the cross-section of the specimen can be smaller than the fracture strain.

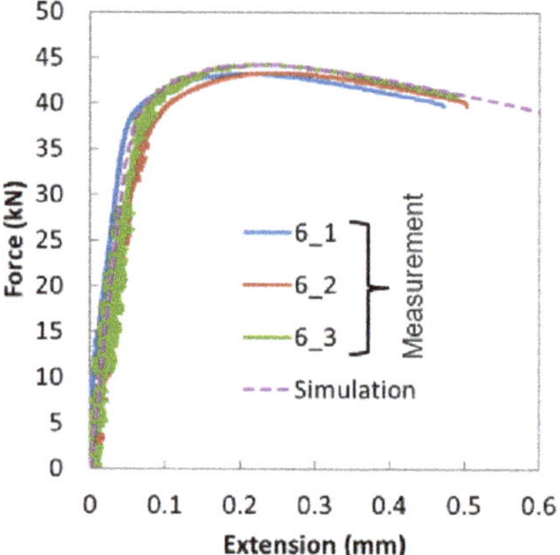

Figure 18. Comparison of forces for notched tensile test.

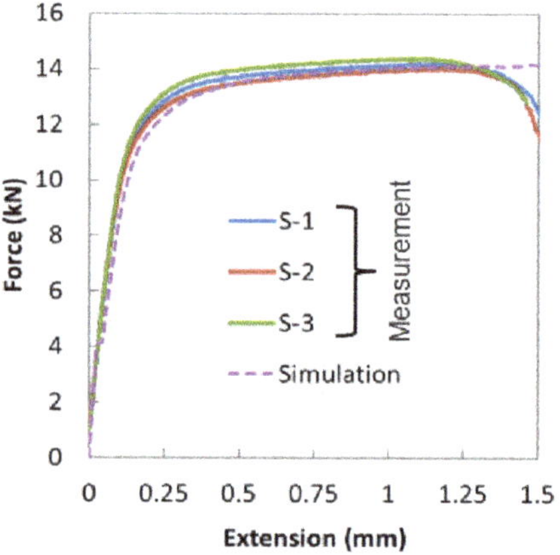

Figure 19. Comparison of forces for shear test.

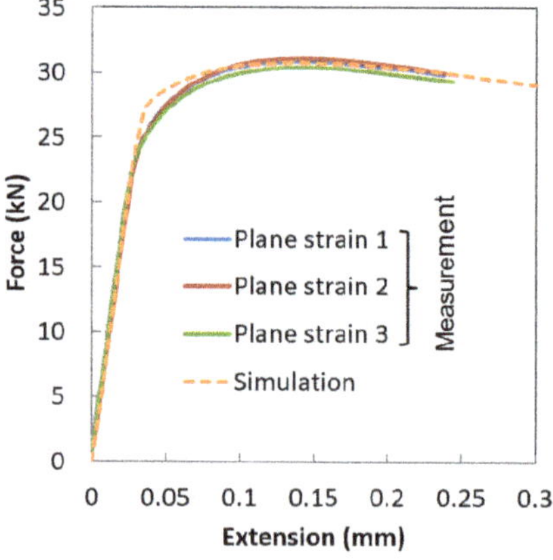

Figure 20. Comparison of forces for plane strain test.

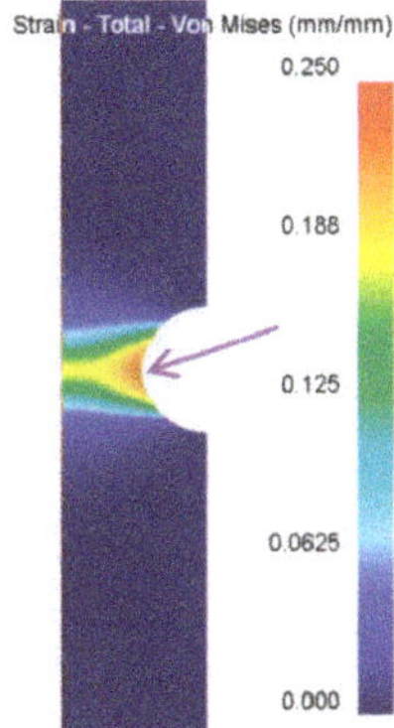

Figure 21. The fracture location in 2D simulation of notched tensile test.

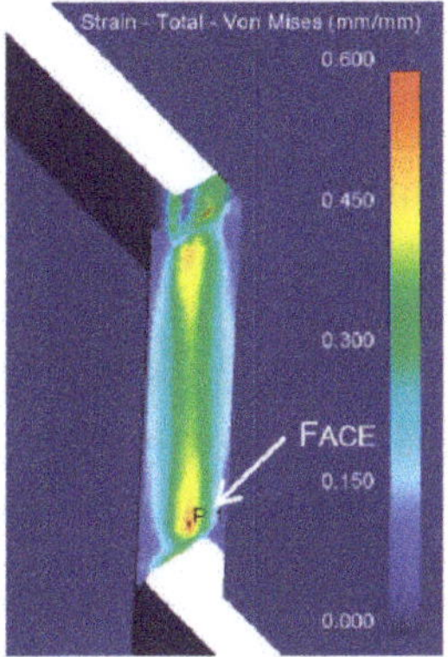

Figure 22. The fracture point in simulation of the shear test.

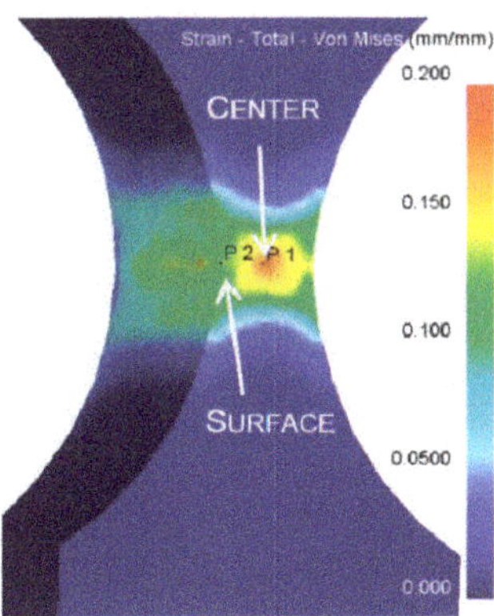

Figure 23. The fracture point in simulation of the plane strain test (longitudinal section).

Fracture strain at crack initiation, as determined from simulations calibrated against the actual test data, are listed in Table 4. Cracking was considered to initiate when the measured force curve either dropped off or began to deviate from the simulated force curve. This time instant and location indicate the critical damage value for the NCL criterion which is characterized by normalized Lode angle and triaxiality. The parameters can be obtained directly in the DEFORM simulation data. The critical damage value is identical, accidentally, for the notched tensile specimen and the plane strain specimen.

Table 4. Critical damage values identified by simulation.

Specimen Type	Fracture Strain (-)	NCL Critical Damage (-)	Average Triaxiality (-)	Normalized Lode Angle (-)
Notched tensile specimen	0.22	0.23	0.40	0.91
Shear test specimen	0.55	0.38	0.06	0.16
Plane strain specimen	0.18	0.23	0.71	0.00

The calculated triaxiality paths at selected points on the specimens are plotted in Figures 24–26. The triaxiality is changing significantly during the stress evolution, so an average value is usually used and recommended. The average value of triaxiality η_{av} is calculated based on the following equation:

$$\eta_{av} = \int_0^{\varepsilon_f} \frac{\eta}{\varepsilon_f} d\varepsilon .$$

(11)

The normalized Lode angle paths at identical points in test specimens are plotted in Figures 27–29. The normalized Lode angle for the notched tensile test is close to 1. In the shear test, the normalized Lode angle is close to 0, similar to in the plane strain test. These and the average stress triaxiality values indicate that the choice of these tests was appropriate to reflect the different stress states in the material. All average triaxiality and Lode angle values are given in Table 2.

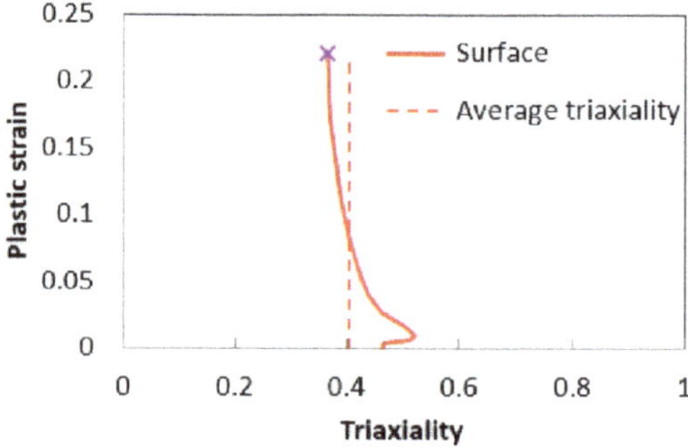

Figure 24. Triaxiality path for notched tensile test.

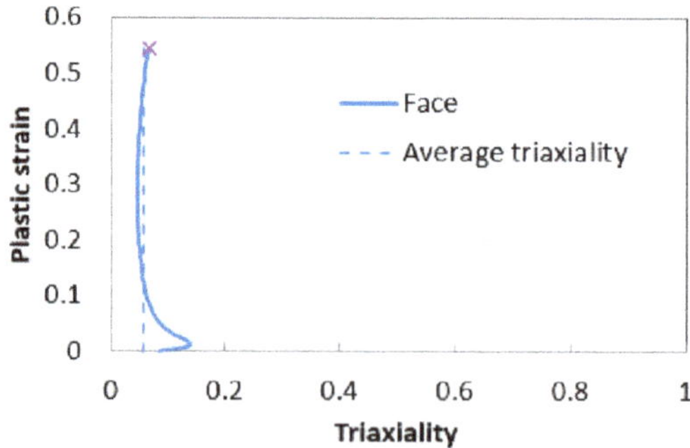

Figure 25. Triaxiality path for shear test.

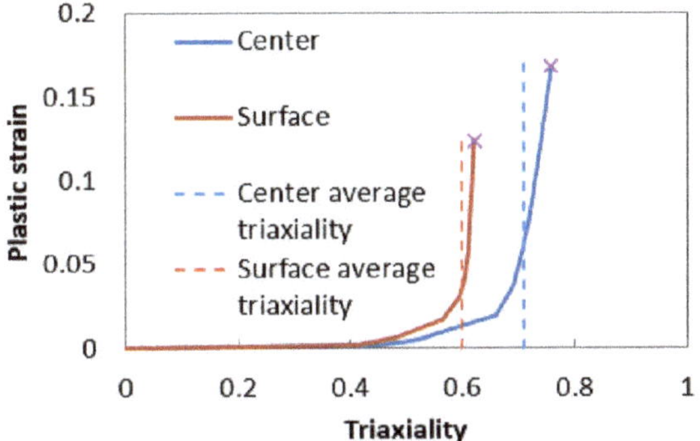

Figure 26. Triaxiality paths for the plane strain test at two points.

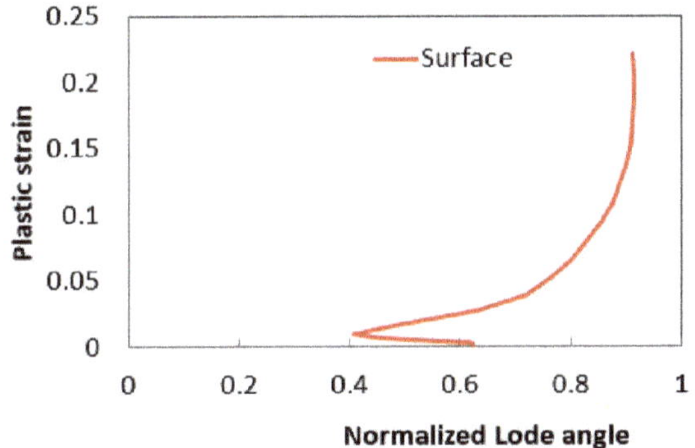

Figure 27. Lode angle for notched tensile test.

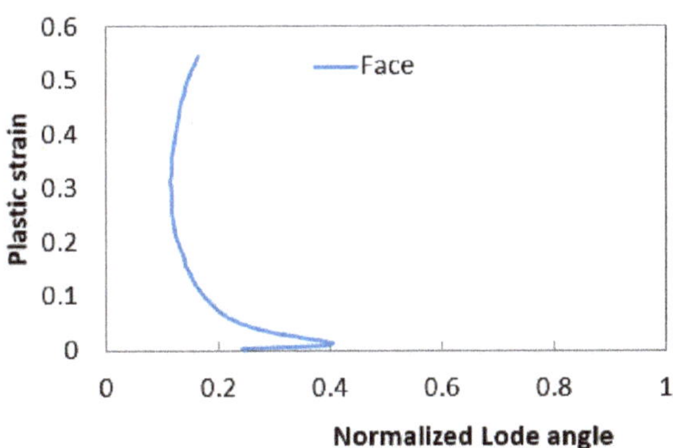

Figure 28. Lode angle for shear test.

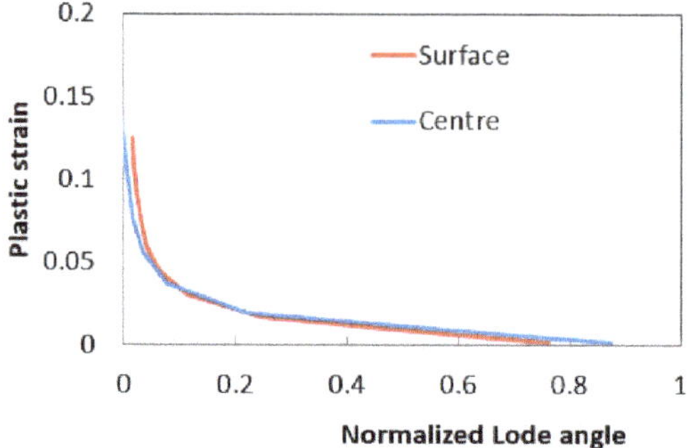

Figure 29. Lode angle for plane strain test.

The Oyane fracture criterion is based on stress triaxiality. The parameters of the Oyane criterion can be evaluated based on literature data [24]. The equation for the criterion can be rewritten as:

$$\varepsilon_f = D_C - \frac{1}{a_0} \int_0^{\varepsilon_f} \frac{\sigma_m}{\sigma_{eq}} d\varepsilon, \tag{12}$$

This expression presents a linear relationship between the fracture strain and the integral:

$$\int_0^{\varepsilon_f} \frac{\sigma_m}{\sigma_{eq}} d\varepsilon, \tag{13}$$

which can be obtained from numerical simulation. The graphic representation of Equation (12) is shown in Figure 30. The material constants for the Oyane criterion can be determined from a linear fit to points for the individual tests. Material constant a_0 can be derived from the slope of the line as its negative inverse value. Dc is in fact the value at the intersection of the line with the ordinate. The values determined in this manner were $Dc = 0.63$ and $a_0 = 0.26$.

When Lode angle is considered for the Oyane criterion, see Figures 27–29, the notched tensile test becomes the least relevant of the three tests. Lode angle for the notched tensile test is far from the other two tests. It seems appropriate to have the values of Lode angle close to each other, see the dependence illustrated in Figure 1. In an alternative course of calculation, the notched tensile test was omitted from the evaluation of material constants. The resulting values $Dc = 0.65$ and $a_0 = 0.27$ do not differ substantially from the previous pair of values.

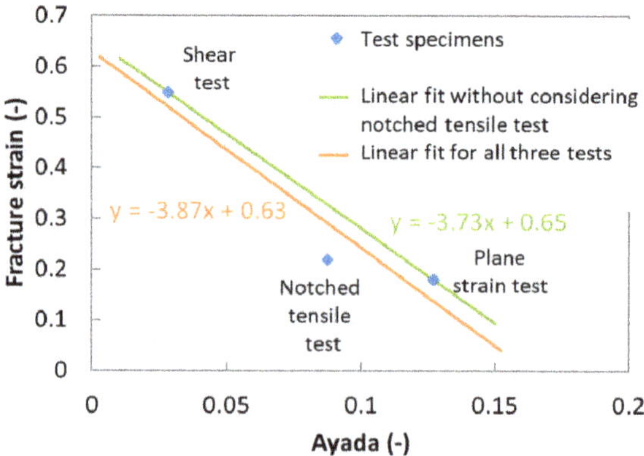

Figure 30. Relationship between $\int_0^{\varepsilon_f} \frac{\sigma_m}{\sigma_{eq}} d\varepsilon$ and ε_f.

Figure 31 shows a graphic representation of the Oyane criterion along with the values from Table 4. in a plot of fracture strain versus triaxiality.

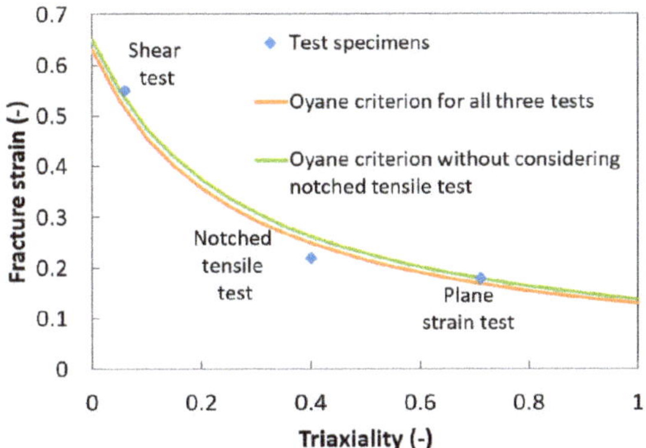

Figure 31. Graphic representation of the Oyane criterion.

The damage coefficients for the Oyane criterion are summarized in Table 5.

Table 5. Critical damage values for the Oyane criterion.

Specimen Type	Critical Damage D_C (-)	Material Coefficient a_0 (-)
All specimens	0.63	0.26
Shear and plane strain only	0.65	0.27

6. Validation of Criteria Coefficients and Discussion

The main purpose of numerical analysis of damage is to predict the failure location and time in a part under load. The validity of the above-calculated critical damage values and material constants can be assessed by comparing numerical simulations with experimental data. This validation was based on the third group of experimental tests: notched and unnotched ring specimens and notched cylinder specimens. All these tests were simulated using NCL and Oyane fracture criteria with accurate replication of the test conditions.

6.1. Radial Compression of Ring Specimens

Two types of ring specimens were tested: notched and unnotched ones. The primary purpose of the notches was to ensure the ring fractures. However, once a simulation of compression of an unnotched ring proved that it would suffer fracture as well, an additional actual test of an unnotched ring was performed to validate the simulation. The simulation models are shown in Figures 32 and 33.

As expected, the numerical analysis confirmed that fracture in the notched ring starts at the notch tip. The fracture propagates inside the ring with the increasing load, opposite the line of contact with the dies. In the unnotched ring, fracture initiates first inside the ring, opposite the outer line of contact with the dies. Then, as loading continues, another fracture occurs on the outer surface where the ring bulges out. Fracture propagation was represented by the softening method implemented in the DEFORM code.

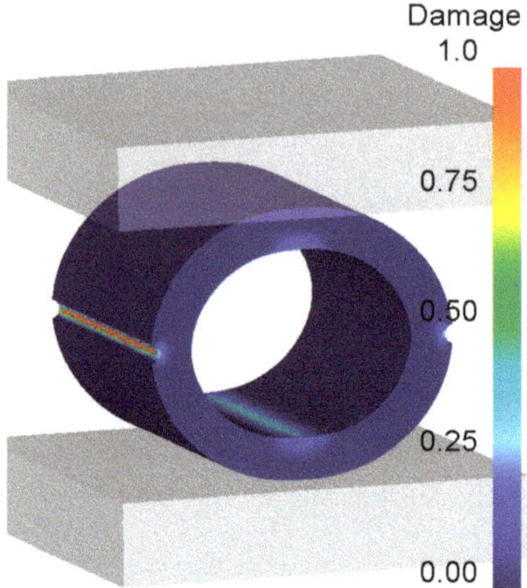

Figure 32. Simulation model of compression of a notched ring.

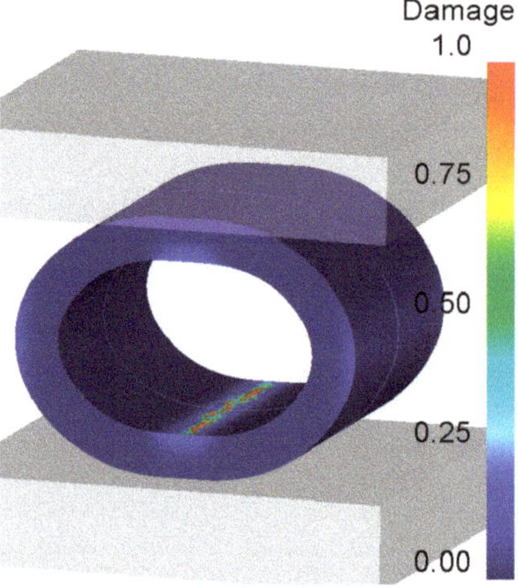

Figure 33. Simulation model of axial compression of an unnotched ring.

The simulation model with the NCL criterion had two versions, each for one of the two critical damage values listed in Table 4. The forces from these simulations and from the experiments are compared in Figure 34. The calculated material constants for the Oyane criterion, as presented in Table 5, were substituted into simulation and the resulting simulated forces were compared with the measurement data, as shown in Figure 35.

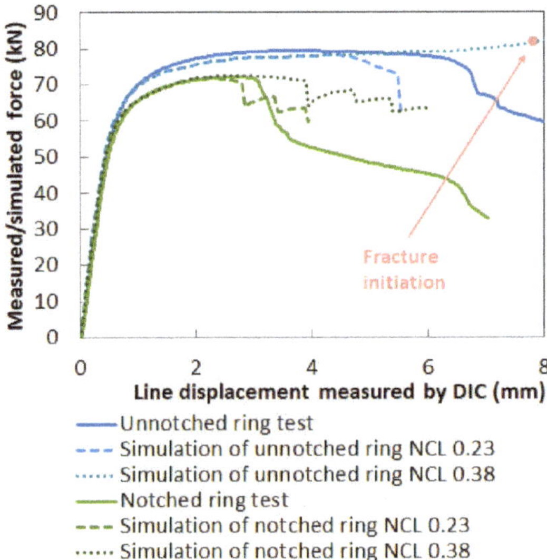

Figure 34. Measured forces in radial compression of rings and the forces simulated with the use of the NCL fracture criterion.

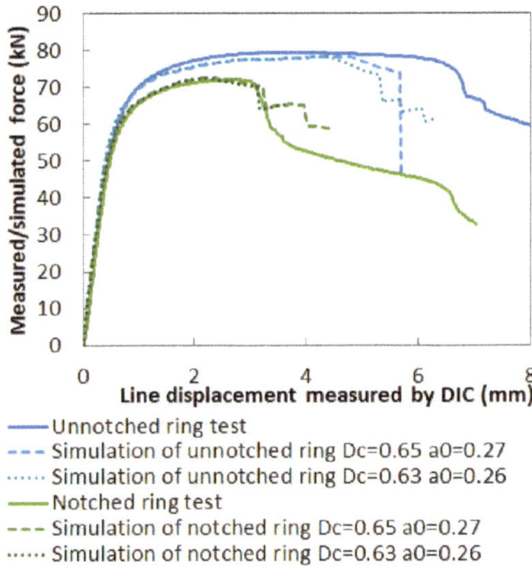

Figure 35. Measured forces for notched and unnotched rings and the forces simulated with the use of the Oyane fracture criterion.

The location of the fracture initiation was predicted correctly by simulation with both the NCL and Oyane fracture criteria. For the notched ring test and the NCL criterion, the simulation model with the value of 0.23 predicted a premature fracture, whereas the one with the value of 0.38 predicted failure later than in the experiment. The actual unnotched ring test and its simulation yielded similar results. However, no drop occurred on the simulated force curve for the critical damage value of 0.38. In the simulations, the time of fracture initiation was understood to be the moment the stress decreased, as indicated by the softening method. The point at which the stress begins to fall is indicated by the red arrow in Figure 34. It illustrates an important finding that although a drop in the simulated force demonstrates that fracture is present, the actual crack initiation might have occurred earlier. During these experiments, the specimen surfaces had to be carefully observed in order to detect fracture. However, the DIC camera was aimed at the specimen face to measure the displacement. The other surfaces were only observed with the naked eye. Hence, the exact time of crack initiation was not determined.

The NCL critical damage value 0.38 in Table 4 applies to the shear test. The value for the notched ring test is the same as the value for the tensile test. The more appropriate value appears to be 0.23, which predicts a slightly earlier failure. The critical damage for the NCL criterion depends significantly on the stress state. This criterion should therefore be used with care.

The simulation of notched ring compression with the Oyane criterion showed very good agreement with the experiment on the notched ring. However, the failure prediction for the unnotched ring was earlier than in the actual test. The calculations with different coefficients gave almost identical results, indicating that the dependence on Lode angle has of little importance for this material with respect to the used triaxiality range from around 0 to 0.7, see Table 4. This result was predictable due to the small difference in the coefficients as shown in graphical representation of Oyane criterion in Figure 31.

6.2. Axial Compression of a Notched Cylinder

Axial compression of a notched cylinder was also used for validating both criteria (Figure 36). The measured and simulated forces for the NCL and Oyane criteria are plotted and compared in Figure 37. In this test, failure is not indicated by a drop in the test force. In the actual tests, the moment of fracture initiation was found by means of DIC measurement. In simulations, the fracture propagation is represented by the softening method. Hence, the crack initiation is identified as a decrease in stress in the relevant region.

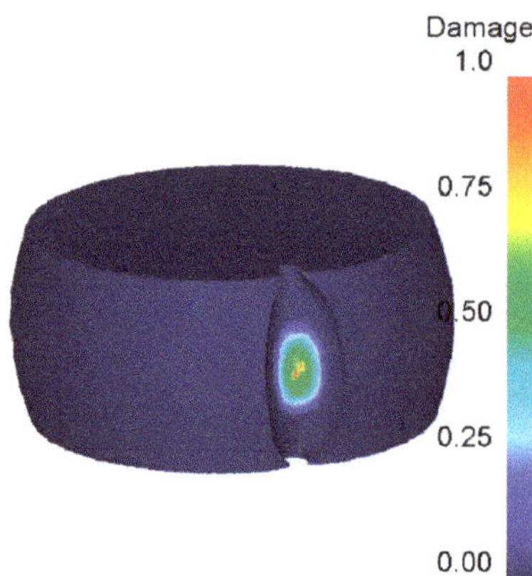

Figure 36. Simulation model of notched cylinder compression.

In the notched cylinder specimen, the NCL criterion predicted cracking rather early for $Dc = 0.23$. The simulation with $Dc = 0.38$ shows better agreement with the actual test data. With reference to the critical values in Table 4, the NCL value from the shear test ($Dc = 0.38$) appears to better predict fracture in this shear dominated process.

Fracture predictions for notched cylinder obtained with the use of Oyane criterion were very accurate, with the fracture being located in the centre of the range of measured fracture displacement values. The simulation with Oyane criterion was performed for two different coefficients, see Figure 31, that were chosen to assess the influence of the Lode angle. As expected, the difference is small for this material and the used triaxiality range, see Table 4.

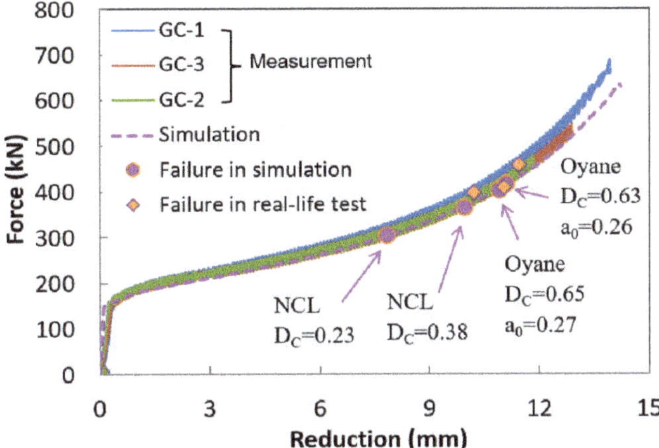

Figure 37. The comparison between measured and simulated compression forces and fracture initiation points for notched cylinder specimens.

Both validation tests showed quite a good applicability of the Oyane criterion, based on fracture parameters from the simple tests performed in this study. The Oyane criterion covers a wider range of stress states and industrial problems using two inserted parameters, as reflects the dependence on stress triaxiality. The NCL model can also predict failure rather accurately but the choice of the critical damage value is strongly dependent on the stress state in the process. The failure prediction in radial compression of rings was more accurate with $Dc = 0.23$, whereas the simulations of axial compression of notched cylinders gave better results for $Dc = 0.38$. In the rings, tensile stress dominated in fracture initiation locations. By contrast, failure in the notched cylinders was shear dominated. Therefore, a different critical damage value was needed for a reliable failure prediction.

7. Conclusions

Tensile and compression tests were conducted to determine plasticity data for DIN 34CrNiMo6 steel. Notched tensile tests, shear tests and plane strain tests were used for finding critical damage values for the Oyane and NCL criteria. Fracture prediction simulations were carried out using DEFORM code. The choice of tests for sufficiently precise fracture prediction was dictated from an engineering point of view. Plasticity and fracture models were validated using simple tests: radial compression of notched and unnotched rings and axial compression of notched cylinders.

Reducing the inaccuracy of the material flow stress is an important issue for increasing accuracy in subsequent predicting the fracture behaviour. Both fracture criteria predicted the failure locations accurately in the validation tests. The time of fracture was predicted more accurately by the Oyane criterion for both tests. The Oyane criterion can be used for failure prediction in industrial problems without any extra experience of and knowledge with measuring the data for material models.

The NCL criterion lacks general applicability, unlike the Oyane criterion. The effect of both triaxiality and Lode angle in the NCL criterion is hidden in maximal principal stress. The NCL criterion does not allow a damage prediction for arbitrary processes but only for similar processes in which the stress state of material does not differ significantly for the critical damage value. The results indicated that it is essential to identify an applicability range for each critical damage value selected for the NCL criterion.

This paper can serve as a guide for fracture evaluation based on a limited number of simple tests. The effectiveness of this procedure was demonstrated above. Using the fracture criteria available in the DEFORM software, one can obtain the desired results without any programming tasks.

Follow-on investigations will involve validating both fracture criteria for other types of specimens. In addition, other fracture criteria will be examined in order to improve failure prediction in industrial conditions.

Author Contributions: Data curation, M.R.; funding acquisition, M.Z.; investigation, I.P.; methodology, M.Z. and J.D.; project administration, J.D.; writing—original draft, I.P. All authors have read and agreed to the published version of the manuscript.

Funding: European Regional Development Fund: CZ.02.1.01/0.0/0.0/17_048/0007350. European Regional Development Fund: CZ.02.1.01/0.0/0.0/16_019/0000836.

Acknowledgments: The paper was developed with a support from the ERDF project Pre-Application Research of Functionally Graduated Materials by Additive Technologies, No. CZ.02.1.01/0.0/0.0/17_048/0007350 and from ERDF Research of advanced steels with unique properties, No. CZ.02.1.01/0.0/0.0/16_019/0000836.

Conflicts of Interest: The authors declare no conflict of interest.

References

1. Zheng, C.; de Sa Cesar, J.M.A.; Andrade Pires, F.M. A comparison of models for ductile fracture prediction in forging processes. *Comput. Methods Mater. Sci.* **2007**, *7*, 389–396.
2. Silva, C.M.A.; Alves, L.M.; Nielsen, C.V.; Atkins, A.G.; Martins, P.A.F. Failure by fracture in bulk metal forming. *J. Mater. Process. Technol.* **2015**, *215*, 287–298. [CrossRef]
3. Yu, S.; Feng, W. Experimental research on ductile fracture criterion in metal forming. *Front. Mech. Eng.* **2011**, *6*, 308–311. [CrossRef]
4. DrieMeier, L.; Moura, R.T.; Machado, I.F.; Alves, M. A bifailure specimen for accessing failure criteria performance. *Int. J. Plast.* **2015**, *71*, 62–86.
5. Bridgman, P.W. *Studies in Large Flow and Fracture*; McGraw Hill: New York, NY, USA, 1952.
6. McClintock, F.A. A criterion for ductile fracture by the growth of holes. *J. Appl. Mech. Trans.* **1968**, *35*, 363–371. [CrossRef]
7. Rice, D.; Tracey, D. On the ductile enlargement of voids in triaxial stress fields. *J. Mech. Phys. Solids* **1969**, *17*, 201–217. [CrossRef]
8. Johnson, G.R.; Cook, W.H. Fracture characteristic of three metals subjected to various strains, strain rates, temperatures and pressures. *Eng. Fract. Mech.* **1985**, *21*, 31–48. [CrossRef]
9. Wierzbicki, T.; Bao, Y.; Lee, Y.W.; Bai, Y. Calibration and evaluation of seven fracture models. *Int. J. Mech. Sci.* **2005**, *47*, 719–743. [CrossRef]
10. Kim, J.; Zhang, G.; Gao, X. Modeling of ductile fracture: Application of the mechanism-based concepts. *Int. J. Solids Struct.* **2007**, *44*, 1844–1862. [CrossRef]
11. Gao, X.; Kim, J. Modeling of ductile fracture: Significance of void coalescence. *Int. J. Solids Struct.* **2006**, *43*, 6277–6293. [CrossRef]
12. Brünig, M.; Chyra, O.; Albrecht, D.; Driemeier, L.; Alves, M. A ductile damage criterion at various stress triaxialities. *Int. J. Plast.* **2008**, *24*, 1731–1755. [CrossRef]
13. Kubík, P.; Šebek, F.; Petruška, J.; Hůlka, J.; Růžička, J.; Španiel, M.; Džugan, J.; Prantl, A. Calibration of Selected Ductile Fracture Criteria Using Two Types of Specimens. *Key Eng. Mater.* **2014**, *592–593*, 258–261. [CrossRef]
14. Španiel, M.; Prantl, A.; Džugan, J.; Růžička, J.; Moravec, M.; Kuželka, J. Calibration of fracture locus in scope of uncoupled elastic–plastic-ductile fracture material models. *Adv. Eng. Softw.* **2014**, *72*, 95–108. [CrossRef]
15. Prantl, A.; Džugan, J.; Konopík, P. Ductile Damage Parameters Identification for Nuclear Power Plants Experimental Part. *COMAT* **2012**.
16. Branco, R.; Costa, J.D.M.; Antunes, F.V.; Perdigão, S. Monotonic and cyclic behaviour of DIN 34CrNiMo6 tempered alloy steel. *Metals* **2016**, *6*, 98. [CrossRef]
17. Bai, Y.; Wierzbicki, T. A comparative study of three groups of ductile fracture loci the 3D space. *Eng. Fract. Mech.* **2015**, *135*, 147–167. [CrossRef]
18. Lou, Y.; Huh, H. Evaluation of ductile fracture criteria in a general three-dimensional stress state considering the stress triaxiality and the lode parameter. *Acta Mech. Solida Sin.* **2013**, *26*. [CrossRef]
19. Stebunov, S.; Vlasov, A.; Biba, N. Prediction of fracture in cold forging with modified Cockcroft-Latham criterion. *Procedia Manuf.* **2018**, *15*, 519–526. [CrossRef]

20. Available online: https://www.gom.com/3d-software/gom-correlate.html (accessed on 14 August 2019).
21. Poláková, I.; Kubina, T. Flow stress determination methods for numerical modelling. In Proceedings of the 22nd International Conference on Metallurgy and Materials, Brno, Czech Republic, 15–17 May 2013; pp. 273–278, ISBN 978-80-87294-41-3.
22. Džugan, J.; Španiel, M.; Prantl, A.; Konopík, P.; Růžička, J.; Kuželka, J. Identification of ductile damage parameters for pressure vessel steel. *Nucl. Eng. Des.* **2018**, *328*, 372–380. [CrossRef]
23. Ling, Y. Uniaxial true stress-strain after necking. *AMP J. Technol.* **1996**, *5*, 37–48.
24. Oyane, M.; Sato, T.; Okimoto, K.; Shima, S. Criteria for ductile fracture and their applications. *J. Mech. Work. Technol.* **1980**, *4*, 65–81. [CrossRef]

Article

Finite Element Analysis on the Temperature-Dependent Burst Behavior of Domed 316L Austenitic Stainless Steel Rupture Disc

Hongbo Zhu [1], Weipu Xu [2], Zhiping Luo [3] and Hongxing Zheng [1,*]

1   Center for Advanced Solidification Technology, School of Materials Science and Engineering, Shanghai University, Shanghai 200444, China; hbzhu3026@163.com
2   Shanghai Institute of Special Equipment Inspection and Technical Research, Shanghai 200333, China; neiltwo@126.com
3   Department of Chemistry, Physics and Materials Science, Fayetteville State University, Fayetteville, NC 28301, USA; zhipingluo@gmail.com
*   Correspondence: hxzheng@shu.edu.cn; Tel.: +86-137-6135-6748

Received: 16 December 2019; Accepted: 5 February 2020; Published: 8 February 2020

Abstract: As a safety device, a rupture disc instantly bursts as a nonreclosing pressure relief component to minimize the explosion risk once the internal pressure of vessels or pipes exceeds a critical level. In this study, the influence of temperature on the ultimate burst pressure of domed rupture discs made of 316L austenitic stainless steel was experimentally investigated and assessed with finite element analysis. Experimental results showed that the ultimate burst pressure gradually reduced from 6.88 MPa to 5.24 MPa with increasing temperature from 300 K to 573 K, which are consistent with the predicted instability pressures acquired by nonlinear buckling analysis using ABAQUS software. Additionally, it was found that a gradual transition from opening ductile mode to cleavage mode happened with increasing temperature due to more cross slips occurring under serious plastic deformation. The equivalent stress, equivalent strain and strain hardening rates acquired by static analysis were effective at rationalizing the temperature-dependent fracture behavior of the domed rupture discs.

Keywords: rupture disc; finite element analysis; burst fracture; mechanical property; austenitic stainless steel

---

1. Introduction

Rupture discs, also known as bursting discs, are widely used in the nuclear power, fire protection and petrochemical industries owing to the advantages provided by their simple structure, high sensitivity and strong venting ability [1–5]. Hydrogen has especially been regarded as a promising alternative energy source; however, safety issues associated with high-pressure hydrogen hinders its application [6]. Several studies related to the spontaneous ignition of high-pressure hydrogen during the utilization of rupture discs have been conducted [7–9]. Presently, a rupture disc can be roughly classified as a domed rupture disc, a reverse domed rupture disc and a flat rupture disc. The domed and flat rupture discs undergo tensile failure of materials, while the reverse domed rupture disc undergoes structure instability failure [10–12].

Schrank [12] investigated the instability behavior of reverse discs by putting them through vacuum cycles to simulate flushing the facility and pressure cycles as emergencies with a sudden rise in pressure, and they found that the tested discs showed no sign of degradation after vacuum and pressure cycling compared to the unstressed discs. As rupture discs are a pressure relief component, one of the critical issues is to accurately predict their ultimate burst pressure (P_b). Lake and Inglis [13] suggested to estimate the burst pressure using

$$P_b = 2.6\sigma_s \times \frac{S}{a}$$

(1)

based on a constant volume assumption, where σ_s is the ultimate tensile strength measured from a conventional uniaxial tensile test, and s and a are the thickness and interdiamater of the domed rupture disc, respectively, as shown in Figure 1. Kanazawa et al. [14] proposed another equation

$$P_b = 4\sigma_s \times \left(\frac{2}{3}\right)^n \times \frac{S}{a}$$

(2)

based on the tensile instability conditions [15], where n is the strain hardening coefficient of the material.

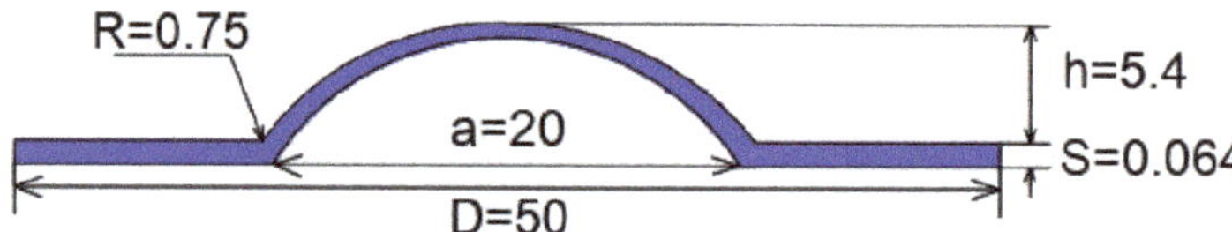

Figure 1. Illustration for the dimensions of the domed rupture disc (unit: mm).

The effect of environmental temperature can, to some extent, be indirectly identified from the tensile strength. Actually, the ultimate burst pressure of a domed rupture disc is closely linked to its relief caliber, thickness and fillet radius with regard to a particular material. Much effort has been made using commercial simulation software [16–20], for example, Jeong et al. [20] carried out a structural analysis of domed stainless-steel rupture discs from the viewpoint of materials failure, and the relationship amongst burst pressure, thickness and superficial groove depth were established; however, no experimental evidence was provided.

In the present study, we used austenitic stainless steel 316L as a model material, to predict its ultimate burst pressure. The equivalent stress, equivalent strain and strain hardening rates were also acquired using the finite element method (ABAQUS/CAE 6.14-4, Dassault Systèmes Simulia Corp., Providence, RI, USA) to better understand the temperature-dependent burst behavior of domed rupture discs.

2. Experimental Procedure

In the present study, a 316L austenitic stainless sheet steel with a thickness of 0.064 mm was adopted to fabricate the domed rupture discs. The chemical composition of the 316L austenitic stainless steel is listed in Table 1. The rupture discs were fabricated by press forming within a die with the dimensions shown in Figure 1. All the rupture discs were annealed at 473 K for 2 h to relieve stress. The bursting tests were conducted in a self-designed rupture set-up, as shown in Figure 2, at 300 K, 373 K, 473 K and 573 K, respectively, where the temperature was controlled by a K-type thermocouple welded between holders with a resolution of ±2 K, and the pressure was applied using compressed air. All the tests for each temperature were performed on at least five samples. The experimental process was as follows: (1) the rupture disc was fixed between the top and bottom holders; (2) the holders were heated to the preset temperature and held for 30 min; and (3) the pressure was applied by filling with compressed air until the fracture happened. The fracture surfaces were taken from the discs and directly examined using a scanning electron microscope (SEM; JSM-6700F, JEOL, Akishima Tokyo, Japan). The finite element analysis procedures are provided in the next section.

Table 1. The chemical composition of the studied steel (wt.%).

C	Si	Mn	P	S	Ni	Mo	Cr	Fe
≤0.03	≤1.00	≤2.00	≤0.035	≤0.03	10.00–14.00	2.00–3.00	16.00–18.00	bal.

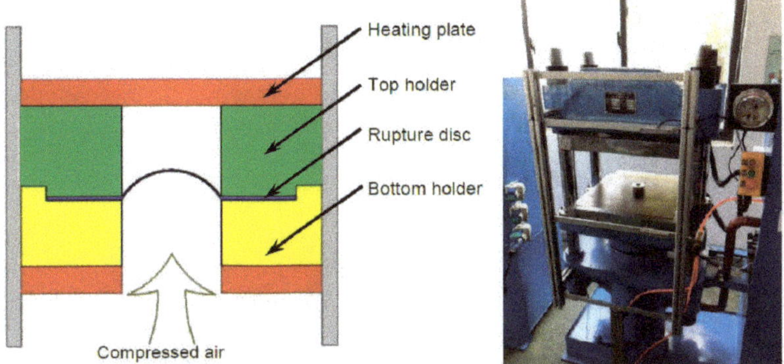

Figure 2. A schematic diagram (left) and the bursting test set-up (right).

3. Results and Discussion

The experimental results showed that all the rupture discs burst at the dome, as shown in Figure 3, and the average ultimate burst pressure was 6.88 MPa at 300 K, gradually decreasing to 6.20 MPa, 5.59 MPa and 5.24 MPa when the temperature rose to 373 K, 473 K and 573 K, respectively. The fracture morphologies are shown in Figure 4. In the case of 300 K, one could observe some parabolic (elongated) dimples and cleavage features besides equiaxed dimples near the inner surface, as marked by the yellow arrows in Figure 4a. Such features were different from the full equiaxed dimples obtained during the conventional uniaxial tensile tests [21,22] and the small punch tests (SPT) [23,24]. It could be inferred that the severe plastic deformation was first initiated by the rapid tearing stress from the outside surface, and then the final fracture changed to an opening mode caused by the direct stress. With rising temperatures, as shown in Figure 4b–d, the proportion of equiaxed/elongated dimples gradually decreased and more cleavage features appeared. The cleavage mode dominated the fracture at 573 K.

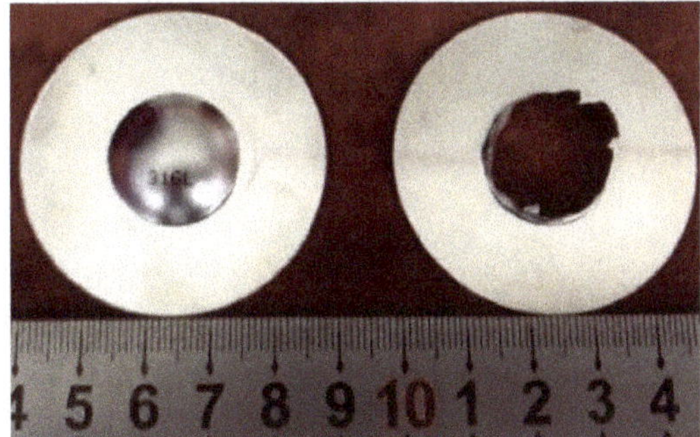

Figure 3. The domed disc before (left) and after (right) bursting.

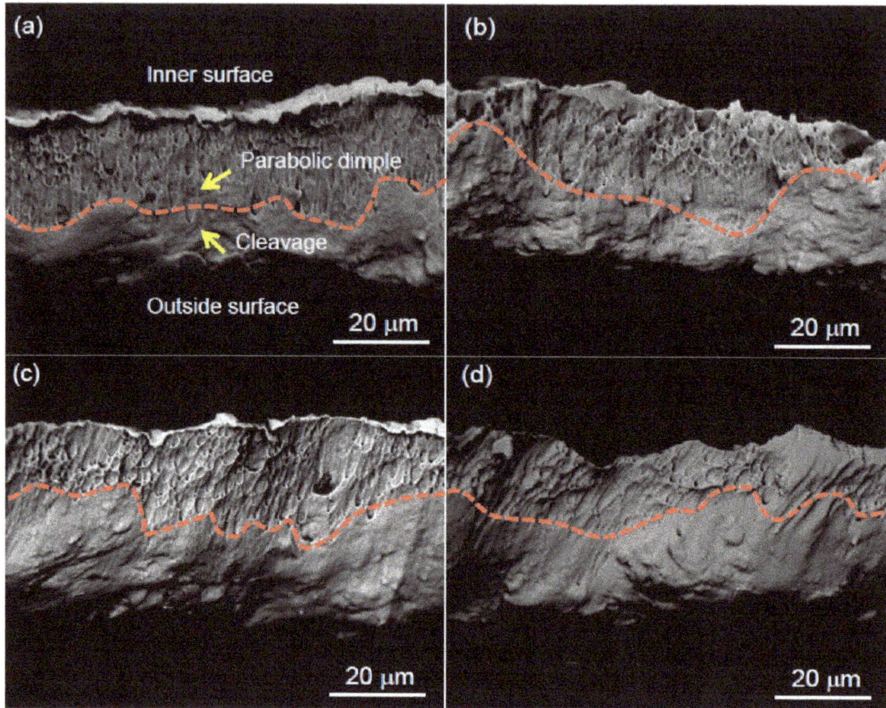

Figure 4. The fracture morphologies of the domed rupture discs at 300 K (**a**), 373 K (**b**), 473 K (**c**) and 573 K (**d**), respectively.

In the present study, finite element analyses, including nonlinear buckling analysis and static analysis during successive loading, were carried out using ABAQUS software. A nonlinear buckling analysis was performed, as the linear elastic constitutive relationship was not applicable for local buckling of the rupture disc. Static analysis was conducted to acquire the equivalent stress, equivalent strain and strain hardening rates to explain the fracture behavior. A Nlgeom option was applied in both the nonlinear buckling analysis and static analysis. The physical properties of the 316L austenitic stainless steel are listed in Table 2 [25,26].

Table 2. The physical properties of the 316L austenitic stainless steel [25,26].

Temperature	300 K	373 K	473 K	573 K
Ultimate tensile strength (MPa)	632	590	492	485
Break elongation (%)	47	36	27	26
Young's modulus (MPa)	195,000	189,000	183,000	176,000
Poisson ratio	0.3	0.3	0.3	0.3

A nonlinear buckling analysis can be used to simulate structural responses after loss of balance, and generally, it is conducted as follows: (1) to analyze eigenvalue buckling to calculate the eigenvalues and obtain buckling modes; (2) to introduce an appropriate number of eigenmodes into the perfect geometry as an initial imperfection; and (3) to obtain a reaction force (i.e., the force acting on the normal direction of the dome surface) versus the arc length curves using Riks method which was used to calculate the post-buckling strength and post-buckling mode [27–30]. A rigid body hemisphere with a radius of 12 mm was applied to provide pressure, and the flat clamped by holders was deformable, with a 360° revolution and preset thickness of 0.064 mm, as shown in Figure 5. The displacement of the hemisphere was set to be 5.4 mm to get a dome of the flat. An encastre option (neither displacement nor rotation) was applied on the edge part, and a displacement without rotation option was applied on the dome part, as shown in Figure 6a. A tet option (element shape) was chosen for the irregular and complex areas. A quadratic was adopted in the geometric order, and other options were subjected to defaults. In addition, we set the maximum load proportionality factor (the ratio of calculated load to applied load) in the stopping criteria (instability criteria) option as one and the DOF value (degree of freedom) as three (the three-dimension mode) on the top of the dome (the node region option). Meanwhile, the Riks analysis would automatically stop once the vault region was fully plastic by presetting the "other" option in the ABAQUS software. C3D10 was a 10-node quadratic tetrahedron for noncontact cases, as shown in Figure 6b. There are 57,153 elements in total and two elements along the wall thickness.

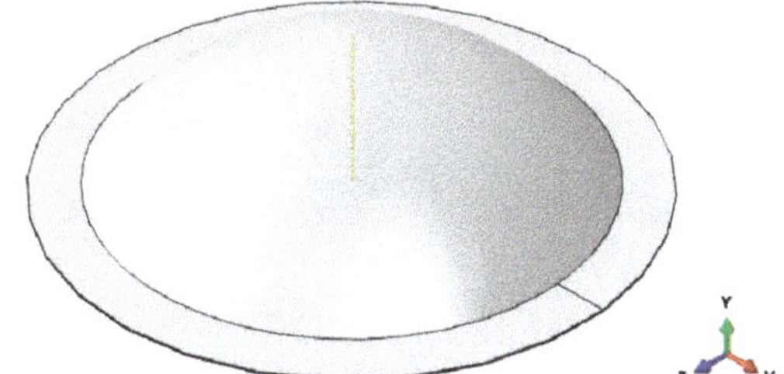

Figure 5. The finite element analysis model of the rupture disc.

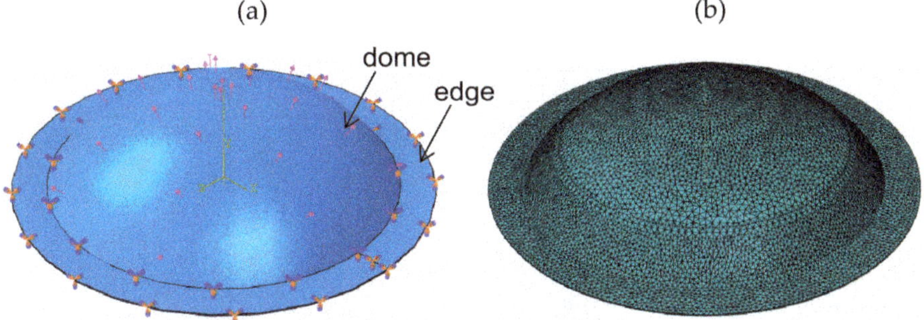

Figure 6. A schematic diagram of the load and boundary conditions set for the finite element analysis (**a**) and the corresponding finite element mesh (**b**).

A series of reaction force (*F*) distributions during successive loading at different temperatures could be obtained. Two representative cloud maps at 300 K were taken as examples, as shown in Figure 7, and the corresponding plastic deformation distributions are shown in Figure 8. One can notice that the maximum reaction force/plastic deformation appeared at the top of the dome, which agreed well with the experimental results, as shown in Figure 3. By extracting the maximum reaction force and the corresponding arc length [31,32], the relationship curves between the reaction force and arc length were plotted in Figure 9a, and then the relationship between the pressure and arc length were obtained, as shown in Figure 9b. According to Figure 9b, the instability pressures were determined to be 7.20 MPa, 6.50 MPa, 5.97 MPa and 5.60 MPa at 300 K, 373 K, 473 K and 573 K, respectively, which were close to the experimental results; here, the instability pressure was defined to be the maximum reaction force divided by the original dome area. Meanwhile, the ultimate burst pressures estimated from Equations (1) [13] and (2) [14] were integrated into Figure 10 for comparison.

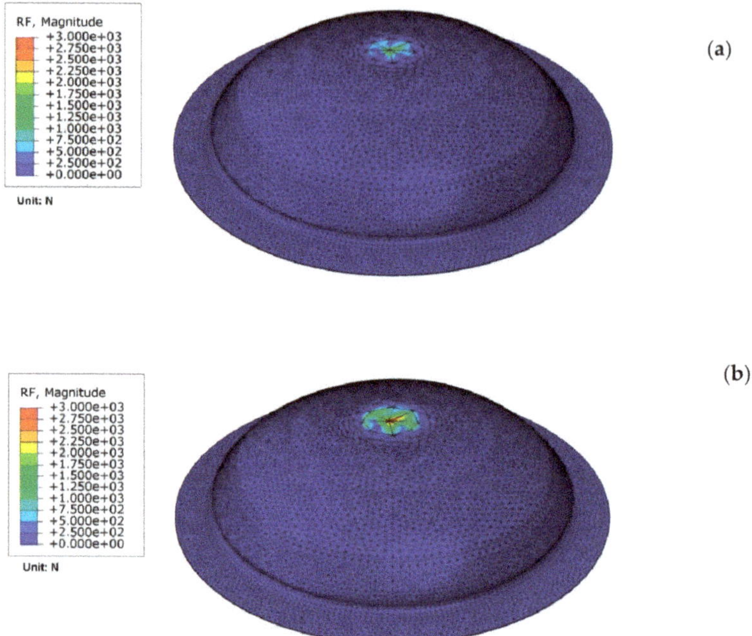

Figure 7. Two representative cloud maps of the reaction force taken from successive loading at 300 K. (**a**) F_{max} = 1500 N and (**b**) F_{max} = 2943 N.

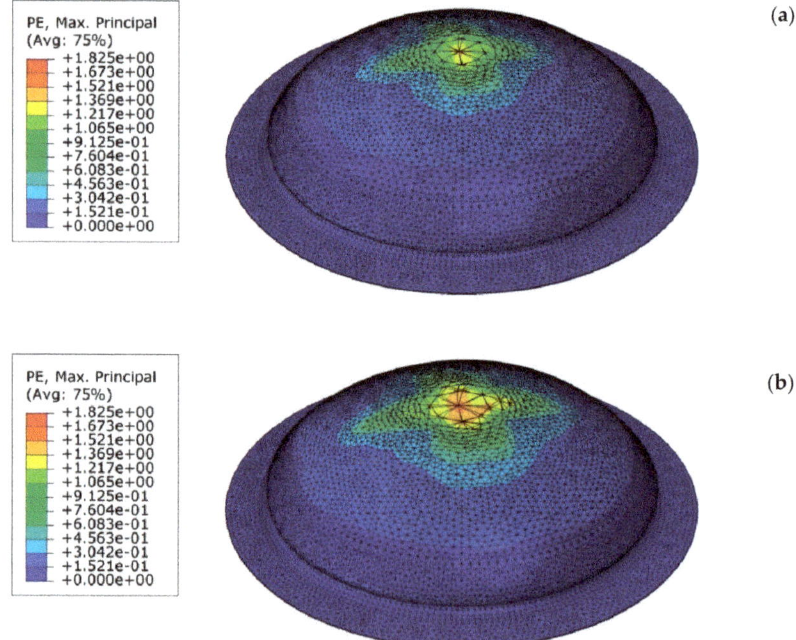

Figure 8. Two representative cloud maps of the plastic deformation distributions taken from successive loading at 300 K. (**a**) F_{max} = 1500 N and (**b**) F_{max} = 2943 N.

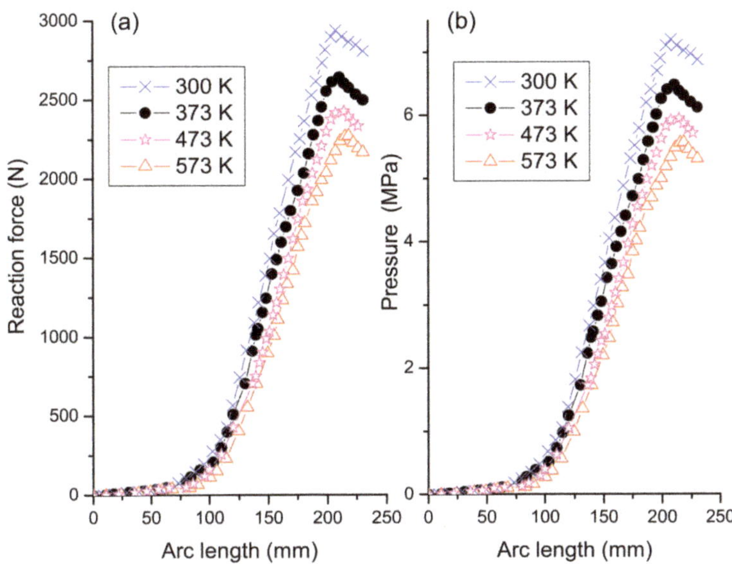

Figure 9. The relationships between the maximum reaction force: (**a**) pressure, and (**b**) arc length, at different temperatures obtained from cloud maps as shown in Figure 7.

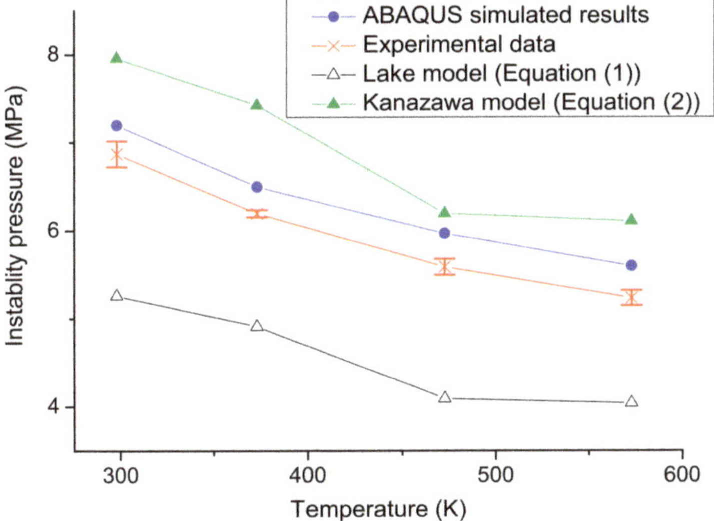

Figure 10. The ultimate burst pressures as a function of temperature for the domed rupture discs. Calculated results using Equations (1) and (2) are integrated for comparison.

According to Figure 10, all the predicted data from the Lake model (Equation (1)) seem to be less than the experimental results, while the data from the Kanazawa model (Equation (2)) are slightly higher. The deviation is explained as follows. Firstly, the Lake model [13] was derived from

$$\sigma = \frac{P \times a}{S} \frac{\left(1 + 4\varphi^2\right)^2}{16\varphi} \tag{3}$$

where σ, p, S, φ and a are equivalent stress, pressure, thickness, relative arch height (vertical displacement divided by interdiameter of the disc) and relief caliber, respectively. The equation was proposed with an assumption that the rupture disc deforms uniformly; however, it is inevitable that the vault of the dome thins the most [33]. Secondly, Kanazawa model was obtained by modifying the equations of

$$\frac{P \times a}{S \times A} = \frac{2^{2+n}\left[\ln\left(1+\varphi^2\right)\right]^n \times \varphi}{\left(1+\varphi^2\right)^3} \tag{4}$$

$$n = \left(5 - \frac{1}{\varphi^2}\right) \times \frac{\ln\left(1+\varphi^2\right)}{2} \tag{5}$$

where A is the hardness coefficient (kg/mm^2) and n is the strain hardening coefficient. Although nonuniform deformation is considered in this model by introducing an arc of a smaller radius, the φ value becomes higher and thus results in an over-predicted value.

Comparatively speaking, the predicted values of the present work using the finite element method are closer to the experimental results, indicating the reasonability of the simulation model described above. $P_{total} = P_0 + \lambda(P_{ref} - P_0)$ was introduced to the ABAQUS software to evaluate the buckling load, where P_0 is a constant value (dead load), P_{ref} is the reference load, and λ is the load proportionality factor. When conducting an ABAQUS/Riks analysis, P_{ref} is altered automatically by the ABAQUS program to obtain the corresponding P_{total} during the analysis process until P_{total} equals P_{ref}, representing that the structure reached stable ($\lambda = 1$). λ changes to maintain P_{total} equals P_{ref} if P_{ref} continues to increase when the structure has lost stability, and the nonlinear bucking strength is expressed as the product of λ and P_0 [27,32].

The equivalent stress (σ_T) and equivalent strain (ε_T) can be further determined by static analysis with the same finite element model. The experimental data of the ultimate burst pressure were applied. The equivalent stress and equivalent strain distributions of the domed rupture disc at 300 K are shown in Figure 11 as an example, and, correspondingly, the relationship between the equivalent stress and equivalent strain can be obtained by extracting data, as shown in Figure 12a. As the temperature rises, the equivalent stress in the dome part gradually decreases. Figure 12b presents the relationship between the strain hardening rate and the equivalent strain by differentiating the curves of Figure 12a. At the low strain stage before yielding, the deformation mode of the 316L austenitic stainless steel is planar slip with a high strain hardening rate [34,35]. As a face-centered cubic structured material, the deformation mode of the 316L austenitic stainless steel gradually changes from planar slip to cross slip with increasing strain, causing saturated dislocation tangles, which may exhaust the strain hardening capability [36]. On the other hand, the thermally activated process could intensify at a higher temperature, as reflected by the enlargement of the cleavage area in Figure 4; that is, the strain hardening capability could be more easily consumed, resulting in a faster burst fracture [37].

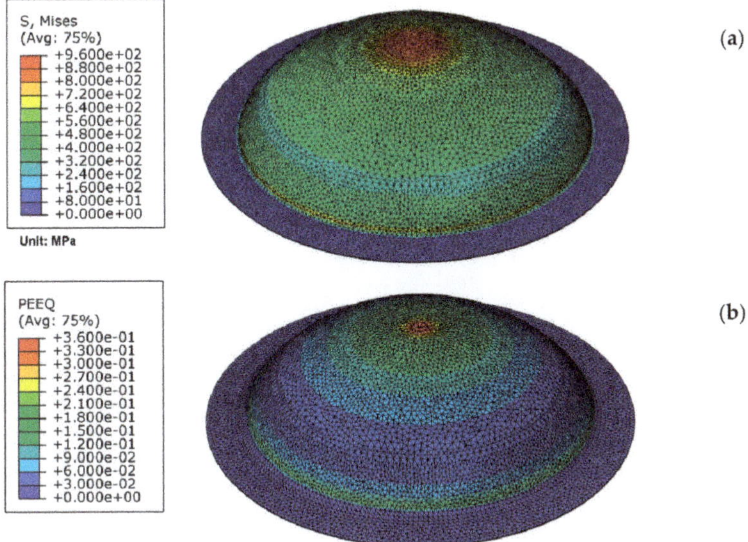

Figure 11. The equivalent stress (**a**) and the corresponding equivalent strain distributions on the domed rupture disc at 300 K (**b**).

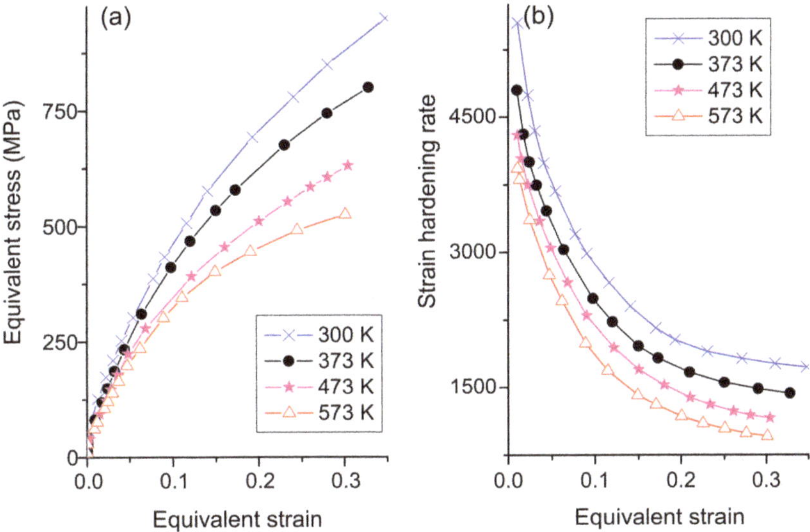

Figure 12. The equivalent strain versus equivalent stress curves (**a**) and the strain hardening rate curves (**b**) at different temperatures.

4. Conclusions

The effect of temperature on the ultimate burst pressure and fracture behavior of the domed 316L austenitic stainless steel rupture discs was investigated. Finite element analysis based on nonlinear buckling analysis and static analysis during successive loading was proven to be an effective way to predict the burst pressure. The following conclusions were obtained:

(1) Experimental results showed that the ultimate burst pressure decreased from 6.88 MPa to 5.24 MPa with increasing temperature from 300 K to 573 K. Instability pressures acquired by nonlinear buckling analysis under different temperatures agreed well with the experimental results.

(2) The fracture morphologies showed a gradual transition from an opening ductile mode to a cleavage mode with increasing temperature due to more cross slips occurring under serious plastic deformation. The equivalent stress, equivalent strain and strain hardening rates acquired by static analysis can be used to effectively rationalize the temperature-dependent fracture behavior of domed rupture discs.

Author Contributions: Data curation, H.Z. (Hongbo Zhu); Formal analysis, H.Z. (Hongbo Zhu), W.X., Z.L. and H.Z. (Hongxing Zheng); Funding acquisition, H.Z. (Hongxing Zheng); Investigation, H.Z. (Hongbo Zhu) and W.X.; Resources, H.Z. (Hongxing Zheng); Software, H.Z. (Hongbo Zhu); Supervision, H.Z. (Hongxing Zheng); Writing—original draft, H.Z. (Hongbo Zhu); Writing—review and editing, Z.L. and H.Z. (Hongxing Zheng). All authors have read and agreed to the published version of the manuscript.

Funding: This research was funded by the National Natural Science Foundation of China (51904182) and the Shanghai Science and Technology Committee (19DZ2270200).

Acknowledgments: The authors are grateful to the National Natural Science Foundation and the Shanghai Science and Technology Committee, which enabled the research to be carried out successfully.

Conflicts of Interest: The authors declare no conflict of interest.

References

1. Zhao, G.B. An easy method to design gas/vapor relief system with rupture disk. *J. Loss Prev. Process Ind.* **2015**, *35*, 321–328. [CrossRef]
2. Mutegi, M.K.; Schmidt, J.; Denecke, J. Sizing rupture disk vent line systems for high-velocity gas flows. *J. Loss Prev. Process Ind.* **2019**, *62*, 103950. [CrossRef]
3. Shannak, B. Experimental and theoretical investigation of gas-liquid flow pressure drop across rupture discs. *Nucl. Eng. Des.* **2010**, *240*, 1458–1467. [CrossRef]
4. Schmidt, J.; Claramunt, S. Sizing of rupture disks for two-phase gas/liquid flow according to HNE-CSE-model. *J. Loss Prev. Process Ind.* **2016**, *41*, 419–432. [CrossRef]
5. More, T.; Johnson, K. A simple, low-cost burst disc for use at cryogenic temperatures. *Cryogenics* **1998**, *38*, 947–948. [CrossRef]
6. Xiao, H.H.; Duan, Q.L.; Sun, J.H. Premixed flame propagation in hydrogen explosions. *Renew. Sustain. Energy Rev.* **2018**, *81*, 1988–2011. [CrossRef]
7. Lee, B.J.; Jeung, I.S. Numerical study of spontaneous ignition of pressurized hydrogen released by the failure of a rupture disk into a tube. *Int. J. Hydrogen Energy* **2009**, *34*, 8763–8769. [CrossRef]
8. Lee, H.J.; Kim, Y.R.; Kim, S.H.; Jeung, I.S. Experimental investigation on the self-ignition of pressurized hydrogen released by the failure of a rupture disk through tubes. *Proc. Combus. Inst.* **2011**, *33*, 2351–2358. [CrossRef]
9. Duan, Q.L.; Xiao, H.H.; Gao, W.; Gong, L.; Wang, Q.S.; Sun, J.H. Experimental study on spontaneous ignition and flame propagation of high-pressure hydrogen release via a tube into air. *Fuel* **2016**, *181*, 811–819. [CrossRef]
10. *GB567.1-567.4-2012. Bursting Disc Safety Devices*; Standards Press of China: Beijing, China, 2013.
11. Gong, L.; Duan, Q.L.; Liu, J.L.; Li, M.; Jin, K.Q.; Sun, J.H. Effect of burst disk parameters on the release of high-pressure hydrogen. *Fuel* **2019**, *235*, 485–494. [CrossRef]
12. Schrank, M. Investigation on the degradation and opening behavior of knife blade burst disks. *J. Loss Prev. Process Ind.* **2014**, *32*, 161–164. [CrossRef]
13. Lake, G.F.; Inglis, N.P. The design and manufacture of bursting disks. *Proc. Inst. Mech. Eng.* **1932**, *142*, 365–378. [CrossRef]
14. Kanazawa, T.; Machida, S.; Miyata, T.; Niimura, Y.; Tanaka, O. Plastic deformation and rupture of thin circular plate under lateral pressure. In *Proceedings of the 11th Japan Congress on Materials Research Testing Methods and Apparatus*; The Society of Material Research: Kyoto, Japan, 1968; Volume 11, pp. 180–183.
15. Weil, N.A. Rupture characteristics of safety diaphragms. *J. Appl. Mech. Trans. ASME* **1959**, *26*, 621–624.

16. Ahmed, M.; Hashmi, M.S.J. Comparison of free and restrained bulge forming by finite element method simulation. *J. Mater. Process. Technol.* **1997**, *63*, 651–654. [CrossRef]

17. Kong, X.W.; Zhang, J.C.; Li, X.Q.; Jin, Z.B.; Zhong, H.; Zhan, Y.; Han, F.J. Experimental and finite element optimization analysis on hydroforming process of rupture disc. *Procedia Manuf.* **2018**, *15*, 892–898. [CrossRef]

18. Xue, L.; Widera, G.E.O.; Sang, Z. Parametric FEA study of burst pressure of cylindrical shell intersections. *J. Press. Vessel Technol. ASME* **2010**, *132*, 031203. [CrossRef]

19. Cherouat, A.; Ayadi, M.; Mezghani, N.; Slimani, F. Experimental and finite element modelling of thin sheet hydroforming processes. *Int. J. Mater. Form.* **2008**, *1*, 313–316. [CrossRef]

20. Jeong, J.Y.; Jo, W.; Kim, H.; Baek, S.H.; Lee, S.B. Structural analysis on the superficial grooving stainless-steel thin-plate rupture discs. *Int. J. Precis. Eng. Manuf.* **2014**, *15*, 1035–1040. [CrossRef]

21. Matsuo, T.; Yamabe, J.; Matsuoka, S. Effects of hydrogen on tensile properties and fracture surface morphologies of type 316L stainless steel. *Int. J. Hydrogen Energy* **2014**, *39*, 3542–3551. [CrossRef]

22. Qin, W.B.; Li, J.S.; Liu, Y.Y.; Kang, J.J.; Zhu, L.N.; Shu, D.F.; Peng, P.; She, D.S.; Meng, D.Z.; Li, Y.S. Effects of grain size on tensile property and fracture morphology of 316L stainless steel. *Mater. Lett.* **2019**, *254*, 116–119. [CrossRef]

23. Peng, J.; Li, K.S.; Dai, Q.; Gao, G.F.; Zhang, Y.; Gao, W.W. Estimation of mechanical strength for pre-strained 316L austenitic stainless steel by small punch test. *Vacuum* **2019**, *160*, 37–53. [CrossRef]

24. Li, K.S.; Peng, J.; Zhou, C.Y. Construction of whole stress-strain curve by small punch test and inverse finite element. *Results Phys.* **2018**, *11*, 440–448. [CrossRef]

25. Shen, W.Z.; Li, C.F.; Wang, P.F.; Chen, G.J. Tensile behavior and flow stress calculation of 316L stainless steel at 25–350 °C. *Trans. Mater. Heat Treat.* **2012**, *33*, 49–54.

26. *GB150.1-GB150.4-2011. GB: Pressure Vessels*; Standards Press of China: Beijing, China, 2012.

27. Luo, G.M.; Hsu, Y.C. Nonlinear buckling strength of out-of-roundness pressure hull. *Thin Walled Struct.* **2018**, *130*, 424–434. [CrossRef]

28. Zhao, Y.S.; Zhai, X.M.; Wang, J.H. Buckling behaviors and ultimate strengths of 6082-T6 aluminum alloy columns under eccentric compression—Part I: Experiments and finite element modeling. *Thin Walled Struct.* **2019**, *143*, 106207. [CrossRef]

29. Verwimp, E.; Tysmans, T.; Mollaert, M.; Berg, S. Experimental and numerical buckling analysis of a thin TRC dome. *Thin Walled Struct.* **2015**, *94*, 89–97. [CrossRef]

30. Pauloa, R.M.F.; Carlone, P.; Valentec, R.A.F.; Teixeira-Dias, F.; Rubino, F. Numerical simulation of the buckling behaviour of stiffened panels: Benchmarks for assessment of distinct modelling strategies. *Int. J. Mech. Sci.* **2019**, *157–158*, 439–445. [CrossRef]

31. Crisfield, M.A. *Non-linear Finite Element Analysis of Solids and Structures*; John Wiley & Sons: Chichester, UK, 1991.

32. Žarkovic, D.; Jovanović, D.; Vukobratović, V.; Brujić, Z. Convergence improvement in computation of strain-softening solids by the arc-length method. *Finite Elem. Anal. Des.* **2019**, *164*, 55–68. [CrossRef]

33. Hill, R. A theory of the plastic bulging of a metal diaphragm by lateral pressure. *Philos. Mag.* **1950**, *41*, 1133–1142. [CrossRef]

34. Byun, T.S.; Hashimoto, N.; Farrell, K. Deformation mode map of irradiated 316 stainless steel in true stress-dose space. *J. Nucl. Mater.* **2006**, *351*, 303–315. [CrossRef]

35. Wu, X.L.; Pan, X.; Mabon, J.C.; Li, M.M.; Stubbins, J.F. The role of deformation mechanisms in flow localization of 316L stainless steel. *J. Nucl. Mater.* **2006**, *356*, 70–77. [CrossRef]

36. Kim, J.W.; Byun, T.S. Analysis of tensile deformation and failure in austenitic stainless steels: Part I—Temperature dependence. *J. Nucl. Mater.* **2010**, *396*, 1–9. [CrossRef]

37. Mishra, S.; Yadava, M.; Kulkarni, K.N.; Gurao, N.P. A new phenomenological approach for modeling strain hardening behavior of face centered cubic materials. *Acta Mater.* **2019**, *178*, 99–113. [CrossRef]

Article

Mathematical Model for Prediction and Optimization of Weld Bead Geometry in All-Position Automatic Welding of Pipes

Baoyi Liao [1,2], Yonghua Shi [1,2,*], Yanxin Cui [1,2], Shuwan Cui [1,2], Zexin Jiang [3] and Yaoyong Yi [4]

[1] School of Mechanical and Automotive Engineering, South China University of Technology, Guangzhou 510640, China; 13430279851@163.com (B.L.); 201810100312@mail.scut.edu.cn (Y.C.); 13597066615@163.com (S.C.)

[2] Guangdong Provincial Engineering Research Center for Special Welding Technology and Equipment, South China University of Technology, Guangzhou 510640, China

[3] Guangzhou Shipyard International Company Limited, Guangzhou 511462, China; wljzxbs@163.com

[4] Guangdong Welding Institute (China-Ukraine E. O. Paton Institute of Welding), Guangzhou 510650, China; yiyy@gwi.gd.cn

*   Correspondence: yhuashi@scut.edu.cn; Tel./Fax.: +86-208-7114-407

Received: 3 September 2018; Accepted: 21 September 2018; Published: 25 September 2018

Abstract: In this study all-position automatic tungsten inert gas (TIG) welding was exploited to enhance quality and efficiency in the welding of copper-nickel alloy pipes. The mathematical models of all-position automatic TIG weld bead shapes were conducted by the response surface method (RSM) on the foundation of central composition design (CCD). The statistical models were verified for their significance and adequacy by analysis of variance (ANOVA). In addition, the influences of welding peak current, welding velocity, welding duty ratio, and welding position on weld bead geometry were investigated. Finally, optimal welding parameters at the welding positions of 0° to 180° were determined by using RSM.

Keywords: all-position automatic tungsten inert gas (TIG) welding; optimal welding parameters; response surface method (RSM); lap joint; weld bead geometry

---

1. Introduction

With the fast development of heavy industries, all-position automatic tungsten inert gas (TIG) welding of pipes had been widely applied to industries such as shipbuilding, nuclear power, chemical industries, and natural gas transportation. Welding plays an important role in joining pipes [1]. Traditional all-position welding of pipes is manual arc welding, but it has low welding efficiency, poor stability, and high cost. Moreover, the manual operation of pipe welding is a challenge to the welder. In order to overcome the aforementioned shortcomings of manual arc welding, an all-position automatic TIG welding process has been developed in recent years. At present, there are four types of automatic TIG butt-welding equipment for pipelines: closed welding heads, open welding heads, orbital welding trolleys, and thick-walled narrow gap TIG welding heads.

In all-position automatic TIG welding of pipes, especially for vertical and overhead position welding, it is a challenge to control the liquid metal flow down from the molten pool under the action of gravity. Therefore, there is a need to adjust welding parameters to keep the droplet transition smooth. In order to obtain high-quality weld beads, selecting optimal weld parameters and controlling the weld bead profile are important. As is well known, weld bead shape has a noteworthy influence on the mechanical properties of lap joints. In all-position automatic TIG welding of pipes, the weld bead profile is impacted not only by the weld parameters such as weld background current, weld peak current, weld speed, weld voltage, but also by the welding position. Therefore, it is hard to obtain several optimal parameters to receive perfect weld bead profiles. To realize this goal, the connection between weld parameters and weld bead shapes need to be established. A lot of research has been undertaken to establish mathematical models to optimize parameters and obtain the relationship [2,3]. Xu et al. [4] optimized the narrow gap all-position gas metal arc (GMA) welding parameters by employing the response surface method (RSM) and building regression models to predict the parameters of weld bead geometry. Rao et al. [5] had obtained the influence of welding parameters and statistical models for the prediction of weld bead shapes in pulsed GMA welding. Koleva [6] researched the connection with electron beam welding parameters and weld bead geometry; moreover, the mathematical model can optimize welding parameters. Karthikeyan and Balasubramanian used RSM to optimize friction stir spot welding procedure parameters and to obtain maximum lap shear strength of the lap joint [7]. The statistical model was built at an invariable position in inchoate research. However, in all-position automatic TIG welding of pipes, the welding position should be used as the input parameter of the statistical model because the weld includes flat, vertical and overhead positions [8], and welding position impacts the weld bead profile.

RSM can be used to predict the weld bead shape and mechanical properties in the welding process [9–12]. The ultimate goal of this paper is to build a new all-position automatic TIG pipes' welding procedure and a mathematical model of weld bead geometry using RSM.

2. Materials and Methods

The schematic diagram of the system for all-position automatic TIG welding of pipelines is shown in Figure 1. The system includes an iOrbital 5000 welding machine, a TOA77 welding torch, a control panel, two argon tanks, a welding fixture and a cool water circulation system. The welding process was as follows:

(1)   Fixing the pipes with a welding fixture.
(2)   Inputting the cool water and argon gas into the TOA77 welding torch.
(3)   The protective gas was inputted into the pipes.
(4)   The welding torch weld around the pipes.

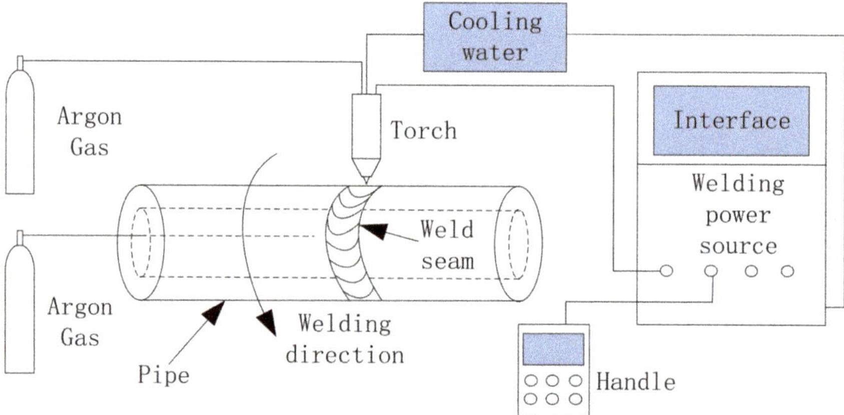

Figure 1. The schematic diagram of the system for all-position automatic tungsten inert gas (TIG) welding of pipes.

The workpieces used in this study are UNS (Unified Numbering System for Metas and Alloys) C70600 with dimensions of Φ57 mm $\times$ 300 mm $\times$ 2 mm, but there is a flaring end of the pipe and the diameter is 62 mm, as shown in Figure 2. The chemical compositions of the parent metal are listed in Table 1. In this study, experiments were executed with one-sided welding with double-sided shaping without an opening groove and welding wire, because the experiments melt the parent metal as filler metals. There is an angle of 30 degrees between the torch and the pipe vertical direction. Before welding, two pipes without a groove were fixed with a gap of 0.5 mm, and they were then lap welded. To ensure the stability of the welding process and avoid weld joints from being influenced by impurities and surface oil contamination, the pipes should be cleaned before welding. Pure argon (99.9% purity) was selected as a shielding gas with a flow rate of 15 L/min on the surface of the pipes and a flow rate of 20 L/min in the pipes to prevent the base metal from oxidizing in the air and enhancing the quality of the weld bead. The tungsten electrode diameter was 3.2 mm and the taper was 60°.

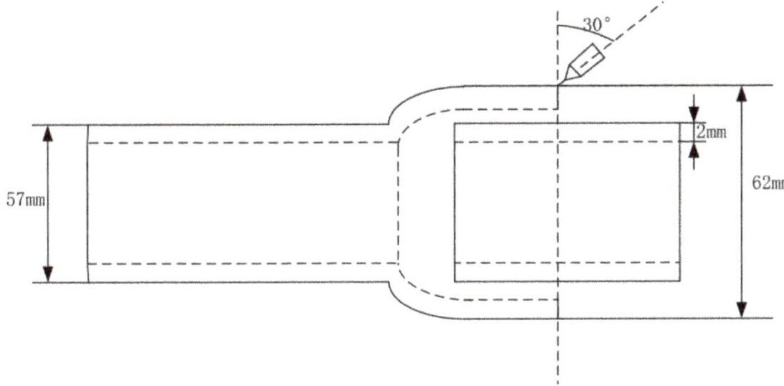

Figure 2. The schematic diagram of the lap joint.

Table 1. Chemical compositions of parent metal (wt%).

Chemical Element	Ni	Fe	Mn	Pb	S	C	Zn	P	Cu
Measured Value	10.66	1.60	0.79	0.01	0.0023	0.0065	0.019	0.01	Bal

In Figure 3, the specimens for weld bead shape observation were cut from the cross-section of the pipes, sealed with bake lite resin, then ground with a series of emery papers (grit size 400, 600, 800, 1200, 1500, 2000 and 2500), polished with a 1.0-μm diamond paste, and etched with an etchant of 5 g ferric chloride + 16 mL hydrochloric acid + 60 mL ethanol. Finally, a high-dynamic camera model NSC1003 (New Imaging Technologies, Paris, France) was used to photograph the weld bead shape and calculate the weld width, weld depth and weld thickness by image processing software. Reducing measurement error is important to the experiments, and therefore each value was measured three times. Then the average of the three measured values was calculated as experimental data.

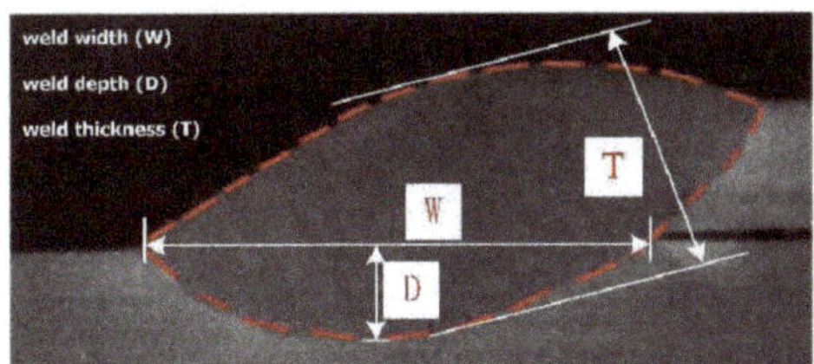

Figure 3. The cross-section of the weld bead. (weld width (W), weld depth (D), weld thickness (T)).

In this investigation, due to the working range of individual factor being extensive, a central composition rotatable four-factor, five-level factorial design matrix was selected. The design software

Design-Expert (V 10. 0. 7, Stat-Ease, MN, USA) was used to establish the design matrix and process the experimental data.

The steps of this investigation are as follows:

(a)   Validation of the primary factors;
(b)   Confirming the working range of the control variables;
(c)   Establishing trial matrix by Design-Expert V10.0.7 software;
(d)   Conducting the experiments as per the design matrix;
(e)   Recording the response parameters;
(f)   Building statistical models;
(g)   Calculating regression coefficients of the multinomial;
(h)   Checking the adequacy of the statistical model;
(i)   Verification of models;
(j)   Receiving, finally, the statistical model;
(k)   Analysis of results;
(l)   Optimizing welding parameters.

3. Creating Mathematical Model

Based on the initial experimental results, four very important welding parameters affecting the weld bead geometry are the welding peak current (I), welding velocity (V), welding duty ratio (d) and welding position (P). The duty ratio is the ratio of welding peak current time to the pulse period. In order to create a mathematical model to describe and forecast the weld bead geometry in all-position automatic welding, I, V, d, and P were chosen as input parameters. The schematic diagrams of the weld bead shape and welding position are shown in Figures 3 and 4, respectively.

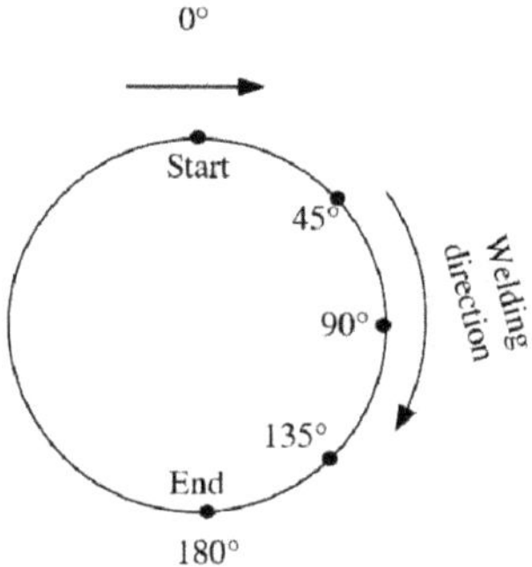

Figure 4. Schematic diagram of welding position.

Before this experiment, the control variable method was used to find out the working range of input parameters. The working range was determined by the steady welding procedure and invisible weld defects. In order to obtain optimized welding parameters, the establishment of a test parameter matrix adopts the center composed design (CCD) in the regression design method. In the CCD design, the upper and lower limit values of the input parameters were coded as $\pm \beta$, the value of β is dependent on the number of input parameters and $\beta = (2^k)^{1/4}$, k is the number of input parameters. The upper level and lower level were coded as ± 1 respectively and the center point was coded as 0. In this research, there are 4 input parameters, so β is 2. Therefore, the upper and lower limit values of the input parameters were coded as ± 2. The codes values for intermediate levels can be calculated by Equation (1):

$$X_i = 2[2X - (X_{max} + X_{min})]/[X_{max} - X_{min}] \tag{1}$$

In Equation (1), X_i is the desired code value of a variable X, and range of values of X is from X_{min} to X_{max}; X_{max} is the upper limits and X_{min} is lower limits of the variable X [13,14].

The factor levels and coded values have been listed in Table 2. The experimental design matrix (Table 3) includes 30 sets of coded situations and constitutes a full replication four-factor factorial design of 16 points, 8 star points and 6 center points [7].

Table 2. Important weld parameters and their levels for copper–nickel alloy pipe.

Factors	Unit	Coded Value				
		−2	−1	0	1	2
I	A	114	117	120	123	126
V	mm/min	57	60	63	66	69
d	%	40	45	50	55	60
P	°	0	45	90	135	180

Table 3. Trial design matrix and response of copper–nickel alloy lap joints.

Std	Run	Coded Variables				Response Parameters		
		I (A)	V (mm/min)	D (%)	P (degree)	W (mm)	D (mm)	T (mm)
1	7	117	60	45	45	1.416	0.29	1.978
2	10	123	60	45	45	1.615	0.363	2.142
3	6	117	66	45	45	1.016	0.199	1.797
4	19	123	66	45	45	1.325	0.272	1.89
5	11	117	60	55	45	2.877	0.577	2.541
6	24	123	60	55	45	4.2	0.851	2.751
7	22	117	66	55	45	1.575	0.315	1.995
8	26	123	66	55	45	2.373	0.462	2.232
9	29	117	60	45	135	3.255	0.637	2.185
10	30	123	60	45	135	4.956	1.02	2.314
11	14	117	66	45	135	2.583	0.525	1.932
12	8	123	66	45	135	3.591	0.706	2.73
13	5	117	60	55	135	4.074	0.835	3.024
14	23	123	60	55	135	5.796	1.298	2.541
15	16	117	66	55	135	3.297	0.606	2.379
16	15	123	66	55	135	5.754	1.179	2.034
17	12	114	63	50	90	2.268	0.441	2.331
18	28	126	63	50	90	3.717	0.762	2.436
19	20	120	57	50	90	3.402	0.697	2.667
20	9	120	69	50	90	2.856	0.462	1.827
21	27	120	63	40	90	2.226	0.356	1.659
22	4	120	63	60	90	5.859	1.201	2.478
23	13	120	63	50	0	0.987	0.202	1.533
24	17	120	63	50	180	4.032	0.826	2.184
25	3	120	63	50	90	3.412	0.699	2.331
26	2	120	63	50	90	3.533	0.724	2.826
27	25	120	63	50	90	3.641	0.528	2.461
28	21	120	63	50	90	2.755	0.597	2.098
29	1	120	63	50	90	2.562	0.74	2.356
30	18	120	63	50	90	2.878	0.742	2.501

In order to better reflect the weld bead geometry, reasonable response parameters need to be selected. The response parameters were named weld width (W), weld depth (D), and weld thickness (T). Figure 3 shows the cross section of weld bead.

On the foundation of central composite design matrix, the value of input parameters, response parameters and the regression model can be built. The connection between measured response and the input parameters could be shown as $y = f(x_1, x_2, \dots, x_i) + \varepsilon$, where y is the value of measured response, x_i is the value of the input parameter, and ε is the systematic error. Y is a power transformation of y [11,15], so the second-order polynomial can be expressed as Equation (2):

$$Y = b_0 + \sum_{i=1}^{4} b_i x_i + \sum_{i=1}^{3} \sum_{j=i+1}^{4} b_{ij} x_i y_j + \sum_{i=1}^{4} b_{ii} x_i^2 \tag{2}$$

where, b_0 is the average of the measured response, the regression coefficients such as b_i, b_{ij} and b_{ii} depend on linear, interaction and squared terms of factors, respectively.

Almost all response surface method problems can be approximated by these polynomials, and the regression coefficients could be obtained by the least squares method.

The regression model adopts the method of stepwise regression. Firstly, the regression model eliminates the insignificant terms and calculates the regression coefficients until the significant terms and the lack-of-fit terms of the regression model meet the requirements of the regression model. Finally, the relational expressions about all-position automatic TIG welding of the pipe within the range of 0–180° were obtained as Equations (3)–(8), which show the relationship between weld geometry shape and input parameters.

The final equations in terms of coded factors are given as follows:

$$W = 3.07 + 0.52\,I - 0.32\,V + 0.73\,d + 0.96\,P + 0.19\,I \times d + 0.27\,I \times P + 0.22\,d^2 - 0.16\,P^2 \tag{3}$$

$$D = 0.64 + 0.12\,I - 0.087\,V + 0.16\,d + 0.20\,P + 0.047\,I \times d + 0.065\,I \times P + 0.034\,d^2 - 0.032\,P^2 \tag{4}$$

$$T = 2.36 + 0.042\,I - 0.17\,V + 0.17\,d + 0.13\,P - 0.098\,I \times d - 0.12\,V \times d - 0.11\,P^2 \tag{5}$$

The final equations in terms of actual factors are given as follows:

$$\begin{aligned} W = {}&100.19346 - 0.64732\,I - 0.10788\,V - 2.29476\,d - 0.20122\,P + \\ &0.012846\,I \times d + 0.00197176\,I \times P + 0.00898719\,d^2 - 0.000783063\,P^2 \end{aligned} \tag{6}$$

$$\begin{aligned} D = {}&22.87088 - 0.15965\,I - 0.028847\,V - 0.47648\,d - 0.050128\,P + \\ &0.00311250\,I \times d + 0.000478241\,I \times P + 0.00134656\,d^2 - 0.0000160301\,P^2 \end{aligned} \tag{7}$$

$$\begin{aligned} T = {}&-62.80003 + 0.34011\,I + 0.34775\,V + 1.32831\,d + 0.012728\,P \\ &- 0.00652083\,I \times d - 0.00811250\,V \times d - 0.0000546879\,P^2 \end{aligned} \tag{8}$$

Analysis of variance (ANOVA) was used to determine the significance and suitability of the regression model. Tables 4–6 show the ANOVA analysis of the weld width, the weld depth and the weld thickness model respectively.

Table 4. Results of analysis of variance (ANOVA) for model of weld width.

Source	Sum of Squares	df *	Mean Square	F Value	p Value (Prob > F)	Significance
Model	47.8	8	5.98	30.86	<0.0001	Significant
I	6.42	1	6.42	33.17	<0.0001	-
V	2.51	1	2.51	12.98	0.0017	-
d	12.69	1	12.69	65.56	<0.0001	-
P	22.04	1	22.04	113.82	<0.0001	-
I × d	0.59	1	0.59	3.07	0.0945	-
I × P	1.13	1	1.13	5.85	0.0247	-
d²	1.44	1	1.44	7.42	0.0127	-
P²	0.72	1	0.72	3.69	0.0683	-
Residual	4.07	21	0.19	-	-	-
Lack of Fit	3.04	16	0.19	0.92	0.5944	Not Significant
Pure Error	1.03	5	0.21	-	-	-
Cor total	51.87	29	-	-	-	-

* Degree of freedom (df), a concept in statics, indicates the number of unconstrained variables in calculating a statistical magnitude. According to the usual definition, df $= n - k$, n is the number of samples and k is the number of constrained variables or conditional number, while k is also the quantity of the other independent statistical magnitude in calculating one statistical magnitude.

Table 5. Results of ANOVA for model of weld depth.

Source	Sum of Squares	df	Mean Square	F Value	*p* Value (Prob > F)	Significance
Model	2.21	8	0.28	31.63	<0.0001	Significant
I	0.33	1	0.33	37.60	<0.0001	-
V	0.18	1	0.18	20.56	0.0002	-
d	0.6	1	0.6	68.85	<0.0001	-
P	0.93	1	0.93	106.39	<0.0001	-
I × d	0.035	1	0.035	3.99	0.0589	-
I × P	0.067	1	0.067	7.63	0.0117	-
d^2	0.032	1	0.032	3.69	0.0685	-
P^2	0.030	1	0.030	3.43	0.0782	-
Residual	0.18	21	0.008744	-	-	-
Lack of Fit	0.14	16	0.009019	1.15	0.4786	Not Significant
Pure Error	0.039	5	0.007863	-	-	-
Cor total	2.40	29	-	-	-	-

Table 6. Results of ANOVA for model of weld thickness.

Source	Sum of Squares	df	Mean Square	F Value	*p* Value (Prob > F)	Significance
Model	2.64	7	0.38	7.89	<0.0001	Significant
I	0.043	1	0.043	0.9	0.3544	-
V	0.72	1	0.72	15.15	0.0008	-
d	0.72	1	0.72	15.15	0.0008	-
P	0.40	1	0.40	8.46	0.0081	-
I × d	0.15	1	0.15	3.20	0.0872	-
V × d	0.24	1	0.24	4.96	0.0365	-
P^2	0.35	1	0.35	7.39	0.0125	-
Residual	1.05	22	0.048	-	-	-
Lack of Fit	0.76	17	0.045	0.78	0.6845	Not Significant
Pure Error	0.29	5	0.058	-	-	-
Cor total	3.69	29	-	-	-	-

Using Design Expert V 10.0.7 Software can calculate the value of coefficient and the significance of each coefficient was confirmed by Student's t test and p values. The values of "Prob > F" less than 0.0500 indicate model terms are significant and values greater than 0.1000 indicate the model terms are not significant [16,17]. Tables 4–6 show the result of ANOVA for the W model, D model and T models, respectively, and the models' F values are 30.86, 31.63 and 7.89. The probability of F (prob > F) is less than 0.0001, in other words, these models are significant. Sometimes, these models show that the test results of lack-of-fit are insignificant relative to the pure error, insignificant lack-of-fit represents that the quadratic model is adequate. According to Table 4, I, d and P are the most important factors of the W model; V, (I × d), (I × P), d^2 and P^2 also could affect W. From Table 5, I, d and P are the most important factors of the D model; V, (I × d), (I × P), d^2 and P^2 also could affect D. I, V, d, P, (I × d), (V × d) and P^2 could affect T as shown in Table 6.

4. Verification of Models

To assure that the established model can predict and control the weld bead shape in actual application, it should test the accuracy of the mathematical model. The test experiments were executed by assigning diverse values for experimental variables within their working limits, but distinguishing them from the values of the design matrix. The values of input parameters, predicted response, actual response and percentage errors are listed in Table 7 respectively. It shows that the percentage errors for any models are less than 9%, and all the percentage errors are within the scope of industrial engineering requirements. Therefore, the statistical models can predict and optimize weld bead shape.

Table 7. Predicted values and actual values of the weld bead geometry.

Designation	Run	1	2	3	4	5
Input Parameters	I (A)	123	122	119	117	115
	V (mm/min)	60	60	60	62	63
	d (%)	55	45	40	60	55
	P (degree)	0	45	90	135	180
Predicted Values	W (mm)	1.977	1.818	2.788	3.958	4.067
	D (mm)	0.426	0.372	0.574	0.783	0.799
	T (mm)	2.071	2.091	2.194	2.707	2.406
Actual Values	W (mm)	1.839	1.978	2.581	3.615	4.256
	T (mm)	0.346	0.346	0.596	0.759	0.954
	W (mm)	2.204	1.991	2.345	2.891	2.278
Percentage Error ** (%)	W	−6.98	8.8	−7.43	−8.67	4.65
	D	−7.04	−6.99	3.38	−4.29	6.88
	T	6.42	−4.78	7.03	6.8	−5.32

$$\text{** Percentage error} = \frac{actual\ value - predicted\ value}{predicted\ value} \times 100.$$

5. Results and Discussion

According to the all models, the prime and interaction influences of input weld parameters on weld bead geometry can be found.

5.1. Influences of Welding Peak Current on Weld Width (W), Weld Depth (D), and Weld Thickness (T)

Considering the influence of the single factor welding peak current on the weld bead geometry, it can be shown from Figure 5 that the welding peak current increases with the increase in weld width, weld depth and weld thickness. This is because with weld peak current increasing leads to heat input increase per unit time, which is good for the weld metal melted and enhancing the deposition efficiency. As the peak current increases, the weld width and depth increase significantly, while weld thickness increases less. The weld thickness increases gradually from 2.331 to 2.436 mm with the increase in welding peak current from 114 to 126 A. This is because as the weld peak current increases, the arc force also increases, and the liquid metal is blown to both sides of the molten pool under the action of arc force, so the weld thickness increases indistinctly.

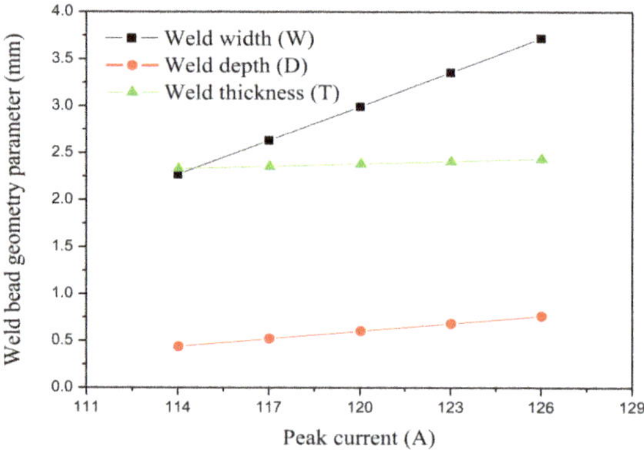

Figure 5. Influence of weld peak current on weld bead. (Welding velocity (V) = 63 mm/min, welding duty ratio (d) = 50%, welding position (P) = 90°).

5.2. Influences of Weld Velocity on W, D and T

Figure 6 indicates that the increase of welding velocity leads to the decrease in the heat input and volume of liquid filler metal per unit time of the weld bead, so the weld width, the weld depth and the weld thickness are correspondingly reduced.

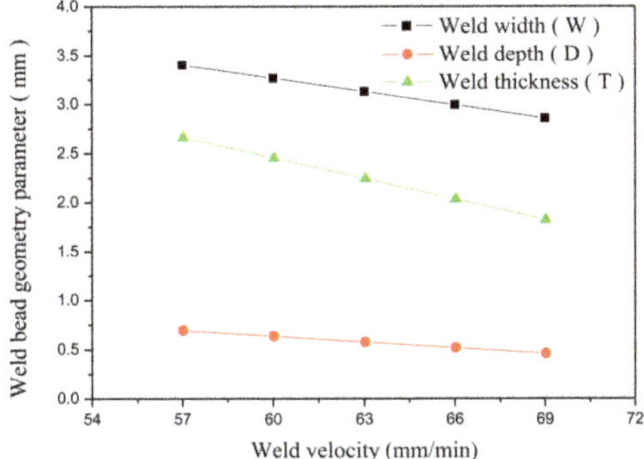

Figure 6. Influence of welding velocity on weld bead. (Welding peak current (I) = 120 A, d = 50%, P = 90°).

5.3. Effects of Duty Ratio on W, D and T

As shown in Figures 7–9, as the duty ratio increases, the weld width, depth and thickness increase correspondingly. This is due to increases in the heat input and volume of the liquid filler metal per unit time of the weld bead. The weld width and depth have a quadratic parabolic relationship with the duty ratio, and the slope of the curve growth is getting bigger and bigger. Because the duty ratio is increased, the weld heat input to the weld bead is larger and the more heat that is accumulated per unit time, the more favorable is the melting of the weld metal so the rate of weld width and weld depth growth is increasing. It can be seen from Figure 9 that the weld thickness increases linearly with increasing duty ratio, although the increase in duty ratio and the accumulation of welding heat contribute to the weld thickness, but excessive heat leads to liquid metal loss on the weld surface.

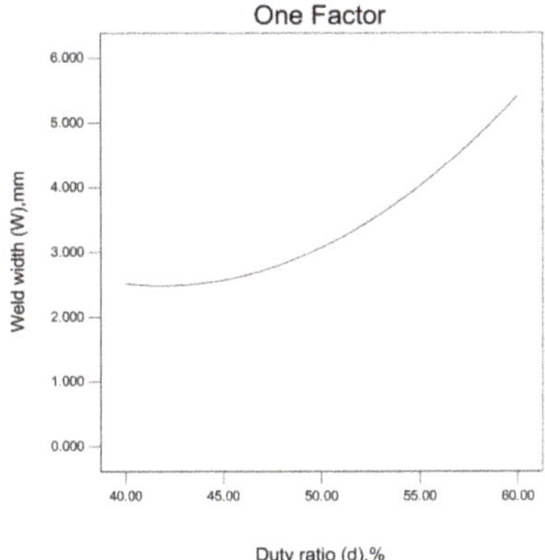

Figure 7. Influence of duty ratio on weld width.

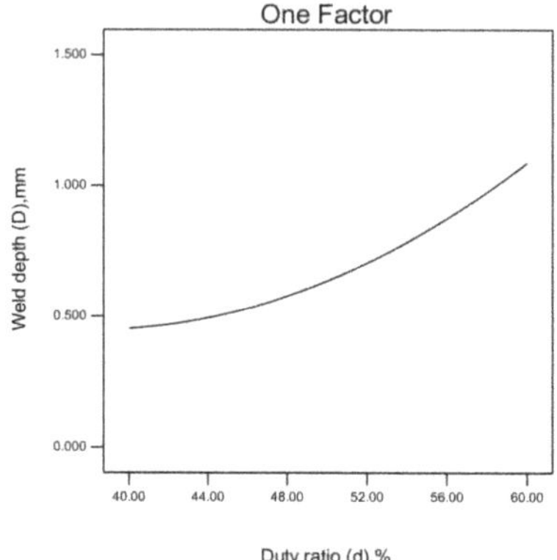

Figure 8. Influence of duty ratio on weld depth.

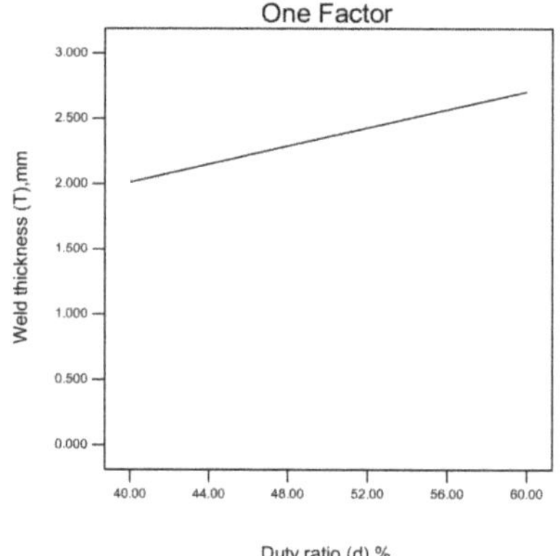

Figure 9. Influence of duty ratio on weld thickness.

5.4. Effects of Welding Position on W, D and T

As shown in Figures 10 and 11, as the degree of welding position increases, the weld width and weld depth increase correspondingly in the welding interval of 0–180°. The weld width and weld depth have a quadratic parabolic relationship to the degree of welding position, and the slope of the curve is getting smaller and smaller due to the gravity of the molten pool and the flow of molten metal along the weld bead during welding. Figure 12 shows the weld thickness increases first and then decreases. The gravity of the molten pool at different welding positions was shown in Figure 13. It was decomposed into a tangential force Gt and a radial force Gr. When welding in the 0–180° interval, the molten pool is subjected to the tangential force Gt, which leads to the molten liquid metal flowing down along the weld bead, and the flowing liquid metal can preheat the remaining weld bead and fill the weld bead, so the shape parameters of the weld width, weld depth and the weld thickness will increase with the increasing of the welding position degree. But when welding is in the 0–90° interval, the molten pool is subjected to the radial force Gr, and Gr = G cosθ. As the degree of the welding position increases, the radial force becomes smaller and smaller, and the direction of the radial force points to the center of the pipe; In the 90–180° interval welding, the molten pool is subjected to the direction of the radial force back to the center of the pipe, and Gr = G sin(θ − 90); with the welding position degree increasing, Gr becomes larger and larger, hindering the increase in weld width, depth and thickness. Therefore, the slope of the curve growth is getting smaller and smaller. Figure 12 shows the peak value of weld thickness for all values of P is received when P is 116°.

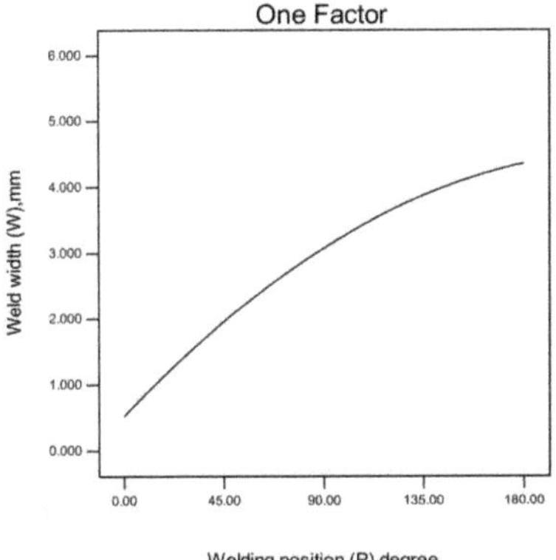

Figure 10. Influence of welding position on weld width.

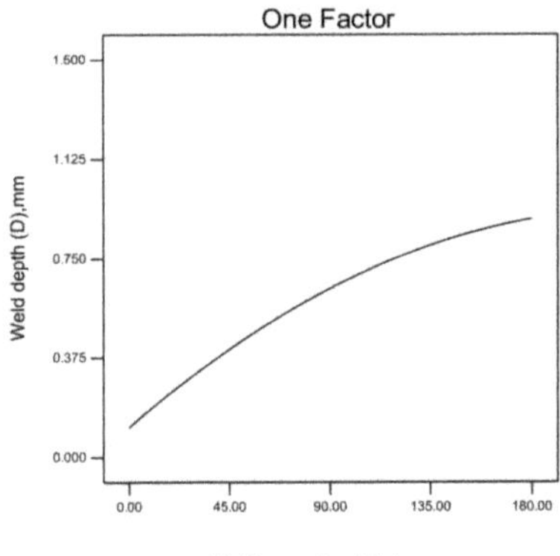

Figure 11. Influence of welding position on weld depth.

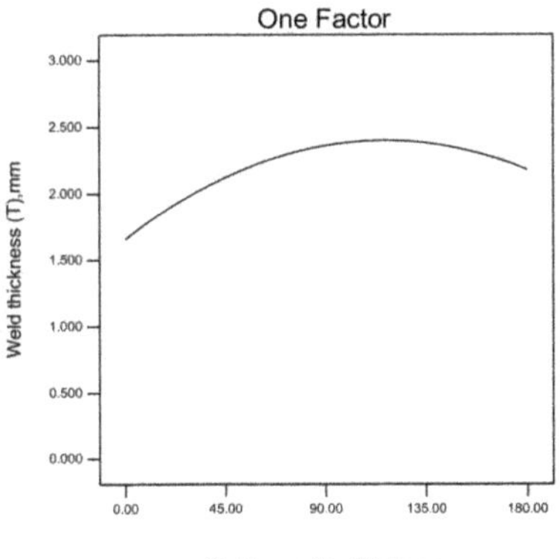

Figure 12. Influence of welding position on weld thickness.

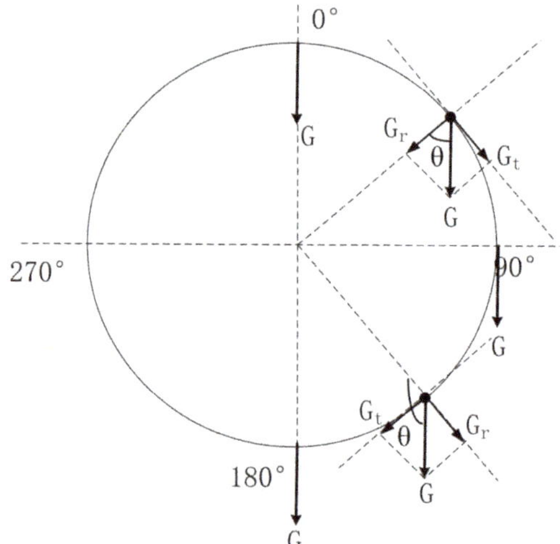

Figure 13. Gravity of molten pool at different welding positions.

5.5. Effects of Two-Factor Interaction on W, D and T

The interaction of I and d are shown in Figures 14–16. The speed of the increase in W and D with the increases in I increases as d increases, and the speed of the increase in W and D with d increases as I increases. This is because that I and d have an active influence on heat input of the weld width and depth. In Figure 16, T increases as I increases when d is less than 50%, while it decreases as I increases when d is more than 50%.

The interaction of I and P are shown in Figures 17 and 18. The rate of the increase in W and D with the increases in I increases gradually as P increases, and the speed of increase in W and D with P increases as I increases.

According to Figure 19, when the welding velocity is small, the T increases with the increase of the duty ratio. As the welding velocity increases, the influence of the duty ratio on T becomes smaller and smaller, because the welding velocity is large and the time of the welding process is short, so the welding heat input and the accumulated welding heat is small. When the duty ratio is small, the welding velocity has little effect on T. When the duty ratio is greater than the critical value, T decreases as the welding velocity increases.

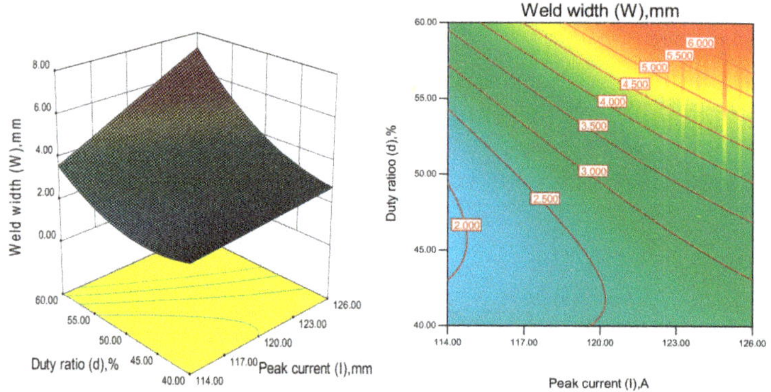

Figure 14. Interaction of peak current and duty ratio on weld width.

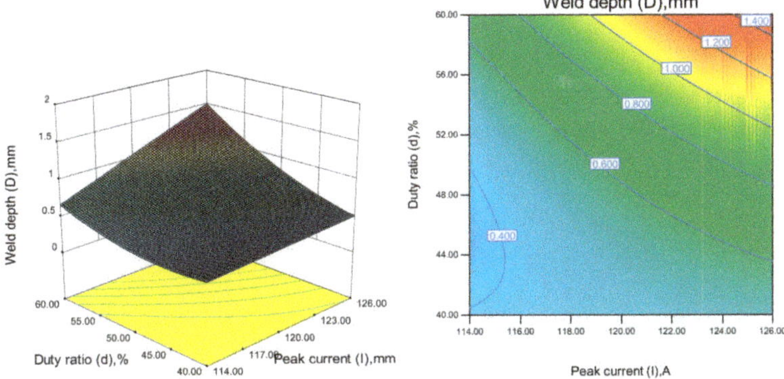

Figure 15. Interaction of peak current and duty ratio on weld depth.

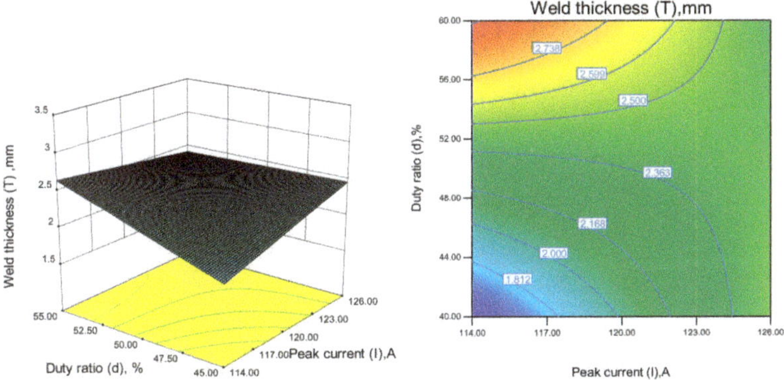

Figure 16. Interaction of peak current and duty ratio on weld thickness.

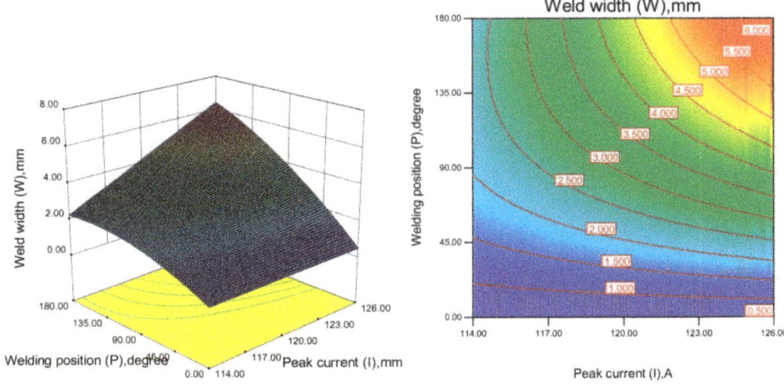

Figure 17. Interaction of peak current and welding position on weld width.

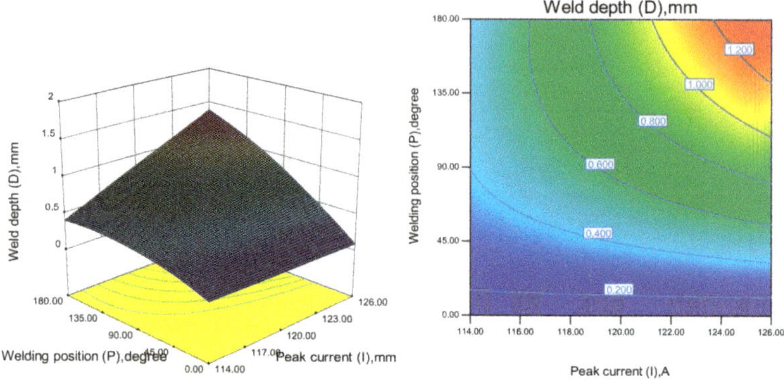

Figure 18. Interaction of peak current and welding position on weld depth.

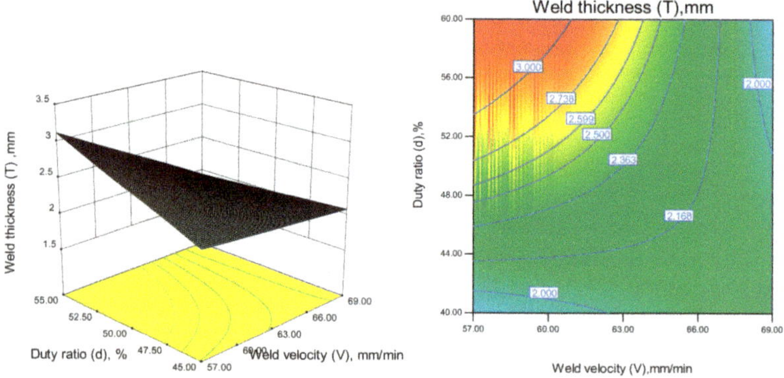

Figure 19. Interaction of weld velocity and duty ratio on weld thickness.

5.6. Optimization of the Welding Parameters by Numerical Method

In order to obtain the ideal weld bead shape and avoid welding defects such as incomplete penetration and weld collapse, numerical analysis was used to optimize the welding parameters. The goal, lower, upper limits and importance for every input and response parameters of the standard are shown in Table 8. Table 9 lists optimal welding parameters at 0, 45°, 90°, 135° and 180° welding position. The Figure 20 has shown the cross sections of weld bead with the optimal parameters at disparate welding position in 0–180°, it shows that the gap between tubes is different for each case; there are two factors lead to this phenomenon. One factor is welding deformation: the welding process generates a lot of heat, which lead to pipes deformation. The other factor is welding sequence: welding experiments weld the bead at the 0° position firstly, so the weld bead was solidified first at the 0° welding position, which results in other gaps between tubes being confirmed.

Table 8. Restraint of numerical optimization.

Name	Goal	Lower	Upper	Importance
I	Is in range	114	126	3
V	Is in range	57	69	3
d	Is in range	40	60	3
P	Is equal to	0	180	3
W	maximize	0.987	5.859	5
D	Is in range	0.8	1.5	5
T	Is in range	1.533	3.204	5

Table 9. Optimal parameters.

I (A)	V (mm/min)	d (%)	P (Degree)
126	57	60	0
125	58	59	45
124	60	58	90
122	60	54	135
119	61	54	180

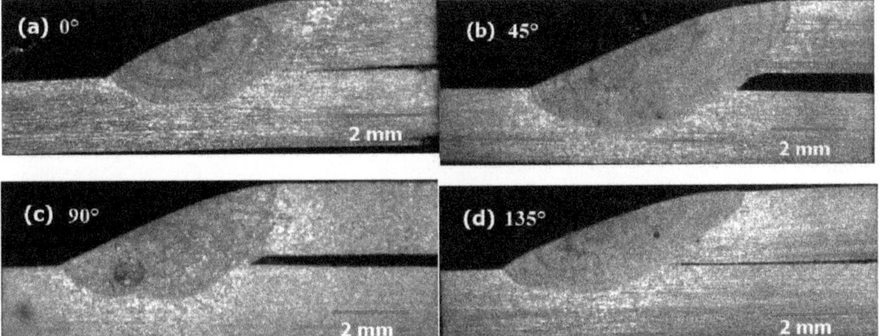

Figure 20. Cross sections of weld bead at different welding position from 0–180°. (**a**) 0° welding position, (**b**) 45° welding position, (**c**) 90° welding position, (**d**) 135° welding position.

6. Conclusions

In this research, the all-position automatic welding of pipes has been researched and statistically analysed. The main conclusions drawn from this research are as follows:

(1) Response surface methodology (RSM) based on center composed design (CCD) can be used to establish a mathematical model, and it can predict weld bead shape in the all-position automatic welding of pipes.

(2) Weld peak current had a prominent active influence on the important weld bead geometry parameters, while welding velocity had a negative influence on the important weld bead geometry parameters. The duty ratio exhibits a quadratic parabolic relationship with W and D, and the slope of the curve increases as the duty ratio increases, while T and the duty ratio increase linearly. The welding position has a quadratic parabolic relationship with the important weld bead geometry parameters, and the slope of the curve decreases as the degree of the welding position increases.

(3) The ideal weld bead geometry can be obtained by choosing the optimal weld parameters with the established statistical models, and this model can be used for the all-position automatic welding of pipes.

Author Contributions: B.L. performed the experiments, analyzed the data, and wrote the original manuscript. Y.S. conceived and designed the experiments, and contributed significantly to analysis and manuscript improvement. Y.C. and S.C. helped perform the analysis with constructive discussions. Z.J. and Y.Y. helped perform the experiments and analyse the data.

Funding: This research was funded by Science and Technology Planning Project of Guangzhou City (Grant No. 201604046026), Science and Technology Planning Project of Guangdong Province (Grant No. 2015B010919005), and National Natural Science Foundation of China (Grant No. 51374111).

Acknowledgments: The authors gratefully acknowledge Guangzhou Shipyard International Company Limited for providing experimental materials.

Conflicts of Interest: The authors declare no conflict of interest.

References

1. Xu, W.H.; Lin, S.B.; Fan, C.L.; Zhuo, X.Q.; Yang, C.L. Statistical modelling of weld bead geometry in oscillating arc narrow gap all-position GMA welding. *Int. J. Adv. Manuf. Technol.* **2014**, *72*, 1705–1716. [CrossRef]
2. Manonmani, K.; Murugan, N.; Buvanasekaran, G. Effects of process parameters on the bead geometry of laser beam butt welded stainless steel sheets. *Int. J. Adv. Manuf. Technol.* **2007**, *32*, 1125–1133. [CrossRef]
3. Kim, D.I.S.; Basu, A.; Siores, E. Mathematical models for control of weld bead penetration in the GMAW process. *Int. J. Adv. Manuf. Technol.* **1996**, *12*, 393–401. [CrossRef]
4. Xu, W.H.; Lin, S.B.; Fan, C.L.; Yang, C.L. Prediction and optimization of weld bead geometry in oscillating arc narrow gap all-position GMA welding. *Int. J. Adv. Manuf. Technol.* **2015**, *79*, 183–196. [CrossRef]
5. Rao, P.S.; Gupta, O.P.; Murty, S.S.N.; Rao, A.B.K. Effect of process parameters and mathematical model for the prediction of bead geometry in pulsed GMA welding. *Int. J. Adv. Manuf. Technol.* **2009**, *45*, 496–505. [CrossRef]
6. Koleva, E. Electron beam weld parameters and thermal efficiency improvement. *Vacuum* **2005**, *77*, 413–421. [CrossRef]
7. Karthikeyan, R.; Balasubramanian, V. Predictions of the optimized friction stir spot welding process parameters for joining AA2024 aluminum alloy using RSM. *Int. J. Adv. Manuf. Technol.* **2010**, *51*, 173–183. [CrossRef]
8. Ii, E.J.L.; Torres, G.C.F.; Felizardo, I.; Filho, F.A.R.; Bracarense, A.Q. Development of a robot for orbital welding. *Ind. Robot* **2005**, *32*, 321–325.
9. Acherjee, B.; Kuar, A.S.; Mitra, S.; Misra, D. Modeling and analysis of simultaneous laser transmission welding of polycarbonates using an FEM and RSM combined approach. *Opt. Laser Technol.* **2012**, *44*, 995–1006. [CrossRef]
10. Gunaraj, V.; Murugan, N. Application of response surface methodology for predicting weld bead quality in submerged arc welding of pipes. *J. Mater. Process. Technol.* **1999**, *88*, 266–275. [CrossRef]
11. Acherjee, B.; Misra, D.; Bose, D.; Venkadeshwaran, K. Prediction of weld strength and seam width for laser transmission welding of thermoplastic using response surface methodology. *Opt. Laser Technol.* **2009**, *41*, 956–967. [CrossRef]
12. Ruggiero, A.; Tricarico, L.; Olabi, A.G.; Benyounis, K.Y. Weld-bead profile and costs optimisation of the CO_2 dissimilar laser welding process of low carbon steel and austenitic steel AISI316. *Opt. Laser Technol.* **2011**, *43*, 82–90. [CrossRef]
13. Gunaraj, V.; Murugan, N. Prediction and comparison of the area of the heat-affected zone for the bead-on-plate and bead-on-joint in submerged arc welding of pipes. *J. Mater. Process. Technol.* **1999**, *95*, 246–261. [CrossRef]
14. Murugan, N.; Gunaraj, V. Prediction and control of weld bead geometry and shape relationships in submerged arc welding of pipes. *J. Mater. Process. Technol.* **2005**, *168*, 478–487. [CrossRef]
15. Rajakumar, S.; Balasubramanian, V. Diffusion bonding of titanium and AA7075 aluminum alloy dissimilar joints—Process modeling and optimization using desirability approach. *Int. J. Adv. Manuf. Technol.* **2016**, *86*, 1–18. [CrossRef]
16. Colombini, E.; Sola, R.; Parigi, G.; Veronesi, P.; Poli, G. Laser quenching of ionic nitrided steel: Effect of process parameters on microstructure and optimization. *Metall. Mater. Trans. A* **2014**, *45*, 5562–5573. [CrossRef]
17. Poli, G.; Sola, R.; Veronesi, P. Microwave-assisted combustion synthesis of NiAl intermetallics in a single mode applicator: Modeling and optimisation. *Mater. Sci. Eng. A* **2006**, *441*, 149–156. [CrossRef]

Article

A Volumetric Heat Source Model for Thermal Modeling of Additive Manufacturing of Metals

Yabin Yang [1,*] and Xin Zhou [2]

[1] School of Materials Science and Engineering, Sun Yat-Sen University, Guangzhou 510275, China
[2] Science and Technology on Plasma Dynamics Laboratory, Air Force Engineering University, Xi'an 710043, China; dr_zhouxin@126.com
* Correspondence: yangyabin@mail.sysu.edu.cn

Received: 27 September 2020; Accepted: 15 October 2020; Published: 22 October 2020

Abstract: In additive manufacturing of metallic materials, an accurate description of the thermal histories of the built part is important for further analysis of the distortions and residual stresses, which is a big issue for additively manufactured metal products. In the present paper, a computationally volumetric heat source model based on a semianalytical thermal modeling approach is proposed. The proposed model is applied to model the thermal response during a selective laser melting (SLM) process. The interaction between the laser and the material is described using a moving volumetric heat source. High computational efficiency can be achieved with considerable accuracy. Several case studies are conducted to examine the accuracy of the proposed model. By comparing with the experimentally measured melt-pool dimensions, it is found that the error between the predictions obtained by the proposed model and the experimental results can be controlled to less than 10%. High computational efficiency can also be achieved for the proposed model. It is shown that for simulating the thermal process of scanning a single layer with the dimension of 2 mm × 2 mm, the calculation can be finished in around 110 s.

Keywords: additive manufacturing; thermal modeling; volumetric heat source; computational efficiency

1. Introduction

Additive manufacturing (AM), defined by ASTM international [1] as "the process of joining materials to make parts from 3D model data, usually layer upon layer, as opposed to subtractive manufacturing and formative manufacturing methodologies", opens up a new era to design novel structural materials with complex geometries. Direct energy deposition and powder-bed fusion are the two main AM approaches for metallic materials [2]. For both approaches, a 3D object is usually first sliced into thousands of 2D layers. Focused thermal energy—such as the laser, electron-beam, or plasma-arc—is then employed on a working plane to melt and fuse the material (coming from wire or powder) locally with the designed path in accordance with the corresponding 2D cross-sectional layout. After one layer is finished, the working plane is lowered for a small distance and the heat source scans the subsequent slice. This process repeats until the complete 3D object is built. Support structures may sometimes be needed to eliminate overhanging. Finally, the support structures are removed and the built part is cut from the baseplate.

Residual stresses and distortions are major issues for the AM process of metals due to the complex set of heat-cooling cycles during the process [3,4]. Part distortions can disqualify precision components with tight dimensional tolerances, and high values of residual stresses can lead to part failure while being built. It is therefore of great interest to optimize the process parameters and part design in order to minimize the undesired residual stresses and distortions. One approach is experimental trial-and-error. A more systematic way to enhance the understanding of the interplay between the processes parameters, design, and final part quality is computational process modeling. Since the transient temperature response is the root cause of part distortions and residual stresses, a computationally efficient thermal model for the AM process of metals is the key step to increase the build quality and repeatability leading to products with superior mechanical properties.

Although the mechanisms for the two AM processes of direct energy deposition and powder-bed fusion are different, the modeling techniques for predicting the thermal transients during theses two AM processes in part scale share many common features [5]. In order to obtain certain computational efficiency, the powder or the wire is usually represented as a continuum having effective thermal properties [2], and the interaction between the focused thermal energy and the material can be simplified as a surface or volumetric heat source applied on the material [6]. The temperature transient is usually obtained by solving the Fourier's law of heat conduction with Neumann and Dirichlet boundary conditions (BCs) using the finite element method (e.g., [7–9]). Phase transitions within the melt pool can also be neglected as a second-order effect to improve the computational efficiency [10,11]. Even with these simplifications, the challenge of the thermal modeling of AM process in part scale is still how to accurately predict the thermal transient within a reasonable amount of time. The modeled part is usually orders of magnitude larger than the heat source, thus resulting in very fine discretizations in both space and time domains to describe the movement of the heat source [2]. It should be noted that even the heat source is modeled as a dimensionless moving point source, the multiscale nature of the problem and the numerical requirements still pose a computational challenge.

One way to address the separation of the scales in the problem is to use an adaptive mesh refinement in the vicinity of the heat source. However, this requires an additional remeshing step within time integration. Zhang et al. [12] developed an adaptive remeshing technique to reduce the computational cost for modeling the heat-transfer process in selective laser melting (SLM, belonging to the family of powder-bed fusion technique). Although the computational efficiency is certainly improved, the influence of the heat source (i.e., the laser) moving path is not fully considered. Instead, the powder-bed deposition is simplified by the scale of an entire layer or fractions of each layer, and each fraction is heated entirely for a certain effective time interval and then cools down. As the thermomechanical response in the metal AM process is very sensitive to the heat source moving path [13], an adaptive remeshing technique which takes into account the influence of the real heat source moving strategy may need to be further developed.

Yang et al. [14,15] proposed a semianalytical thermal approach based on the superposition principle and applied it in modeling the thermal response of the SLM process. The total temperature is decoupled as the superposition of an analytical field and a numerical field. The analytical field corresponds to the temperature caused by the moving heat source in a semi-infinite space, for which a closed-form expression exits. The numerical field is employed to account for the BCs and solved numerically. In the semianalytical approach, there is no need to give fine discretization for the heat source. The mesh size in solving the numerical field scales with the dimension of the modeled part and hence, the computational efficiency can be improved. In the model developed by Yang et al. [14,15], the moving heat source is discretized as a number of dimensionless point sources. Considerable accuracy of the thermal predictions can be achieved at a distance of 100 μm away from the point heat source [14]. However, predictions of the near-field temperature evolution, especially in the vicinity of the center of the melting zone, are less accurate [16,17]. Furthermore, in AM process, the nonaxisymmetric melt-pool observations with respect to the moving direction of the heat source suggest the heat energy distributes nonuniformly over a certain volume, and this cannot be captured by the dimensionless point heat source, for which the energy diffuses isotropically in the space domain. Therefore, it would be more appropriate to model the heat source as volumetric. Li et al. [18] compared the effect of using surface and volumetric heat sources in modeling the laser melting of ceramic materials and found the model incorporating the volumetric heat source increased the accuracy of melt-pool predictions.

In the present paper, taking the SLM process as an example, a volumetric heat source model based on the semianalytical thermal approach proposed by Yang et al. [14,15] is developed to predict the thermal histories of the built part in SLM. Compared to the point heat source model [14,15], the proposed volumetric heat source model is able to capture the nonaxisymmetric nature of the melt pool and give more accurate temperature predictions in the vicinity of the heat source. Meanwhile, the computational efficiency of the proposed volumetric heat source model is not impaired. The predicted melt-pool dimensions by modeling the heat source in SLM as point and volumetric sources are compared. The accuracy of the proposed volumetric heat source model is evaluated by comparing the corresponding simulation results with experiments. Section 2 introduces the volumetric heat source model based on the semianalytical thermal approach. Three case studies are investigated in Section 3 to evaluate the accuracy of the proposed model. The article concludes with a reiteration of the most salient points of the study.

2. Model Description

The semianalytical thermal model developed by Yang et al. [14,15] is briefly introduced in Section 2.1 The SLM process is taken as an example to explain how the semianalytical approach is utilized to model the thermal transients. The volumetric heat source model is then detailed in Section 2.2.

2.1. Semianalytical Thermal Model

Consider a 3D body V that has already been built on the baseplate, as shown in Figure 1a. The top, lateral, and bottom surfaces of the body V are represented by ∂V_{top}, ∂V_{lat}, and ∂V_{bot}, respectively. In SLM, a thin layer is laid on the top surface ∂V_{top}. The bottom surface ∂V_{bot} is bonded to the baseplate and the lateral surface ∂V_{lat} is in contact with the powder. Since the mean conductivity of the solid body V is much larger than that of the powder [19], it can be assumed that there is no heat transfer between body V and the powder, and hence, the lateral surface ∂V_{lat} is thermally insulated. The uppermost layer of powder is also neglected since its overall heat capacity is also negligible. The moving laser is modeled as a moving heat source enforced on the top surface ∂V_{top}. During SLM, the baseplate is usually preheated and keeps a relatively constant temperature [20,21]. This is considered as prescribing a fixed temperature T_c on the bottom surface ∂V_{bot}. Consequently, the temperature of body V caused by the moving heat source on ∂V_{top} is governed by the heat equation

$$\frac{\partial T}{\partial t} = \alpha \nabla^2 T + \frac{\dot{Q}_v}{\rho c_p} \tag{1}$$

with the BCs

$$\frac{\partial T}{\partial x_i} n_i = 0, \quad \text{on} \quad \partial V_{\text{top}} \quad \text{and} \quad \partial V_{\text{lat}}, \quad i = 1, 2, 3 \tag{2}$$

$$T = T_c, \quad \text{on} \quad \partial V_{\text{bot}}. \tag{3}$$

The initial condition is

$$T = T_{\text{ini}}, \quad \text{at} \quad t = 0. \tag{4}$$

The time is represented by t and the thermal diffusivity $\alpha = k/\rho c_p$, where k is the conductivity, ρ is the density and c_p is the specific heat. The heat generation rate is represented by $\dot{Q}_v$. The Cartesian coordinate system is illustrated in Figure 1a. The normal to the boundary surfaces of ∂V_{top} and ∂V_{top} are denoted as n_i, and T_{ini} is the initial temperature.

It should be noted that Equation (1) is linear by assuming the thermal properties k, ρ, and c_p are temperature independent. Although for most materials, the specific heat c_p and conductivity k are both temperature dependent. However, it has been demonstrated by many studies [15,22,23] that by choosing appropriate effective thermal constants c_p and k, the linear governing Equation (1) can still result in accurate prediction of the temperature field. The energy losses due to convection, radiation, and phase transitions are not explicitly accounted for and instead an effective laser power is later employed to implicitly characterize these energy losses.

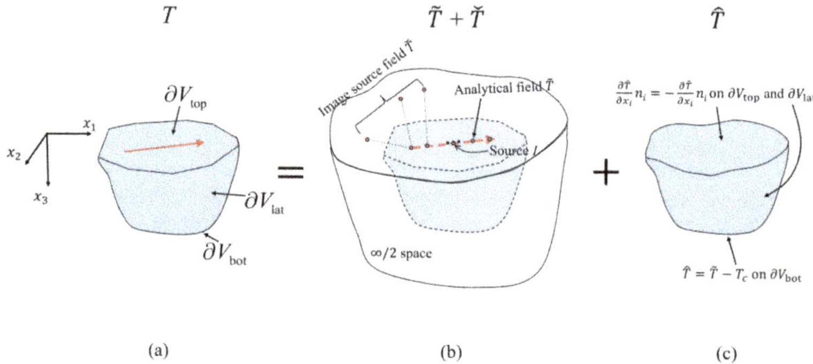

Figure 1. (**a**) A laser scanning is applied on the top surface of the body V. (**b**) The scanning line is discretized by a finite number of heat sources. Image sources by mirroring the original heat sources are added with respect to the boundaries. In a semi-infinite space, the temperature caused by the original discretized heat sources is $\tilde{T}$, and the temperature caused by the image sources is $\check{T}$. (**c**) Temperature filed $\hat{T}$ is employed to account for the boundary conditions (BCs). The total temperature T can be decomposed as the superposition of $\tilde{T}$, $\check{T}$, and $\hat{T}$.

Since the governing Equation (1) and the BCs expressed by Equations (2) and (3) are all linear, the temperature T can be decomposed as

$$T = \tilde{T} + \check{T} + \hat{T}, \tag{5}$$

where $\tilde{T}$ is the analytical temperature field caused by the moving heat source in a semi-infinite space, for which the boundary surface coincides with the top surface ∂V_{top} of the body V, as shown in Figure 1b. By discretizing the moving heat source as a finite number of individual heat sources, as shown in Figure 1b, the analytical field $\tilde{T}$ can be expressed as

$$\tilde{T} = \sum_{I=1}^{N} \tilde{T}^{(I)}, \tag{6}$$

where $\tilde{T}^{(I)}$ is the temperature field caused by the Ith heat source in the semi-infinite space and N is the total number of heat sources. The source density, which represents the number of heat sources per unit length, is given by

$$\rho_s = \frac{1}{v\Delta t}, \tag{7}$$

where v is the heat source moving speed and Δt is the duration between the activation of two consecutive sources.

In Equation (6), the $\check{T}$ and $\hat{T}$ fields are employed to account for the BCs. The $\check{T}$ is the image source field, which is the temperature field caused by the image sources in the semi-infinite space (see Figure 1b) and can also be obtained analytically, and $\hat{T}$ is the complementary field being solved numerically (see Figure 1c). The image sources are added by mirroring the original heat sources with respect to the boundary surfaces. As illustrated in Figure 2, image source $J_1^{(1)}$ is added by mirroring the heat source I with respect to boundary $\partial B^{(1)}$. Consider that the power associated with heat source I is P. If the power of image source $J_1^{(1)}$ is P, the heat flux on boundary $\partial B^{(1)}$ caused by heat source

I and image source $J_1^{(1)}$ would be 0; while if the power of image source $J_1^{(1)}$ is $-P$, the temperature on boundary $\partial B^{(1)}$ caused by heat source I and image source $J_1^{(1)}$ would be 0. Therefore, the no heat flux and prescribed temperature BCs shown in Equations (2) and (3) can be satisfied by adding the image sources. However, take Figure 2 as an example, image source $J_1^{(1)}$ would affect the heat flux and temperature on boundary $\partial B^{(2)}$, and thus, a second-order image source $J_2^{(2)}$ (the subscript denotes the order of the image source and the superscript represents the boundary by which the image source is mirrored) needs to be added. It is obvious that for the parallel boundaries shown in Figure 2, an infinite number of image sources will finally be needed. Therefore, to reduce the computational cost, only a finite number of image sources are added and the BCs are finally accounted for by the complementary field $\hat{T}$. The $\check{T}$ is expressed as

$$\check{T} = \sum_{J=1}^{M} \check{T}^{(J)},\tag{8}$$

where M is the total number of image sources. The analytical expression of $\check{T}^{(J)}$ is the same as $\tilde{T}^{(I)}$, which is detailed discussed in Section 2.2.

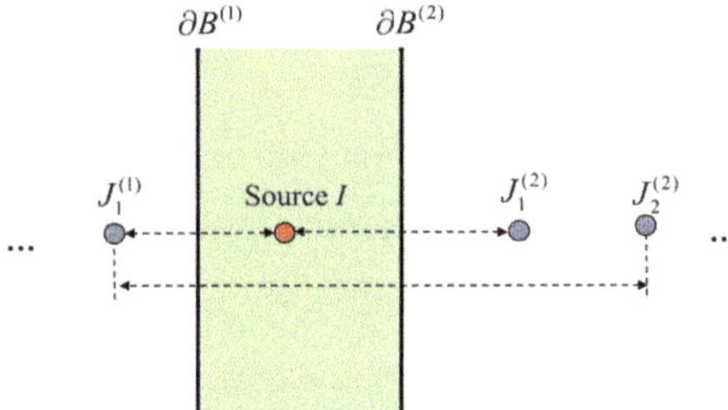

Figure 2. The strategy of adding image sources. The original heat source is represented by the red source and the image source is represented by the blue source. Image source $J_1^{(1)}$ is added by mirroring the original source with respect to boundary $\partial B^{(1)}$, and image source $J_1^{(2)}$ is added by mirroring the original source with respect to boundary $\partial B^{(2)}$. Image source $J_2^{(2)}$ is added by mirroring image source $J_1^{(1)}$ with respect to boundary $\partial B^{(2)}$. The subscript of $J_2^{(2)}$ denotes the order of the image source, and the superscript represents the boundary by which the image source is mirrored. For the parallel boundaries $\partial B^{(1)}$ and $\partial B^{(2)}$, an infinite number of image sources need to be added.

The complementary temperature field $\hat{T}$ is obtained by solving

$$\frac{\partial \hat{T}}{\partial t} = \alpha \nabla^2 \hat{T}\tag{9}$$

with the BCs

$$\frac{\partial \hat{T}}{\partial x_i} n_i = -\frac{\partial \tilde{T}}{\partial x_i} n_i - \frac{\partial \check{T}}{\partial x_i} n_i, \quad \text{on} \quad \partial V_{\text{top}} \quad \text{and} \quad \partial V_{\text{lat}},\tag{10}$$

$$\hat{T} = -\tilde{T} - \check{T} + T_c, \quad \text{on} \quad \partial V_{\text{bot}},\tag{11}$$

and initial condition

$$\hat{T} = -\tilde{T} - \check{T} + T_{\text{ini}}, \quad \text{at} \quad t = 0,\tag{12}$$

where Equations (10) to (12) are a direct consequence of Equations (2) to (4).

The analytical field $\tilde{T}$ serves to partially capture the steep temperature gradient in the vicinity of the heat source, and the image source field $\check{T}$ is employed to describe the steep temperature gradient when the heat source is close to the boundary. As a result, the temperature gradient in $\hat{T}$ field is relatively smooth, and thus, a relatively coarse mesh can be applied in solving $\hat{T}$.

It should be noted that in the semianalytical approach, as there is the need to add image sources, body V needs to be a convex polyhedron (where any two points within the polyhedron can be connected by a line). For arbitrary complex geometries, the semianalytical approach can still be utilized by using the $\hat{T}$ field only to account for the BCs. Then, without the image source field and since the temperature gradient will become very steep when the heat source is close to the boundary, a fine discretization in the vicinity of the boundary may be necessary for solving the $\hat{T}$ field to obtain certain accuracy. The computational efficiency and accuracy for using the $\hat{T}$ field only to account for BCs to model the nonconvex part are detailed discussed in [15]. In the present paper, for the purpose of validating the accuracy of the proposed volumetric heat source model, only convex body V is considered so that the accuracy of the model is not very sensitive to the mesh size with the assistance of the image sources.

2.2. Volumetric Heat Source

In the model developed by Yang et al. [14,15], sources $I = 1$ to N and the corresponding images sources are modeled as dimensionless point sources. In the present paper, for individual source I, it is described as a half-ellipsoidal volumetric source (see Figure 3), and the corresponding analytical solutions for $\tilde{T}$ and the gradient of $\tilde{T}$ are developed.

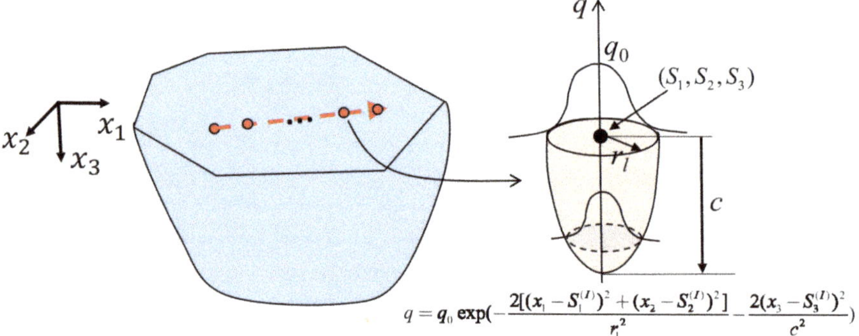

Figure 3. The discretized heat source is modeled as a half-ellipsoidal volumetric source.

Consider the energy distribution for the heat source I, shown in Figure 3, expressed as

$$q(x_i) = q_0 \exp\left(-2\frac{(x_1 - S_1^{(I)})^2 + (x_2 - S_2^{(I)})^2}{r_I^2} - 2\frac{(x_3 - S_3^{(I)})^2}{c^2}\right), \quad i = 1, 2, 3, \tag{13}$$

where q_0 is the maximum energy density at the position of $x_i = S_i^{(I)}$. The parameters r_I and c are illustrated in Figure 3. In SLM, the r_I can be considered as the laser spot radius and the parameter c can be seen as the optical penetration depth (OPD) of the laser. Equation (13) satisfies that the energy density reduces to q_0/e^2 at the surface of the ellipsoidal source.

The total energy of the heat source is Q, and conservation energy requires that

$$Q = \int_0^\infty \int_{-\infty}^\infty \int_{-\infty}^\infty q(x_1, x_2, x_3)\,\mathrm{d}x_1\mathrm{d}x_2\mathrm{d}x_3. \tag{14}$$

Evaluating Equation (14) yields

$$q_0 = \frac{4\sqrt{2}Q}{\pi\sqrt{\pi}cr_I^2}. \tag{15}$$

If heat source I is considered as a single dimensionless point source, the corresponding temperature $\tilde{T}_p^{(I)}$ in a semi-infinite space is given by [24]

$$\tilde{T}_p^{(I)}(x_i^{(P)}, t) = 2QG(x_i^{(P)}, t), \tag{16}$$

where

$$G(x_i^{(P)}, t) = \frac{1}{\rho c_p(\pi\alpha\eta)^{3/2}} \exp\left(-\frac{(x_1 - x_1^{(P)})^2 + (x_2 - x_2^{(P)})^2 + (x_3 - x_3^{(P)})^2}{\eta}\right). \tag{17}$$

The $\eta = 4\alpha(t - t_0^{(I)})$ and $x_i^{(P)}$ is the coordinates for the point of interest. Source I is activated at $t = t_0^{(I)}$. Function $G(x_i^{(P)}, t)$ is the Green function, which is a fundamental solution of Equation (1). Due to the boundary effect caused by the boundary surface of the semi-infinite space, there is a factor 2 in Equation (16).

Due to the linearity of Equation (1), the temperature for point $x_i^{(P)}$ at moment t induced by a heat source with the energy density of $q(x_i)$ is equal to $q(x_i)G(x_i^{(P)}, t)$. In addition, we need to consider the boundary effect caused by the boundary surface of the semi-infinite space. This can be solved by adding an image source with respect to the boundary surface of the semi-infinite space. Consequently, the temperature $\tilde{T}^{(I)}$ caused by a volumetric heat source shown in Figure 3 is expressed as

$$\tilde{T}^{(I)} = \int_{-\infty}^{\infty} \int_{-\infty}^{\infty} \int_{-\infty}^{\infty} q(x_1, x_2, x_3)G(x_1^{(P)}, x_2^{(P)}, x_3^{(P)}, t)dx_1dx_2dx_3. \tag{18}$$

Finally, the temperature $\tilde{T}^{(I)}$ is given by

$$\tilde{T}^{(I)} = B\frac{\exp(-(I_{xy} + I_z))}{D}, \tag{19}$$

where

$$
\begin{aligned}
B &= \frac{2\sqrt{2}Q}{\rho c_p \pi^{3/2}}, \\
I_{xy} &= -\frac{2[(x_1^{(P)} - S_1^{(I)})^2 + (x_2^{(P)} - S_2^{(I)})^2]}{2\eta + r_l^2}, \\
I_z &= -\frac{2(x_3^{(P)} - S_3^{(I)})^2}{2\eta + c^2}, \\
D &= (2\eta + r_l^2)\sqrt{2\eta + c^2}.
\end{aligned}
\tag{20}
$$

It can be seen that if r_l and c both become 0, source I would become a dimensionless point source, and Equation (19) would be equivalent to the temperature caused by a point source as shown in Equation (16). For any given heat source, the total energy $Q = PA\Delta t$, where P is the power and A is a nondimensional coefficient which implicitly accounts for the energy losses during the SLM process, such as due to laser absorptivity, convection, radiation, and phase transitions. The expression for any given image source J is the same as Equation (19) but with different source positions $S_i^{(J)}$. Therefore, with Equation (19), the analytical field $\tilde{T}$ and image source field $\check{T}$ can be finally obtained using Equations (6) and (8), respectively.

The gradients of $\tilde{T}$ and $\check{T}$ used for solving the $\hat{T}$ field are given by

$$\frac{\partial \tilde{T}}{\partial x_i} = -\frac{4x_i^{(P)}}{2\eta + \lambda^2}\tilde{T},$$

$$\frac{\partial \check{T}}{\partial x_i} = -\frac{4x_i^{(P)}}{2\eta + \lambda^2}\check{T},$$

(21)

where $\lambda = r_l$ for $i = 1, 2$ and $\lambda = c$ for $i = 3$.

3. Results and Discussions

Three numerical examples are investigated in the present study. For the first example, a single laser scan is applied on a very large baseplate to mimic the scanning of the first layer. The baseplate can be assumed to be a semi-infinite space and thus, the total temperature field T simply becomes $\tilde{T}$ in the absence of any BCs. The predicted melt-pool width and depth by the proposed volumetric heat source model are compared with the experiments reported in [25]. The material used is stainless steel (SS) 316L. For the second example, a single laser scan is applied on a baseplate with the cubic profile and thus, the BCs need to be considered. The material Ti6Al4V is considered in this example to demonstrate that the proposed model can be applied to different materials. The accuracy of the proposed model is further validated by comparing the predicted melt-pool width and depth with the experiments [8]. A third example is presented to demonstrate the ability of the proposed model to simulate the multiple laser scanning process with various heat source moving paths in the AM process.

The material properties of SS316L and Ti6Al4V are tabulated in Table 1. As suggested by Yang et al. [14], the values quoted in Table 1 are representative for a temperature which is slightly higher than the melting point (T_m) of SS 316L and Ti6Al4V, respectively. The thermal properties for SS316L and Ti6Al4V in Table 1 correspond to the temperatures of 1727 °C [26] and 2227 °C [27], respectively. The temperature fields $\tilde{T}, \check{T}$, and $\hat{T}$ are all calculated using an in-house Matlab code. The $\tilde{T}$ and $\check{T}$ fields are directly calculated using Equations (6) and (8), and the $\hat{T}$ is solved with an explicit finite difference scheme, centered in space and forward in time.

Table 1. Material properties.

	T_m (°C)	k (W/mK)	ρc_p (MJ/Km^3)
SS316L [26]	1427	18.97	3.8
Ti6Al4V [27]	1650	42	4.38

3.1. A Single Laser Scan on a Semi-Infinite Space

A single laser scan is applied on a semi-infinite space as schematically illustrated in Figure 3 to mimic scanning the first SS 316L powder layer on a very large baseplate [25]. The initial temperature is set to be 25 °C. The laser spot radius r_l is 27 μm and the OPD c is set as 60 μm, which is close to the diameter of the powder (54 μm).

A convergence study with respect to the source density ρ_s is first investigated. The maximum temperature experienced by a point with a distance of 20 μm to the laser scanning line with various source density ρ_s is calculated. This distance is close to the laser spot radius. The source density $\rho_s = 100 \times 10^4$ Wms^{-1} is taken as the reference. As shown in Figure 4, the error $|\tilde{T}(\rho_s) - \tilde{T}_{\text{ref}}|/\tilde{T}_{\text{ref}}$ for two different P/v is plotted as a function the source density ρ_s. It can be seen that the error quickly reduces to a value close to 0 and converges at $\rho_s = 10 \times 10^4$ Wms^{-1}. Hence, for the following calculations, the source density $\rho_s = 10 \times 10^4$ Wms^{-1} is used.

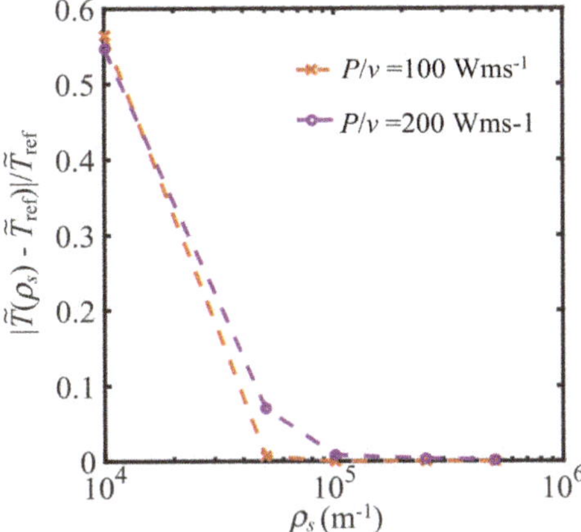

Figure 4. The temperature convergence of $\tilde{T}$ as a function of the source density ρ_s.

The predicted melt-pools by the volumetric heat source model within the $x_1 - x_2$ plane and $x_2 - x_3$ plane are shown in Figure 5a,b, respectively. The power is 150 W and the speed is 0.8 m/s. The melt-pool is shown as the white area in Figure 5 and determined as the material points heated above the melting point. A regular hexahedral grid with the size of 10 µm is employed to plot Figure 5. The melt-pool width and depth along the scanning line at $P = 150$ W and $v = 0.8$ m/s are shown in Figure 6. By examining the temperature of the material points in the vicinity of the scanning line (along the width and depth directions, respectively), the boundary of the melt-pool is determined to be the first material point with the temperature below the melting point. The resolution for determining the melt-pool width and depth in this example is 1 µm. It can be seen that the melt-pool width and depth are both small at the beginning of the laser scan, but quickly reach a higher steady-state value. The melt-pool width and depth then decrease towards the completion of the laser scan. Hence, the calculated width and depth of the melt-pool shown in the following are determined by the steady-state value along the laser scanning line.

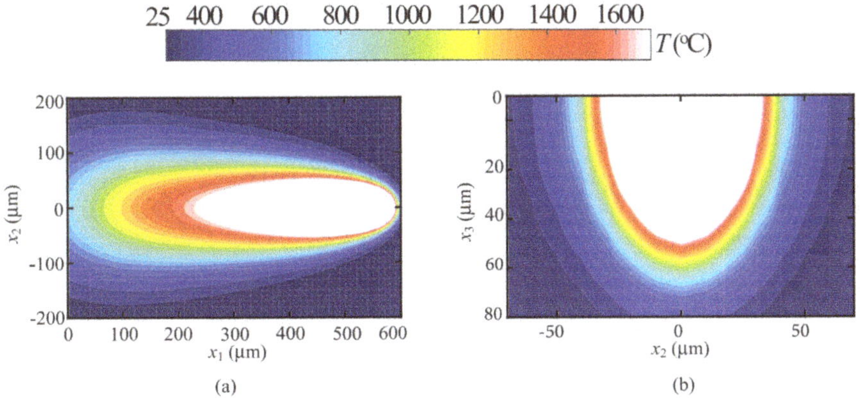

Figure 5. The temperature profile within the (**a**) $x_1 - x_2$ plane and (**b**) $x_2 - x_3$ plane for the power of 150 W and the speed of 0.8 m/s. The melt-pool is shown as the white area.

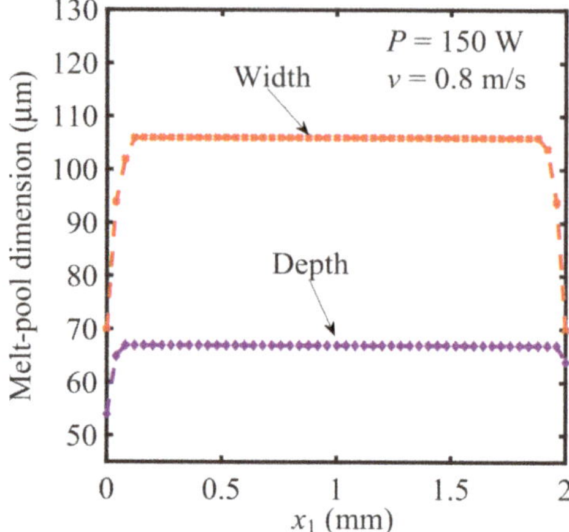

Figure 6. The melt-pool width and depth along the laser scanning line at the power of 150 W and the speed is 0.8 m/s.

The melt-pool width and depth by the proposed volumetric heat source model for various laser powers and speeds, and the corresponding experimental measurements are reported in [25] are plotted in Figure 7a,b. The coefficient A is set to be 0.43 to result in the best agreement for the volumetric heat source model. The simulation results by the point heat source model are also plotted in Figure 7.

The uncertainty of the experimental measurements is 5 μm. As explained in [25], the experimental uncertainty is mainly because that the plasma/metal vapor plume during the SLM process can change the laser absorptivity, which causes fluctuations of the effective power. Incorporating a variable coefficient A in the proposed model to accurately account for the variance of the effective power during the SLM would be quite difficult and thus, a more common approach [25] to take the coefficient A as a constant is employed. It can be observed from Figure 7 that the melt-pool width and depth increase with the increasing P/v. The value of P/v for the power of 200 W and speed of 1.2 m/s is the same as that for the power of 300 W and speed of 1.8 m/s, and thus, close values of melt-pool width and depth can be observed in Figure 7 both for the experimental and simulation results.

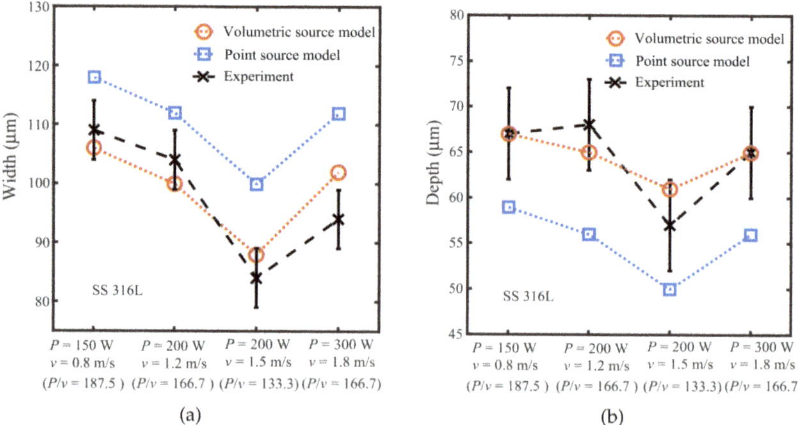

Figure 7. The predicted melt-pool (**a**) width and (**b**) depth and the experimental measurements under various laser powers and scanning speeds.

In Figure 7, it can also be seen that the predicted melt-pool width and depth by the volumetric heat source model agree well with the experimental measurements for all the four sets of laser power and speed. In comparison, the predicted melt-pool width by the point heat source model is overestimated (see Figure 7a) and the depth is underestimated (Figure 7b). The predicted melt-pool depth by the point heat source model is actually half of the predicted width as the heat energy diffuses isotropically in the space domain for the point heat source. If we tune the coefficient A to make the width calculated by the point heat source model agree well with the experiments, the calculated depth will also reduce accordingly. Therefore, the difference between the half width and depth of the melt-pool cannot be captured by the point heat source model. In the volumetric heat source model, by adjusting the parameter c to characterize the OPD of the laser, the melt-pool with shallow or deep depth can be conveniently captured.

The error of the predicted results in Figure 7 with respect to the average experimental measurements is evaluated by

$$\varepsilon = \frac{|Y - Y_{\exp}|}{Y_{\exp}}, \tag{22}$$

where Y represents the predicted melt-pool width or depth, and $Y_{\exp}$ represents the averaged experimental melt-pool width or depth. Figure 8 shows the errors of the predicted melt-pool width and depth for both the point and volumetric heat source models. The rectangular bars represent the average error $\bar{\varepsilon}$ for the four sets of laser power and speed, and the error bar corresponds to the maximum and minimum error, respectively, in the four sets of laser power and speed. It can be seen from Figure 8 that the maximum errors of the predicted melt-pool width and depth for the point heat source model are close to 20%, while the corresponding maximum errors for the volumetric heat source model are less than 10%. The average errors of the predicted melt-pool width and depth for the point heat source model are around 15%, and the corresponding average errors for the volumetric heat source model are around 5%.

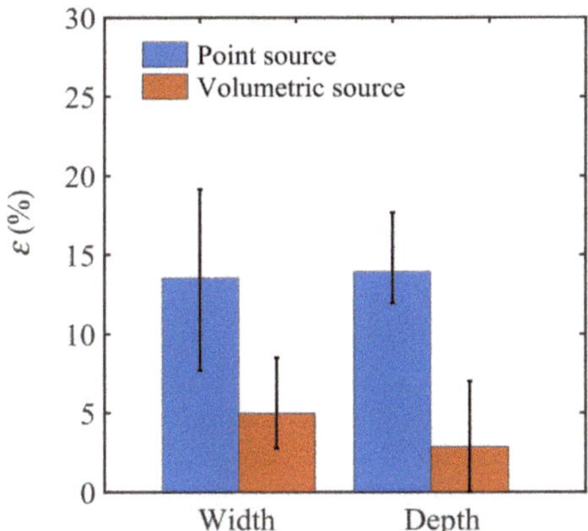

Figure 8. The errors $\varepsilon = |Y - Y_{\exp}|/Y_{\exp}$ of the predicted width and depth with respect to the experiments.

3.2. A Single Laser Scan on a Finite Space

In this example, a single scan experiment reported in literature [8] is simulated using the proposed model. As shown in Figure 9, the baseplate has a dimension of 4 mm × 2 mm × 0.5 mm and the material is Ti6Al4V. The bottom surface of the baseplate is set to be at 20 °C during the whole process. The laser spot radius is 26 μm and the OPD c is set to be 40 μm, which is also close to the powder diameter (35 μm). The coefficient A is set to be 0.77 as suggested in [8]. The laser speed is 0.2 m/s and various laser powers are investigated. A total number of 64 cubic finite difference cells is used to discretize the baseplate, resulting in a cell size of 1 mm $\times$ 0.5 mm $\times$ 0.125 mm. The $\tilde{T}$ and $\check{T}$ fields can be readily obtained for any point of interest $x_i^{(P)}$ within the body analytically, while the $\hat{T}$ field can only be calculated at the nodes of the finite difference cells. Therefore, a linear interpolation of known $\hat{T}$ values is utilized to estimate the value of $\hat{T}(x_i^{(P)})$ on any given material point, which is given by

$$\hat{T}(x_i^{(P)}) = \sum_{q=1}^{N_q} \Phi^{(q)} \hat{T}(x_i^{(q)}), \tag{23}$$

where $q = 1, 2, \ldots, N$ is the index for the grid point with $\hat{T}(x_i^{(q)})$ and the total number of grid points of the finite difference cell is denoted as N_q. For an 8-node hexahedral finite difference cell ($N_q = 8$) used in this example, the function $\Phi^{(q)}$ is expressed as

$$\Phi^{(q)} = \frac{1}{8}(1 + \xi_i \xi^{(q)}), \tag{24}$$

where ξ_i is the normalized position of $x_i^{(P)}$ in a right-handed local coordinate system having an origin located at the center of the cell and $\xi^{(q)}$ is the normalized position of $x_i^{(q)}$ given in the same coordinate system. The resolution for determining the melt-pool width and depth in this example is 1 µm.

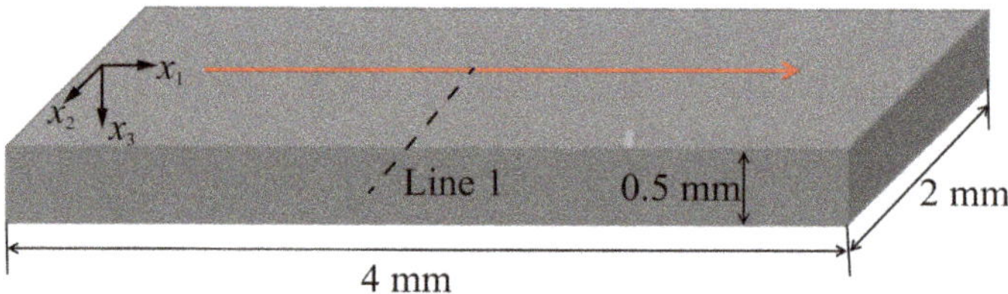

Figure 9. A single laser scan is applied on a finite space.

The predicted melt-pool width and depth obtained by the point and volumetric heat source models as well as the experimental measurements are shown in Figure 10a,b, respectively. The experimental results were measured for different places along the scanning line by optical microscopy based on the solidified microstructure [8]. As explained in [8], because the boundaries of the melt-pool were difficult to characterize, some large error bars can be observed in Figure 10. It can be seen from Figure 10 that the trend of the predicted melt-pool width and depth obtained by the point and volumetric heat source models agree well with that of the experimentally obtained results. The melt-pool width and depth increase with the increasing P/v. The melt-pool widths predicted by the point and volumetric heat source models both agree reasonably well with the experiments. However, in Figure 10b, it can be seen that the melt-pool depth is underestimated by the point heat source model at the powers of 20 W and 40 W. In comparison, the melt-pool depths predicted by the volumetric source model have good agreements with the experimentally measured values at all the four sets of laser power. This is because the different heat energy distributions within the laser scanning plane (i.e., the $x_1 - x_2$ plane) and along the depth direction can be well captured by the volumetric heat source model. In comparison, for the point heat source model, both the heat energy diffusion within the laser scanning plane and the the phenomenon of penetration along the depth direction cannot be accurately described. For the powers of 60 W and 80 W, the melt-pool depths happen to be approximately half of the melt-pool width, and thus, the predicted depths obtained by the point heat source model agree well with the experimental results.

A nonlinear thermal analysis using the finite element method was also conducted in [8], in which the nonlinear thermal properties of Ti6Al4V were employed. The temperature distributions along the width direction (see line 1 in Figure 9) at the powers of 40 W and 80 W at a certain time calculated by the proposed volumetric heat source model were compared with the results obtained by the nonlinear model in [8], as shown in Figure 11. Good agreements can be observed between the nonlinear model and the proposed approach. The disparities of the temperature obtained by the two models shown in Figure 11 are characterized by $\kappa = |T_{vol} - T_{non}|/T_{non}$, where T_{vol} is the temperature obtained by the proposed volumetric heat source model and T_{non} is the temperature obtained by the nonlinear model proposed in [8]. It is found the κ is less 10%. The small disparities are expected, as a nonlinear set of thermal properties which is temperature dependent is used in the nonlinear model while constant thermal properties are employed in the proposed model.

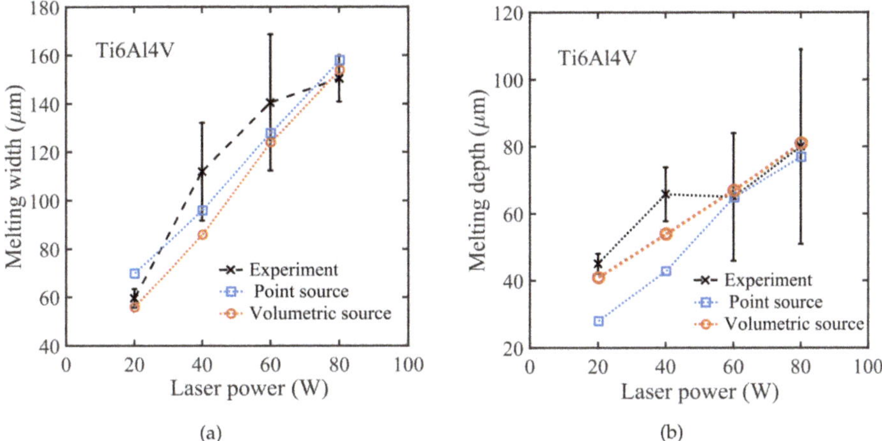

Figure 10. The predicted melt-pool (**a**) width and (**b**) depth with the experimental measurements.

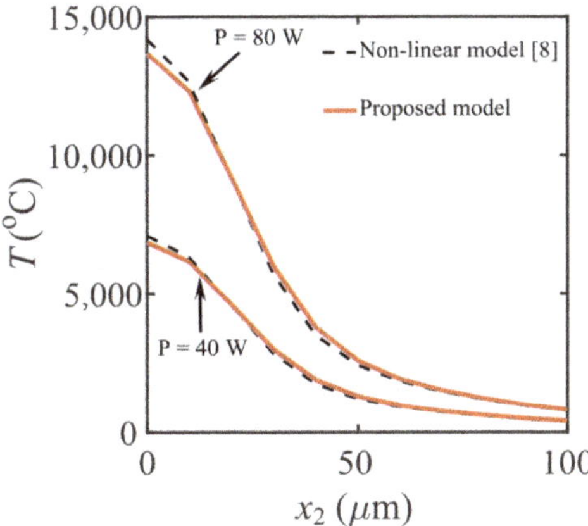

Figure 11. The temperature along the width direction (see line 1 in Figure 9) obtained by the proposed volumetric heat source model and the nonlinear model reported in [8].

3.3. Building a New Layer with Multiple Laser Scans

Finally, a numerical example for scanning a new layer with multiple scans is conducted. Two scanning strategies are considered in the proposed model. The material of Ti6Al4V is employed in this example. As shown in Figure 12a, a cube with the dimension of $l = 2$ mm is assumed already built, and two scanning strategies (unidirectional, see Figure 12b; and alternating, see Figure 12c) are considered to build the new layer. A total number of 64 cubic finite difference cells is used to discretize the body. The temperature of the bottom surface of the cube is set to be 20 °C during the whole process. An arbitrary set of process parameters is chosen to investigate the temperature histories of point G1 and G2 (see Figure 12a). The laser spot radius is set to be 50 μm and the OPD is 60 μm. The power of the laser is set to be 40 W with a speed of 0.8 m/s. The hatching distance between adjacent tracks is 90 μm and a total number of 21 tracks are applied on the top surface ∂V_{top}. The thermal response is calculated by the volumetric heat source model.

Figure 13a shows the temperature histories of point G1 under the unidirectional and alternating scanning strategies. The temperature evolutions under the two scanning strategies are well captured. The first peak temperature for the alternating scanning strategy appears earlier than that for the unidirectional scanning strategy. This is because compared with the unidirectional scanning strategy, the first track in the alternating scanning strategy arrives earlier at point G1. The second laser track is the same in both alternating and unidirectional scanning strategies, and thus, the second peak temperatures of point G1 for both scanning strategies occur at the same moment. It also demonstrates that the temperature history of point G1 is sensitive to the scanning strategy. The temperature field $\tilde{T}$ without considering the BCs under the unidirectional scanning strategy is also plotted in Figure 13a. The peak temperature of $\tilde{T}$ is only 1439 °C, while the peak temperature of T is 4493 °C. It can be observed that after including the BCs, the total temperature T is reasonably much higher than the $\tilde{T}$, which indicates it is essential to consider the BCs in the thermal modeling of the AM process.

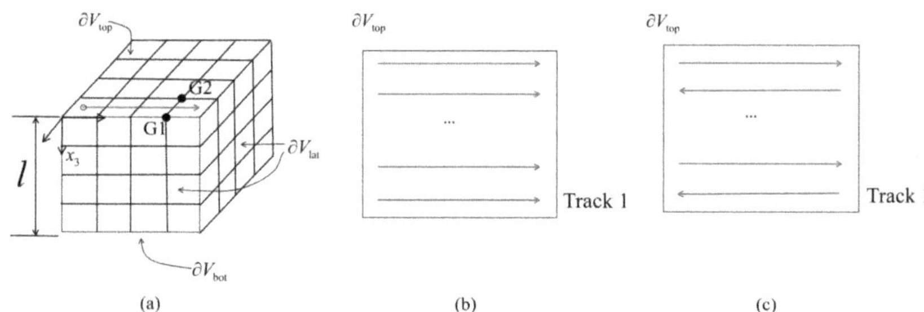

Figure 12. (**a**) A total of 21 laser tracks are applied on the top surface ∂V_{top} of the cube, which is discretized by 64 finite difference cells. The temperature histories of point G1 (1.5, 0, 0) mm and G2 (1.5, −0.5, 0) mm are investigated. Two scanning strategies, (**b**) unidirectional and (**c**) alternating, are employed.

The temperature histories of point G2 under the two different scanning strategies are compared in Figure 13b. It can be seen that the peak temperature of point G2 is reasonably lower than that of point G1 because point G1 is at the boundary. The peak temperature in Figure 13b is around 1000 °C, which is lower than the melting point 1650 °C. This indicates that the set of process parameters employed may cause the lack of fusion. It is also important to note that scanning strategies can modify the temperature histories which, in turn, have an impact on the resulting residual stresses.

The calculation time for this example by the volumetric heat source model is around 110 s. The calculation associated with $\tilde{T}$ field is only 0.45 s. The high efficiency is because the steep temperature gradient is captured by the $\bar{T}$ and $\hat{T}$ fields, which can be calculated analytically, and thus, coarse mesh can be applied in solving $\tilde{T}$. With this high computational efficiency, it would be very efficient to investigate the thermal histories of an AM part and optimize the process parameters. The calculation is performed using a single-core Intel i7 6600U quad-core processor with a clock speed of 2.60 GHz and 8 GB RAM.

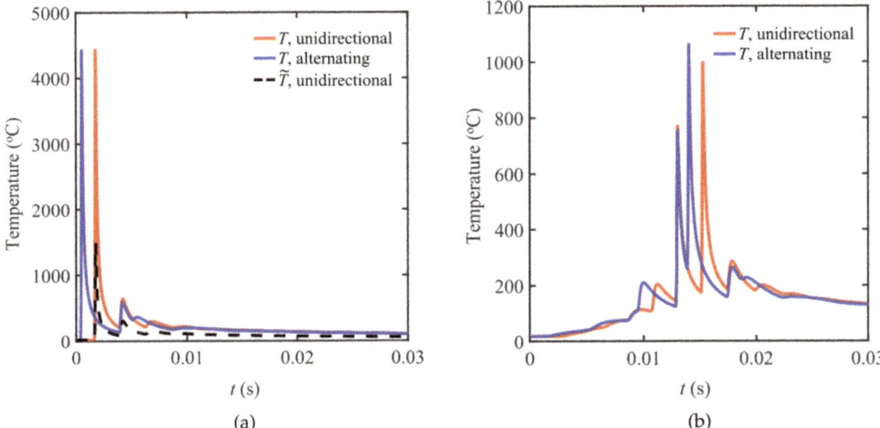

Figure 13. Temperature histories of (**a**) point G1 and (**b**) point G2.

4. Limitations and Future Extensions

In the proposed model, a half-ellipsoid heat source is employed to describe the interaction between the laser and the materials in SLM. Since the laser spot radius is usually small in SLM, it is appropriate to assume a circle heat source profile within the laser scanning plane ($x_1 - x_2$ plane). When the size of heat source becomes large—such as in wire arc additive manufacturing, for which the characteristic size of the heat source is usually in mm scale—a more complex heat source profile, such as the Goldak double ellipsoidal heat source [16], may need to be applied. Therefore, suitable heat source profiles for different metal AM processes with the corresponding analytical solutions of $\tilde{T}$ and Grad ($\tilde{T}$) need to be further developed. Moreover, in the half-ellipsoid heat source, the OPD of the laser is characterized by the constant parameter c, which is independent of the laser power and speed. However, the OPD may vary for different processing parameters, and this cannot be accounted for by the present model. Hence, a thermal model which is capable of characterizing the variance of OPD as a function of the process parameters is also of interest to investigate.

5. Conclusions

A computationally efficient volumetric heat source model based on the semianalytical thermal approach is proposed to simulate the thermal response of the produced part in the AM process. It has shown that the asymmetry of the melt pool in width and depth directions can be well captured by the proposed volumetric heat source model. Two materials of SS 316L and Ti6Al4V are employed to validate the accuracy of the proposed model. As the steep temperature gradients are mainly captured by the analytical field $\tilde{T}$ and the image source field $\check{T}$, which can be obtained analytically, high computational efficiency can be achieved in solving $\hat{T}$ field. Compared with the point heat source model, the volumetric heat source model has higher accuracy and in the meantime, the computational cost of the proposed volumetric heat source model is not impaired because of the concise expressions of $\tilde{T}$ and Grad($\tilde{T}$). Moreover, the computational efficiency of the proposed model is independent of the laser spot radius and OPD, indicating that various laser profiles can be easily modeled. The high computational efficiency of the proposed model suggests it is promising to be incorporated in an optimization process (optimizing the topology of the built part or process parameters) to reduce the residual stresses and distortions.

Author Contributions: Conceptualization, Y.Y.; methodology, Y.Y.; validation, Y.Y.; resources, X.Z.; writing–review and editing, Y.Y., X.Z. All authors have read and agreed to the published version of the manuscript.

Funding: This research was funded by the Science and Technology on Plasma Dynamics Laboratory, Air Force Engineering University, Xi'an, China (Project No. 614220206021808).

Conflicts of Interest: The authors declare no conflict of interest.

References

1. *Additive Manufacturing General Principles Terminology*; Standard, International Organization for Standardization: Geneva, Switzerland, 2015.
2. Francois, M.M.; Sun, A.; King, W.E.; Henson, N.J.; Tourret, D.; Bronkhorst, C.A.; Carlson, N.N.; Newman, C.K.; Haut, T.; Bakosi, J.; et al. Modeling of additive manufacturing processes for metals: Challenges and opportunities. *Curr. Opin. Solid State Mater. Sci.* **2017**, *21*, 198–206. [CrossRef]
3. Parry, L.; Ashcroft, I.; Wildman, R.D. Understanding the effect of laser scan strategy on residual stress in selective laser melting through thermo-mechanical simulation. *Addit. Manuf.* **2016**, *12*, 1–15. [CrossRef]
4. Zaeh, M.F.; Branner, G. Investigations on residual stresses and deformations in selective laser melting. *Prod. Eng.* **2010**, *4*, 35–45. [CrossRef]
5. Ding, J. Thermo-Mechanical Analysis of Wire and Arc Additive Manufacturing Process. Ph.D. Thesis, Cranfield University, Cranfield, Bedfordshire, UK, 2012. .
6. Hodge, N.; Ferencz, R.; Solberg, J. Implementation of a thermomechanical model for the simulation of selective laser melting. *Comput. Mech.* **2014**, *54*, 33–51. [CrossRef]
7. Chin, R.; Beuth, J.; Amon, C. Successive deposition of metals in solid freeform fabrication processes, Part 1: Thermomechanical models of layers and droplet columns. *J. Manuf. Sci. Eng.* **2001**, *123*, 623–631. [CrossRef]
8. Fu, C.; Guo, Y. 3-dimensional finite element modeling of selective laser melting ti-6al-4v alloy. In Proceedings of the Solid Freeform Fabrication Symposium 2014 Proceedings, Austin, TX, USA, 4–6 August 2014; pp. 1129–1144.
9. Ding, J.; Colegrove, P.; Mehnen, J.; Ganguly, S.; Almeida, P.S.; Wang, F.; Williams, S. Thermo-mechanical analysis of Wire and Arc Additive Layer Manufacturing process on large multi-layer parts. *Comput. Mater. Sci.* **2011**, *50*, 3315–3322. [CrossRef]
10. Bai, X.; Zhang, H.; Wang, G. Improving prediction accuracy of thermal analysis for weld-based additive manufacturing by calibrating input parameters using IR imaging. *Int. J. Adv. Manuf. Technol.* **2013**, *69*, 1087–1095. [CrossRef]
11. Zhao, H.; Zhang, G.; Yin, Z.; Wu, L. A 3D dynamic analysis of thermal behavior during single-pass multi-layer weld-based rapid prototyping. *J. Mater. Process. Technol.* **2011**, *211*, 488–495. [CrossRef]
12. Zhang, Y.; Guillemot, G.; Bernacki, M.; Bellet, M. Macroscopic thermal finite element modeling of additive metal manufacturing by selective laser melting process. *Comput. Methods Appl. Mech. Eng.* **2018**, *331*, 514–535. [CrossRef]
13. Kruth, J.P.; Deckers, J.; Yasa, E.; Wauthlé, R. Assessing and comparing influencing factors of residual stresses in selective laser melting using a novel analysis method. *Proc. Inst. Mech. Eng. Part B J. Eng. Manuf.* **2012**, *226*, 980–991. [CrossRef]
14. Yang, Y.; Knol, M.; van Keulen, F.; Ayas, C. A Semi-Analytical Thermal Modelling Approach for Selective Laser Melting. *Addit. Manuf.* **2018**, *21*, 284–297. [CrossRef]
15. Yang, Y.; van Keulen, F.; Ayas, C. A computationally efficient thermal model for selective laser melting. *Addit. Manuf.* **2020**, *31*, 100955. [CrossRef]
16. Goldak, J.; Chakravarti, A.; Bibby, M. A new finite element model for welding heat sources. *Metall. Trans. B* **1984**, *15*, 299–305. [CrossRef]
17. Flint, T.; Francis, J.; Smith, M.; Balakrishnan, J. Extension of the double-ellipsoidal heat source model to narrow-groove and keyhole weld configurations. *J. Mater. Process. Technol.* **2017**, *246*, 123–135. [CrossRef]
18. Li, J.; Li, L.; Stott, F. Comparison of volumetric and surface heating sources in the modeling of laser melting of ceramic materials. *Int. J. Heat Mass Transf.* **2004**, *47*, 1159–1174. [CrossRef]
19. Gusarov, A.; Yadroitsev, I.; Bertrand, P.; Smurov, I. Model of radiation and heat transfer in laser-powder interaction zone at selective laser melting. *J. Heat Transf.* **2009**, *131*, 072101. [CrossRef]
20. Kempen, K.; Vrancken, B.; Thijs, L.; Buls, S.; Van Humbeeck, J.; Kruth, J.P. Lowering thermal gradients in Selective Laser melting by pre-heating the baseplate. In Proceedings of the Solid Freeform Fabrication Symposium Proceedings, Austin, TX, USA, 12–14 August 2013.

21. Sato, Y.; Tsukamoto, M.; Masuno, S.; Yamashita, Y.; Yamashita, K.; Tanigawa, D.; Abe, N. Investigation of the microstructure and surface morphology of a Ti6Al4V plate fabricated by vacuum selective laser melting. *Appl. Phys. A* **2016**, *122*, 439. [CrossRef]
22. Yang, Y.; Ayas, C. Computationally efficient thermal-mechanical modelling of selective laser melting. AIP Conference Proceedings. *AIP Publ.* **2017**, *1896*, 040005.
23. Childs, T.; Berzins, M.; Ryder, G.; Tontowi, A. Selective laser sintering of an amorphous polymer—Simulations and experiments. *Proc. Inst. Mech. Eng. Part B J. Eng. Manuf.* **1999**, *213*, 333–349. [CrossRef]
24. Carslaw, H.S.; Jaeger, J.C. *Conduction of Heat in Solids*, 2nd ed.; Clarendon Press: Oxford, UK, 1959.
25. Khairallah, S.A.; Anderson, A.T.; Rubenchik, A.; King, W.E. Laser powder-bed fusion additive manufacturing: Physics of complex melt flow and formation mechanisms of pores, spatter, and denudation zones. *Acta Mater.* **2016**, *108*, 36–45. [CrossRef]
26. Khairallah, S.A.; Anderson, A. Mesoscopic simulation model of selective laser melting of stainless steel powder. *J. Mater. Process. Technol.* **2014**, *214*, 2627–2636. [CrossRef]
27. Wits, W.W.; Bruins, R.; Terpstra, L.; Huls, R.A.; Geijselaers, H. Single scan vector prediction in selective laser melting. *Addit. Manuf.* **2016**, *9*, 1–6. [CrossRef]

Publisher's Note: MDPI stays neutral with regard to jurisdictional claims in published maps and institutional affiliations.

Article

Thermomechanical Simulations of Residual Stresses and Distortion in Electron Beam Melting with Experimental Validation for Ti-6Al-4V

Fawaz M. Abdullah [1],*, Saqib Anwar [1] and Abdulrahman Al-Ahmari [1,2]

[1] Industrial Engineering Department, College of Engineering, King Saud University,
P. O. Box 800, Riyadh 11421, Saudi Arabia; sanwar@ksu.edu.sa (S.A.); alahmari@ksu.edu.sa (A.A.-A.)
[2] Raytheon Chair for Systems Engineering (RCSE Chair), King Saud University,
P.O. Box 800, Riyadh 11421, Saudi Arabia
* Correspondence: fmuthanna@ksu.edu.sa; Tel.: +00966-553989419

Received: 13 July 2020; Accepted: 20 August 2020; Published: 25 August 2020

Abstract: Electron beam melting (EBM) is a relatively new process in three-dimensional (3D) printing to enable rapid manufacturing. EBM can manufacture metallic parts with thin walls, multi-layers, and complex internal structures that could not otherwise be produced for applications in aerospace, medicine, and other fields. A 3D transient coupled thermomechanical finite element (FE) model was built to simulate the temperature distribution, distortion, and residual stresses in electron beam additive manufactured Ti-6Al-4V parts. This research enhances the understanding of the EBM-based 3D printing process to achieve parts with lower levels of residual stress and distortion and hence improved quality. The model used a fine mesh in the layer deposition zone, and the mesh size was gradually increased with distance away from the deposits. Then, elements are activated layer by layer during deposition according to the desired material properties. On the top surface, a Gaussian distributed heat flux is used to model the heat source, and the temperature-dependent properties of the powder and solid are also included to improve accuracy. The current simulation has been validated by comparing the FE distortion and temperature results with the experimental results and other reported simulation studies. The residual stress results calculated by the FE analysis were also compared with the previously reported simulation studies on the EBM process. The results showed that the finite element approach can efficiently and accurately predict the temperature field of a part during the EBM process and can easily be extended to other powder bed fusion processes.

Keywords: additive manufacture; Ti-6Al-4V; finite element analysis; temperature distribution; distortion; residual stress; experimental validation

---

1. Introduction

Due to current rapid changes in technology, manufacturing engineering is receiving a lot of attention. Manufacturing methods and designs have been improved to make life easier and more straightforward, especially concerning technologies that allow for accurate and fast manufacturing. Electron beam melting (EBM) is a relatively new additive manufacturing technology based on the powder bed fusion process [1]. The components can be produced on a layer-by-layer basis by melting the powder metal using an electron beam. The energy density of an electron beam is high enough to melt a wide variety of metals and alloys. The material then cools and solidifies to form a fully dense geometry [2]. A unique feature of EBM is its ability to fabricate complex geometries and structures (e.g., meshed, porous, cellular). Schwerdtfeger et al. used electron beam melting to manufacture a unique structure that exhibited "auxetic behavior" a negative Poisson's ratio, and high impact and shear resistance [3]. EBM processes have the potential to work with many material classes, for example, aluminum alloys [4], steel (H13) [5], cobalt-based super-alloys [6], and titanium alloys [7]. The material used in this study is Ti-6Al-4V, which has many applications, such as medical and aerospace, due to its properties, including good biocompatibility and corrosion resistance, low density, human allergic response, and mechanical strength [8].

Despite a great interest in obtaining part accuracy using the electron beam melting process [9], there have been only a few studies reporting on the geometric aspects of electron beam melting. The accuracy of the EBM parts is greatly affected by the residual stresses and thermal distortions resulting from rapid heating and cooling cycles [10], which are not well understood. These are important issues because they severely degrade the dimensional accuracy and mechanical performance of the components. It is crucial to study, understand, and control distortion and residual stresses through the simulation of the additive manufacturing process. Some methods have been applied to decrease distortion and residual stresses in parts produced by similar processes (e.g., welding). These methods include the use of appropriate design approaches, such as presetting/offsetting, mechanical restraint, preheating, limiting heat input, controlling process parameters, and applying mechanical stress relief techniques [11].

Research work has been undertaken to investigate the thermal phenomena of the EBM process with the help of finite element (FE) modeling. Some studies have been carried out on this issue, such as Denlinger et al. [12] developed a FE model for predicting the in situ thermomechanical response of Ti-6Al-4V during the electron beam melting process. The 3D thermomechanical analysis is performed to model the distortion, residual stress in the workpiece, and experimental temperature in situ. The distortion measurements were performed during the deposition of a single-bead-wide of 16 layers to validate the model. Both in situ deformation and post-process residual stress measurements indicate that stress relaxation occurs during Ti-6Al-4V deposition. The results showed that failure to apply stress relaxation in the constituent model leads to errors in residual stress and situ deformation predictions of more than 500% when compared to experimental measurements. Cheng et al. [13] developed a FE model to simulate the thermomechanical process in EBM to build overhanging parts. A two-dimensional thermomechanical model was developed by using ABAQUS to simulate residual stress distribution, temperature, and overhanging part model distortions. The process parameters included in the FE model were beam scanning speed, beam diameter, and beam current. Thermomechanical characteristics such as thermal gradients and thermal stresses around the overhang build were evaluated and analyzed. The main results showed that the overhang areas have a higher maximum temperature, higher tensile stress, and greater distortion than the areas above the solid base plate.

Ninggang [14] utilized a 3D FE model in ABAQUS to study the complex thermal process of Electron-beam additive manufacturing (EBAM). The finite element model was developed to simulate the transfer of transit heat in one part during EBAM under a moving heat source. The selected parameters in the FE model included efficiency coefficient, voltage, current, penetration, and beam diameter. The key findings include the following: (1) concerning the state of the powder layer, the size of the melt pool is greater with higher maximum temperature compared to a solid layer, which indicates the importance of taking powder into account for the accuracy of the model. (2) The diameter of the larger electron beam will reduce the maximum temperature in the melting trough, and temperature gradients can be much smaller, which gives a lower cooling rate. In another study [15], a 3D transient fully coupled thermomechanical model was built to study the deformation and residual stress in the EBM of Ti-6Al-4V plates. Single-layer, 6-layers, and 11-layers were optimized using both simulations and experiments to ensure the model accuracy and provide confidence in using simulations to optimize the process. The process parameters used in this study were current, voltage, and speed. The results showed that preheating at least twice is an effective way to reduce both distortion and residual stresses. The results also showed that, instead of distortion accumulating in a monotonous manner as more layers are added, the distortion increases as the first 2–3 layers are deposited, and then decreases with the deposition of additional layers.

Umer et al. [16] used a two-dimensional thermomechanical FE model to predict the stresses and distortions associated with the manufacture of overhang structures by EBM for Ti-6Al-4V alloys. The EBM process parameters were (Beam current, speed, voltage and etc.). Various support structure geometries were designed and evaluated. The key findings of this study are as follows. (1) Temperature predictions during each accumulated layer could be useful in determining the phase transform from powder to the liquid and from liquid to solid. (2) The residual stress patterns of the deposited layers with supports were found to be completely different from the unsupported model. Also, the overhang regions showed fluctuating stresses from tensile to compression.

It can be seen from the reviewed literature that very few papers have reported on the modeling of the electron beam melting process. It is worth mentioning that all the reported FE models on the EBM are focused on thick parts (>1 mm). At the same time, thin-walled parts (<1 mm) show different behavior during EBM as compared to the thick parts. Furthermore, none of the reported models consider the combined effect of real conditions, such as preheating temperature, the energy of the electron beam, cooling time, beam scanning path, and beam speed, when applying the FE modeling approach. The current study takes into consideration the real conditions, which were not incorporated into the previous finite element based model for analyzing the EBM process. The objective of this work is to develop a finite element model to study and understand the thermomechanical characteristics of the thin-walled Ti-6Al-4V parts during the electron beam melting process. According to previous studies, the thin-walled EBM parts have good stability to control the surface roughness [17,18]. The thin wall EBM parts were selected due to their significant applications such as biomedical implants with a very limited filling area [19]. A three-dimensional (3D) fully coupled temperature-displacement finite element model is developed to simulate the thermomechanical behavior in the electron beam melting of thin-walled Ti-6Al-4V components.

2. Materials and Methods

2.1. Finite Element Modeling

Finite element analysis was run using the commercial software ABAQUS 6.13 (Dassault Systèmes, Paris, France) to simulate the temperature history, residual stresses distribution, and distortion in multi-layer raster scanning and deposition during EBM of thin-walled Ti-6Al-4V parts. The thermomechanical analysis was performed using a 3D transient, fully coupled temperature-displacement approach. The finite element analysis package Abaqus/Standard was used to simulate the electron beam melting process.

The model was divided into two sections; the stainless-steel base-plate (also known as the substrate) and the powder layers of Ti-6Al-4V. Figure 1 shows a view of the FE model of the electron beam melting process consisting of a base-plate and some layers of Ti-6Al-4V. The dimensions of the base plate are 10 mm (length) × 3.6 mm (width) × 1.5 mm (height). The dimensions of each powder layer are 5 mm (length) × 0.6 mm (width) × 0.05 mm (thickness). It should be noted that, after the deposition of several layers, the width of the layers will become the thickness of the final part. Further, the minimum thickness of a part that can be produced in the ARCAM AB machine is 0.6 mm. Furthermore, the thickness of 0.05 mm (50 μm) is the recommended layer deposition thickness from ARCAM AB for producing Ti-6Al-4V parts. The 5 mm length of each layer was based on the previous FE modeling studies reported on the EBM of Ti-6Al-4V [20]. The FE model was developed in two configurations for studying the thermomechanical behavior of the thin-walled parts. In the first configuration, the FE model consists of five EBM layers and was used to study the residual stresses and temperature distributions in the deposited layers. In the second configuration, the FE model comprises of fifteen EBM layers to study the distortion and variation of thermal gradient in the deposited layers. It should be noted that the initial plan was to run the second configuration of the FE model with fifty EBM layers so that the distortion simulated in the FE model can be directly compared with the 2.5 mm height EBM fabricated parts (50 layers).

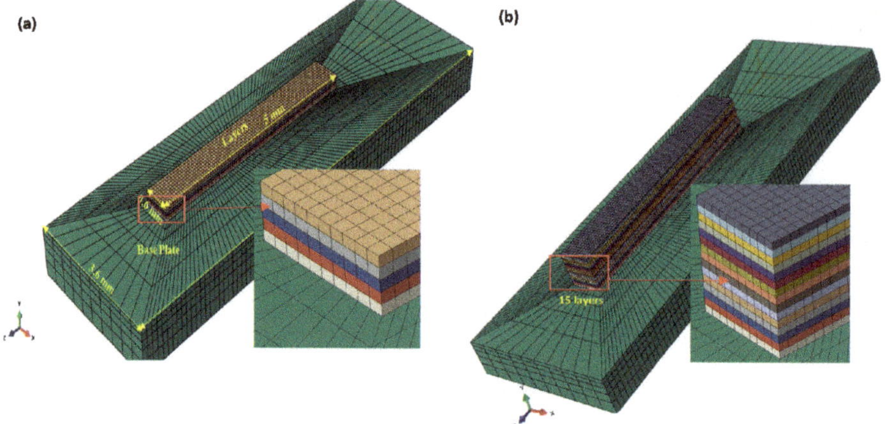

Figure 1. FE model of the EBM process, (**a**) with five layers, (**b**) with fifteen layers.

2.1.1. Material Properties

It should be noted that the EBM process is a thermomechanical process, so the temperature- dependent properties of the material must be used. In the current model, the temperature-dependent thermal and mechanical properties of the base plate and Ti-6Al-4V are considered, as listed in Tables 1 and 2, respectively. It should be noted that the main difference in the powder and solid Ti-6Al-4V is between the thermal conductivity (see Table 2), other properties are almost the same as mentioned in [21]. Also, it can be noticed in Table 2 that the thermal conductivity at T = 1950 K is quite high because the Ti-6Al-4V transform to the liquid phase and exists as a liquid between 1950 < T < 2700 K, as reported by [22]. The thermal conductivity of the liquid Ti-6Al-4V is quite high as compared to the solid-state [22–24]. It should be noted that the melt pool dynamics depend on the EBM process parameters as well as on the properties of the material used. However, the melt the pool is most sensitive to the thermal conductivity for Ti-6Al-4V, as explained by [25].

Table 1. Material properties for the 304 stainless steel base plate [26].

Temperature (K)	Density (kg/m^3)	Heat Capacity (J/(kg K))	Thermal Conductivity (W/(m K))	Thermal Expansion Coefficient (m/(m K))	Yield Strength (MPa)	Young's Modulus (GPa)	Poisson's Ratio
273	0.0079	462	0.0146	1.7×10^{-5}	265	198.5	0.294
373	0.00788	496	0.0151	1.7×10^{-5}	218	193	0.295
473	0.00783	512	0.0161	1.8×10^{-5}	186	185	0.301
573	0.00779	525	0.0179	1.9×10^{-5}	170	176	0.31
673	0.00775	540	0.018	1.9×10^{-5}	155	167	0.318
873	0.00766	577	0.0208	2.0×10^{-5}	149	159	0.326
1073	0.00756	604	0.0239	2.0×10^{-5}	91	151	0.333
1473	0.00737	676	0.0322	2.1×10^{-5}	25	60	0.339
1573	0.00732	692	0.0337	2.1×10^{-5}	21	20	0.342
1773	0.00732	935	0.12	2.2×10^{-5}	10	10	0.388

Table 2. Material properties for Ti-6Al-4V [15,27].

Temperature (K)	Density (kg/m^3)	Heat Capacity (J/(kg K))	Thermal Conductivity (W/(m K)) Solid	Thermal Conductivity (W/(m K)) Powder [21]	Thermal Expansion Coefficient (m/(m K))	Yield Strength (MPa)	Young's Modulus (GPa)	Poisson's Ratio
293	4420	546	7	0.2	9×10^{-6}	850	102	0.345
400	4402	567	7.8	-	9.16×10^{-6}	720	101	0.35
500	4391	591	8.9	-	9.31×10^{-6}	680	95	0.355
600	4376	611	10.5	-	9.46×10^{-6}	630	91	0.36
700	4361	636	11.7	-	9.61×10^{-6}	590	85	0.365
800	4345	656	13	-	9.76×10^{-6}	540	80	0.37
900	4331	679	14.5	-	9.90×10^{-6}	490	75	0.375
1000	4319	699	16.2	-	1.01×10^{-5}	450	70	0.385
1100	4303	719	18.4	-	1.02×10^{-5}	400	65	0.395
1200	4289	733	20.1	-	1.04×10^{-5}	360	60	0.405
1300	4278	647	19.7	-	1.05×10^{-5}	315	35	0.43
1400	4264	664	21.7	-	1.06×10^{-5}	268	20	0.43
1950	4189	790	72	28.3	1.10×10^{-5}	20	10	0.43

2.1.2. Initial and Boundary Conditions

The preheating temperature of 1003 K was applied as an initial condition to the base plate at t = 0 s as shown in Figure 2a. The preheating temperature was 1003 K, as reported in previous studies [13]. Moreover, all the external surfaces of the base plate and the deposited layers were assigned radiation boundary conditions. Equation (1) shows the heat loss by radiation surrounding the layers

$$Q = -A{\cdot}\sigma{\cdot}\varepsilon \left(T^4 - T_a^4\right) \tag{1}$$

where A is the surface area, σ is the Stefan–Boltzmann constant, ε is the emissivity, and T_a is the ambient temperature. The negative sign indicates the heat loss due to radiation. The transient temperature field T (x y z t) throughout the domain was obtained by solving the 3D heat conduction equation, Equation (2)

$$\rho C\frac{\partial T}{\partial t} = \frac{\partial}{\partial x}\left(k\frac{\partial T}{\partial x}\right) + \frac{\partial}{\partial y}\left(k\frac{\partial T}{\partial y}\right) + \frac{\partial}{\partial z}\left(k\frac{\partial T}{\partial z}\right) + Q \tag{2}$$

where T is the temperature, ρ is the density, C is the specific heat, k is the heat conductivity, and Q is the internal heat generation per unit volume. For thermoelasticity, the materials expand or contract as temperature changes; therefore thermal strain, which depends on current and initial temperatures, is an important part of the total strain. The total strain in the component, $\varepsilon^{\text{Total}}$ can be represented as

$$\varepsilon^{\text{Total}} = \varepsilon^P + \varepsilon^e + \varepsilon^T \tag{3}$$

where $\varepsilon^{\text{Total}}, \varepsilon^P, \varepsilon^e$, and ε^T are the total strain, plastic strain, elastic strain, and thermal strain, respectively. However, the plastic strain (ε^P) was not computed in the current model, as the FE model was set to perform the elastic thermal analysis during the EBM process. The constitutive equation for the stress can be written as

$$a = C\left(\varepsilon^e + \varepsilon^T\right) \tag{4}$$

where a is the stress and C is the fourth-order material stiffness tensor. To better illustrate the constitutive law, the equation can be written in indicial notation as follow:

$$Q_{ij} = \frac{E}{(1+V)(1-2V)}\left(V\delta_{ij}\varepsilon_{kk} + (1+v)\varepsilon_{jj} - (1+v)\alpha\Delta T\delta_{ij}\right) \tag{5}$$

where (i, j, k = 1, 2, 3), v is Poisson's ratio, E is Young's modulus, and δ_{ij} is the Kronecker delta. The elastic behavior was modeled using the isotropic generalized Hooke's law.

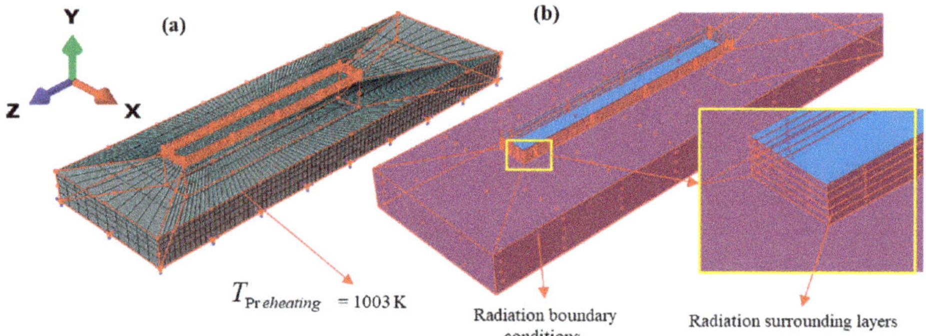

Figure 2. (**a**) Preheating temperature for the whole model. (**b**) Radiation surrounding the base plate and deposited layers.

The temperature-dependent material properties have been used to improve accuracy. Thermal radiation is considered as the boundary condition, described by Newton's law of cooling and the Stefan-Boltzmann law, as the part is produced by EBM in a full vacuum. The ARCAM AB vacuum system provides a basic pressure of 5×10^{-5} mbar or better throughout the entire building cycle. During the process, the partial pressure of He is 4×10^{-3} mbar. This confirms a clean and controlled building environment, which is important for maintaining the chemical specifications of building materials. However, the convection between the environment and the powder layers due to the vacuum is not included. Hence, the radiation is taken to be the heat transfer between part/powder and the surroundings. Radiation from the side of the base plate, sides of layers, and any free surfaces on the model have been used, as shown in Figure 2b.

2.1.3. Electron Beam Modeling

The heat distribution from the electron beam was simulated as a conical moving heat source with a Gaussian distributed intensity at each depth [14]. In this research, the heat source was modeled as Equation (6). The intensity decreases with an increase in penetration depth.

$$S(x, y, z) = f(z) \frac{8 \eta U I_b}{\pi \Phi_E^2} exp \left\{ - \frac{8(x - x_s)^2 + (y - y_s)^2}{\Phi_E^2} \right\} \tag{6}$$

whereas *f(z)* is given by Equation (7)

$$f(z) = \frac{2}{h}\left(1 - \frac{z}{h}\right) \tag{7}$$

The terms used in Equations (3) and (4) are defined as follows: $(x,,z)$ is the volumetric heat source along X, Y, and Z axes, efficiency coefficient η (0.9), voltage U (60 kV), beam current I_b (0.002 mA), beam penetration h (0.05 mm), and beam diameter Φ_E (0.3 mm). x_s and y_s are the positions of the heat source (electron beam) center.

A user subroutine DFLUX coded in FORTRAN was used to apply the heat source in the FE model. The subroutine DFLUX was called under the interaction module so that contact could be established between the heat source (electron beam spot) and layers, as shown in Figure 3. It is called at the beginning of each iteration and reads the simulation time to determine the position of the heat source center so that the domain of the volumetric heat flux can be established. In this study, the penetration of the beam was chosen to be the same as the thickness of the layer, as adopted by [28].

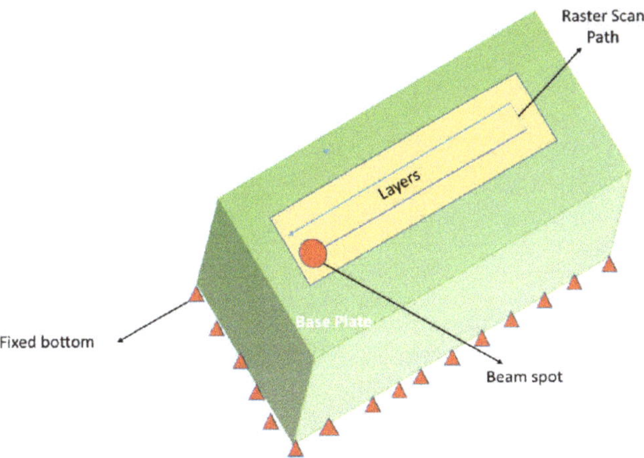

Figure 3. Electron beam movement.

2.2. Contact

Contact between the base plate material and powder layers was assumed to be rigid. The tie constraint in ABAQUS was used to transfer the heat from the first layer to the base plate, and also from the nodes of the top layers to the bottom layers, as shown in Figure 4. The tie constraint locks the temperature and deformation degrees of freedom between the nodes such that the slave nodes (bottom nodes) follow the master nodes (top nodes). The characteristics of the tie constraints are:

- The deformation effect will be transferred from one layer to another layer through the tied nodes.
- Heat degree of freedom will be transferred from one layer to another through the tied nodes.

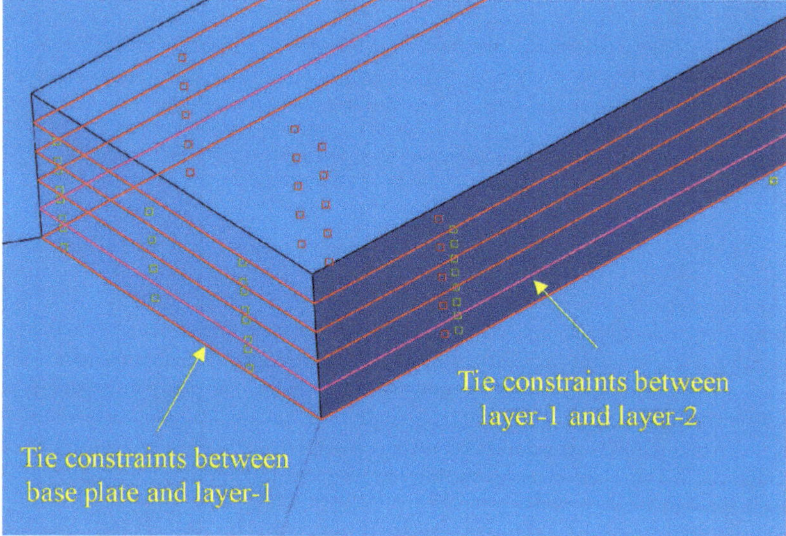

Figure 4. Contact between the base plate material and the powder layers, and between successive layers.

2.3. FE Mesh

The mesh size for the model was selected after several trials. The model was run with different mesh sizes, and the results from each model were compared. Hence, fine meshes were used in the layers deposition zone, and the mesh size was gradually increased with distance from the deposits. In regions more separated from the heat-affected area, coarser meshes were utilized. The C3D8T coupled thermal-displacement elements were used to mesh the base plate and the powder layers. Mesh sensitivity analysis was carried out to ensure that the used mesh (element) size was neither too time-consuming nor leading to discretization errors. Eight node coupled thermal-displacement elements (C3D8T) were selected to mesh the layers with an element size dimension of 0.06 mm × 0.06 mm × 0.05 mm. The same element type was used in the base plate. The size of the elements in the fine mesh region in the base plate was also the same as for the layers. Figure 5 shows the mesh used in this model.

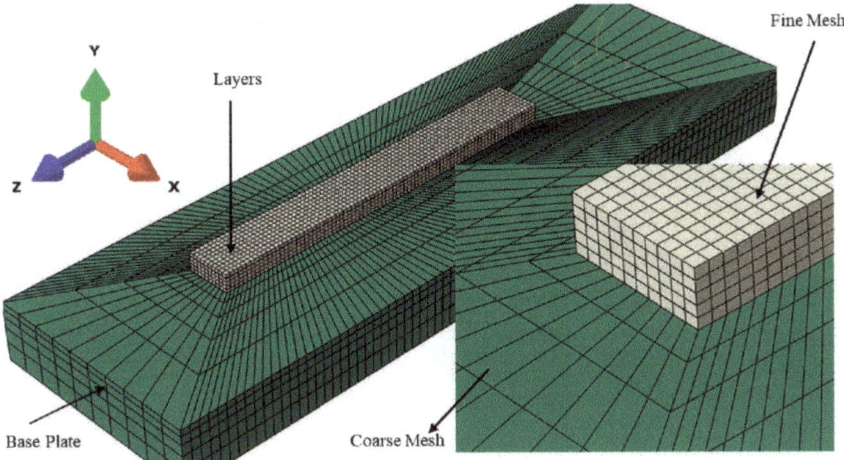

Figure 5. Meshed FE model.

2.4. Powder Addition in a Layer by Layer Fashion

To add layers one by one in the FE model, the model change option in ABAQUS was used. At a step in time, a set of elements was added onto the base plate by activating the layer to form rectangular deposits along the centerline of the base plate by using the model change option in the interaction module. However, at the first step, all layers were deactivated except layer one (L1). After the melting phase in the L1, the cooling phase (Step-2 in Figure 6) was followed in this layer. The time for the cooling step was assumed to be 3 s [29]. After the cooling time ended, the second layer (L2) was activated on top of the previously melted and cooled L1, and so on for the remaining layers, as shown in Figure 6. Once a layer is activated, it remains active for the remaining analysis to take part in the heating, conduction, and radiation phenomena.

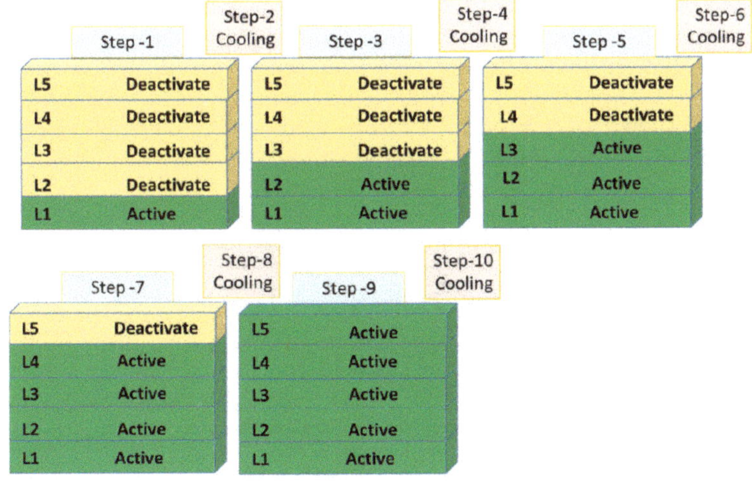

Figure 6. Using model change to activate the deactivated layers during the EBM process.

The whole model (base plate and powder layers) was assigned with a uniform temperature distribution of T-preheat as the initial thermal condition. During the simulation, the electron beam traveled along the X and Z-axes on the top surface of each powder layer. After the melting phase, the phase was changed from powder to solid phase, and then the simulation of the cooling phase followed. The field variables in ABAQUS were used to transform the martial properties from powder to solid-state. During the electron beam scanning in Step-1 (melting phase), the Ti-6Al4V properties were maintained as powder by activating the field variable 1. After the electron beam scan was accomplished, Ti-6Al4V properties were changed to solid by activating the field variable 2 in Step-2 (cooling phase). This process was repeated in the subsequent melting and cooling steps. The field variable can be controlled in the ABAQUS within the property module, step module, and edit keywords options. The time for the cooling step was assumed to be 3 s, as shown in Table 3. Five layers were built with ten steps (transient fully coupled temperature-displacement) to handle the model. The last cooling step was 1100 s, which was the time needed for the temperature of the EBAM part to drop to a room temperature of 293 K.

Table 3. Model transformation phases.

Layer	Phase	Step	Description	Time (s)	Total Time (s)
Complete model	Preheating	Initial	Preheating phase for the whole model at 1003k	0	0
L1	Melting	1	Electron beam scanning along X-Z direction	0.0265	0.0265
	Cooling	2	Transformation of material from powder to solid-state then cooling phase	3	3.0265
L2	Melting	3	Activation of layer-2 and scanning of electron beam along X-Z direction	0.0265	3.0795
	Cooling	4	Transformation of material from powder to solid-state then cooling phase	3	6.0795
L3	Melting	5	Activation of layer-3 and scanning of electron beam along X-Z direction	0.0265	6.106
	Cooling	6	Transformation of material from powder to solid-state then cooling phase	3	9.106
L4	Melting	7	Activation of layer-4 and scanning of electron beam along X-Z direction	0.0265	9.1325
	Cooling	8	Transformation of material from powder to solid-state then cooling phase	3	12.1325
L5	Melting	9	Activation of the next layer and scanning of electron beam along X-Z direction	0.0265	12.159
	Cooling	10	Transformation of material from powder to solid-state then cooling phase	1100	112.16
Total time					1112.16

2.5. Fabrication of the Thin-Walled EBM Parts

To validate the FE model results, thin-walled Ti-6Al-4V parts were produced via the electron beam melting machine from ARCAM A2 machine (ARCAM AB, Mölndal, Sweden). The particle size of the Ti-6Al-4V powder was between 10 μm and 50 μm in which the particle diameter of 37 μm occupies the highest proportion, and the layer thickness of 50 μm was used [30]. The thickness of the thin-walled parts was set at 0.6 mm, and the length of the parts was set as 5 mm. Two types of thin-walled parts were produced; one with a height of 5 mm (100 EBM layers), and second with the height of the 2.5 mm (50 EBM layers). The layout of the thin-walled parts fabricated on the stainless steel base plate is shown in Figure 7. Parts were fabricated by using the ARCAM AB recommended process parameters, as listed in Table 4. It should be noted that the 0.6 mm thickness is the minimum width of the parts that can be produced with the selected Ti-6Al-4V powder using the ARCAM AB recommended process parameters (see Table 4). A profile projector was used to measure the distortion in the thin-walled parts, as shown in Figure 8. The heights of the EBM parts were more than the heights of the parts simulated in the FE model due to the size constraints of the distortion measuring equipment (profile projector). The distortion in the bigger height parts can be measured more conveniently and reliably as compared to the small height parts, i.e., 0.75 mm as produced in simulations due to computational constraints.

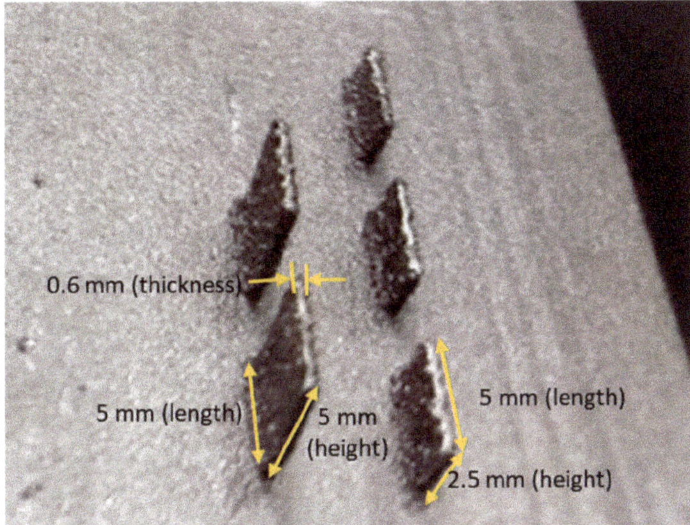

Figure 7. EBM printed thin-walled parts for model validation.

Table 4. Parameters used in the simulations and for producing the EBM parts [29].

Parameters	Values
Electron beam diameter, Φ (mm)	0.3
Scan speed, v (mm/s)	400
Acceleration voltage, U (kV)	60
Beam current, Ib (mA)	0.002
Powder layer thickness, $tlayer$ (mm)	0.05
Beam penetration depth, dP (mm)	0.05
Preheat temperature, $Tpreheat$ (k)	1003

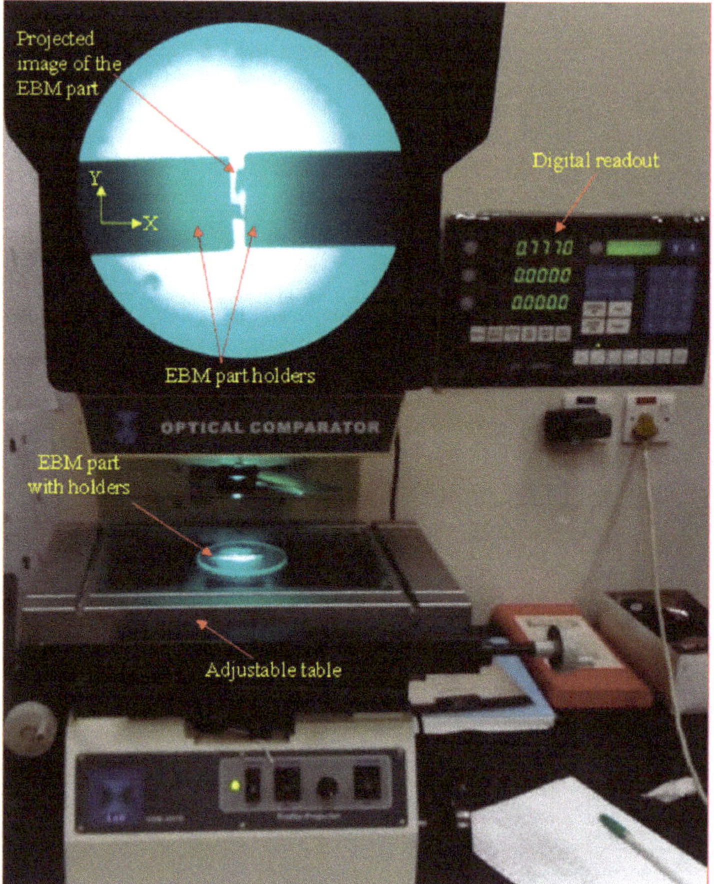

Figure 8. Profile projector setup for measuring the distortion in the EBM parts.

A white light interferometer based 3D profilometer Contour GTK from Bruker (Tucson, AZ, USA) was used to measure the distortion profile on the base plate after the removal of the thin-walled parts, as shown in Figure 9. A 5× interferometer lens with a fixed field of view 2.2 mm × 1.7 mm was used to capture the distortion profile on the base plate.

Furthermore, the microstructures of the thinned walled parts were also revealed on the side faces (0.6 mm thickness side). The samples were first ground with SiC papers P180, P300, P600, P800, P1000, and P1500 and then polished with Al_2O_3 suspension followed by etching with Kroll's agent. The microstructures were obtained primarily to link the microstructural variations of the alpha and beta phases along the thin side faces with the recorded temperature distributions from the FE simulations. This is because the percentage of the alpha and beta phases and the size of these phases are significantly affected by the temperature input. The developed FE model provided the opportunity to study the temperature–time history, and to link it with the achieved microstructures.

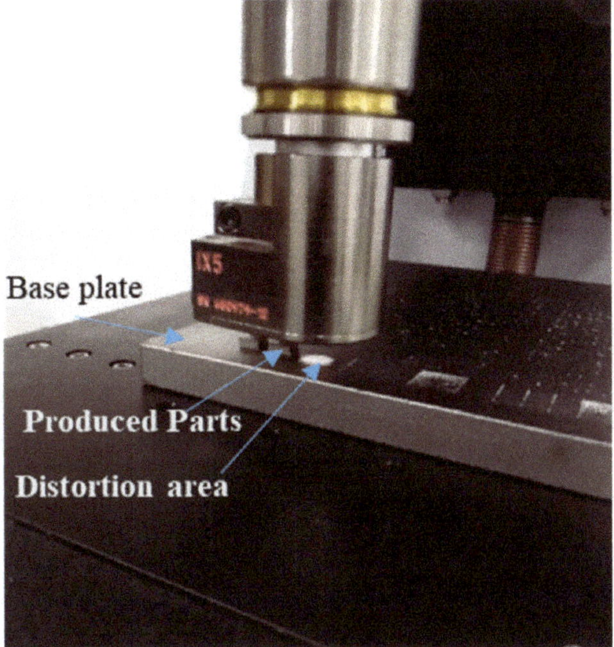

Figure 9. The Contour GTK 3D profilometer from Bruker.

3. Results and Discussion

The results and discussion are arranged as follows: temperature field, residual stresses, thermal distortion, experimental validation, and microstructure evolution.

3.1. Temperature Field

Figure 10 shows the contour and cross-section plots of the temperature field of the melt pool and surrounding areas at different times. The cross-sections aim to show the effect of heat from the electron beam on the EBM layers and base plate. Figure 10 also shows that the peak temperature during the process in the first layer was 2702 K, while the highest temperature during the layer-5 deposition was 3456 K. It can be concluded that the temperature increases from one layer to another due to heat accumulation between layers as the process proceeds. It should be noted that, in some regions in the model, the lowest temperature was below the preheating temperature because the radiation boundary conditions were defined on all the external surfaces of the base plate and the deposited layers. Furthermore, the preheating temperature was defined as an initial temperature, which was subject to change as the model computes. Therefore, in some regions in the model, the temperature goes below the initial preheat temperature due to the radiation effect. Also, in some regions, the temperature goes higher due to the heat input from the electron beam. The same methodology was adopted in other studies as well reported by Cheng, Lu, and Chou [31,32]. They showed that after the cooling step, the temperature goes below the preheating temperature due to the radiation effect.

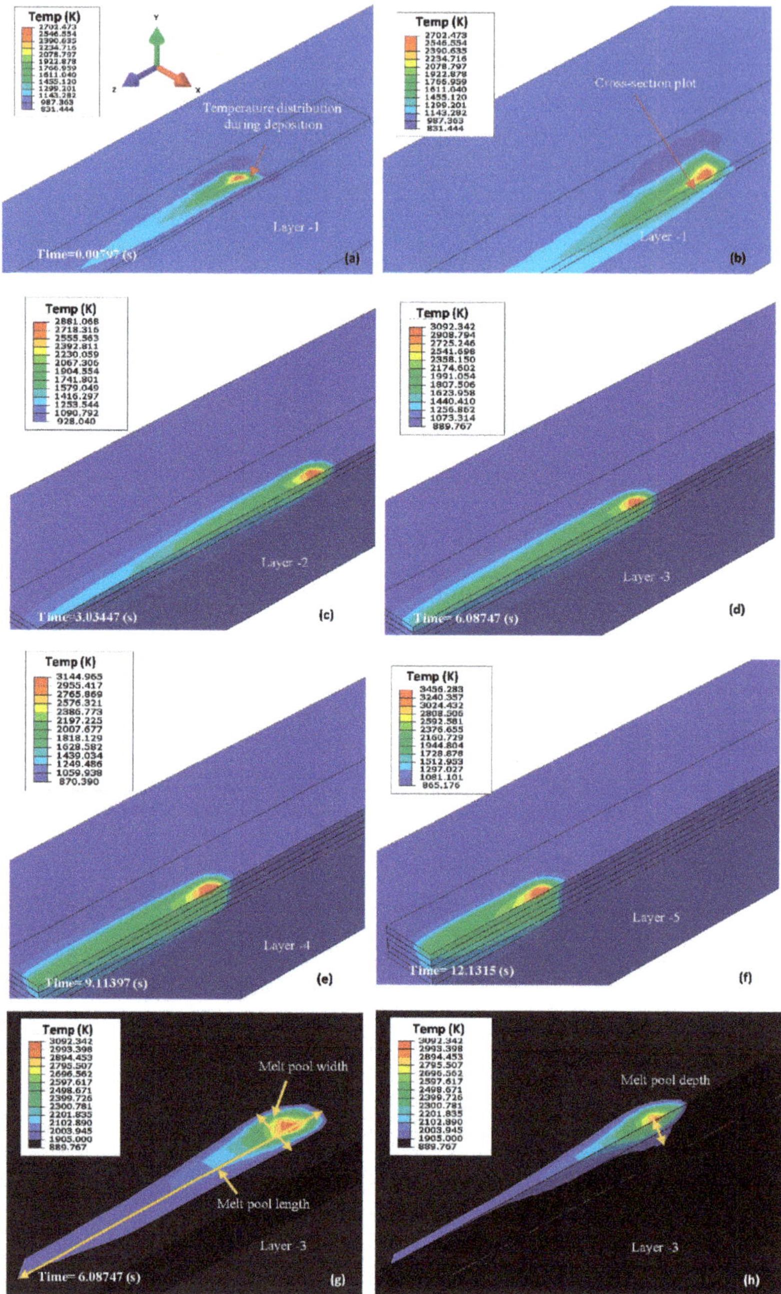

Figure 10. Contour and cross-section plots of temperature field of the melt pool and surrounding areas at different times: (**a**) Layer-1 (contour plot), (**b**) Layer-1 (cross-section plot), (**c**) Layer-2 (cross-section plot), (**d**) Layer-3 (cross-section plot), (**e**) Layer-4 (cross-section plot), (**f**) Layer-5 (cross-section plot), (**g**) Layer-3 (melt pool contour plot), and (**h**) Layer-3 (melt pool cross-section).

To show the dimensions of the melt pool, Figure 10g,h are presented with the lower limit of the temperature as the melting point of the Ti-6Al-4V, so that only the melt pool would be visible. Figure 10g,h show the contour and cross-sectional views of the melt pool size and shape as the beam traveled during the deposition of layers at different times. The melt pool shape and size depend on the material properties and the EBM parameters. The depth of the melt pool can be influenced by the movement of temperature-dependent fluid within the melt pool [33]. However, the size and shape of a melt pool plays a crucial role in the determination of the microstructure in manufactured metals [33].

3.1.1. Temperature Gradient

The temperature gradients were analyzed along the beam trailing direction (Z-direction) and below the beam center point (Y-direction), as shown in Figure 11a,c. The Z-direction nodes were along the top surface of the layers, while the Y-direction nodes were along the height of the deposited layers. Figure 11b presents the thermal gradients along the Z-axis, which have reached a maximum temperature of 3030 K from the focused electron beam. The temperature gradually decreases along the Z-axis to the preheating temperature. Figure 11d shows the temperature curve for thermal gradients along the Y-axis (height) of the selected side cross-section view layers, which have reached a maximum temperature of 2716 K from the focused electron beam. The temperature gradually decreases along the Y direction. Since the electron beam melting process is a hot process, during which the powder is kept at high temperatures throughout the entire melting process. The melting of the electron beam can easily reach temperatures 2000 K and more to melt the materials such as titanium alloys. The melting point for titanium alloys is 1900 K [34]. Figure 11 shows that the temperature is high enough to melt the selected material that is difficult to dissolve through traditional technologies.

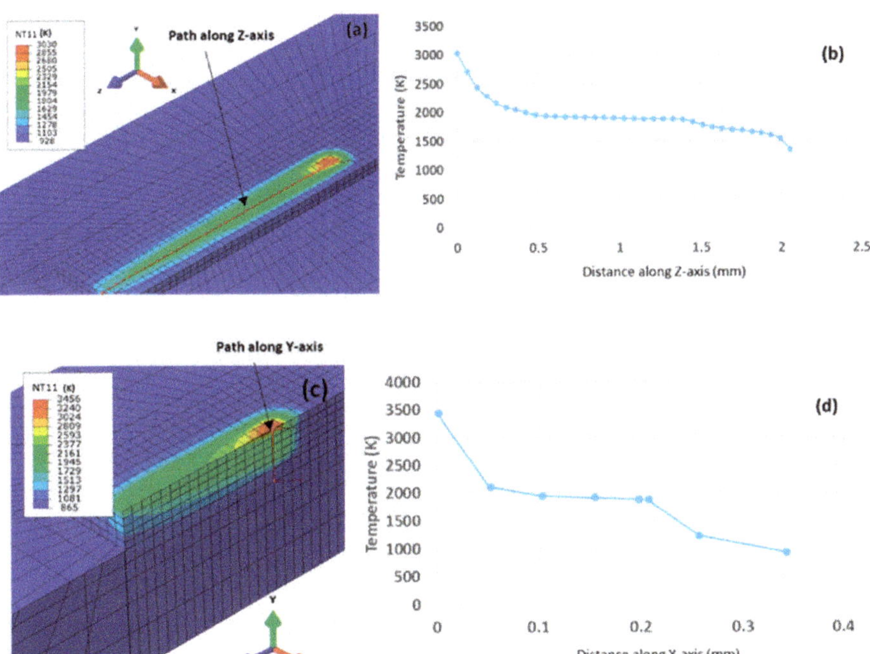

Figure 11. (**a**) Nodal path along the Z-axis (**b**) The plot for the temperature variation along the Z-axis nodal path, (**c**) nodal path along Y-axis, (**d**) the plot of the temperature variation along the Y-axis nodal path.

3.1.2. Model Validation

The results obtained from the EBM process model have been validated by comparing them with the results from the reported studies. Figure 12 presents the simulation results of temperature history after final cooling from Shen and Chou [14] and the current model simulation results on the electron beam melting process (ARCAM AB) with the same parameters for both models. Figure 12 shows that the range of the temperature gradient during deposition is almost the same. The maximum temperature (Shen & Chou) [14] is 3007 °C, and 3015 °C for the simulated result. The temperature dropped to room temperature (25 °C) after final cooling on both models. The temperature rose from 750 °C (preheating temperature) to 3015 °C in 0.3 s and dropped again to 750 °C in about 3 s.

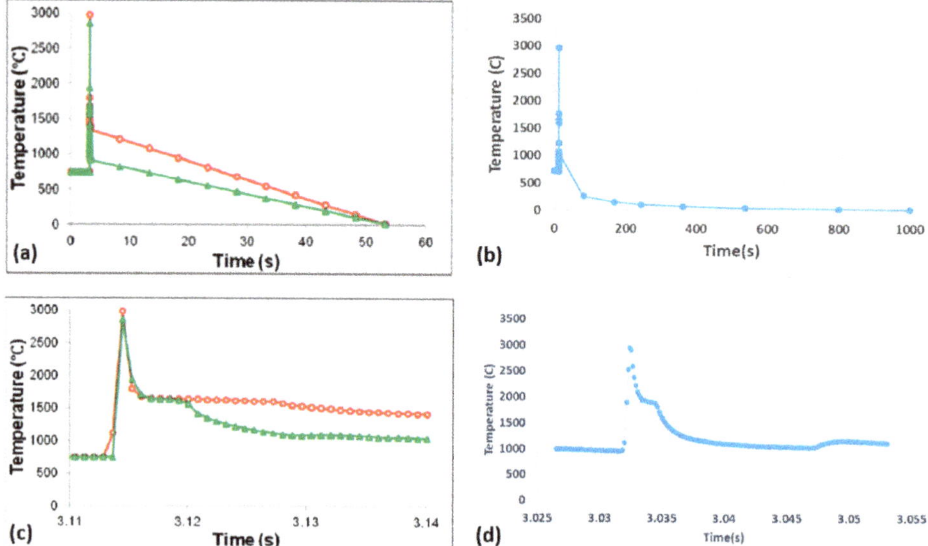

Figure 12. (**a**) Simulation results of temperature history after final cooling (Shen & Chou [14]) (**b**) Zoomed-in view of simulation results of temperature history after final cooling (Shen & Chou [14]) cooling (**c**) Current model simulation results of temperature history after final cooling (**d**) Zoomed-in view of current model simulation results of temperature history after final.

Figure 13 shows a comparison of the maximum temperature for the electron beam melting process when producing thin-wall parts. The current model results are compared with the experimental and simulation results reported by Shen and Chou [14]. The histogram shows that the maximum temperature for the current model and experimental results (Shen and Chou) are almost the same, with a minimal difference of around 170 °C. It can also be seen from Figure 13 that the maximum temperatures for the experimental, simulation (Shen and Chou), and current simulated results are 2800 °C, 2950 °C, and 2970 °C, respectively.

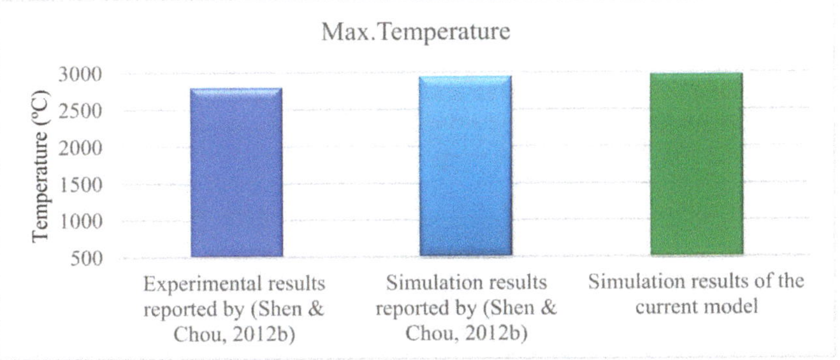

Figure 13. The Maximum temperatures compared with the experimental results.

3.2. Residual Stresses

The residual stress is the stress resident inside a structure after all applied forces have been removed. The magnitude and nature of residual stresses occurring in the final deposits affect the integrity of the whole structure. In general, compressive residual stresses are beneficial because they increase the load resistance and prevent the growth of cracks. However, tensile residual stresses are disadvantageous because they reduce the load resistance and accelerate the growth of cracks.

The von Mises stress distribution in the electron beam melting process of the five-layer model after the final cooling stage shows that the maximum stress occurs at the end of the electron beam scanning location and has a magnitude of about 840 MPa. After the EBM process started, the von Mises stress rapidly increased to 442 MPa. During the deposition process, it maintained a value between 333 MPa and 442 MPa, and after final cooling, it rose to 840 MPa, as shown in Figure 14b.

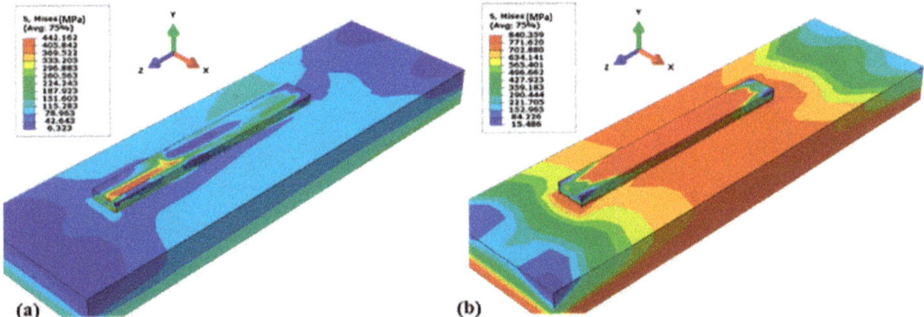

Figure 14. (**a**) Von Mises stress during deposition at t = 9.125 s, layer-4. (**b**) Von Mises stress after final cooling.

The residual stress distribution during deposition is shown in Figure 14a. It shows the stresses along with the three axes. Figure 14a,b indicate that the residual stresses in the lower part were mostly tensile stress due to the cooling phase of the molten layers [29]. After deposition, the re-melted base of the deposits began to shrink. This shrinkage was restricted by the underlying material and hence encouraged the tensile stresses.

The components of normal stress along the X-Axis and Z-Axis are stresses along the scan direction, which are indicated as S11 and S33, respectively. S33 are the normal stresses along the beam scan direction along the length of the layers, while the S11 are the normal stresses along the beam scan direction along the width of the layers. In contrast, S22 defines the normal residual stresses along the beam penetration direction (Y-axis). Such stress components reflect the highest possible magnitude of the residual stress at the given point being selected. Von mises stress and normal stresses S11, S22, and S33 along three spatial directions are shown in Figure 15. The residual stress distribution after the final cooling step is shown in Figure 15a–d (a cross-sectional view is used to show the internal residual stress). The maximum tensile stress of 592 MPa and the maximum compressive stress of 353 MPa can be observed in Figure 15. Although these residual stress values are lower than the yield point of the Ti-6Al-4V, still such EBM parameters should be used that the residual stresses in the produced parts should be minimum for better and safe performance [14]. The FE model allows us to explore the EBM parameters, which can lead to lower residual stresses, as discussed later in Section 4.

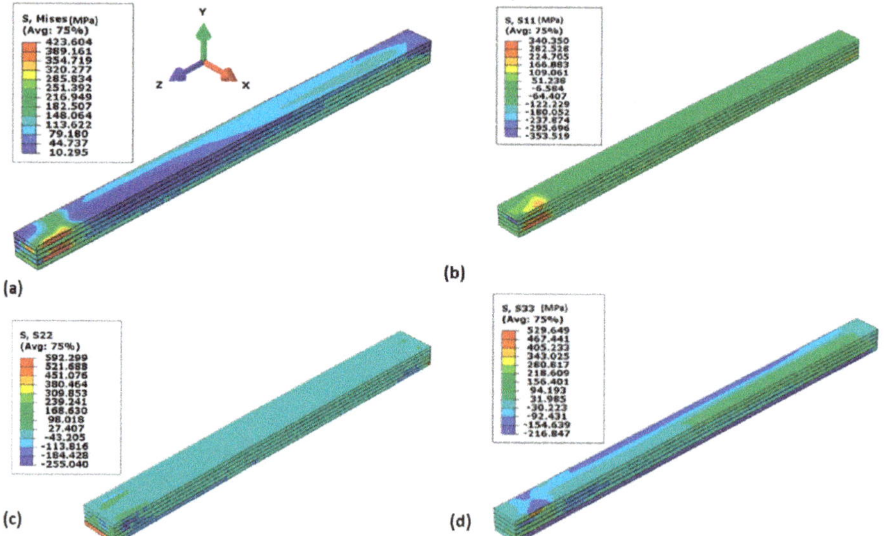

Figure 15. Contour plots of residual stress within deposition (cross-section). (**a**) Von mises stress. (**b**) Normal stresses along the beam scan direction (S11, along X-axis), (**c**) normal stresses along the beam penetration direction (S22, along Y-axis), and (**d**) normal stresses along the beam scan direction (S33, along Z-axis).

3.2.1. Instantaneous Stress

The instantaneous von Mises stress within the deposits during the electron beam melting process is shown in Figure 16. As the EBM process started, the von Mises stress rapidly increased to 820 MPa. The maximum von Mises stress existed at the end of the electron beam scanning location with a value of around 830 MPa. The von Mises stress after the deposition process had a greater magnitude than that during the deposition process.

Three paths are drawn on the top surface of the deposits to present the average distribution and magnitude of residual stresses in part. Along the z-direction, the middle part of the top surface was compressed with a stress magnitude of approximately 700 MPa, while at the edges, it was lower, as shown in Figure 17. The trend of residual stress on each path was relatively uniform along the longitudinal direction in the left and right sides. For the normal stresses along z, tensile stresses with a magnitude of approximately 700 MPa existed near the center part, and residual stresses ranging from 300 to 400 MPa existed at both edges.

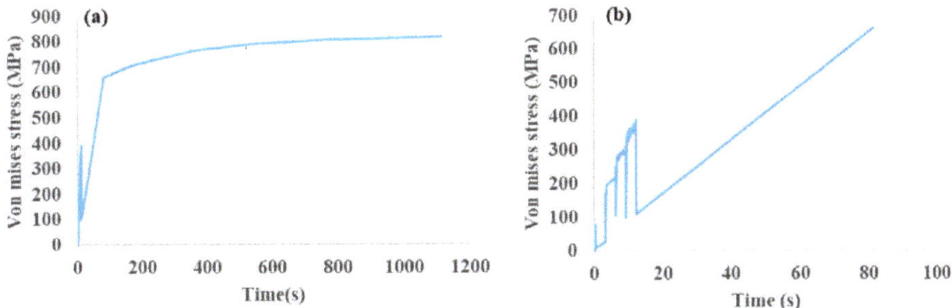

Figure 16. Von Mises stress plot (**a**) During deposition (MPa) (**b**) Zoomed-in view at the first 15 s.

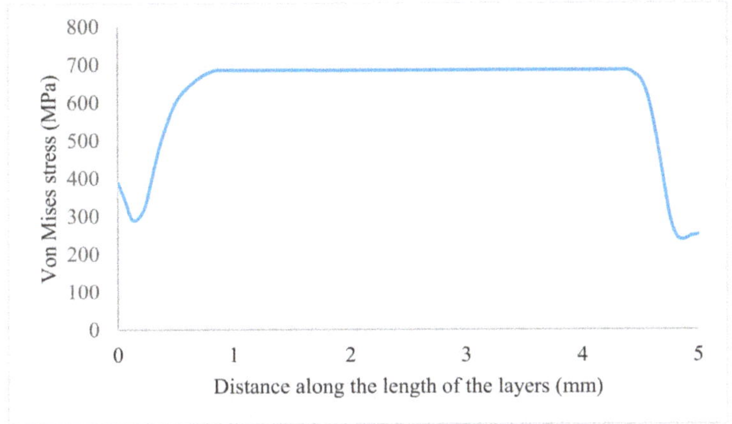

Figure 17. Von Mises stress after final cooling along the length of the deposited layers.

3.2.2. Model Results Versus Reported Results

Figure 18 compared the normal stress from the current model after cooling with the results of [15] along the beam scan direction (along the length of the layers). The residual stress levels are approximately the same (5%), with minimal differences due to the location of the path chosen on each model. It can be seen that the maximum residual stress in [15] after the cooling is 350 MPa, which is almost the same (5% error) as simulated in the current model.

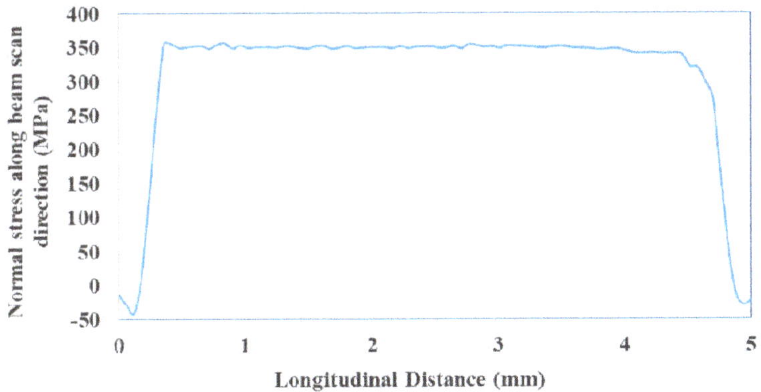

Figure 18. Current model simulation results of residual stress along with beam scan direction (length of the layers).

3.3. Thermal Distortion

Heating and cooling during the EBM process produce non-uniform thermal expansion, which results in a problematic distribution of residual stresses in the heat-affected zone and unexpected distortion. These residual stresses may increase fatigue and fractures during deposition. Deformation is determined from the dimensional accuracies of the structures. It is therefore very important to predict the behavior of the material after the EBM process, and to study manufacturing parameters, to control deformation and residual stresses. During EBM, the base plate will continuously shrink and expand, leading to a deformed shape. In this research, deformation along the X-direction was the main focus under consideration, and Figure 19a illustrates the base plate deflection during the electron beam melting process. For each deposition layer, the base plate first bent down due to thermal expansion on the top surface and then bent up due to thermal shrinkage during the cooling process. The curve of base plate distortion along the X-axis is shown in Figure 19b. After completely cooling down, the base plate maintained its deformed shape. In Figure 19a–d, it can be concluded that more layers lead to higher distortion in the base plate.

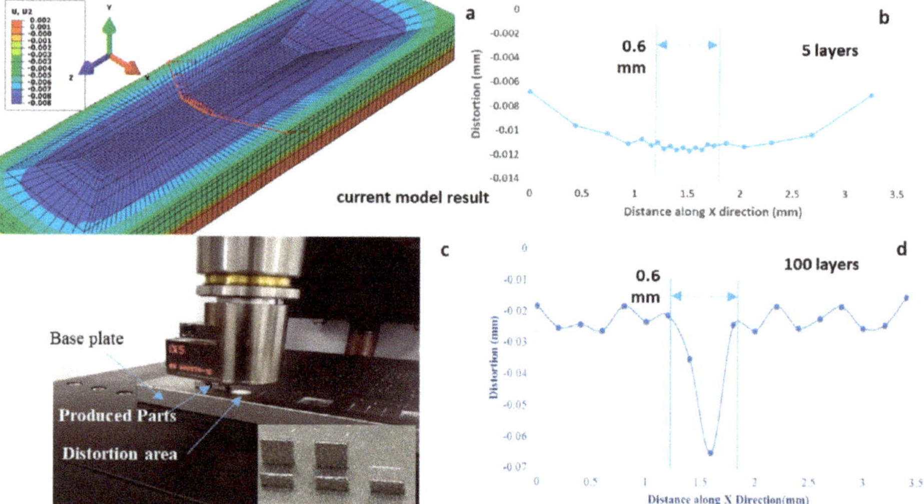

Figure 19. (**a**) Thermal distortion of the base plate along the X-axis. (**b**) Thermal distortion curve of the base plate along the X-axis. (**c**) Thermal distortion of the experimental base plate along the X-axis. (**d**) Thermal distortion curve of the experimental base plate along the X-axis.

3.4. Experimental Validation

The profile projector machine was used to measure the distortion results of the parts produced by EBM. The profile of the EBM parts can be seen on the screen, as shown in Figure 20. The projector magnifies the profile of the specimen and displays this on the built-in projection [1] screen on this screen. There is typically a grid that can be rotated 360 degrees so that the X-Y axis of the screen can be aligned with a straight edge of the part to be examined or measured. This projection screen displays the profile of the piece, which is magnified for ease of measurement. The distortion measurement from both sides for the 50-layer part (2.5 mm height) produced with EBM/Ti-6Al-4V are shown in Figure 20a–d. The distortion seems to increase in the base plate, and the first few layers as more layers are added. The profiles of the distortion in the right and left sides can be clearly seen in Figure 20.

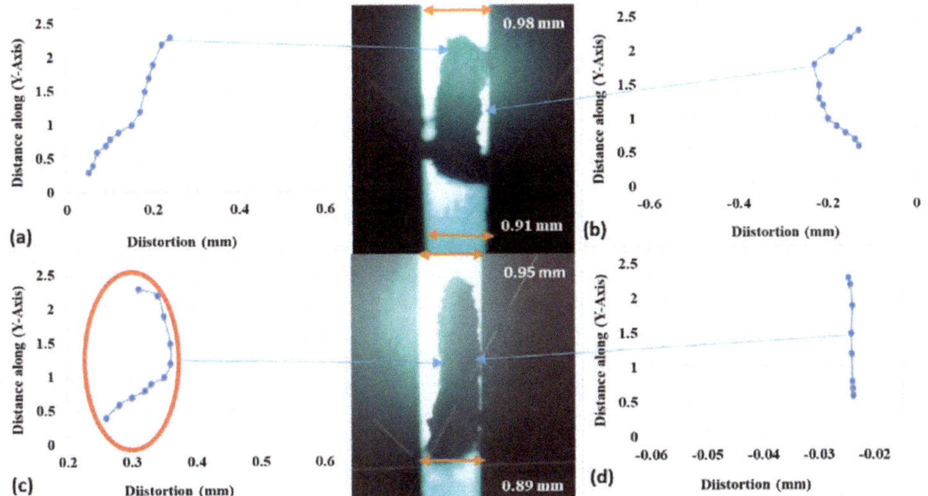

Figure 20. Distortion along the Y-axis for the 50-layers (all sides) for 2.5 mm parts. (**a**) Left side from the front view. (**b**) Right side from the front view. (**c**) Left side from the bottom view. (**d**) Right side from the bottom view

The distortion measurement from both sides is shown in Figure 21a–d for the 100 layer part (5 mm height), produced with EBM/Ti6Al-4V. The distortion increases in the base plate and the first few layers as more layers are added. The profile of distortion can be seen more clearly than in the previous profile of 50 layers part, as shown in Figure 21.

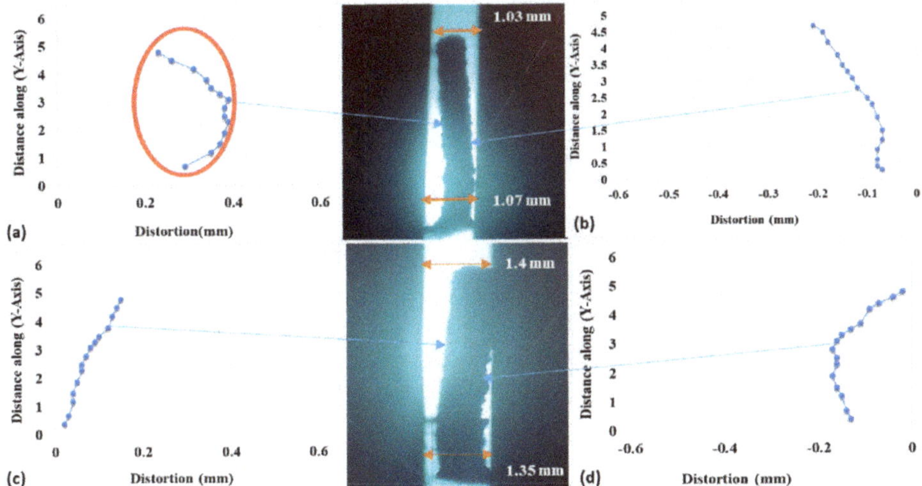

Figure 21. Distortion along the Y-axis for the 100-layers (all sides) for 5 mm height parts. (**a**) Left side from the front view. (**b**) Right side from the front view. (**c**) Left side from the bottom view. (**d**) Right side from the bottom view

An important observation is made from Figures 20 and 21 that the overall total distortion in the EBM produced parts remains almost the same, independent of height. For example, the total distortion in the 2.5 mm height part was measured to be around 0.15 mm, and the total distortion measured in the 5 mm height part was approximately 0.17 mm, which is a very similar value.

The comparison of distortion in the built layers between the experimental and simulation results is presented in Figure 22a–e. Although inherently capable of performing calculations on 50 or 100 layers, limitations of the computer used meant that simulation was stopped at 15 layers, as shown in Figure 22. The distortion in the EBM produced parts was experimentally measured. The experimental results showed that the overall distortion was independent of the part height produced. The simulated pattern of the distortion in the FE model results was qualitatively compared with experimental data and showed a similar trend. When more layers are added, the first few layers are bent down due to thermal expansion on the top surface and then bent up due to thermal shrinkage during the cooling process. After complete cooling down, the layers maintained their profile shape.

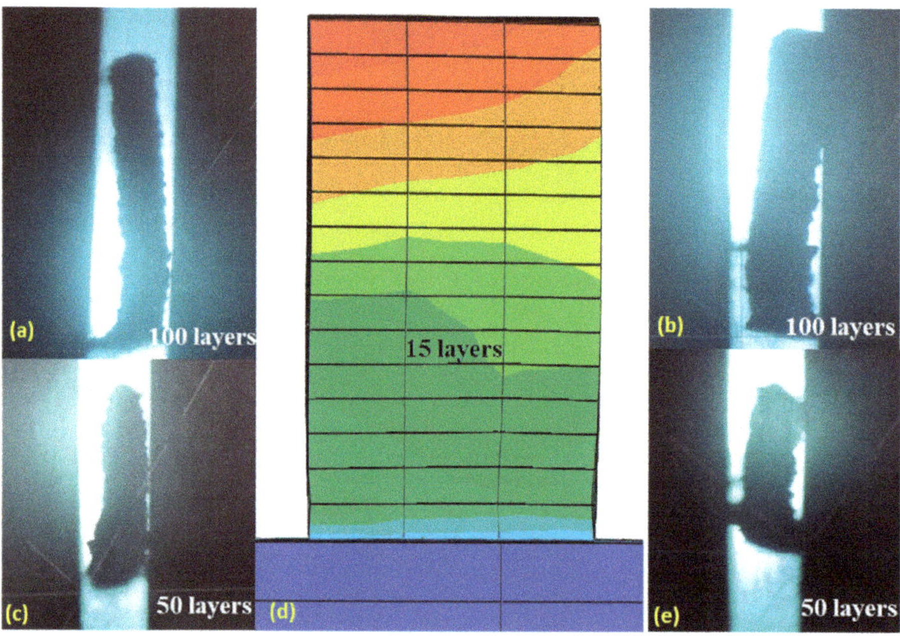

Figure 22. Comparison between current model result and measured distortion within the layers. (**a**) Left side from the front view (100 layers). (**b**) Right side from the bottom view (100 layers). (**c**) Left side from the front view (50 layers). (**d**) The distortion in the current model result (**e**) Right side from the bottom view (50 layers).

3.5. Microstructure Evolution

The samples were first grinding with SiC papers P180, P300, P600, P800, P1000, and P1500, and then polishing with Al_2O_3 suspension followed by etching with Kroll's agent. The microstructures were obtained primarily to see the variation of the alpha and beta phases along the thin side faces.

Figure 23a,b presents the microstructure of the EBM produced Ti-6Al-4V parts. It can be noted that the columnar grains (α and β) are found in the bottom (almost first three layers) of the produced part due to the cumulative heat deposited in the part near the bottom, as shown in Figure 23a. The result shows that the coarse structure has been found above the base plate, fine structure at the mid of the building layers, and finer structure near the top of the produced layers. This is because the bottom of the simulated model has more heat cycles due to the cumulative heat deposited when compared with the middle region of the part which receives fewer heat cycles. For instance, in Figure 23c, it can be seen that layer-2 receives several heat cycles due to electron beam scans on subsequent layers as compared to layer-8, which only receives 2 heat cycles (see Figure 23d). Usually, the higher heat accumulation and rapid extraction of heat favor the higher concentration of the beta phase and vice versa for the alpha phase [35,36]. Therefore, in Figure 23a, more β phase and wider α grains could be observed. In contrast, in Figure 23b, a refined lamellar α + β structure is produced because this region (middle region) receives fewer heat cycles (see Figure 23d) and experiences less heat accumulation as compared to the bottom region.

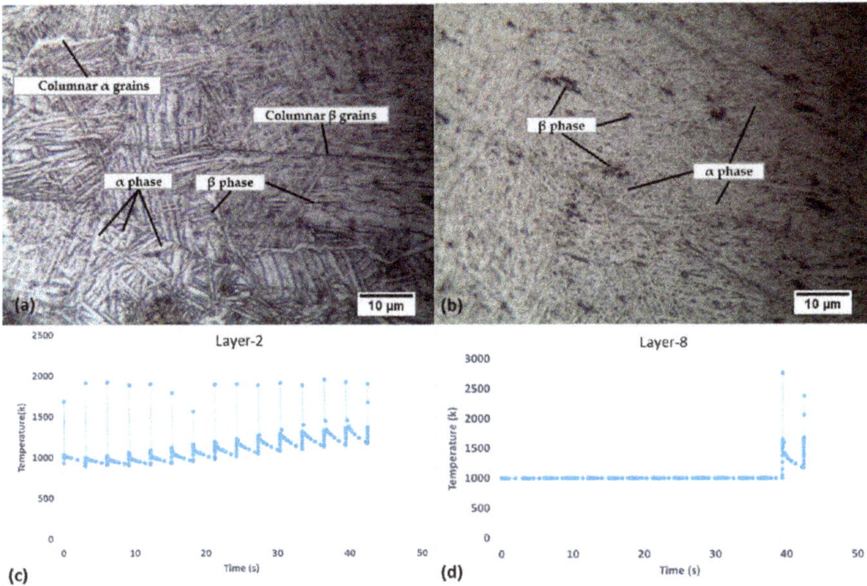

Figure 23. Microstructures obtained by microscope for Ti-6Al-4V alloys vs. current simulated model temperature profiles. (**a**) At 0.075 mm above the base plate (layer-2). (**b**) At 0.375 mm above the base plate (layer-8). (**c**) Temperature profile for layer-2. (**d**) Temperature profile for layer-8.

4. The Effects of Different Electron Beam Parameters

This section shows the effects of EBM processing parameters on producing a thin part of Ti-6Al-4V. Three parameters are studied: beam speed, beam current, and voltage. These three parameters were selected as they are easily adjusted and are believed to be the ones having the most influence on the objects produced. Beam current is the current due to the electrons in the electron beam. Beam speed is the electron beam scan speed. In the study, Ti6Al-4V parts were modeled using different processing parameters.

Process parameter values were chosen according to the machine's parameter range in order to achieve the largest difference between the high and low values of each process parameter. Using a wide range in each parameter increases the probability that significant effects will be detected. As mentioned above, Cheng and Chou reported the EBM parameters used for producing a part by ARCAM AB machine, which were 400 mm, 0.002 mA, and 60 kV. However, in this section, nine combinations have been used to study the effect of real conditions such as beam speed, current, and voltage for a selected geometry.

4.1. The Effect of Scan Speed on the EBM Process

Table 5 shows the combination of parameters used with varying scan speed with other parameters fixed. The scan speed has been varied within the range reported for the ARCAM AB Machine, which is capable of scan speeds up to 8000 m/s (ARCAM AB, 2016).

Table 5. The levels of scan speed changes.

Parameter	Level-1	Level-2	Level-3
Speed (mm/s)	300	400	500
Current (mA)	0.002	0.002	0.002
Voltage (kV)	60	60	60

Figure 24 shows the scan speed versus temperature with other parameters fixed to study the effect of scan speed on the electron beam melting process. The results conclude that the increase in scan speed leads to a decrease in temperature. The first combination of 300 mm/s, 0.002 mA, and 60 kV, has the highest temperature of 3977 K. The second combination has a temperature of 3193 K, and the third has a lower value of 2765 K. Figure 24 presents the scan speed versus von Mises stress to study the effect of scan speed on stress in the electron beam melting process. The results show that an increase in scan speed leads to an increase in the residual stresses deposited in the part produced by EBM. The first combination of 300 mm/s, 0.002 mA, and 60 kV received the least residual stress with a value of 665 MPa. The second combination had a value of 676 MPa, and the third had the highest residual stress value of 705 MPa. It can be concluded that, as the speed increases, the time of contact becomes less, which leads to a decrease in the temperature and an increase in the residual stress.

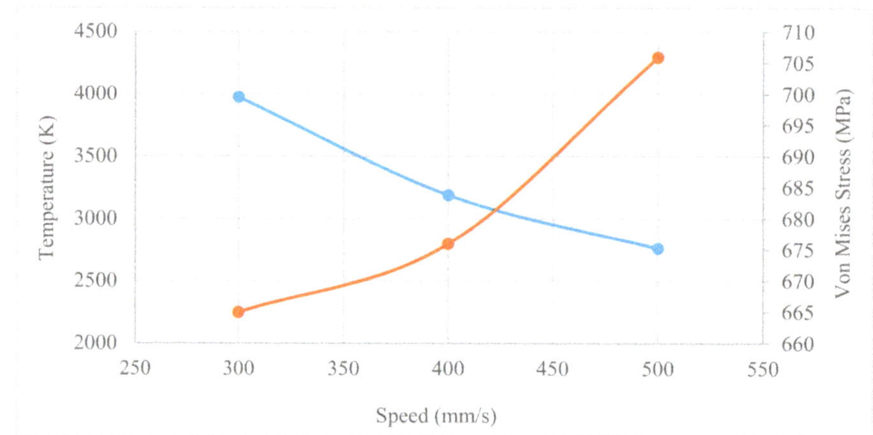

Figure 24. Scanning speed vs. temperature & stress.

4.2. The Effect of Current on the EBM Process

Table 6 shows the combination of parameters used with varying current, and other parameters fixed. The current has been varied within the range reported H. Gong, Rafi, Starr, & Stucker [37] for the ARCAM AB machine (0.001, 0.002, and 0.003 mA).

Table 6. The levels of current changes.

Parameter	Level-1	Level-2	Level-3
Speed (mm/s)	400	400	400
Current (mA)	0.001	0.002	0.003
Voltage (kV)	60	60	60

Figure 25 shows the current versus temperature to study the effect of current on the temperature in the electron beam melting process. The results conclude that the increase in current leads to a rise in the temperature of the electron beam melting process. The first combination of 300 mm/s, 0.001 mA, and 60 kV, has the lowest temperature of 1926 K. The second combination has a temperature of 3192 K, and the third has the highest value of 4357 K. Figure 25 presents the current versus von Mises stress at different applied currents, to study the effect of current on stress in the electron beam melting process. It is concluded that the increase in current leads to an increase in the residual stress deposited in the part produced by EBM. The first combination of the three parameters has the least residual stress, with a value of 666 MPa. The second combination has a value of 676 MPa, and the third has the highest residual stress with a value of 709 MPa. It can be concluded that, as the current increases, the spot size is decreasing the beam energy, so it leads to an increase in the temperature and residual stress.

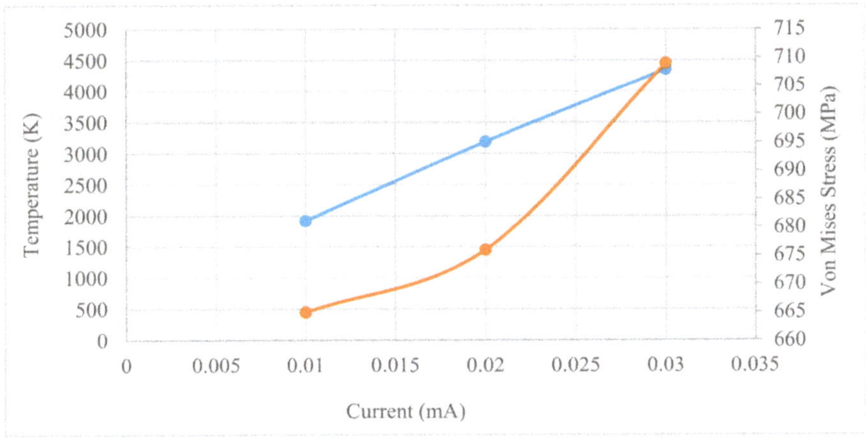

Figure 25. Current vs. temperature& Von Mises Stress.

4.3. The Effect of Voltage on the EBM Process

The combination of parameters used with varying voltage and other parameters fixed is shown in Table 7. The voltages used are 50, 60, and 70 kV.

Table 7. The levels of Voltage changes.

Parameter	Level-1	Level-2	Level-3
Speed (mm/s)	400	400	400
Current (mA)	0.002	0.002	0.002
Voltage (kV)	50	60	70 k

Figure 26 shows the voltage versus temperature at various set voltages to study the effect of voltage on the temperature in the EBM process. The results conclude that the increase in voltage leads to a rise in the temperature of the EBM process. The first combination of scan speed, current, and voltage has the lowest temperature of 2780 K. The second combination has a temperature of 3192 K, and the third has the highest value of 3693 K. Figure 26 presents the voltage versus von Mises stress at different set voltages to study the effect of voltage on stress in the EBM process. It can be concluded that the increase in voltage leads to lower residual stresses deposited in the part produced by EBM. The first combination of the three parameters has the highest residual stress with a value of 705 MPa. The second combination has a value of 676 MPa, and the third has the least residual stress with a value of 661 MPa. The results conclude that as the voltage increases as the energy increases, leading to an increase in the temperature and a decrease in the residual stress due to the potential difference between the two nodes.

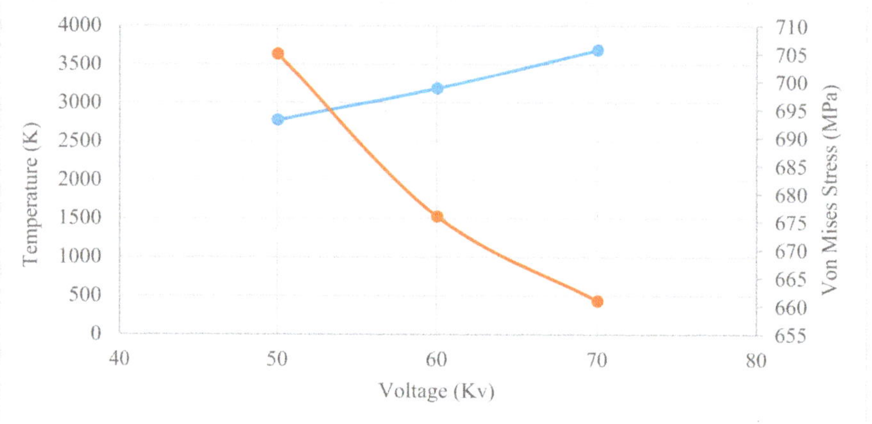

Figure 26. Voltage vs. temperature & Von Mises Stress.

5. Conclusions

The thermomechanical finite element model was developed for the multi-layer EBM process of Ti-6Al-4V, to investigate the thermal and mechanical behavior of deposited materials involved in the electron beam melting process. In this study, the thermomechanical analysis was performed using a 3D transient, fully coupled temperature-displacement model.

The developed FE model study the effect of real conditions such as preheating temperature, the energy of the electron beam, cooling time, and beam scanning path and speed for a selected geometry. The simulated results have been compared with other reported simulated results on EBM, and the experimental results. The results revealed the characteristics of temperature distribution, residual stress, and deformation within the deposited layers and the base plate. Based on the results and discussion, several conclusions have been drawn and can be stated as follows:

- The FEA can predict the thermomechanical behavior of products fabricated by the electron beam melting process or similar processes with localized heat sources such as laser sintering, laser cladding, and welding.
- During the deposition of layers with EBM, high residual stresses resulted after final cooling, i.e., the stresses in the deposited layers increased with cooling due to thermal contraction. It is highly recommended to validate the FE residual stress results by conducting the experimental observations and measurements as well.
- The distortion in the EBM produced parts was experimentally measured. The experimental results showed that the overall distortion was independent of the part height produced. The simulated pattern of the distortion in the results from the FE model was found to be in close agreement with the experimental data.
- The maximum predicted temperatures and the temperature profiles were found to be in close agreement with the reported experimental and simulated results.

The effects of EBM processing parameters on the production of a thin part of Ti-6Al-4V have been studied to understand the most influential parameters. Findings are as follows:

- An increase in scan speed leads to a decrease in temperature.
- An increase in scan speed leads to an increase in the residual stresses within the part produced by EBM.
- An increase in current leads to a rise in temperature of the electron beam melting process.
- A decrease in current leads to a reduction in the residual stresses within the part produced by EBM.

○   An increase in voltage leads to a rise in the temperature of the electron beam melting process.

○   An increase in voltage leads to lower residual stresses within the part produced by EBM.

For the EBM machine, characteristics of the beam, and materials used in this study, the best combination of parameters was 400 mm/s, 0.002 mA, and 60 kV speed, current, and voltage, respectively, to achieve parts with low levels of residual stress and distortion and hence improved quality.

Author Contributions: F.M.A.: Data curation, formal analysis, investigation, methodology, software, visualization, writing—original draft; S.A.: conceptualization, data curation, formal analysis, investigation, methodology, software, supervision, validation, visualization, writing–original draft, writing—review & editing; A.A.-A.: funding acquisition, resources, supervision, writing—review & editing. All authors have read and agreed to the published version of the manuscript.

Funding: This study received funding from the Raytheon Chair for Systems Engineering. The authors are grateful to the Raytheon Chair for Systems Engineering for funding.

Acknowledgments: The authors thank the Deanship of Scientific Research and RSSU at King Saud University for their technical support.

Conflicts of Interest: The authors declare no conflict of interest.

References

1. Ameen, W.; Al-Ahmari, A.; Mohammed, M.K. Self-supporting overhang structures produced by additive manufacturing through electron beam melting. *Int. J. Adv. Manuf. Technol.* **2019**, *104*, 2215–2232. [CrossRef]
2. Ameen, W.; Al-Ahmari, A.; Mohammed, M.; Mian, S. Manufacturability of overhanging holes using electron beam melting. *Metals* **2018**, *8*, 397. [CrossRef]
3. Schwerdtfeger, J.; Heinl, P.; Singer, R.; Körner, C. Auxetic cellular structures through selective electron-beam melting. *Phys. Status Solidi B* **2010**, *247*, 269–272. [CrossRef]
4. Yu, P.; Qian, M.; Tomus, D.; Brice, C.A.; Schaffer, G.B.; Muddle, B.C. Electron beam processing of aluminium alloys. In *Materials Science Forum*; Trans Tech Publications Ltd.: Stafa-Zurich, Switzerland, 2009; Volume 618, pp. 621–626.
5. Cormier, D.; Harrysson, O.; West, H. Characterization of H13 steel produced via electron beam melting. *Rapid Prototyp. J.* **2004**, *10*, 35–41. [CrossRef]
6. Gaytan, S.M.; Murr, L.E.; Martinez, E.; Martinez, J.L.; Machado, B.I.; Ramirez, D.A.; Medina, F.; Collins, S.; Wicker, R.B. Comparison of microstructures and mechanical properties for solid and mesh cobalt-base alloy prototypes fabricated by electron beam melting. *Metall. Mater. Trans. A* **2010**, *41*, 3216–3227. [CrossRef]
7. Parthasarathy, J.; Starly, B.; Raman, S.; Christensen, A. Mechanical evaluation of porous titanium (Ti6Al4V) structures with electron beam melting (EBM). *J. Mech. Behav. Biomed. Mater.* **2010**, *3*, 249–259. [CrossRef]
8. Leyens, C.; Peters, M. *Titanium and Titanium Alloys: Fundamentals and Applications*; John Wiley & Sons: Hoboken, NJ, USA, 2003.
9. Liu, W.; Li, L.; Kochhar, A. A method for assessing geometrical errors in layered manufacturing. Part 1: Error interaction and transfer mechanisms. *Int. J. Adv. Manuf. Technol.* **1998**, *14*, 637–643. [CrossRef]
10. Lim, G.; Lau, K.; Cheng, W.; Chiang, Z.; Krishnan, M.; Ardi, D. Residual Stresses in Ti-6Al-4V Parts Manufactured by Direct Metal Laser Sintering and Electron Beam Melting. *Brit. Soc. Strain Meas.* **2017**, *71*, 348–353.
11. Weman, K. *Welding Processes Handbook*; Elsevier: Amsterdam, The Netherlands, 2011.
12. Denlinger, E.R.; Heigel, J.C.; Michaleris, P. Residual stress and distortion modeling of electron beam direct manufacturing Ti-6Al-4V. *Proc. Inst. Mech. Eng. Part B J. Eng. Manuf.* **2015**, *229*, 1803–1813. [CrossRef]
13. Cheng, B.; Chou, K. Thermal stresses associated with part overhang geometry in electron beam additive manufacturing: Process parameter effects. *Proc. Annu. Int. Solid Free. Fabr. Symp.* **2014**, *25*, 1076.
14. Shen, N.; Chou, K. Thermal modeling of electron beam additive manufacturing process: Powder sintering effects. In Proceedings of the ASME 2012 International Manufacturing Science and Engineering Conference Collocated with the 40th North American Manufacturing Research Conference and in participation with the International Conference on Tribology Materials and Processing, Notre Dame, IN, USA, 4–8 June 2012; pp. 287–295.

15. Cao, J.; Gharghouri, M.A.; Nash, P. Finite-element analysis and experimental validation of thermal residual stress and distortion in electron beam additive manufactured Ti-6Al-4V build plates. *J. Mater. Process. Technol.* **2016**, *237*, 409–419. [CrossRef]
16. Umer, U.; Ameen, W.; Abidi, M.H.; Moiduddin, K.; Alkhalefah, H.; Alkahtani, M.; Al-Ahmari, A. Modeling the Effect of Different Support Structures in Electron Beam Melting of Titanium Alloy Using Finite Element Models. *Metals* **2019**, *9*, 806. [CrossRef]
17. Algardh, J.K.; Horn, T.; West, H.; Aman, R.; Snis, A.; Engqvist, H.; Lausmaa, J.; Harrysson, O. Thickness dependency of mechanical properties for thin-walled titanium parts manufactured by Electron Beam Melting (EBM)®. *Addit. Manuf.* **2016**, *12*, 45–50. [CrossRef]
18. Isaev, A.; Grechishnikov, V.; Pivkin, P.; Ilyukhin, Y.; Kozochkin, M.; Peretyagin, P. Structure and machinability of thin-walled parts made of titanium alloy powder using electron beam melting technology. *J. Silic. Based Compos. Mater.* **2016**, *68*, 46.
19. Cormier, D.; West, H.; Harrysson, O.; Knowlson, K. Characterization of Thin Walled Ti-6Al-4V Components Reduced via Electron Beam Melting 440. In Proceedings of the 2004 International Solid Freeform Fabrication Symposium, Austin, TX, USA, 4 August 2004.
20. Everhart, W.; Dinardo, J.; Barr, C. The effect of scan length on the structure and mechanical properties of electron beam-melted Ti-6Al-4V. *Metall. Mater. Trans. A* **2017**, *48*, 697–705. [CrossRef]
21. Fu, C.; Guo, Y. 3-dimensional finite element modeling of selective laser melting Ti-6Al-4V alloy. In Proceedings of the 25th Annual International Solid Freeform Fabrication Symposium, Tuscaloosa, AL, USA, 4 August 2014; pp. 1129–1144.
22. Boivineau, M.; Cagran, C.; Doytier, D.; Eyraud, V.; Nadal, M.H.; Wilthan, B.; Pottlacher, G. Thermophysical properties of solid and liquid Ti-6Al-4V (TA6V) alloy. *Int. J. Thermophys.* **2006**, *27*, 507–529. [CrossRef]
23. Liu, S.; Shin, Y.C. Additive manufacturing of Ti6Al4V alloy: A review. *Mater. Des.* **2019**, *164*, 107552. [CrossRef]
24. Abbes, B.; Anedaf, T.; Abbes, F.; Li, Y. Direct energy deposition metamodeling using a meshless method. *Eng. Comput.* **2020**. [CrossRef]
25. Ahmed, S.H.; Mian, A. Influence of material property variation on computationally calculated melt pool temperature during laser melting process. *Metals* **2019**, *9*, 456. [CrossRef]
26. Seyedian, C.M.; Haghpanahi, M.; Sedighi, M. *Investigation of the Effect of Clamping on Residual Stresses and Distortions in Butt-Welded Plates*; Scientia Iranica. Transaction B: Mechanical Engineering: Tehran, Iran, 2010.
27. Ameen, W.; Al-Ahmari, A.; Mohammed, M.; Abdulhameed, O.; Umer, U.; Moiduddin, K. Design, finite element analysis (FEA), and fabrication of custom titanium alloy cranial implant using electron beam melting additive manufacturing. *Adv. Prod. Eng. Manag.* **2018**, *13*, 267–278. [CrossRef]
28. Gong, X.; Cheng, B.; Price, S.; Chou, K. Powder-bed electron-beam-melting additive manufacturing: Powder characterization, process simulation and metrology. In Proceedings of the Early Career Technical Conference, Birmingham, AL, USA, 2–3 November 2013; pp. 55–66.
29. Cheng, B.; Lu, P.; Chou, K. Thermomechanical investigation of overhang fabrications in electron beam additive manufacturing. In Proceedings of the ASME 2014 International Manufacturing Science and Engineering Conference Collocated with the JSME 2014 International Conference on Materials and Processing and the 42nd North American Manufacturing Research Conference, Detroit, MI, USA, 9–13 June 2014.
30. He, J.; Li, D.; Jiang, W.; Ke, L.; Qin, G.; Ye, Y.; Qin, Q.; Qiu, D. The Martensitic Transformation and Mechanical Properties of Ti6Al4V Prepared via Selective Laser Melting. *Materials* **2019**, *12*, 321. [CrossRef] [PubMed]
31. Shen, N.; Chou, K. Thermomechanical investigation of overhang fabrications in electron beam additive manufacturing. In Proceedings of the International Manufacturing Science and Engineering Conference, Detroit, MI, USA, 9–13 June 2014; p. V002T02A024.
32. Shen, N.; Chou, K. Numerical thermal analysis in electron beam additive manufacturing with preheating effects. In Proceedings of the 23rd solid freeform fabrication symposium, Tuscaloosa, AL, USA, 6 August 2012; pp. 774–784.
33. Guo, Q.; Zhao, C.; Qu, M.; Xiong, L.; Escano, L.I.; Hojjatzadeh, S.M.; Parab, N.D.; Fezzaa, K.; Everhart, W.; Sun, T.; et al. In-situ characterization and quantification of melt pool variation under constant input energy density in laser powder bed fusion additive manufacturing process. *Addit. Manuf.* **2019**, *28*, 600–609. [CrossRef]

34. Cezairliyan, A.; Miiller, A. Melting point, normal spectral emittance/at the melting point/, and electrical resistivity/above 1900 K/ of titanium by a pulse heating method. *J. Res.* **1977**, *82*, 119–122. [CrossRef]
35. Raghavan, S.; Nai, M.L.S.; Wang, P.; Sin, W.J.; Li, T.; Wei, J. Heat treatment of electron beam melted (EBM) Ti-6Al-4V: Microstructure to mechanical property correlations. *Rapid Prototyp. J.* **2018**, *24*, 774–783. [CrossRef]
36. Singh, A.P.; Yang, F.; Torrens, R.; Gabbitas, B. Heat Treatment, Impact Properties, and Fracture Behaviour of Ti-6Al-4V Alloy Produced by Powder Compact Extrusion. *Materials* **2019**, *12*, 3824. [CrossRef]
37. Gong, H.; Rafi, K.; Starr, T.; Stucker, B. The effects of processing parameters on defect regularity in Ti-6Al-4V parts fabricated by selective laser melting and electron beam melting. In Proceedings of the 24th Annual International Solid Freeform Fabrication Symposium—An Additive Manufacturing Conference, Austin, TX, USA, 12 August 2013; pp. 424–439.

 *metals*

Article

Process Parameters Decision to Optimization of Cold Rolling-Beating Forming Process through Experiment and Modelling

Long Li, Yan Li *, Mingshun Yang and Tong Tong

School of Mechanical and Precision Instrument Engineering, Xi'an University of Technology, Xi'an 710048, China; lilong678@stu.xaut.edu.cn (L.L.); yangmingshun@xaut.edu.cn (M.Y.); 2170221141@stu.xaut.edu.cn (T.T.)
* Correspondence: jyxy-ly@xaut.edu.cn; Tel.: +86-029-82312776

Received: 8 March 2019; Accepted: 31 March 2019; Published: 2 April 2019

Abstract: The cold roll-beating forming (CRBF) process is a particular cold plastic bulk forming technology for metals that is adequate for shaping the external teeth of important parts. The process parameters of the CRBF process were studied in this work to improve the process performance. Of the CRBF process characteristics, the forming forces, tooth profile angle, surface roughness, and forming efficiency were selected as the target indices to describe the process performance. Single tooth experimental tests of ASTM 1045 steel were conducted with different roll-beating modes, spindle rotation speeds, and feed speeds. Using analysis of variance (ANOVA) and regression analysis, the influence of the process parameters in each index was investigated, and regression models of each index were established. Then, the linear weighted sum method and compound entropy weight method were used to determine the process parameters for multi-objective optimization. The results show that the impact capacity and optimum value range of the process parameters vary in different indices, and that, to achieve the comprehensive optimum effect of a small forming force, high product quality, and high forming efficiency, the optimal process parameter combination is the up-beating mode, a spindle rotation speed of 801 r/min, and a feed speed of 960 mm/min.

Keywords: plasticity forming; cold roll-beating forming; process parameter; multi-objective optimization

---

1. Introduction

As important mechanical products, gears, spline shafts, and other transmission parts are widely used in various mechanical products [1,2], the rapid development of major industries, such as automobile, aircraft, special engineering machinery, and wind and nuclear power equipment, has led to a huge demand for transmission parts with higher performance [3,4]. In the production of mechanical parts, objectives such as creating a lightweight design, establishing short process chains, and improving material and energy efficiency are difficult to meet using the current cutting methods to shape the external teeth of transmission parts [3,5,6]. Plastic forming, as a near-net shape forming technology, is a promising method to solve some of the problems in traditional cutting. As such, the new process and novel production equipment have received attention [7,8]. Among the varieties of plastic forming technologies, cold roll-beating forming (CRBF) is an incremental metal bulk forming technology for forming the external teeth of high-load transmission parts, and it has the advantages of environmental protection, flexibility, and lower cost. It has a wide range of application prospects in the automobile industry, the aerospace sector, and equipment manufacturing [9–11].

Some studies have been carried out on the CRBF process, and some research results have been published. On the basis of the kinematics of the CRBF process, Fengkui Cui et al. [12,13] analyzed and studied the forming process of involute spline cold roll-beating in which the workpiece was continuously indexed. In view of the forming error caused by the continuous indexing motion of workpiece, a design and modification method of the rolling wheel were put forward, and a technological scheme of roll wheel manufacturing was provided that improved the geometric accuracy of involute spline CRBF. Mingshun Yang et al. [14] established a simplified mathematical model for describing the residual height of the formed tooth bottom under the assumption of rigid plastic deformation, and they discussed the influence of process parameters on the residual height of the tooth bottom. These studies illustrate the basic kinematic characteristics of the CRBF process. On the basis of these studies, there have been some numerical simulation and experimental studies dedicated to the metal deformation characteristics and mechanism of the CRBF process. Fengkui Cui [15] and Xinqin Gao [16] established a simplified finite element model of the CRBF process and simulated the material behavior and changes in the principal stress, hydrostatic pressure, and principal strain of the deforming area. Then, the metal deformation characteristics of the CRBF process were given. On the basis of the deformation characteristics, a constitutive model of the material for the CRBF process was further studied [17–19]. The results of these studies provide guidance for simulating the CRBF process and explaining the forming mechanism. Through the simulation of a complete tooth groove CRBF process, Xiaoming Liang et al. [20] analyzed changes in the radial and tangential forming force throughout the whole forming process and discussed the influence of the roll-beating mode on forming forces. Fengkui Cui et al. [21] linked the plastic deformation of metal to the surface residual stress that occurs during the CRBF process by observing the grains changing in the spline tooth profile section, and they explained the residual stress generation in terms of the forming mechanism. Zhiqi Liu et al. [11] performed an experimental study on the CRBF mechanism from the microcosmic point of view. The depth of the microhardness layer and the model of grain evolution during the CRBF process were established. Fengkui Cui et al. [10,22,23] measured the residual stress, hardness, and roughness of a spline surface fabricated by CRBF, and a series of empirical expressions were obtained to describe the effects of the roll-beating speed and feed rate on these indices.

On the basis of clear characteristics and the forming mechanism of CRBF, there is a need for further quantitative discussion about the effect of process parameters on the CRBF process to promote the practical application of CRBF. In particular, to improve the process performance of CRBF, it is necessary to clarify the influence of process parameters on the forming force, forming quality, and forming efficiency. In view of this situation, ASTM 1045 steel was tested by a CRBF experiment on the outer teeth. The empirical formulas of the indices of the forming force, product quality, and forming efficiency under different process parameters were established according to the experimental results. The aim of this study is to determine the influence of different process parameters on CRBF and provide a method to realize the comprehensive optimization of the forming force, forming quality, and forming efficiency by judiciously selecting the appropriate process parameters.

2. Materials and Methods

2.1. Description of CRB Process

CRBF is a special intermittent free cold forging process. It is different from the quasi-static forming process of die forging, drawing, rolling, etc. During the CRBF process, the metal is subjected to intermittent impacts at a certain frequency, and the deformation produced by each impact is small. The deformation zone continuously migrates through the feed of the workpiece, and the amount of deformation gradually accumulates to accomplish the final forming purpose. For different types of parts, the process of forming the external tooth profile in CRBF can be described by the single tooth groove forming process. A schematic diagram is presented in Figure 1. The roller is assembled eccentrically on the spindle, and the geometric profile of the roller is designed according to the profile of the target tooth. The spindle rotation drives the roller to roll-beat the workpiece, and the roller can rotate around the roller shaft as it beats and rolls the workpiece. During the forming process, the workpiece is fed continuously to accumulate the plastic deformation caused by each roll-beating of the roller. Finally, a tooth groove consistent with the profile of the target tooth is formed on the workpiece. When the feed direction of the workpiece is clear, we can change the rotation direction of the spindle to choose the mode of the CRBF process. In order to express this clearly in a way similar to the milling process, CRBF can be defined by a coordinate system in which the y-axis is the feed direction of the workpiece and the z-axis is perpendicular to the workpiece to be processed. As such, when the angular velocity direction of the spindle rotation is consistent with the x-axis, it is an up-beating mode; when it is inconsistent, it is a down-beating mode. Through the above description of the CRBF process, it can be understood that the main processing parameters are spindle rotation speed, feed speed, and roll-beating mode.

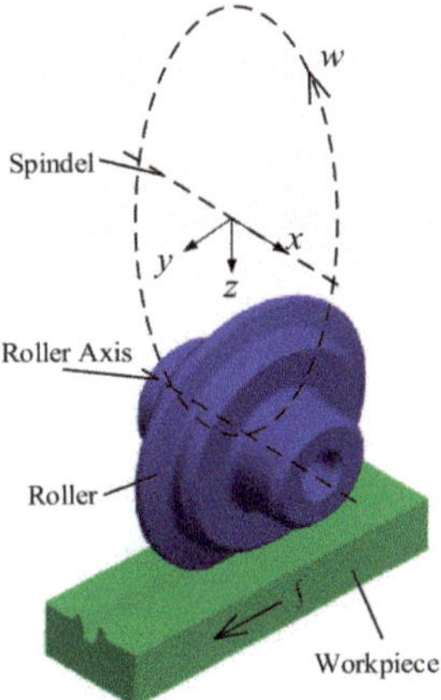

Figure 1. Schematic of the cold roll-beating (CRB) principle of an external tooth groove.

2.2. Material Characterization

The material of the workpieces was ASTM 1045 steel for the CRBF experiment because it has good mechanical properties and manufacturability, and it is widely used to manufacture functional parts. The main chemical composition of the material is shown in Table 1.

Table 1. Chemical composition of ASTM 1045 steel (mass fraction, %).

Element	C	Si	P	S	Cr	Mn
wt.%	0.46	0.32	0.035	0.04	0.22	0.62

The main deformation of materials during the forming process is compression deformation. After normalization, a compression test was carried out by a 100-kN mechanical testing machine (MTS Inc., Eden Prairie, MN, USA). The size of the sample was $\Phi 6$ mm $\times$ 10 mm, and the down-pressure velocity was 0.5 mm/min. Figure 2 illustrates the experimental result.

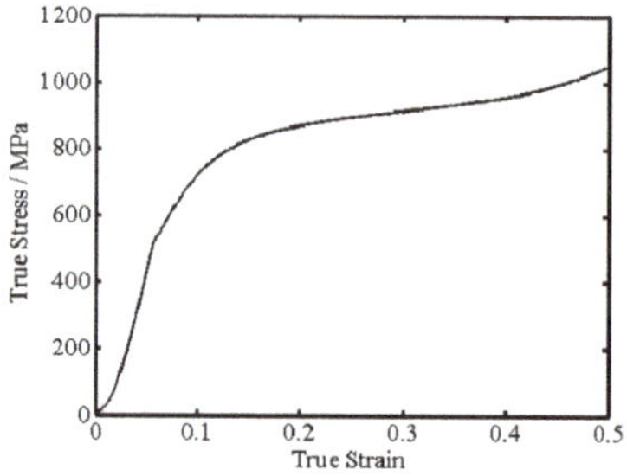

Figure 2. Stress-strain curve of ASTM 1045 steel.

For the CRBF experiment, the material was made into workpieces by wire-electrode cutting and milling. The length along the feed direction of the workpiece was 45 mm. The surface hardness and roughness of the workpieces were HV 180–220 and Ra 0.8–1.6, respectively.

2.3. Experimental Setup

According to the principle of the external tooth groove CRBF, a CRBF device was designed and realized by a horizontal milling machine to carry out the forming experiment, as shown in Figure 3. In this CRBF device, the roller is eccentrically mounted on the spindle through a roller shaft and two bearings (INA NKXR20Z, INA Inc., Nuremberg, Germany), and the axial clearance is adjusted by two sets of gaskets. The radius of the trajectory of the roller edge is 74 mm, and the roller is made of 20CrMnTi. After quenching and tempering treatment, the surface of the roller is conditioned and Rockwell hardness number reaches 58–64 HRC. The dimensions and tooth detail of the roller are shown in Figure 4.

Figure 3. CRBF experimental equipment.

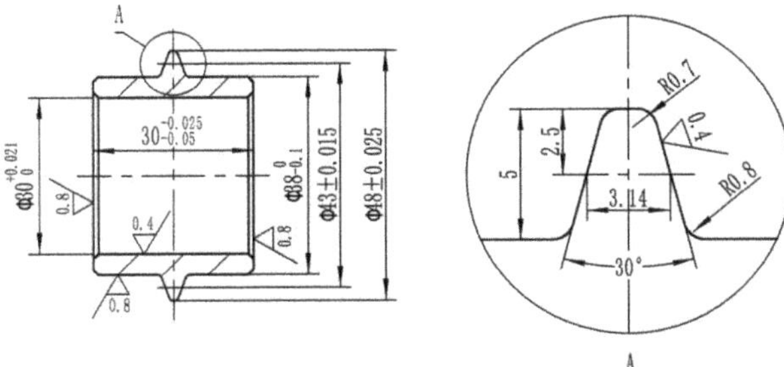

Figure 4. Dimensions and tooth detail of the roller (dimensions in mm).

The spindle rotation speed, feed speed, and roll-beating mode were chosen as the research factors. The spindle rotation speeds are 475 r/min, 950 r/min, and 1500 r/min. The feed speeds of the workpiece are 30 mm/min, 60 mm/min, 120 mm/min, 240 mm/min, 480 mm/min, and 960 mm/min. The different roll-beating modes of the CRBF process were realized by changing the rotation direction of the spindle. The experiments were carried out with all combinations of all process parameters at all levels. The experimental implementation scheme is shown in Table 3. The roll-beating depth was set at 2.5 mm, and the workpiece was coated with lubricating oil.

During the forming process, a piezoelectric three-direction force sensor PCB261A03 (PCB Inc., Buffalo, NY, USA) was used to obtain forming force data. After removing oil from the surface of the experimental tooth grooves with a metal cleaning agent, the three-dimensional macroscopic geometrical data of the surface geometry of the tooth grooves were measured and obtained by the super 3-D microscope system VHX-5000 (Keyence Inc., Osaka, Japan). The measuring accuracy of this instrument is 0.01 mm. The surface roughness of the tooth wall was measured by laser confocal microscopy using the DCM 3D (Leica Inc., Wetzlar, Germany).

3. Results

For metal bulk forming technology, forming forces, forming quality, and forming efficiency are the important objectives that we are interested in optimizing.

In the coordinate system described in Figure 1, the forming force in the x-direction was low during the forming process because of the geometric symmetry of the tooth shape of the roller. The forming force in the z-direction F_z is the main forming force in the CRBF process, and the forming force in the y-direction F_y determines the main shaft torque load and the workpiece feed system load of the forming equipment. F_z and F_y directly affect the design of the forming equipment and the implementation of the forming process.

Figure 5 shows F_z and F_y measured in the up-beating and down-beating modes for a spindle rotation speed of 475 r/min and a feed speed of 240 mm/min. For the same forming equipment and workpiece, the total time of the CRBF process is mainly related to the feed speed. Figure 5 shows that the actual roll-beating lasts nearly 16.5 s when the feed speed is 240 mm/min and that the forming forces of CRBF form a pronounced sharp pulse. This is consistent with the characteristics of intermittent impact loading in the CRBF process. In either mode, the F_z peak of each roll-beating gradually increases in the early stage of the forming process; then, it reaches the maximum and remains stable, after which it decreases gradually until the end of the forming process. Therefore, F_{zam}—the average value of the F_z peak of each roll-beating in the stability region—is used to characterize F_z. The F_y peak of each roll-beating during the CRBF process also has increasing, stable, and decreasing regions. However, the F_y peak of each roll-beating in the stable region is not the maximum of the whole forming process, differing from the pattern for F_z and varying for different roll-beating modes: In the up-beating mode, the maximum F_y is obtained between the increasing and stable regions, and the maximum F_y in the down-beating mode occurs before the end of the stable region. For practical applications, the maximum

F_y of the whole forming process is worth more than the average value of the F_y peak of each roll-beating in the stable region. As such, to express the maximum F_y of the whole forming process, F_{ymax} is used to characterize F_y in the CRBF process.

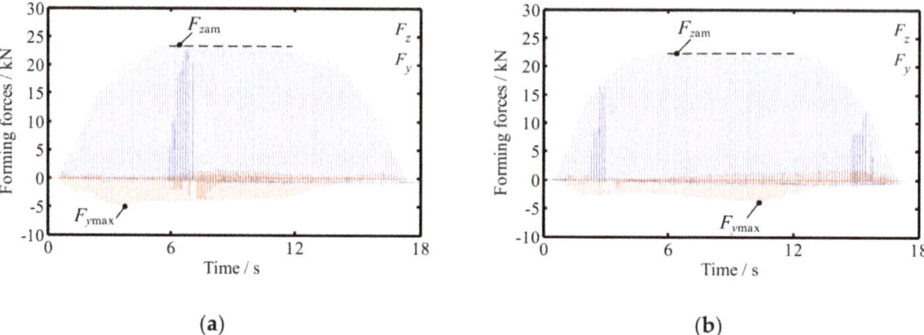

Figure 5. Forming forces of the CRBF forming process: (**a**) Up-beating; (**b**) down-beating.

For product quality attributes, this paper mainly considers geometric accuracy, surface roughness, and hardening. The geometry accuracy of the single tooth profile is usually described by profile error and lead error. The lead error is primarily determined by the feeding straightness of the feed system of the forming equipment and the axial positioning accuracy of the spindle rotation. The lead error is not sensitive to the process parameters analyzed in this paper. There is an obvious tooth profile error in the tooth groove formed by CRBF because, in the real forming process, the material has elastic recovery, which produces profile errors. These profile errors mainly appear as angle errors in the tooth profile, as shown in Figure 6. When the roller roll-beats the workpiece, the metal extends along the tooth profile. After the roller leaves the workpiece, the metal flowing along the tooth profile has elastic recovery, which ultimately makes the angle of the tooth groove wall smaller than the tooth profile angle of the roller. The above discrepancy in the tooth angle is defined as the angle error of the tooth profile and is represented by β. For a different roll-beating mode, roll-beating speed, and feed speed, the stress field and hardening of metals are different, and this variation affects the elastic recovery. As such, β is obviously affected by the process parameters. Therefore, β is used to characterize the geometry accuracy in this paper.

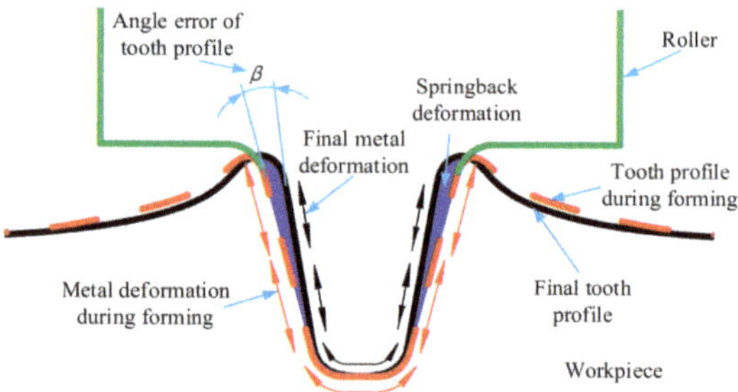

Figure 6. Schematic of profile error generation.

Transmission parts mainly transfer power through tooth wall meshing. Reducing the surface roughness of the tooth wall directly improves the service life and transfer efficiency of the parts and reduces the working vibration. Therefore, the surface roughness is an important standard of surface quality. To facilitate measurement, each groove formed in the experiment was divided into two parts by the middle of the tooth bottom using wire-electrode cutting. The length of the specimen was 10 mm along the tooth's lead direction, as shown in Figure 7. It can be seen that the desired surface finish of the tooth wall can be achieved by grinding.

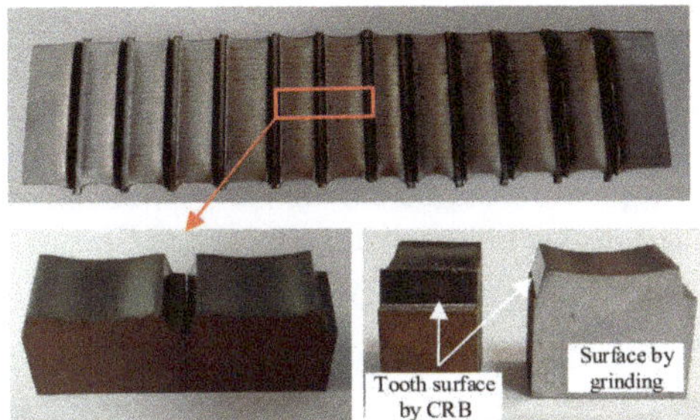

Figure 7. Experimental forming of parts and specimens.

From the measurement of the surface roughness of the tooth wall, it was observed that the micro-morphology of the tooth wall formed by CRBF is an irregular fish-scale pattern, as shown in Figure 8. This phenomenon is caused by the multiple rolling of the surface of the workpiece and the rotation of the roller in the roll-beating process. Therefore, surface roughness Sa of the tooth wall is used to characterize the surface quality.

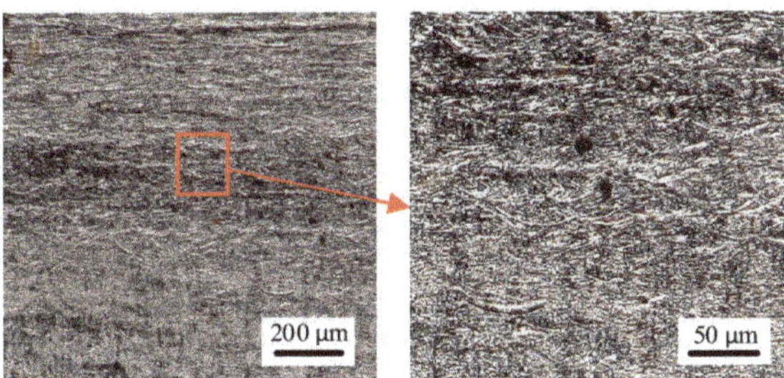

Figure 8. Surface morphology of tooth wall.

The metallographic structure of the metal after CRBF is shown in Figure 9. As a result of the strong plastic deformation of the metal during the forming process, the grain was refined and highly fibrous on the tooth wall and tooth bottom. Because of the impact loading of CRBF, the grain refinement and fibrosis are mainly concentrated in the surface layer of the metal. Comparing the metallographic structure of the tooth top, tooth wall, and tooth bottom, it can be found that the grain refinement degree of the tooth bottom near the tooth wall is the highest, and the grain refinement degree is the weakest at the tooth top.

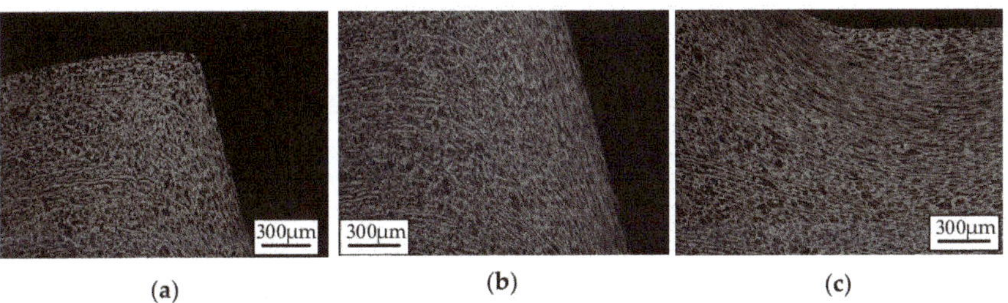

(a)   (b)   (c)

Figure 9. Metallographic structure of different parts of formed tooth groove: (**a**) Tooth top; (**b**) tooth wall; (**c**) tooth bottom.

The hardness of each part was measured. The hardness at the position of the tooth top, tooth wall, and tooth bottom near the tooth wall is 256, 292, and 310 HV, respectively. The hardness of the internal metal, which does not change in structure, is 210 HV. This is consistent with the change in grain refinement. The hardness of the tooth wall resulting from the use of different process parameters was measured. The results show that the hardening rate is between 135% and 145%, and the hardening degree is not very sensitive to the change in process parameters. This is because the change in processing parameters has little effect on the metal deformation after final forming. Therefore, workpiece hardening is not considered as an index of forming quality in this study.

For the forming efficiency, the feed speed is directly used to characterize the forming efficiency. The research objectives and their corresponding indices are shown in Table 2.

Table 2. Research objectives and their corresponding indices.

Object	Forming Forces		Forming Quality		Forming Efficiency
	F_z	F_y	Geometry Accuracy	Surface Roughness	
Characteristic	The form of sharp pulse		Obvious tooth angle error	Irregular fish-scale pattern	Positive correlation with feed speed
Index	F_{zam}	F_{ymax}	β	Sa	f

The experimental results of the above indices and corresponding process parameters are listed in Table 3, in which the results for β and Sa are the average of the two sides of the tooth wall. In addition, in order to facilitate mathematical representation and analysis, the variable m is used to denote the roll-beating mode: $m = 1$ represents up-beating and $m = -1$ represents down-beating.

Table 3. CRBF experimental results.

Serial Number	Roll-Beating Mode, m	Spindle Rotation Speed, w (r/min)	Feed Speed, f (mm/min)	F_{zam} (N)	F_{ymax} (N)	β (°)	Sa (μm)
1	1	475	30	17,051	3457	1.18	0.079
2	1	475	60	18,540	3996	1.07	0.087
3	1	475	120	20,752	4528	1.20	0.124
4	1	475	240	23,175	5125	1.36	0.135
5	1	475	480	27,656	5625	1.51	0.099
6	1	475	960	31,515	6154	1.65	0.085
7	1	950	30	18,391	3534	1.07	0.108
8	1	950	60	20,747	4087	0.92	0.113
9	1	950	120	22,028	4529	0.82	0.155
10	1	950	240	25,928	5022	0.55	0.165
11	1	950	480	27,436	5595	0.61	0.138
12	1	950	960	31,702	6288	1.09	0.109
13	1	1500	30	18,361	3689	0.62	0.165
14	1	1500	60	20,048	4509	0.59	0.242
15	1	1500	120	22,494	5330	0.46	0.253
16	1	1500	240	24,219	5659	0.32	0.265
17	1	1500	480	27,781	6632	0.42	0.280
18	1	1500	960	33,179	7415	1.03	0.232
19	−1	475	30	16,592	2958	0.77	0.075
20	−1	475	60	19,392	3173	0.26	0.090
21	−1	475	120	20,933	3526	0.16	0.110
22	−1	475	240	23,185	3626	0.08	0.112
23	−1	475	480	26,707	3590	1.12	0.088
24	−1	475	960	30,204	3561	1.81	0.073
25	−1	950	30	17,636	3914	0.80	0.093

Table 3. *Cont.*

Serial Number	Roll-Beating Mode, m	Spindle Rotation Speed, w (r/min)	Feed Speed, f (mm/min)	F_{zam} (N)	F_{ymax} (N)	β (°)	Sa (μm)
26	−1	950	60	19,201	4284	0.48	0.140
27	−1	950	120	21,470	4361	0.12	0.147
28	−1	950	240	23,501	4461	0.09	0.166
29	−1	950	480	26,194	4419	0.60	0.174
30	−1	950	960	29,737	4321	1.33	0.117
31	−1	1500	30	16,222	4407	0.55	0.172
32	−1	1500	60	18,807	4984	0.50	0.261
33	−1	1500	120	21,210	5349	0.34	0.266
34	−1	1500	240	22,936	5784	0.19	0.296
35	−1	1500	480	26,577	5722	0.48	0.296
36	−1	1500	960	30,288	5576	1.22	0.200

4. Discussion

4.1. Significance Analysis of Process Parameters to Objectives

ANOVA was used to test the significance of each process parameter, and the results are shown in Table 4. The probability of $F_{(0.05)} > F$ is represented by the P value of the right-sided test. When the F value of the objective is greater than $F_{(0.05)}$ and the P value is less than 0.05, the factor has a significant effect on the objective [24,25].

Table 4. Significance analysis of process parameters to objectives.

Source	$F_{(0.05)}$	F_{zam}		F_{ymax}		β		Sa	
		F	P	F	P	F	P	F	P
m	4.139	60.18	$<10^{-4}$	410.64	$<10^{-4}$	36.32	$<10^{-4}$	0.34	0.5746
w	3.295	8.33	0.0074	453.89	$<10^{-4}$	27.17	$<10^{-4}$	482.2	$<10^{-4}$
f	2.545	817.55	$<10^{-4}$	253.6	$<10^{-4}$	27.43	$<10^{-4}$	33.7	$<10^{-4}$
$m \times w$	3.295	8.76	$<10^{-4}$	94.49	$<10^{-4}$	11.57	0.0025	2.42	0.1390
$m \times f$	2.545	2.28	0.1249	98.27	$<10^{-4}$	6.47	0.0062	1.23	0.3621
$w \times f$	2.255	2.38	0.0943	6.00	0.0045	2.28	0.1055	4.21	0.0164

The results of ANOVA show that for the ranges of the tested levels, the three process parameters have a significant influence on F_{zam}, for which the most important factor is feed speed, the second most important is the roll-beating mode, and the least important is the spindle rotation speed. The interaction item $m \times w$ is significant for F_{zam}, and $m \times f$ is considered statistically non-significant. The F value of $w \times f$ is larger than $F_{(0.05)}$, but the P value is bigger than 0.05. These values indicate that, although the influence of $w \times f$ on F_{zam} is very weak, it still has some impact.

All the tested sources of variation in F_{ymax} are significant, and the order of importance of those sources is $w > m > f > m \times f > m \times w > w \times f$.

Each process parameter has a significant influence on β, and m is the most important factor. The parameters w and f almost have the same effect on β. Of the interactive terms, $m \times w$ and $m \times f$ are significant. $P > 0.05$, the $w \times f$ interaction is considered statistically non-significant, but its F value is bigger than $F_{(0.05)}$, so its impact cannot be ignored.

For Sa, m and the interaction terms containing m are not significant. So, from a statistical view, it can be asserted that the change in roll-beating mode does not affect the roughness of the tooth wall. All the other sources are significant, and w is the most important.

4.2. Regression Models

Because m is not a continuous variable—i.e., it only takes values of 1 or −1—a piecewise function, denoted by $T(w, f)$, that consists of two polynomial models is used to characterize the response of F_{zam},

$F_{y\max}$, and β to the influences of the different parameters. Meanwhile, given that m is irrelevant to Sa, $T(w, f)$ is directly used to express Sa. The functional form of $T(w, f)$ is a polynomial model which is widely used in regression analysis and empirical modeling. The function is shown in Equation (1). In order to ensure regression accuracy, the degrees of freedom of w and f in the function $T(w, f)$ are 2 and 3, respectively, and the interaction between w and f is accounted for.

$$T(w, f) = a_{00} + a_{10}w + a_{01}f + a_{20}w^2 + a_{11}wf + a_{02}f^2 + a_{21}w^2 f + a_{12}wf^2 + a_{03}f^3 \tag{1}$$

Table 5 gives the final regression model of each index. The applicable range of these regression models is within the range of the experimental parameters. The term R^2 denotes the prediction capability of the regression equations. Generally, an R^2 value greater than 0.9 indicates that the regression equation is acceptable for fitting the experimental data, and an R^2 value greater than 0.95 indicates that all predicted values are reliable and quite close to the actual values. Apart from the R^2 value of β's regression model in the down-beating mode being slightly less than 0.95, the R^2 values of all the other regression models are higher than 0.95. Furthermore, the P values of all regression models are less than 10^{-4}, as shown in Table 4.

Table 5. Regression models and validation results.

Regression Model	R^2		P
	$m = 1$	$m = -1$	
$F_{zam} = \begin{cases} 11060 + 11.82w + 58.81f - 0.00506w^2 - 0.02087wf - 0.07666f^2 \\ \quad + 7.724 \times 10^{-6}w^2 f + 5.563 \times 10^{-6}wf^2 + 4.027 \times 10^{-5}f^3, \quad m = 1 \\ \\ 14440 + 4.462w + 50.55f - 0.002426w^2 - 0.009655wf - 0.07253f^2 \\ \quad + 5.319 \times 10^{-6}w^2 f - 4.689 \times 10^{-7}wf^2 + 4.076 \times 10^{-5}f^3, \quad m = -1 \end{cases}$	0.9926	0.9886	$<10^{-4}$
$F_{y\max} = \begin{cases} 3600 - 1.13w + 13.02f + 0.0007435w^2 - 0.00103wf - 0.02282f^2 \\ \quad + 1.377 \times 10^{-6}w^2 f - 7.974 \times 10^{-7}wf^2 + 1.357 \times 10^{-5}f^3, \quad m = 1 \\ \\ 1711 + 2.599w + 7.194f - 0.0005536w^2 - 0.000828wf - 0.01656f^2 \\ \quad + 1.659 \times 10^{-6}w^2 f - 2.078 \times 10^{-6}wf^2 + 1.139 \times 10^{-5}f^3, \quad m = -1 \end{cases}$	0.9818	0.9849	$<10^{-4}$
$\beta = \begin{cases} 1.418 - 0.0004332w + 0.001788f + 2.494 \times 10^{-9}w^2 - 5.875 \times 10^{-9}wf \\ \quad + 2.469 \times 10^{-6}f^2 + 1.634 \times 10^{-9}w^2 f + 2.569 \times 10^{-9}wf^2 - 2.609 \times 10^{-9}f^3, \quad m = 1 \\ \\ 0.5644 + 0.0005123w - 0.005542f - 1.711 \times 10^{-7}w^2 - 3.932 \times 10^{-6}wf \\ \quad + 2.249 \times 10^{-5}f^2 + 1.259 \times 10^{-9}w^2 f + 6.398 \times 10^{-10}wf^2 - 1.479 \times 10^{-8}f^3, \quad m = -1 \end{cases}$	0.9567	0.9480	$<10^{-4}$
$Sa = 0.09151 - 0.0001148w + 0.0004310f + 1.124 \times 10^{-7}w^2 + 3.031 \times 10^{-7}wf$ $\quad -1.407 \times 10^{-6}f^2 + 4.014 \times 10^{-13}w^2 f - 2.876 \times 10^{-10}wf^2 + 1.003 \times 10^{-9}f^3$		0.9563	$<10^{-4}$

From the established regression models, the predicted values of the indices resulting from the process parameters in Table 3 were obtained. Then, we plotted these points to the coordinates that define the experimental values compared with the predicted values, as seen in Figure 10. The distribution of these points is not related to the form and coefficient of the regression equation but only to the fitting error. Figure 10 shows that the points are close to the straight line of y = x, and the error of the predicted values versus experimental values is distributed uniformly. These results indicate that the regression models have high fitting ability.

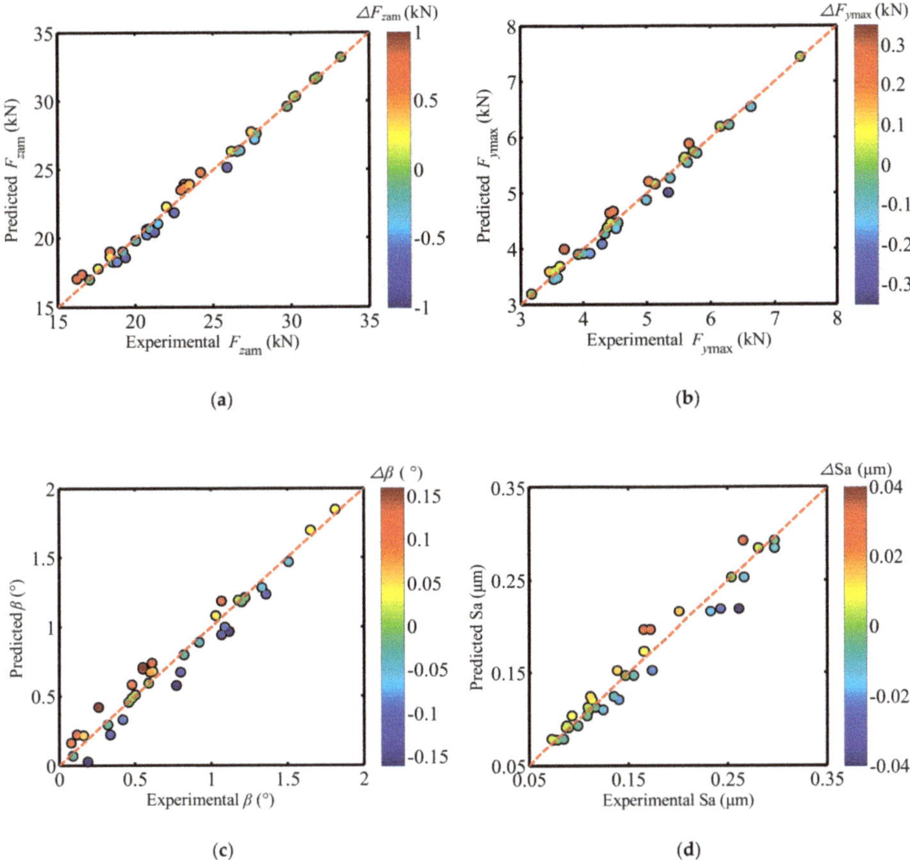

Figure 10. Correlation of predicted values versus experimental values: (**a**) F_{zam}; (**b**) F_{ymax}; (**c**) β; and (**d**) Sa.

As such, according to the results of the above analysis, it is proved that the established regression models can effectively and reliably describe the influence of the tested process parameters on F_{zam}, F_{ymax}, β, and Sa.

4.3. Influence of Process Parameters on Each Index

In accordance with the regression equations listed in Table 5, Figures 11–14 illustrate the relation surface between each index and process parameter.

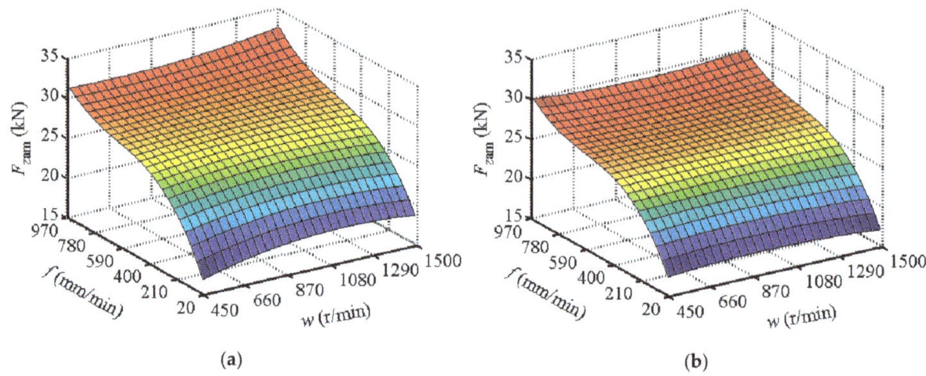

Figure 11. Surface graphs of F_{zam} regression model: (**a**) Up-beating ($m = 1$); (**b**) Down-beating ($m = -1$).

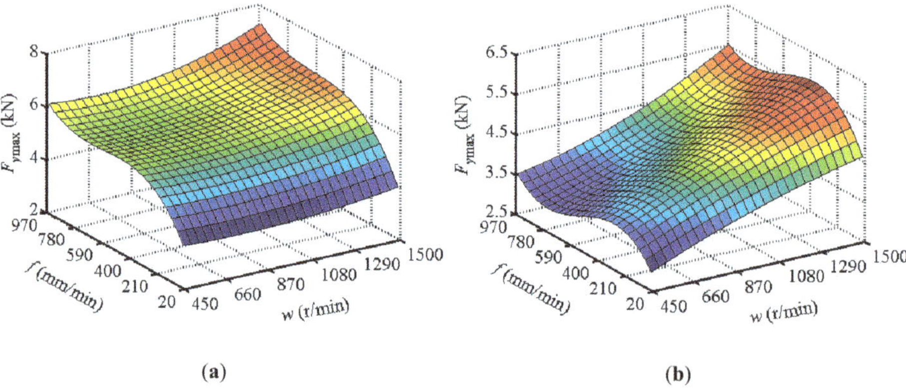

Figure 12. Surface graphs of F_{ymax} regression model: (**a**) Up-beating ($m = 1$); (**b**) Down-beating ($m = -1$).

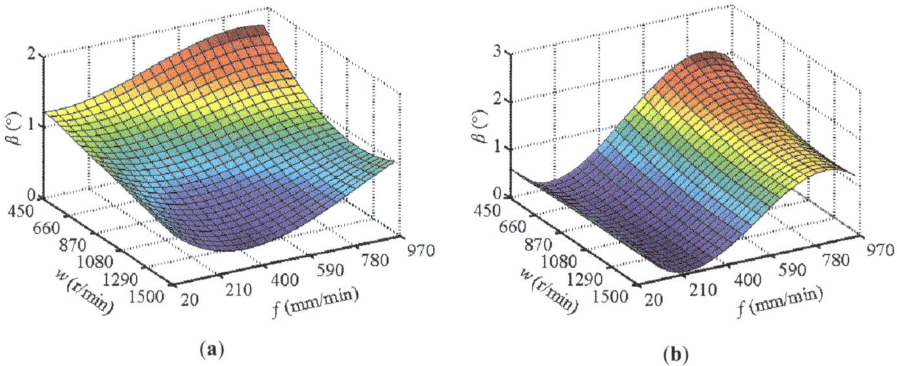

Figure 13. Surface graphs of β regression model: (**a**) Up-beating ($m = 1$); (**b**) Down-beating ($m = -1$).

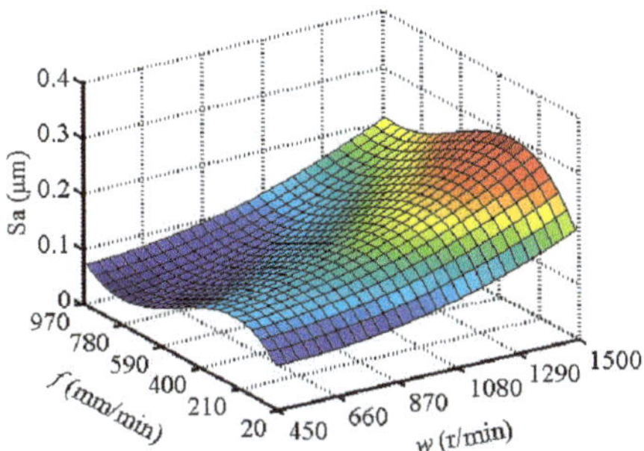

Figure 14. Surface graphs of Sa regression model.

Figure 11 shows that in the different roll-beating modes, the change rules of F_{zam} with w and f are consistent, but the change in F_{zam} in the up-beating mode is larger. The value of f has a great influence on F_{zam}, where lower f means smaller F_{zam}. The degree of this effect is greater when f is less than 400 mm/min, below which F_{zam} decreases rapidly with decreasing f. The effect of w on F_{zam} varies in different f regions. When f is in the lower range, F_{zam} increases first and then decreases with the increase in w. At higher f values, F_{zam} decreases first and then increases when w increases. However, the range of F_{zam} changes is limited by only changing w. Therefore, the down-beating mode and low f are enough to significantly reduce F_{zam}.

As shown in Figure 12, the change trend of F_{ymax} resulting from f and w varies for the different roll-beating modes, and the change range of F_{ymax} in the up-beating mode is larger. For up-beating, with an increase in w, F_{ymax} decreases first and then increases. Compared with w, f has a greater influence on F_{ymax}. Increasing f leads to an obvious increase in F_{ymax}. When f is small and w is near 900 r/min, F_{ymax} is the smallest. In the down-beating mode, F_{ymax} is more affected by w than f. F_{ymax} and w are proportional to each other. In response to increasing f, F_{ymax} increases rapidly and then decreases. As such, the down-beating mode, low w, and small f or f close to 780 mm/min are advantageous for reducing F_{ymax}.

The whole change range of β is smaller in the up-beating mode. β decreases with the increase in w. In the low-w region, increasing f causes β to increase. In the high-w region, β decreases first and then increases when f grows, and the minimum β value is obtained around $f = 450$ mm/min. In the down-beating mode, β is weakly affected by w and decreases slightly with an increase in w. β is strongly influenced by f. As a result of increasing f, β decreases first and reaches the minimum; when f is near 240 mm/min, β increases and reaches the maximum near $f = 840$ mm/min. Figure 13 shows that in different roll-beating modes, setting w to 1280–1500 r/min and setting f to 30–590 mm/min allows β to be controlled at its lower level.

As shown in Figure 14, Sa increases when w increases. With increasing f, Sa increases rapidly first and then decreases gradually. Sa is the maximum at $w = 1500$ r/min and $f = 350$ mm/min. When w is within 450–870 r/min and f is within 600–960 mm/min, Sa is controlled under 0.1 μm.

In addition, for the single tooth forming process, a higher f value means higher forming efficiency.

4.4. Optimize

From the above analysis, it can be seen that the influence of each process parameter on the forming force, forming quality, and forming efficiency indices has its own characteristics. There are some contradictions between the ranges of the process parameters, and these conflicts result in each index having its own optimal value at the same time; in order to determine the process parameters that allow each index to achieve comprehensive optimization, the linear weighted sum method was used to establish the comprehensive evaluation function E for F_{zam}, F_{ymax}, β, Sa, and forming efficiency, as shown in Equation (2).

$$E = \sum_{i=1}^{n} C_i e_{ij} \tag{2}$$

In Equation (2), i denotes the ordinal number of the index (in this case, $i = 1, 2, ..., 5$); j is the number of the process parameter combination; e_{ij} represents the evaluation value of the ith index obtained under the jth process parameter combination; C is the weight coefficient, and the sum of the weight coefficients is 1. The function E allows corresponding weight coefficients to be assigned to each index according to the importance of each index, and it includes the linear combination of the multiple indices. It reduces the multi-objective optimization problem to the numerical optimization of the function E, which realizes the comprehensive optimization of the multiple objectives.

To solve the above, the first step is to get the evaluation value of each index so that they have the same order of magnitude. The Min-Max normalization function is used to convert the index data into the evaluation value. So, all the evaluation values of each objective range from 0 to 1. This Min-Max function takes two forms, as shown by Equations (3) and (4), in which O represents the data of each index. In this case, F_{zam}, F_{ymax}, β, and Sa are normalized by Equation (3), and f is normalized by Equation (4).

$$e_{ij} = \frac{maxO_{ij} - O_{ij}}{maxO_{ij} - minO_{ij}} \tag{3}$$

$$e_{ij} = \frac{O_{ij} - minO_{ij}}{maxO_{ij} - minO_{ij}} \tag{4}$$

In the range of process parameters, m = $[-1, 1]$, w $\in$ [475, 1500], f $\in$ [30, 960], and the maximum and minimum values of each index are solved. The results are shown in Table 6.

Table 6. The $maxO_{ij}$ and $minO_{ij}$ values of each index.

Index	i	$maxO_{ij}$	$minO_{ij}$
F_{zam}	1	33,163	16,987
F_{ymax}	2	7441	2968
β	3	2.1013	0.0282
Sa	4	0.3007	0.0476
f	5	960	30

The entropy weight method was used to determine the weight coefficients, as it has strong objectivity and is widely applicable to practical engineering optimal problems [26]. This method uses the evaluation data of a sample to calculate the entropy weight coefficient of each index as the weight coefficient of each index. In order to calculate the entropy weight coefficient, it is first and foremost necessary to determine a large enough sample of evaluation values. Given the range of the process parameters, with the interval steps of w and f set to 5, there are 77,044 combinations of the different process parameters. Then, the evaluation values for each index under each group of process parameters are calculated by the regression model in Section 4.2, Equations (3)–(4), and Table 6. As such, we get a $5 \times 77,044$ evaluation matrix as the sample data. Then, the entropy weight coefficient of each index is obtained by Equations (5) and (6), in which H_i represents the entropy value of the ith index, and c_i is the entropy weight coefficient for the ith index. During the calculation, it is important to note that when f is 30, calculating H_5 is meaningless. To avoid this, the value of 30 for f is replaced by 30.00001 in these calculations.

$$H_i = -\frac{\sum\limits_{j=1}^{k} \frac{e_{ij}}{\sum\limits_{j=1}^{k} e_{ij}} \ln \frac{e_{ij}}{\sum\limits_{j=1}^{k} e_{ij}}}{\ln k} \tag{5}$$

$$c_i = \frac{1 - H_i}{n - \sum_{i=1}^{n} H_i} \tag{6}$$

However, the weight coefficient obtained by this method has a lack of horizontal comparison among the objectives. Accordingly, to reflect the importance of the different objectives, an additional subjective weight coefficient λ is added to the entropy weight coefficient to get a composite weight coefficient, which is the C in Equation (2). The composite weight coefficient can be calculated by Equation (7). The subjective weight coefficient is given according to the importance of each objective. For the CRBF process, the first considered object is the forming quality, followed by the forming force and forming efficiency. Consequently, the subjective weight coefficients of β and Sa in the evaluation of tooth profile angle error and tooth wall roughness are large. Among them, considering that the tooth wall roughness is maintained at a certain level under the different process parameters, the subjective weight coefficient of Sa is smaller than that of β. For the indices used to evaluate the forming force, compared with F_{zam}, F_{ymax} is more important since it directly reflects the torque required of the spindle and the load of the feed system during the forming process. A smaller F_{ymax} implies lower energy consumption. Therefore, the subjective weight coefficient of F_{ymax} is slightly higher than that of F_{zam}. The forming efficiency and F_{zam} are a set of contradictory indices for CRBF. High efficiency means a large feed speed. However, with a large feed speed, F_{zam} is larger, as shown in Figure 11. A larger F_{zam} means that the forming equipment is subjected to a greater load, which speeds up the wear of the equipment parts and increases the cost of the forming process. As such, it is appropriate to assign F_{zam} and f the same subjective weight coefficients. The results of the weight coefficient calculations are listed in Table 7.

$$C_i = \frac{\lambda_i c_i}{\sum_{i=1}^{n} \lambda_i c_i}; \sum_{i=1}^{n} \lambda_i = 1 \tag{7}$$

Table 7. The weight coefficients of objectives.

Objective Number, i	Entropy Weight Coefficient, c_i	Subjective Weight Coefficient, λ_i	Composite Weight Coefficient, C_i
1	0.1786	0.15	0.1413
2	0.1240	0.175	0.1145
3	0.1578	0.275	0.2289
4	0.1672	0.25	0.2205
5	0.3724	0.15	0.2947

From Equations (2)–(4), the regression model of the indices, Table 6, and Table 7, the extended expression of the comprehensive evaluation function E is obtained, as shown in Equation (8).

$$E = \begin{cases} 0.5397 + 7.351 \times 10^{-5} w - 1.103 \times 10^{-3} f - 7.303 \times 10^{-8} w^2 + 5.934 \times 10^{-7} wf + 2.207 \times 10^{-6} f^2 \\ \quad - 2.835 \times 10^{-10} w^2 f - 6.132 \times 10^{-11} wf^2 - 1.285 \times 10^{-9} f^3, \quad m = 1 \\ \\ 0.6527 - 6.206 \times 10^{-5} w - 7.234 \times 10^{-5} f - 4.367 \times 10^{-8} w^2 + 2.757 \times 10^{-7} wf - 2.002 \times 10^{-7} f^2 \\ \quad - 2.283 \times 10^{-10} w^2 f + 2.372 \times 10^{-10} wf^2 + 1.1175 \times 10^{-10} f^3, \quad m = -1 \end{cases} \tag{8}$$

For the process parameter ranges, $m = [-1, 1]$, $w \in [475, 1500]$, and $f \in [30, 960]$, and the surface graph of the comprehensive evaluation function E is obtained and shown in Figure 15. The coordinates of each point on the surface represent a process parameter combination and the evaluation values resulting from this process parameter combination. A higher evaluation value corresponds to a more reasonable combination of process parameters in order to achieve the comprehensive process effect of a small forming force, high forming quality, and high forming efficiency. It can be seen that in the up-beating and down-beating modes, the evaluation value of the process parameter combination is higher in the ranges of $w \in [475, 1200]$ and $f \in [600, 960]$, respectively. In the case of w and f, with a minimum interval of 1, the vertex coordinates of the comprehensive evaluation function surface at $m = -1$ and $m = 1$ are (850, 882, 0.6354) and (801, 960, 0.6664), respectively. Subsequently, it is concluded that for CRBF, the optimum process parameters according to the comprehensive consideration of forming forces, forming quality, and forming efficiency are the following: Up-beating mode, a spindle rotation speed of 801 r/min, and a feed speed of 960 mm/min.

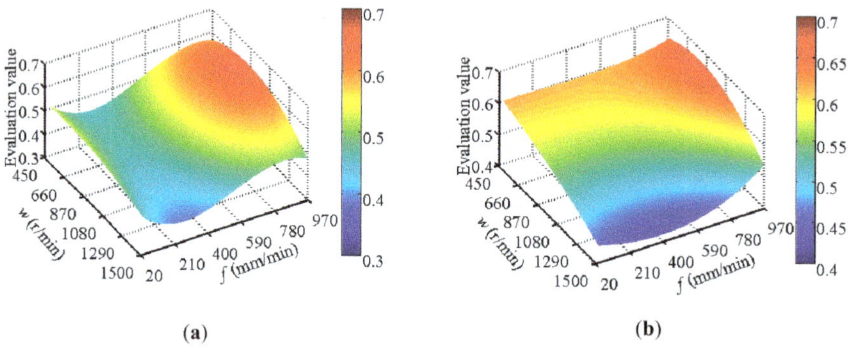

(a) **(b)**

Figure 15. Surface graphs of comprehensive evaluation function: (a) Up-beating ($m = 1$); (b) down-beating ($m = -1$).

With the optimal process parameters, a confirmation experiment was repeated three times, and the average values of F_{zam}, F_{ymax}, β, and Sa from the experiments are shown in Table 8. A comparison of predicted and experimental results reveals that the experimental results are close to the optimal solution obtained from the predicted model. The percentage errors between the prediction results and the confirmation experimental results are less than 7%.

Table 8. Optimal solution and confirmation experiment results.

Item	F_{zam} (kN)	F_{ymax} (kN)	β (°)	Sa (μm)	f (mm/min)	Evaluation
Optimal solution	29,700	4010	1.411	0.09714	960	0.6664
Experimental	29,814	4108	1.51	0.091	960	0.6568
Percentage error	0.55	2.44	6.99	6.32	0	1.42

Next, taking the experiments in Table 3 as control experiments and using the measured data of each index, the comprehensive evaluation value of each experiment was obtained by Equations (2)–(4). Comparing the comprehensive evaluation values from the confirmation experiment and the control experiments, it can be seen that the control experimental results are not higher than the confirmation experimental evaluation values, as shown in Figure 16. Therefore, it can be asserted that the regression models for F_{zam}, F_{ymax}, β, and Sa are correct, and the method of process parameter selection for multi-objective optimization of the CRBF process in this paper is feasible.

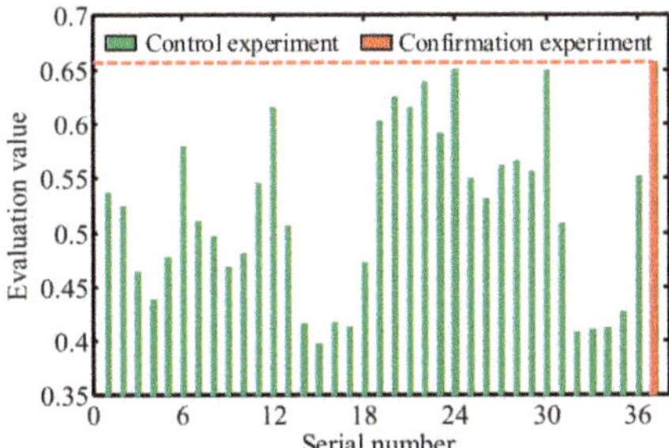

Figure 16. Comprehensive evaluation value of control experiments and confirmation experiment.

5. Conclusions

In this paper, the influences of several process parameters on the CRBF process were studied by forming an external tooth with ASTM 1045 material. The purpose was to derive a method for selecting process parameters that result in a balance between the forming forces, product quality, and forming efficiency. The following conclusions can be drawn:

(1) The feed speed plays a leading role in the peak value of the main forming force during the CRBF process, and the roll-beating mode has the second most important impact; the impact of the spindle rotation speed is minimal. For the forming force in the feed direction, the most influential parameter is the spindle rotation speed, followed by the roll-beating mode and feed speed. The down-beating mode, low feed speed, and low spindle rotation speed are conducive to reducing forming forces.

(2) The influence of the CRBF process on the geometric accuracy of the tooth profile primarily manifests in the tooth profile angle error. The most influential parameter is the roll-beating mode, followed by the spindle rotation speed and feed speed. A high spindle rotation speed and low feed speed bring the profile of the formed tooth groove closer to the tooth profile of the roller.

(3) The surface roughness of the tooth wall obtained by CRBF can be similar to that of a ground tooth wall. The most important impact parameter is the spindle rotation speed, followed by the feed speed. A low spindle rotation speed and high feed speed are good for reducing the surface roughness of the tooth wall.

(4) The regression models of different indices were established and showed that the judicious selection of process parameters is an important issue in CRBF. The linear weighted sum method and the compound entropy weight method were used to determine the process parameters that result in a good balance between the forming forces, product quality, and forming efficiency. The results show that to get the comprehensive optimum effect of a small forming force, high product quality, and high forming efficiency, one should set the roll-beating mode, spindle rotation speed, and feed speed to up-beating, 801 r/min, and 960 mm/min, respectively. In addition, given a multi-objective optimization problem, the prediction model of the forming forces, tooth angle error, and surface roughness of the tooth wall for CRBF and the process parameter selection method can be used reliably in similar experiments and theoretical studies.

Author Contributions: Conceptualization, L.L. and Y.L.; data curation, L.L. and T.T.; funding acquisition, L.L. and Y.L.; investigation, L.L.; methodology, M.Y.; project administration, Y.L.; writing—original draft, L.L.; writing—review & editing, M.Y.

Funding: This research was funded by National Natural Science Foundation of China (Grant No. 51475366, 51805433), Natural Science Basic Research Plan in Shaanxi Province of China (Grant No. 2016JM5074), and Ph.D. Innovation fund projects of Xi'an University of Technology (Grant No. 310-252071601).

Acknowledgments: The authors are grateful to National Natural Science Foundation of China, Natural Science Basic Research Plan in Shaanxi Province of China, and Ph.D. Innovation fund projects of Xi'an University of Technology, which enabled the research to be carried out successfully.

Conflicts of Interest: The authors declare no conflict of interest.

References

1. Gröbel, D.; Schulte, R.; Hildenbrand, P.; Lechner, M.; Engel, U.; Sieczkarek, P.; Wernicke, S.; Gies, S.; Tekkaya, A.E.; Behrens, B.A.; et al. Manufacturing of functional elements by sheet-bulk metal forming processes. *Prod. Eng.* **2016**, *10*, 63–80. [CrossRef]
2. Karpuschewski, B.; Beutner, M.; Köchig, M.; Wengler, M. Cemented carbide tools in high speed gear hobbing applications. *CIRP Ann.* **2017**, *66*, 117–120. [CrossRef]
3. Tekkaya, A.E.; Allwood, J.M.; Bariani, P.F.; Bruschi, S.; Cao, J.; Gramlich, S.; Groche, P.; Hirt, G.; Ishikawa, T.; Löbbe, C.; et al. Metal forming beyond shaping: Predicting and setting product properties. *CIRP Ann. Manuf. Technol.* **2015**, *64*, 629–653. [CrossRef]
4. Tekkaya, A.E.; Khalifa, N.B.; Grzancic, G.; Hölker, R. Forming of lightweight metal components: Need for new technologies. *Procedia Eng.* **2014**, *81*, 28–37. [CrossRef]
5. Jeswiet, J.; Geiger, M.; Engel, U.; Kleiner, M.; Schikorra, M.; Duflou, J.; Neugebauer, R.; Bariani, P.; Bruschi, S. Metal forming progress since 2000. *CIRP J. Manuf. Sci. Technol.* **2008**, *1*, 2–17. [CrossRef]
6. Pusavec, F.; Kramar, D.; Krajnik, P.; Kopac, J. Transitioning to sustainable production—Part II: Evaluation of sustainable machining technologies. *J. Clean. Prod.* **2010**, *18*, 1211–1221. [CrossRef]
7. Groche, P.; Fritsche, D.; Tekkaya, E.A.; Allwood, J.M.; Hirt, G.; Neugebauer, R. Incremental bulk metal forming. *CIRP Ann. Manuf. Technol.* **2007**, *56*, 635–656. [CrossRef]
8. Yang, D.Y.; Bambach, M.; Cao, J.; Duflou, J.R.; Groche, P.; Kuboki, T.; Sterzing, A.; Tekkaya, A.E.; Lee, C.W. Flexibility in metal forming. *CIRP Ann.* **2018**, *67*, 743–765. [CrossRef]
9. K, L. Modern metal forming technology for industrial production. *J. Mater. Process. Technol.* **1997**, *71*, 2–13. [CrossRef]
10. Cui, F.; Su, Y.; Xu, S.; Liu, F.; Yao, G. Optimization of the physical and mechanical properties of a spline surface fabricated by high-speed cold roll beating based on taguchi theory. *Math. Probl. Eng.* **2018**, *2018*, 1–12. [CrossRef]
11. Niu, T.; Li, Y.-T.; Liu, Z.-Q.; Qi, H.-P. Forming mechanism of high-speed cold roll beating of spline tooth. *Adv. Mater. Sci. Eng.* **2018**, *2018*, 1–6. [CrossRef]
12. CUI, F.; Li, Y.; Zhou, Y.; Yang, J.; Zhou, Z.; Li, C. Cad system of roller for involute spline and simulation of grinding process. *Chin. J. Mech. Eng.* **2005**, *41*, 210–215. [CrossRef]
13. Cui, F.; Li, Y.; Zhou, Y.; Zhou, Z.; Zhang, F. Technologic research on rolling involute spline axis. *Trans. Chin. Soc. Agric. Mach.* **2006**, *37*, 189–192. [CrossRef]
14. Yang, M.; Li, Y.; Dong, H.; Li, Y. Scale-like texture defect of slab metal cold roll-beating. *Mater. Res. Innov.* **2015**, *19*, S5–911. [CrossRef]
15. Cui, F.; Wang, X.; Zhang, F.; Xu, H. Metal flowing of involute spline cold roll-beating forming. *Chin. J. Mech. Eng.* **2013**, *26*, 1056–1062. [CrossRef]
16. Gao, X.; Yang, M.; Wei, F.; Li, Y.; Yuan, Q.; Cui, F. Analysis of material behaviour in slab cold roll-beating forming operations. *J. Balkan Tribol. Assoc.* **2016**, *22*, 637–651.
17. Cui, F.K.; Xie, K.G.; Xie, Y.F.; Hou, L.M.; Dong, X.D.; Li, Y. Constitutive model of cold roll-beating forming of 40cr. *Mater. Res. Innov.* **2015**, *19*, S8–284. [CrossRef]
18. Li, Y.; Li, Y.; Yang, M.; Yuan, Q.; Cui, F. Analyzing the thermal mechanical coupling of 40cr cold roll-beating forming process based on the johnson-cook dynamic constitutive equation. *Int. J. Heat Technol.* **2015**, *33*, 51–58. [CrossRef]
19. Kui, C.F.; Fei, L.Y.; Xue, X.J.; Peng, H.K.; Ge, X.K.; Yu, W.; Yan, L. The dynamic characteristics and constitutive model of 1020 steel. *Emerg. Mater. Res.* **2017**, *6*, 124–131. [CrossRef]
20. Liang, X.; Li, Y.; Cui, L.; Yang, M.; Xiao, J.; Cui, F. The effect of different roll-beating methodson deformation forces of rack cold roll-beating. *Rev. Facult. Ingenieria* **2016**, *31*, 164–174. [CrossRef]
21. Cui, F.; Su, Y.; Xie, K.; Xiaoqiang, W.; Ruan, X.; Liu, F. Analysis of metal flow behavior and residual stress formation of complex functional profiles under high-speed cold roll-beating. *Adv. Mater. Sci. Eng.* **2018**, *2018*, 1–10. [CrossRef]
22. Ding, Z.H.; Cui, F.K.; Liu, Y.B.; Li, Y.; Xie, K.G. A model of surface residual stress distribution of cold rolling spline. *Math. Probl. Eng.* **2017**, *2017*, 1–21. [CrossRef]

23. Cui, F.K.; Liu, F.; Su, Y.X.; Ruan, X.L.; Xu, S.K.; Liu, L.B. Surface performance multiobjective decision of a cold roll-beating spline with the entropy weight ideal point method. *Math. Probl. Eng.* **2018**, *2018*, 1–7. [CrossRef]
24. Asiltürk, İ.; Neşeli, S.; İnce, M.A. Optimisation of parameters affecting surface roughness of co28cr6mo medical material during cnc lathe machining by using the taguchi and rsm methods. *Measurement* **2016**, *78*, 120–128. [CrossRef]
25. Colmenero, A.N.; Orozco, M.S.; Macías, E.J.; Fernández, J.B.; Muro, J.C.S.-D.; Fals, H.C.; Roca, A.S. Optimization of friction stir spot welding process parameters for al-cu dissimilar joints using the energy of the vibration signals. *Int. J. Adv. Manuf. Technol.* **2018**. [CrossRef]
26. Cui, F.; Su, Y.; Wang, X.; Yu, X.; Ruan, X.; Liu, L. Surface work-hardening optimization of cold roll-beating splines based on an improved double-response surface-satisfaction function method. *Adv. Mech. Eng.* **2018**, *10*, 1–8. [CrossRef]

Article

Achievements of Nearly Zero Earing Defects on SPCC Cylindrical Drawn Cup Using Multi Draw Radius Die

Rudeemas Jankree and Sutasn Thipprakmas *

Department of Tool and Materials Engineering, King Mongkut's University of Technology Thonburi, 126 Prachautid Rd., Bangmod, Thungkru, Bangkok 10140, Thailand; rudeemas.jan@gmail.com
* Correspondence: sutasn.thi@kmutt.ac.th

Received: 3 July 2020; Accepted: 27 August 2020; Published: 8 September 2020

Abstract: In recent years, the old-fashioned cylindrical cup shapes are still widely used, and there are many defects which could not be solved yet. In the present research, the classical earing defects, which are mainly caused by the material mechanical property of the anisotropic property of the material (R-value), are focused on. The multi draw radius (MDR) deep drawing die is applied and investigated to achieve nearly zero earing defects by encountering the R-value during the deep drawing process. Based on the experiments, in different directions in the sheet plane, the somewhat concurrent plastic deformation could be controlled, and the uniform elongated grain microstructure and uniform strain distributions on the cup wall could be achieved. Therefore, on the basis of these characteristics, the earing defects could be prevented, and the nearly zero earing defects could be achieved. However, to achieve the nearly zero earing defects, the suitable MDR die design relating to the R-value should be strictly considered. In the present research, to apply the MDR die for the medium carbon steel sheet grade SPCC cylindrical drawn cup, the following was recommended: the large draw radius positioned at 45° to the rolling direction and the small draw radius positioned along the plane and at 90° to the rolling direction. Therefore, in the present research, it was originally revealed that the nearly zero earing defects could be successfully performed on the process by using the MDR die application.

Keywords: cylindrical cup; deep drawing; draw radius; earing; microstructure

---

1. Introduction

Sheet-metal products are increasingly fabricated to serve in various manufacturing industries especially the aerospace industry, electronics industry, and automobile industry. The complex shapes with high precision are also increasingly required in recent years. These products are commonly fabricated by sheet-metal forming processes such as bending, stamping, and deep drawing processes. Therefore, based on the experiments and finite element method (FEM) techniques, many studies have been performed and also reported by researchers and engineers to develop these sheet-metal forming processes and meet the mentioned requirements. Several researchers have focused on improving the quality of sheet-metal products as well as manufacturing the complex shapes with high precision through that associated with the experiments and FEM techniques [1–4]. In contrast, in terms of old-fashioned cylindrical cup shapes, the cylindrical cups are still widely used in various sheet-metal manufacturing industries. In general, almost all of them are conventionally fashioned by deep drawing process. On the basis of experimental and FEM works, the developments on this process have been continuously reported in many past studies [5–9]. For examples, Mahdavian and Tui Mei Yen [5] determined the effect of different punches' head profiles on the deep drawing of 5005H34 aluminium circular blanks to manufacture a cup-shape product. Nei et al. [6] investigated the deep drawing at various temperatures and studied the microstructure evolutions of three typical regions including bottom, corner, and wall taken from the drawn cylindrical part with the largest limiting drawing ratio (LDR). Liu et al. [7] showed the increase in the formability and quality of the pure aluminium spherical bottom cylindrical parts (SBCP) by using magnetic medium-assisted sheet metal drawing process. Bassoil et al. [8] studied the effects of a draw bead working on an Al6014-T4 strip according to assigned industrial conditions by the handy draw bead simulator (DBS). Sezek et al. [9] educated the effects of the die radius on blankholder forces and drawing ratio. The results also showed that the major parameters affected cup wall thickness were blankholder force, die radius, and lubricant use. However, in terms of earing defects as shown in Figure 1, there are few studies that have carried out the investigation and prevention these defects [10–15]. Marton et al. [10] investigated the prediction of the earing defect on 0.3 and 3.0 mm thick cold rolled 1050 type aluminium sheets subjected to annealing heat treatments for different time intervals to promote the recrystallization processes and obtain different earing behaviors. It was shown that the proposed method was able to predict the type and magnitude of earing with satisfactory results for both 0.3 and 3.0 mm sheet thicknesses. Kishor and Kumar [11] studied the earing problem in deep drawing of flat bottom cylindrical cups using the FEM (LSDYNA). The optimization of the initial blank shape was proposed to meet the smallest earing defects. Walde and Riedel [12] investigated the earing defects on the magnesium alloy AZ31 that the crystallographic texture and plastic anisotropy were usually pronounced during the rolling process. Based on the FEM in conjunction with a viscoplastic self-consistent texture model, the results showed that not only on the initial texture of sheet metal but also the evolution of texture during drawing process caused the earing defects. Izadpanah et al. [13] studied the deformation behaviors of sheet metals using advanced anisotropic yield criteria by FEM simulation. The results revealed that the earing profile and thickness distribution obtained from experiments well corresponded with FEM simulations. Cazacu et al. [14] proposed the anisotropy plasticity CPB06 theory, and Singh et al. [15] applied and implemented it in an FE model to predict the nonuniform material flow characteristics, earing defects, and thickness distributions successfully. Some studies were performed to prevent earing defects by making the material property into isotropic material. Olaf Engler et al. [16] studied the microstructure evolutions resulting in earing profiles on the Al alloy (AA8011A) during the down-stream process. The results showed that the earing defects could be controlled. Tang et al. [17] studied the microstructure and texture, tensile mechanical properties in terms of strength and elongation, and the anisotropy of conventional unidirectional rolling (UR) and cross rolling (CR) sheets at room temperature. The results showed that the CR sheet produced a deeper drawn cup than that of the UR sheet due to its lower normal anisotropy $\bar{R}$ value and layer elongation compared with those of the UR sheet. However, these techniques cause the increases in additional rolling operations resulting in the increases in production cost and time consuming. Next, by using the FEM, the studies have been focused on the applications of models, while the proposal of new anisotropy models is not covered. In addition, Phanitwong and Thipprakmas [18], by using multidraw radius (MDR), showed the interesting results of material flow characteristics during the deep drawing process in which the nonaxisymmetric material flow characteristic due to the anisotropy property of the material could be encountered by using the MDR die, and the axisymmetric material flow characteristic could be formed. Therefore, in the present research, the new idea of MDR application for the earing prevention during the deep drawing process is proposed. Specifically, the different draw radius positioned in different directions in the sheet plane was designed to encounter the material mechanical property of the anisotropy property of the material, and then the earing could be prevented. The MDR die seemed to be difficult to design and fabricate due to the complicated three-dimensional shape of draw radius (die shoulder radius or hole edge corner radius). However, based on the computer-aided design (CAD) and computer-aided manufacturing (CAM) technologies nowadays, this MDR die could be designed and fabricated in general. In addition, by comparing with other techniques applied for the deep drawing process such as draw bead application, this MDR die application shows the less process parameters to be concerned. Therefore, as the benefits of this application, it is easier to set the process parameters, and it can be performed on the deep drawing operation with no any additional operations. This results in the decreases in the production cost and time consumption. However, the conceptual design of MDR deep drawing die is strictly designed related to the R-value. As the results, they originally revealed that the MDR deep drawing die could be useful to reduce earing defects as well as to meet the nearly zero earing defects with no additional operations.

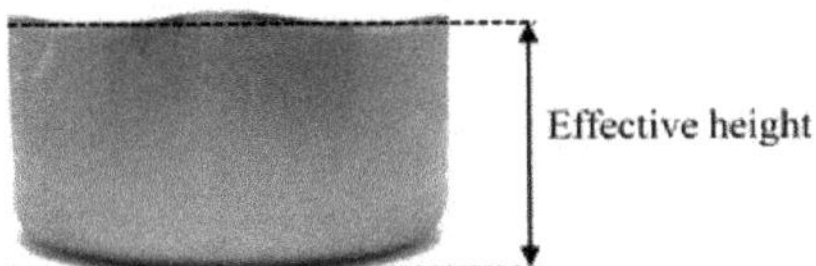

Figure 1. Earing defects and effective height of deep drawn parts.

2. Materials and Method

In the present research, the medium carbon steel sheet grade SPCC (JIS) with the thickness of 0.975 mm was used as a workpiece material. The chemical compositions were listed in Table 1. The mechanical properties were also examined by the tensile testing technique, and they were listed in Table 2. As a major material property that affected the earing defects, the R-value along the plane, at 45 and 90° to the rolling direction were examined, and they were 2.1, 1.9, and 2.6, respectively. Table 2 also shows the other material properties of Young's modulus, Poisson's ratio, and ultimate tensile strength. The microstructures along the plane, at 45 and 90° to the rolling direction were examined as well. The sheet materials were also sectioned and further processed by subsequent mounting, polishing, and etching with 2% nital etchant. Optical microscopy was used to observe and capture microstructure images for microscopic examinations. The images of examined microscopic along the plane, at 45 and 90° to the rolling direction are, respectively, shown in Figure 2a–c. In addition, the grain sizes were also examined on both horizontal and vertical (thickness) directions as they were also reported in Figure 2. The cylindrical cup of 60.0 mm in diameter, 26.2 mm in height, and 5.0 mm in cup bottom radius, as shown in Figure 3, was used as a model of cylindrical deep drawn cup. To fabricate this cup, the initial blank size of 100.0 mm was calculated based on the deep drawing theory [19]. The initial blank was prepared by using a wire electrical discharge machine (Wire-EDM, Sodick Model AQ325L, Yokohama, Kanagawa, Japan). Figure 4a shows the punch and die designed based on the deep drawing theory [19]. As listed in Table 3, the punch diameter of 57.8 mm and die diameter of 60.0 mm were set. The die radius of 5.0 mm was designed. The deep drawing clearance of 1.1 mm was set. Figure 4b shows the press machine, which includes a universal sheet metal testing machine (JT TOHSI INC., Model SAS-350D, Minato-ku, Tokyo, Japan). After the deep drawing process, the obtained cylindrical cups were sectioned by a wire-EDM machine for material thickness examinations. The material thickness was measured using a digital micrometer (Insize, serie IS13108, Loganville, GA, USA). The cup height was measured using a vernier height gauge (Mitutoyo, Model CD-6″ ASX, Kawasaki, Kanagawa, Japan), and the earing defects were calculated. Five samples from each deep drawing condition were used to inspect the obtained cylindrical cups. The amount of material thickness and earing defects were calculated based on these obtained cylindrical cups, and the average material thickness and earing defect values were reported.

Table 1. Chemical compositions of SPCC steel (JIS).

C	Si	Mn	P	S	Fe
0.04	0.01	0.16	0.014	0.006	Bal.

Table 2. Mechanical properties of SPCC steel (JIS).

Young's Modulus [GPa]	Ultimate Tensile Strength [MPa]	Elongation at Break [%]	Poisson's Ratio [v]	Hardness [HV]	Anisotropy Parameters		
					R0°	R45°	R90°
208	317	51	0.33	110	2.1	1.9	2.6

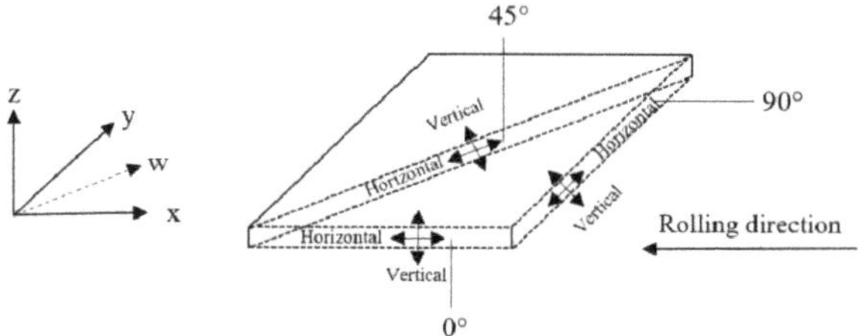

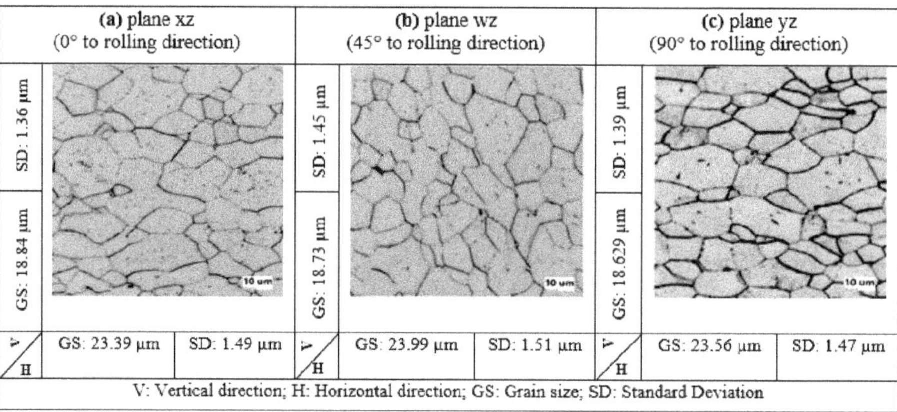

(a) plane xz (0° to rolling direction)		(b) plane wz (45° to rolling direction)		(c) plane yz (90° to rolling direction)	
SD: 1.36 µm / GS: 18.84 µm		SD: 1.45 µm / GS: 18.73 µm		SD: 1.39 µm / GS: 18.629 µm	
GS: 23.39 µm	SD: 1.49 µm	GS: 23.99 µm	SD: 1.51 µm	GS: 23.56 µm	SD: 1.47 µm
V: Vertical direction; H: Horizontal direction; GS: Grain size; SD: Standard Deviation					

Figure 2. Microstructure examinations of SPCC steel (JIS): (**a**) plane xz (0° to rolling direction); (**b**) plane wz (at 45° to rolling direction); (**c**) plane yz (at 90° to rolling direction).

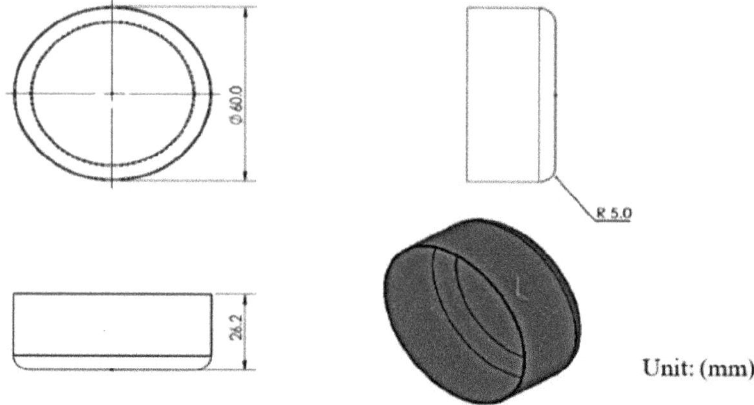

Figure 3. Model of cylindrical deep drawn part.

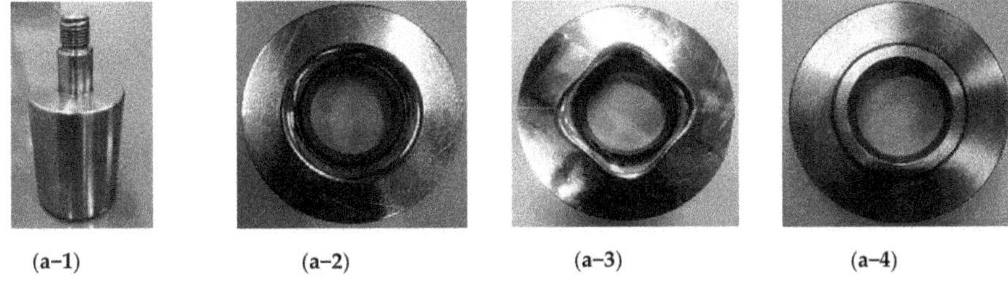

(**a–1**)   (**a–2**)   (**a–3**)   (**a–4**)

(**a**) Die set: (**a–1**) punch; (**a–2**) conventional die; (**a–3**) multi draw radius (MDR) die; (**a–4**) blank holder.

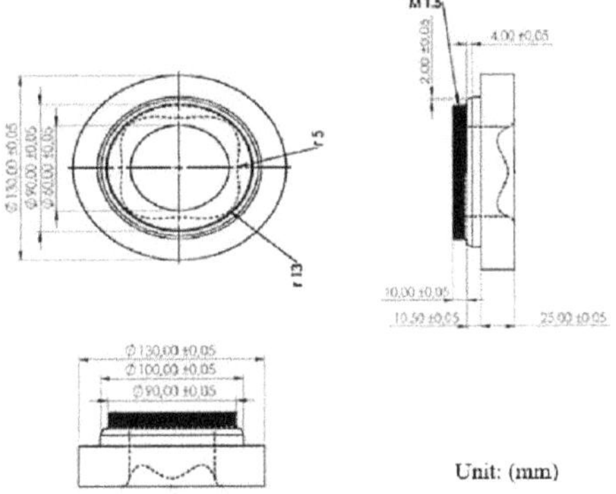

(**b**) Detail drawing of MDR die.

(**c**) Press machine

Figure 4. Die set for experiment and press machine: (**a**) die set; (**b**) detail drawing of MDR die; (**c**) press machine.

Table 3. Experimental conditions.

Sheet material			SPCC (JIS)
Punch diameter (mm)			57.8
Punch radius (mm)			5.0
Punch velocity (mm/s)			1.0
Die diameter (mm)			60.0
Die radius (mm)	Conventional die		5.0 and 13.0
	MDR die	Small radius	5.0
		Large radius	6.0, 8.0, 10.0, 12.0, 13.0 and 14.0
Blankholder force (kN)			3
Sheet thickness (mm)			0.975
Initial sheet diameter (mm)			100.0
Drawing ratio (initial sheet diameter/punch diameter)			1.67
Clearance (mm)			1.1
Lubricant			Paraffinic mineral oil with extreme pressure agent (Iloform (TDN81))

3. Results and Discussion

3.1. Analysis of the Microstructure Evolutions, Strain Distributions, and Thickness Distributions on Deep Drawn Parts Using Conventional Dies

As per the deep drawing theory [19], the cylindrical cup is initially formed after the blank is drawn over the draw radius. This characteristic resulted in the microstructure evolutions and generated strain distributions on the cup. Specifically, the crystal grain deformed adjusting to macroscopic deformation, and so the deformation of crystal grain would naturally correspond to macroscopic strain. On the basis of the fundamentals of the deep drawing mechanism, the material at the cup bottom zone was subjected to equi-biaxial tension and the stretching flange deformation was generated based on the equi-biaxial elongation. This deformation characteristic commonly results in the material thinning. The stretching flange characteristic also commonly formed at the cup bottom radius zone. The bending characteristic was formed at this zone, in addition. Therefore, in general, the material thinning was greatest, and the possibility of breakage was also the highest at the cup bottom radius zone. Next, to form a cup wall, the material at flange portion moves in the direction to the center of die, and then this material must shrink in circular direction. As these characteristics, the tension and compression stresses are commonly generated in radial and circular directions, respectively. Therefore, the material elongates and shrinks in radial and circular directions, respectively. Based on these characteristics, after the cup wall completely formed, the earing defects were simultaneously generated. Therefore, relating to the *R*-value, the analysis of the microstructure evolutions and strain distributions on the cup wall zone was focused on in the present research to clarify the earing defects as well as to use as fundamental information for a new die design to prevent earing defects. In comparison to the initial blank sheet, the microstructure evolutions along the plane at 45 and 90° to the rolling direction at the middle and near edge of the cup walls in the case of a draw radius of 5.0 and 13.0 mm are shown in Figure 5. The examined elongated grains in the horizontal and vertical directions were reported as well. Based on these examined grain sizes, the logarithmic strain in each direction was calculated based on Equation (1). Therefore, by the same token, the logarithmic strains in radial and thickness directions could be calculated as well. Next, by using the volume constancy law, the logarithmic strain in the circular direction could be calculated following Equation (2), and they were reported in the Figure 6.

$$\text{Logarithmic strain} = \ln (l/l_0) \tag{1}$$

when l_0 is the initial grain size and l is the deformed grain size.

$$\varepsilon_\theta = -(\varepsilon_r + \varepsilon_t) \tag{2}$$

when ε_θ is the logarithmic strain in the circular direction, ε_r is the logarithmic strain in the radial direction, and ε_t is the logarithmic strain in the thickness direction.

The same manner of strain distribution analysis in the cases of a draw radius of 5.0 and 13.0 mm could be observed. In terms of thickness strain distribution, it was negative in all directions in the sheet plane. The large negative thickness strain was generated in the direction along the plane and at 90° to the rolling direction, and the small negative one was generated in the direction at 45° to the rolling direction. In comparison to the middle cup wall zone, at the near edge of the cup wall, the positive thickness strain was generated instead of the negative one. Again, the large thickness strain was still generated in the direction at 45° to the rolling direction. Therefore, the material thinning was formed at the middle of cup wall, whereas the material thickening was formed at the near edge of the cup wall. To validate these results, the thickness examinations were carried out, and the thickness distributions are shown in Figure 7. There was a small change of material thickness on the cup bottom zone due to the stretching flange deformation characteristic. However, the maximum material thinning was generated on the cup bottom radius because the bending was added to the stretching flange deformation characteristics, and then the large plastic deformation was generated in that area. It was also observed that the changes in material thickness with respect to draw radii were somewhat at the same level. The effects of the rolling direction on the material thickness of the cup bottom zone and cup bottom radius zone were very small. With the cup wall zone, the results showed that the material thinning was decreased along the cup wall height. The effects of the rolling direction on material thickness were clearly illustrated, especially on the cup wall zone. These thickness distribution results agreed well with the thickness strain distribution results calculated on the basis of the deformed grain size. These results also corresponded well with the deep drawing theory and literature [18,19]. Next, in terms of radial strain and circular strain distributions, the radial strain was positive, but the circular strain was negative. The large radial strain was generated in the direction along the plane and at 90° to the rolling direction, and the small one was generated in the direction at 45° to the rolling direction. Again, the large circular strain was generated in the direction along the plane and at 90° to the rolling direction, and the small one was generated in the direction at 45° to the rolling direction. These results could be explained that, owing to the R-value as listed in Table 2, the plastic deformation would occur earlier in the direction along the plane and at 90° to the rolling direction due to the large R-value. Therefore, the material would flow into these directions from other portions, i.e., the portion at 45° to the rolling direction. Therefore, the nonconcurrent plastic deformation was generated and then caused the restriction of material flown into the die. In addition, the highly excessive elongating material flowed outward in the direction to edge of the blank sheet. Therefore, the earing defects were formed by the peak of the earing profile that was along the plane and at 90° to the rolling direction, whereas the bottom of the earing profile was at 45° to the rolling direction. Figure 8a,b show the earing defects in the case of a draw radius of 5.0 and 13.0 mm, respectively. The obtained earing defects were approximately of 1.7 mm in both cases of the draw radius of 5.0 and 13.0 mm. It was also observed that the peak of the earing profile formed along the plane and at 90° to the rolling direction, whereas the bottom of the earing profile formed at 45° to the rolling direction. They agreed well with radial and circular strain distributions calculated on the basis of the deformed grain size. These results also corresponded well with the deep drawing theory and literature [18,19]. As a result, based on the microstructure evolutions and strain distributions, the occurrence of earing defects could be clearly clarified. These data are valuable information for supporting MDR die design and development to achieve the nearly zero earing defects.

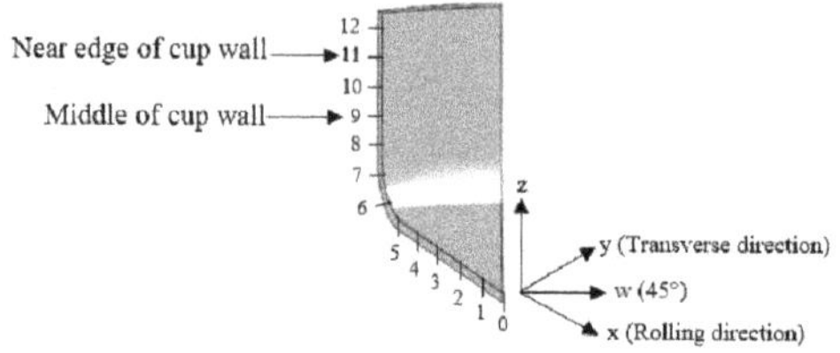

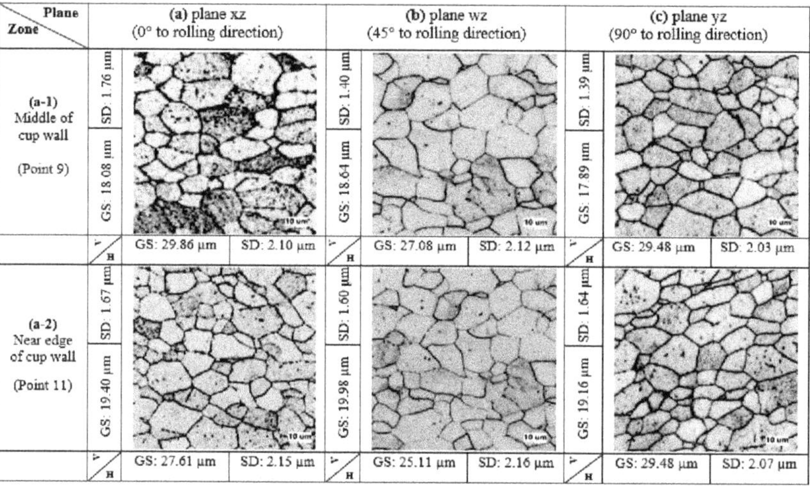

Plane / Zone	(a) plane xz (0° to rolling direction)	(b) plane wz (45° to rolling direction)	(c) plane yz (90° to rolling direction)
(a-1) Middle of cup wall (Point 9)	SD: 1.76 µm GS: 18.08 µm	SD: 1.40 µm GS: 18.64 µm	SD: 1.39 µm GS: 17.89 µm
	GS: 29.86 µm SD: 2.10 µm	GS: 27.08 µm SD: 2.12 µm	GS: 29.48 µm SD: 2.03 µm
(a-2) Near edge of cup wall (Point 11)	SD: 1.67 µm GS: 19.40 µm	SD: 1.60 µm GS: 19.98 µm	SD: 1.64 µm GS: 19.16 µm
	GS: 27.61 µm SD: 2.15 µm	GS: 25.11 µm SD: 2.16 µm	GS: 29.48 µm SD: 2.07 µm

(**a**) Draw radius 5.0 mm.

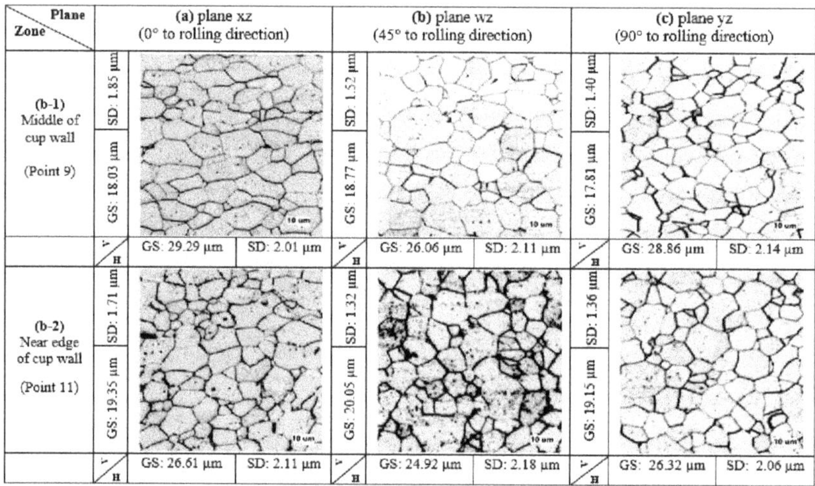

Plane / Zone	(a) plane xz (0° to rolling direction)	(b) plane wz (45° to rolling direction)	(c) plane yz (90° to rolling direction)
(b-1) Middle of cup wall (Point 9)	SD: 1.85 µm GS: 18.03 µm	SD: 1.52 µm GS: 18.77 µm	SD: 1.40 µm GS: 17.81 µm
	GS: 29.29 µm SD: 2.01 µm	GS: 26.06 µm SD: 2.11 µm	GS: 28.86 µm SD: 2.14 µm
(b-2) Near edge of cup wall (Point 11)	SD: 1.71 µm GS: 19.35 µm	SD: 1.32 µm GS: 20.05 µm	SD: 1.36 µm GS: 19.15 µm
	GS: 26.61 µm SD: 2.11 µm	GS: 24.92 µm SD: 2.18 µm	GS: 26.32 µm SD: 2.06 µm

(**b**) Draw radius 13.0 mm.

Figure 5. Comparison of microstructure evolutions on deep drawn parts with respect to draw radii: (**a**) draw radius 5.0 mm; (**b**) draw radius 13.0 mm.

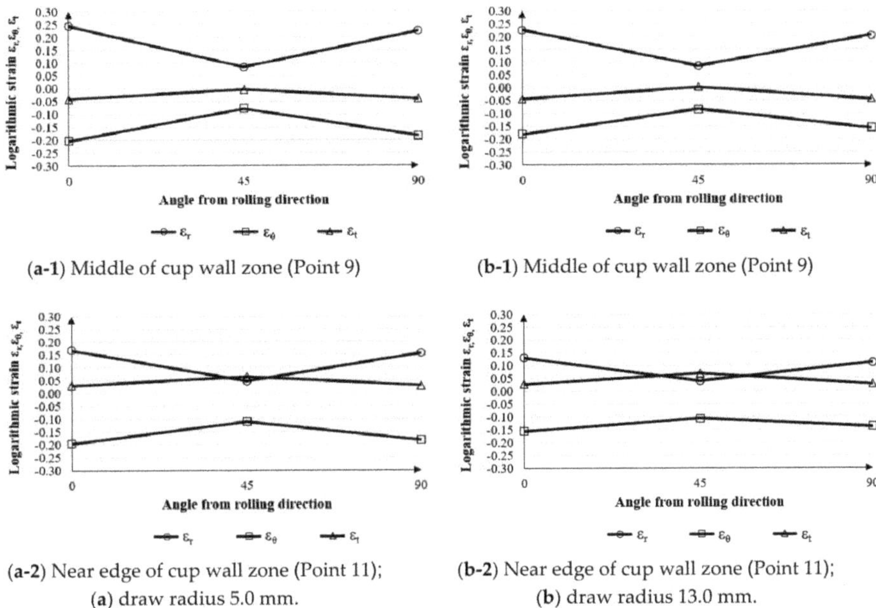

(**a-1**) Middle of cup wall zone (Point 9)

(**b-1**) Middle of cup wall zone (Point 9)

(**a-2**) Near edge of cup wall zone (Point 11);
(**a**) draw radius 5.0 mm.

(**b-2**) Near edge of cup wall zone (Point 11);
(**b**) draw radius 13.0 mm.

Figure 6. Strain distributions on cup wall zone with respect to draw radii: (**a**) draw radius 5.0 mm; (**b**) draw radius 13.0 mm.

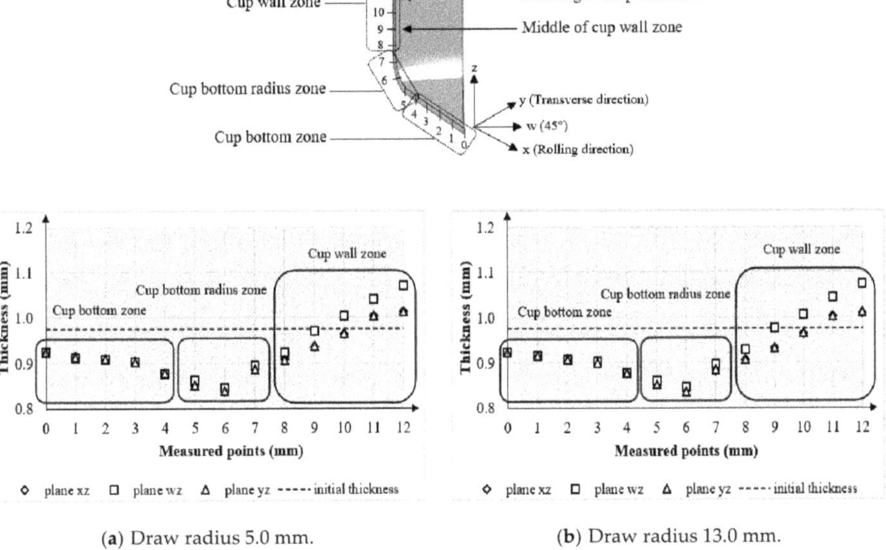

(**a**) Draw radius 5.0 mm.

(**b**) Draw radius 13.0 mm.

Figure 7. Thickness distributions on deep drawn parts with respect to draw radii: (**a**) draw radius 5.0 mm; (**b**) draw radius 13.0 mm.

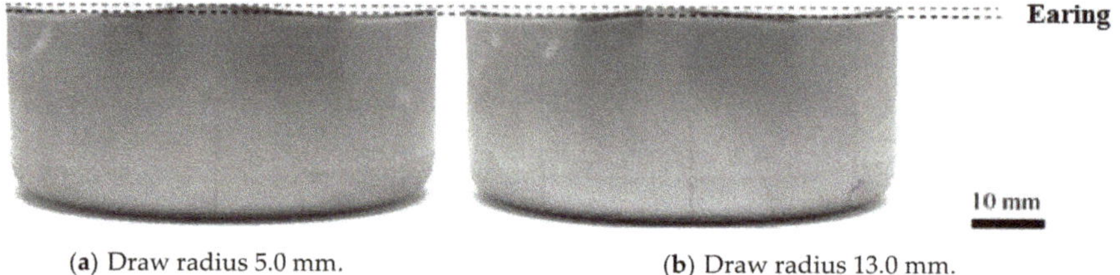

(a) Draw radius 5.0 mm.   (b) Draw radius 13.0 mm.

Figure 8. Earing defects with respect to draw radius obtained from the conventional die application: (**a**) draw radius 5.0 mm; (**b**) draw radius 13.0 mm.

3.2. Conceptual Design of Multi Draw Radius (MDR) Deep Drawing Die

As per the author past research [18], the multi draw radius (MDR) deep drawing die is mainly proposed to increase the limiting drawing ratio of cylindrical deep drawn parts (LDR) by reducing the nonaxisymmetric material flow during deep drawing process. In the present research, this MDR die was proposed and applied to prevent the earing defects. Figure 9 shows the conceptual design of the MDR die. By using the conventional die, as shown in Figure 9a, the nonconcurrent plastic deformation characteristic was formed during the deep drawing process because of the effects of the R-value. As aforementioned, the plastic deformation would occur more early in the direction along the plane and at 90° to the rolling direction due to the large R-value. This resulted in the restriction of material flown into the die due to the highly excessive elongating material flow into these directions from other portions. Then, the nonaxisymmetric material flow characteristic on the flange portion during the deep drawing process was generated [18]. These plastic deformation and material flow characteristics caused the material in the direction along the plane and at 90° to the rolling direction were easier to stretch compared to that in direction at 45° to the rolling direction, and then, the earing defects were formed. To encounter these characteristics and prevent earing defects, the later plastic deformation generated in the direction at 45° to the rolling direction should be driven to generate earlier as it generated in the direction along the plane and at 90° to the rolling direction. Therefore, the plastic deformation in different directions in the sheet plane during the deep drawing process could be concurrently generated. As per the past research [18], by using the MDR die, the axisymmetric material flow characteristic could be achieved. Therefore, the MDR die might be able to solve the above-mentioned plastic deformation characteristic and earing defects. As suggested in the past research, the larger draw radius was positioned at 45° to the rolling direction and the smaller draw radius positioned along the plane and at 90° to the rolling direction. This resulted in the plastic deformation generating in the direction at 45° to the rolling direction, which could be made to occur earlier, and then the concurrent plastic deformation characteristic could be controlled, as shown in Figure 9b. The uniform material stretching in each direction in the sheet plane could be generated, and the uniform strain distribution in each direction in the sheet plane could be obtained. Concerning these plastic deformation mechanisms, by using the MDR die, the achievement of nearly zero earing defects by the process encountering the material mechanical property of the anisotropy property of the material could be met. However, the draw radius in each direction in the sheet plane should be positioned as follows: the larger draw radius positioned at 45° to the rolling direction and the smaller draw radius positioned along the plane and at 90° to the rolling direction. In the present research, the different MDR dies were investigated to achieve nearly zero earing defects, and the results are shown in Figure 10. Specifically, the small draw radius of 5.0 mm, as recommended for conventional deep drawing die, was positioned along the plane and at 90° to the rolling direction. The large draw radius of 6.0–14.0 mm was positioned at 45° to the rolling direction. The results showed the decreases, and again, the increases in the earing defects as the large draw radius increased. The smallest earing defects approximately of 0.1 mm could be generated with the large draw radius of 13.0 mm applied.

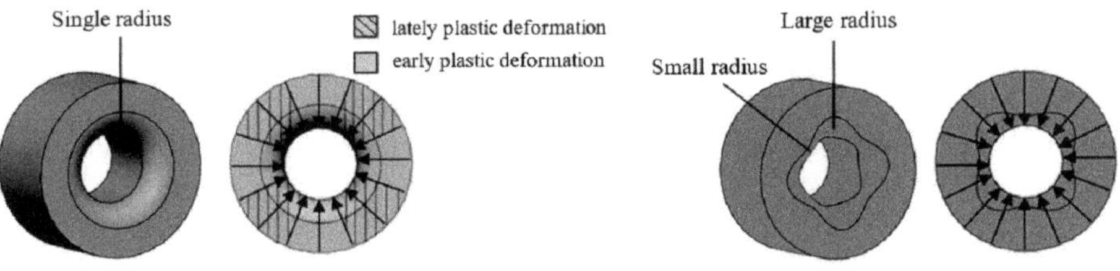

Figure 9. Illustration of plastic deformation characteristics with respect to die types: (**a**) conventional die application; (**b**) MDR die application.

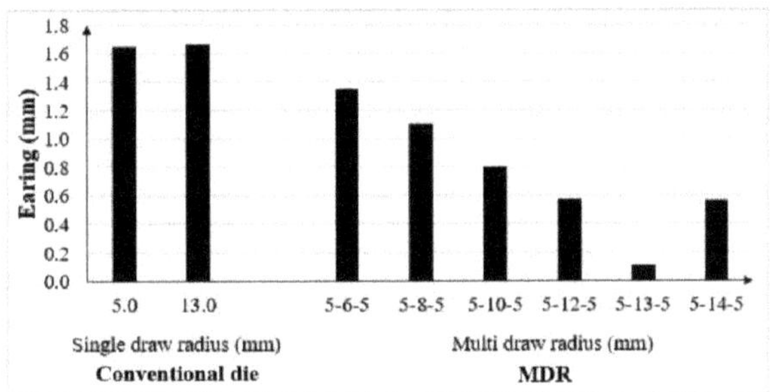

Figure 10. Comparison of earing defects between conventional and MDR die applications.

3.3. Application of Multidraw Radius (MDR) Deep Drawing Die

As aforementioned, the small draw radius of 5.0 mm, as recommended for conventional deep drawing die, was positioned along the plane and at 90° to the rolling direction, and the large draw radius of 13.0 mm positioned at 45° to the rolling direction was designed to achieve nearly zero earing defects. As this MDR die was applied, the nearly uniform elongated grain microstructure evolutions in each direction in the sheet plane could be achieved and the earing defects could be prevented. The examined microstructure evolutions are shown in Figure 11b, and the examined elongated grains in horizontal and vertical directions were reported as well. By comparing the conventional die use, these results revealed that by using the MDR die, the microstructure evolutions in each direction in the sheet plane during the deep drawing process, especially on the cup wall zone, could be encountered by the multi draw radius, and the nearly uniform elongated grain in each direction in the sheet plane on each cup wall height, i.e., the middle and near edge of cup walls, could be achieved. In terms of strain distribution, as shown in Figure 12b, by comparing with the conventional die use, the changes in radial, circular, and thickness strains in the direction at 45° to the rolling direction were clearly identified. Specifically, the radial strain was increased in the positive direction, but the circular and thickness strains were increased in the negative direction. In addition, the strain distributions also showed the interesting results that the nearly uniform strain distribution in each direction in the sheet plane on each cup wall height, i.e., the middle and near edge of cup walls, could be achieved. In terms of thickness distribution, as shown in Figure 13b, the thickness distributions showed the good agreement with the strain distribution results. As the thickness strain in the direction at 45° to the rolling direction increased in the negative direction, the thickness distribution showed the decreases in material thickness compared to that in the case of conventional die use. Moreover, the nearly uniform thickness distribution in each direction in the sheet plane on each cup wall height could be achieved. Based on these microstructure evolutions, strain distributions, and thickness distributions, in terms of earing defects, as shown in Figure 14, the results revealed that the earing defects could be reduced, and the nearly zero earing defects could be achieved. The results showed that, by using the MDR die, the earing defects of approximately 0.1 mm were formed. To more clearly clarify the MDR die use, the MDR die of large die radius of 12.0 mm was also designed and investigated. Again, the results of microstructure evolutions, strain distributions, and thickness distributions are shown in Figure 11b, Figure 12b, and Figure 13b, respectively. The same characteristics of those MDRs with the large die radius of 13.0 mm use could be observed. However, it could be noted that the increases in radial, circular, and thickness strains were smaller than those in the case of MDR with a die radius of 13 mm use. Therefore, the earing defects were larger. These results confirmed that, by using the MDR die, the nearly zero earing defects could be achieved. However, the suitable MDR die design relating to the *R*-value of material should be strictly considered.

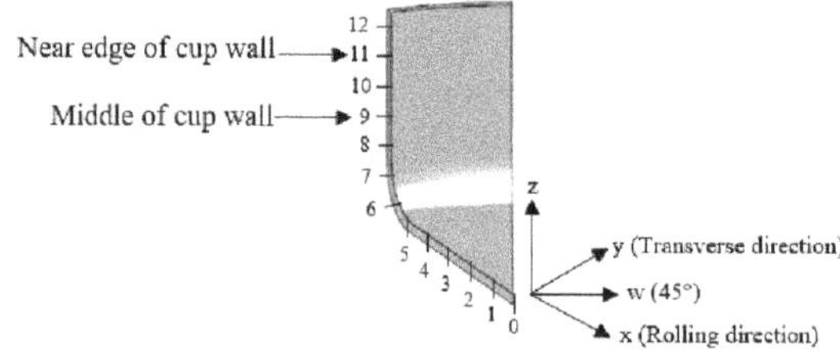

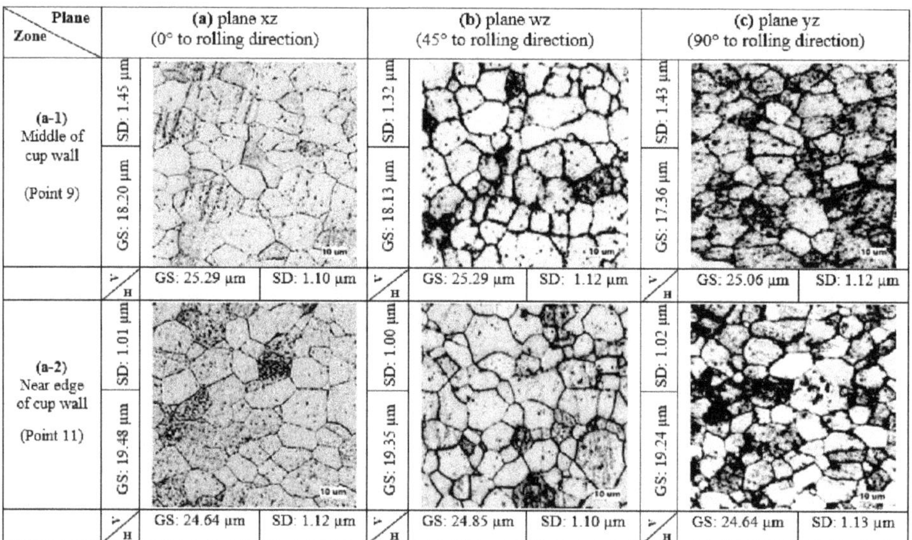

Zone / Plane	(a) plane xz (0° to rolling direction)		(b) plane wz (45° to rolling direction)		(c) plane yz (90° to rolling direction)	
(a-1) Middle of cup wall (Point 9)	SD: 1.45 μm GS: 18.20 μm		SD: 1.32 μm GS: 18.13 μm		SD: 1.43 μm GS: 17.36 μm	
	GS: 25.29 μm	SD: 1.10 μm	GS: 25.29 μm	SD: 1.12 μm	GS: 25.06 μm	SD: 1.12 μm
(a-2) Near edge of cup wall (Point 11)	SD: 1.01 μm GS: 19.48 μm		SD: 1.00 μm GS: 19.35 μm		SD: 1.02 μm GS: 19.24 μm	
	GS: 24.64 μm	SD: 1.12 μm	GS: 24.85 μm	SD: 1.10 μm	GS: 24.64 μm	SD: 1.13 μm

(**a**) MDR die 5-13-5: large radius 13.0 mm; small radius 5.0 mm.

Figure 11. *Cont.*

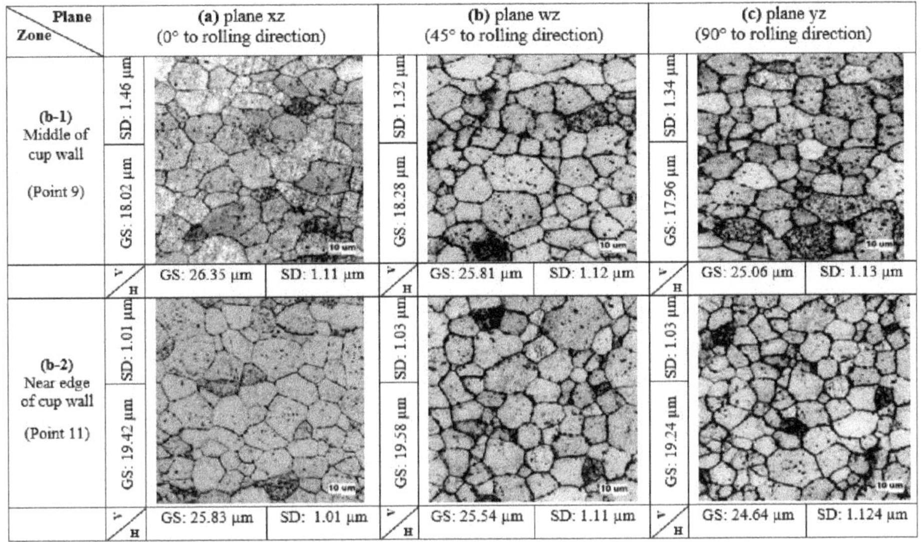

Plane / Zone		(a) plane xz (0° to rolling direction)	(b) plane wz (45° to rolling direction)	(c) plane yz (90° to rolling direction)
(b-1) Middle of cup wall (Point 9)	SD: 1.46 µm / GS: 18.02 µm		SD: 1.32 µm / GS: 18.28 µm	SD: 1.34 µm / GS: 17.96 µm
		GS: 26.35 µm SD: 1.11 µm	GS: 25.81 µm SD: 1.12 µm	GS: 25.06 µm SD: 1.13 µm
(b-2) Near edge of cup wall (Point 11)	SD: 1.01 µm / GS: 19.42 µm		SD: 1.03 µm / GS: 19.58 µm	SD: 1.03 µm / GS: 19.24 µm
		GS: 25.83 µm SD: 1.01 µm	GS: 25.54 µm SD: 1.11 µm	GS: 24.64 µm SD: 1.124 µm

(**b**) MDR die 5-12-5: large radius 12.0 mm; small radius 5.0 mm.

Figure 11. Comparison of microstructure evolutions on deep drawn parts with respect to MDR dies: (**a**) MDR die 5-13-5; (**b**) MDR die 5-12-5.

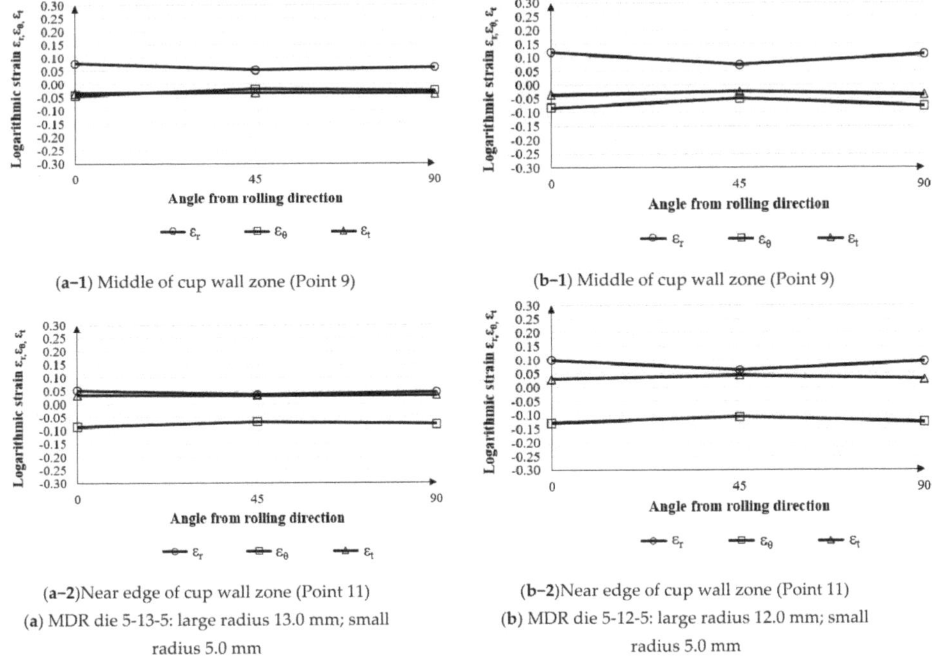

(**a–1**) Middle of cup wall zone (Point 9)

(**b–1**) Middle of cup wall zone (Point 9)

(**a–2**) Near edge of cup wall zone (Point 11)

(**b–2**) Near edge of cup wall zone (Point 11)

(**a**) MDR die 5-13-5: large radius 13.0 mm; small radius 5.0 mm

(**b**) MDR die 5-12-5: large radius 12.0 mm; small radius 5.0 mm

Figure 12. Comparison of strain distributions on cup wall zone with respect to MDR dies: (**a**) MDR die 5-13-5; (**b**) MDR die 5-12-5.

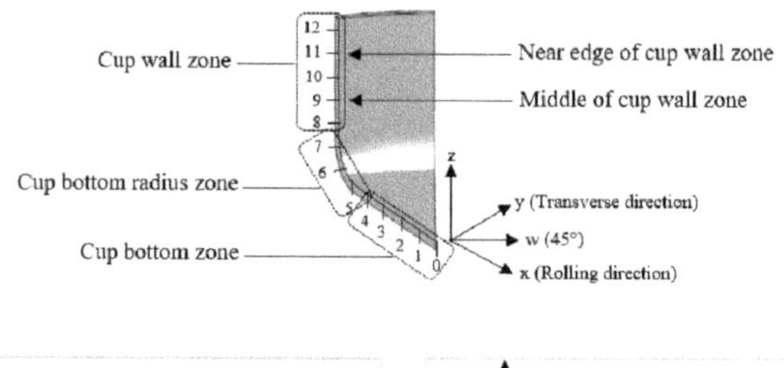

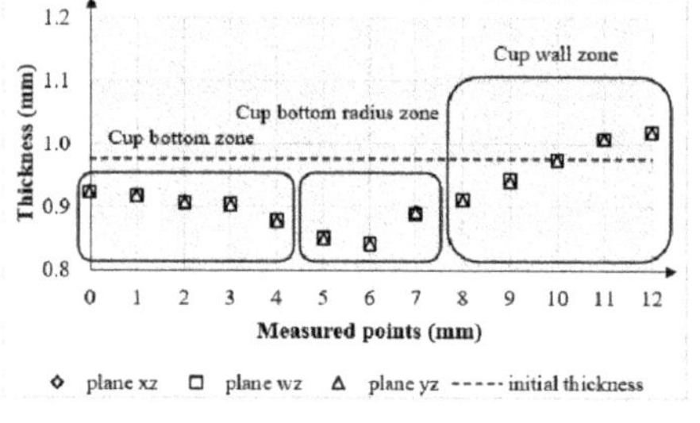

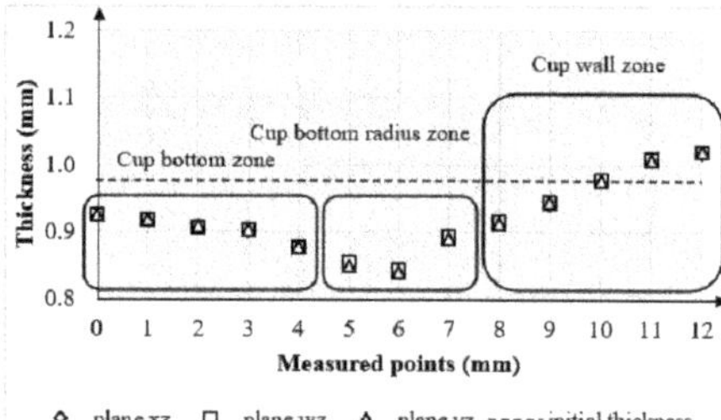

(**a**) MDR die 5-13-5: large radius 13.0 mm; small radius 5.0 mm.

(**b**) MDR die 5-12-5: large radius 12.0 mm; small radius 5.0 mm.

Figure 13. Illustration of thickness distributions on deep drawn parts with respect to MDR dies: (**a**) MDR die 5-13-5; (**b**) MDR die 5-12-5.

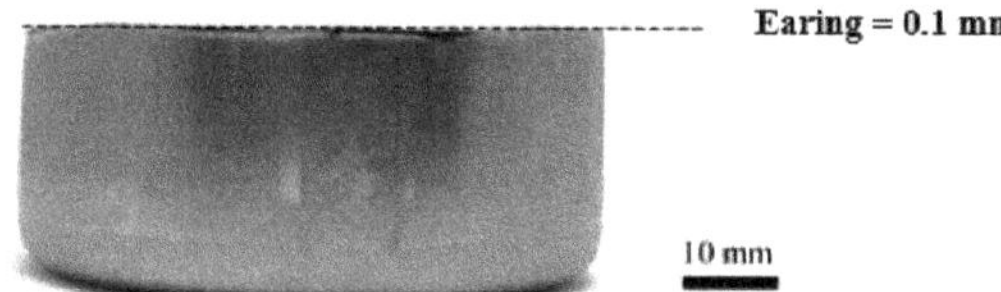

Figure 14. Illustration of earing defects on deep drawn part by using the MDR die.

4. Conclusions

In the present research, the earing defects as the major problem in the deep drawing process, especially for cylindrical drawn parts, were focused on. The MDR die application was proposed for reducing the earing defects as well as for achieving the nearly zero earing defects in the cylindrical deep drawing process. First, the effects of the draw radius on microstructure evolutions, strain distributions, thickness distributions, and earing defects were investigated. The results confirmed that, based on the material mechanical property of the R-value, the draw radius significantly caused the microstructure evolutions in different directions in the sheet plane. In addition, the changes in these microstructure evolutions significantly affected the strain and thickness distributions in different directions in the sheet plane, and then, the earing defects were formed. However, the changes in the draw radius rarely affected these microstructure evolutions, strain and thickness distributions, and earing defects. Next, the MDR die was applied and investigated to prevent earing defects. It was designed based on the principle of that, during the deep drawing process, the multi draw radius could encounter the effects of the R-value on microstructure evolutions and strain distributions in each direction in the sheet plane, especially on the cup wall zone. The plastic deformation generated in each direction in the sheet plane should be concurrently performed, and the axisymmetric material flow characteristic could be achieved. Therefore, the nearly uniform elongated grains in each direction in the sheet plane on each cup wall height could be achieved, and the nearly uniform strain and thickness distributions, especially on cup wall zone in each direction in the sheet plane, could be formed. Based on these microstructure evolutions, strain distributions and thickness distributions, the earing defects could be reduced, and the nearly zero earing defects could be achieved by using the MDR die. However, to achieve the nearly zero earing defects, the suitable MDR die design relating to the *R*-value should be strictly considered. In the present research, it was suggested that the larger draw radius should be positioned at 45° to the rolling direction and the smaller draw radius positioned along the plane and at 90° to the rolling direction.

Author Contributions: Conceptualization, R.J. and S.T.; data curation, R.J. and S.T.; funding acquisition, R.J. and S.T.; investigation, R.J.; methodology, R.J. and S.T.; project administration, S.T.; supervision, S.T.; writing—original draft, R.J.; writing—review and editing, S.T. All authors have read and agreed to the published version of the manuscript.

Funding: This research was funded by the "Petchra Pra Jom Klao Master's Degree Research Scholarship" from King Mongkut's University of Technology Thonburi. The APC was also funded by King Mongkut's University of Technology Thonburi.

Acknowledgments: The authors would like to express their gratitude to Wiriyakorn Phanitwong, Department of Industrial Engineering, Rajamangala University of Technology Rattanakosin, and Arkarapon Sontamino, Department of Mechanical Engineering Technology, College of Industrial Technology, King Mongkut's University of Technology North Bangkok for their advice on experiments in this present research.

Conflicts of Interest: The authors declare no conflict of interest.

References

1. Cao, Q.; Du, L.; Li, Z.; Lai, Z.; Li, Z.; Chen, M.; Li, X.; Xu, S.; Chen, Q.; Han, X.; et al. Investigation of the Lorentz-force-driven sheet metal stamping process for cylindrical cup forming. *J. Mater. Process. Technol.* **2019**, *271*, 532–541. [CrossRef]

2. Hashemi, A.; Gollo, M.H.; Seyedkashi, S.M.H. Study of Al/St laminated sheet and constituent layers in radial pressure-assisted hydrodynamic deep drawing. *Mater. Manuf. Process.* **2017**, *32*, 54–61. [CrossRef]

3. Thipprakmas, S. Finite element analysis of sided coined-bead technique in precision V-bending process. *Int. J. Adv. Manuf. Technol.* **2013**, *65*, 679–688. [CrossRef]

4. Phanitwong, W.; Thipprakmas, S. Centered coined-bead technique for precise U-bent part fabrication. *Int. J. Adv. Manuf. Technol.* **2016**, *84*, 2139–2150. [CrossRef]

5. Mahdavian, S.; Fion, T.M.Y. Effect of Punch Geometry in the Deep Drawing Process of Aluminium. *Mater. Manuf. Process.* **2007**, *22*, 898–902. [CrossRef]

6. Nie, H.; Chi, C.; Chen, H.; Li, X.; Liang, W. Microstructure evolution of Al/Mg/Al laminates in deep drawing process. *J. Mater. Res. Technol.* **2019**, *8*, 5325–5335. [CrossRef]

7. Liu, Y.; Li, F.; Li, C.; Xu, J. Enhancing formability of spherical bottom cylindrical parts with magnetic medium on deep drawing process. *Int. J. Adv. Manuf. Technol.* **2019**, *103*, 1669–1679. [CrossRef]

8. Bassoli, E.; Sola, A.; Denti, L.; Gatto, A. Experimental approach to measure the restraining force in deep drawing by means of a versatile draw bead simulator. *Mater. Manuf. Process.* **2019**, *34*, 1286–1295. [CrossRef]

9. Sezek, S.; Savas, V.; Aksakal, B. Effect of Die Radius on Blank Holder Force and Drawing Ratio: A Model and Experimental Investigation. *Mater. Manuf. Process.* **2010**, *25*, 557–564. [CrossRef]

10. Benke, M.; Hlavacs, A.; Piller, I.; Mertinger, V. Prediction of earing of aluminium sheets from {h00} pole figures. *Eur. J. Mech. A Solids* **2020**, *81*, 103950. [CrossRef]

11. Kishor, N.; Kumar, D.R. Optimization of initial blank shape to minimize earing in deep drawing using finite element method. *J. Mater. Process. Technol.* **2002**, *130*, 20–30. [CrossRef]

12. Walde, T.; Riedel, H. Simulation of earing during deep drawing of magnesium alloy AZ31. *Acta Mater.* **2007**, *55*, 867–874. [CrossRef]

13. Izadpanah, S.; Ghaderi, S.H.; Gerdooei, M. Material parameters identification procedure for BBC2003 yield criterion and earing prediction in deep drawing. *Int. J. Mech. Sci.* **2016**, *115*, 552–563. [CrossRef]

14. Cazacu, O.; Plunkett, B.; Barlat, F. Orthotropic yield criterion for hexagonal closed packed metals. *Int. J. Plast.* **2006**, *22*, 1171–1194. [CrossRef]

15. Singh, A.; Basak, S.; P.S., L.P.; Roy, G.G.; Jha, M.N.; Mascarenhas, M.; Panda, S.K. Prediction of earing defect and deep drawing behavior of commercially pure titanium sheets using CPB06 anisotropy yield theory. *J. Manuf. Process.* **2018**, *33*, 256–267. [CrossRef]

16. Engler, O.; Aegerter, J.; Calmer, D. Control of texture and earing in aluminium alloy AA 8011A-H14 closure stock. *Mater. Sci. Eng. A* **2020**, *775*, 138965. [CrossRef]

17. Tang, W.; Huang, S.; Li, D.; Peng, Y. Mechanical anisotropy and deep drawing behaviors of AZ31 magnesium alloy sheets produced by unidirectional and cross rolling. *J. Mater. Process. Technol.* **2015**, *215*, 320–326. [CrossRef]

18. Phanitwong, W.; Thipprakmas, S. Multi Draw Radius Die Design for Increases in Limiting Drawing Ratio. *Metals* **2020**, *10*, 870. [CrossRef]

19. Lange, K. *Handbook of Metal Forming*; McGraw-Hill Inc.: New York, NY, USA, 1985.

Article

Multi Draw Radius Die Design for Increases in Limiting Drawing Ratio

Wiriyakorn Phanitwong [1,*] and Sutasn Thipprakmas [2]

[1] Department of Industrial Engineering, Rajamangala University of Technology Rattanakosin, Nakhon Pathom 73170, Thailand

[2] Department of Tool and Materials Engineering, King Mongkut's University of Technology Thonburi, Bangkok 10140, Thailand; sutasn.thi@kmutt.ac.th

* Correspondence: wiriyakorn.pha@rmutr.ac.th; Tel.: +662-4416000 (ext. 2681); Fax: +662-4416000 (ext. 2650)

Received: 22 May 2020; Accepted: 24 June 2020; Published: 30 June 2020

Abstract: As a major sheet metal process for fabricating cup or box shapes, the deep drawing process is commonly applied in various industrial fields, such as those involving the manufacture of household utensils, medical equipment, electronics, and automobile parts. The limiting drawing ratio (LDR) is the main barrier to increasing the formability and production rate as well as to decrease production cost and time. In the present research, the multi draw radius (MDR) die was proposed to increase LDR. The finite element method (FEM) was used as a tool to illustrate the principle of MDR based on material flow. The results revealed that MDR die could reduce the non-axisymmetric material flow on flange and the asymmetry of the flange during the deep drawing process. Based on this material flow characteristic, the cup wall stretching and fracture could be delayed. Furthermore, the cup wall thicknesses of the deep drawn parts obtained by MDR die application were more uniform in each direction along the plane, at 45°, and at 90° to the rolling direction than those obtained by conventional die application. In the present research, a proper design for the MDR was suggested to achieve functionality of the MDR die as related to each direction along the plane, at 45°, and at 90° to the rolling direction. The larger draw radius positioned for at 45° to the rolling direction and the smaller draw radius positioned for along the plane and at 90° to the rolling direction were recommended. Therefore, by using proper MDR die application, the drawing ratio could be increased to be 2.75, an increase in LDR of approximately 22.22%.

Keywords: deep drawing; limiting drawing ratio (LDR); draw radius; anisotropy; finite element method

1. Introduction

In recent years, the available sheet metal components are able to serve almost all manufacturing industries, such as is the case for sheet metal components used in automobile and aerospace. The fabrication of such sheet metal components by means of sheet metal die is commonly classified according to the utilization of die bending, die deep-drawing, and die cutting processes [1]. Based on these processes, and through the associated experiments and finite element method (FEM) techniques, many studies have been carried out by researchers and engineers to overcome the major defects that occur on these sheet metal components [2–6] as well as to improve the formability for each sheet metal forming process [7–10]. In the present research, we focused on the deep drawing process. This process is extensively applied to fabricate various consumer products such as household utensils, medical equipment, electronics, and automobile parts. This process is also cost-effective because it is characterized by high production rates and gives finished parts of good quality without additional operations. Nowadays, for sheet metal components manufactured by deep drawing process, a more complex profile and higher formability are required [11–15]. The optimization of process parameters to enhance the formability of AA 5182 alloy in the deep drawing of square cups by hydroforming was carried out [11]. A bead optimization algorithm was developed to increase the efficiency of the bead design while simultaneously considering manufacturability. The influences of the deep drawing depth, initial specimen geometry, and bead height on formability were subsequently investigated by means of sensitivity analysis [12]. Abe et al. proposed the technique of local work hardening with punch indentation to improve sheet metal formability [13]. The most important process parameter affecting thinning was the peak pressure, whereas the pressure path had the least effect on formability. The square deep draw steel/carbon fiber reinforced plastic (CFRP) hybrid composite material was investigated. The effects of fiber orientation on formability were also investigated [14]. In addition, the micro deep drawn parts are also focused on [16–19]. C. J. Wang. et al. carried out the research on the micro deep drawing process of a conical part with ultra-thin copper foil using a multi-layered DLC film-coated die [16]. The grain size effect on multi-stage micro deep drawing of a micro cup with domed bottom was investigated by W. T. Li et al. [17]. However, in terms of old-fashioned cylindrical cup shapes, deep drawing is also a conventional sheet metal forming process for wide application in industry at a very high production rate. The products of cylindrical cup shapes are also still widely used in various sheet metal manufacturing industries. Therefore, the developments on this process have been continuously reported in many previous studies on the basis of experimental and FEM works [20–29]. Some previous studies were carried out to prevent major defects, such as wrinkles, earing, and fracture defects [20–23]. In addition, as a common formability indicator of cylindrical cup forming, the limiting drawing ratio (LDR) has been also investigated in many previous studies [24–29]. Many techniques have been proposed to increase in LDR. A new technique for deep drawing of elliptic cups through a conical die without blankholder or draw beads was proposed to increase LDR [24]. The LDR of aluminum tailored friction stir welded blanks could be increased using a modified conical tractrix die technique. By using this technique, the improvements in LDR of approximately 27% and 14% were recorded, respectively, for the dissimilar grade and the dissimilar gauge aluminum tailor friction stir welded blanks [26]. Bandyopadhyay, K. et al. showed that the LDR of tailor welded blanks (TWBs) fabricated using two dissimilar material combinations of dual phase (DP) and interstitial free (IFHS) steels could be improved with restricted weld movement by shifting the initial weld line position [27]. However, in the previous studies, the increases in LDR were still limited. In addition, the way to increase LDR was also complex as additional operations were applied. In the present research, therefore, a new approach to increase in LDR is proposed and investigated. Based on the mechanical property of plastic strain ratio (R-value), the material flow on flange along the perpendicular differed, and the resulting cup wall stretching and fracture as well as the drawing ratio was limited. This new technique of multi draw radius, termed MDR, is proposed in the present research to encounter the material property of plastic strain ratio and generate the same manner of material flow on the flange along the perpendicular. Specifically, the MDR die could reduce the non-axisymmetric material flow on flange and the asymmetry of the flange during the deep drawing process. Furthermore, on the basis of this technique, the deep drawing process could be applied without additional operations, resulting in production cost and time being saved. However, for the proper design of MDR die, we recommend that the larger draw radius be positioned at 45° to the rolling direction and the smaller draw radius positioned along the plane and at 90° to the rolling direction. The application of the MDR was compared with the conventional die, and the results illustrated that the LDR could be increased.

2. Proposed Multi Draw Radius (MDR) Die and Its Principle

As per the deep drawing theory [1], the LDR is the maximum ratio of initial blank diameter to punch diameter in which the cylindrical cup could be formed without any fractures. The LDR also depends upon the type of material used as well as relates to the mechanical property of material being of common concern in the deep drawing process. In terms of fracture, the fracture is commonly generated on the basis of two deep drawing characteristics [1]. Specifically, the first is that the wrinkle defects are generated on the flange due to the overly low blankholder pressure applied which causes the obstacle of material flow into the die. This manner resulted in the generation of cup wall stretching and fractures. Next, the second is that the flange is tightly clamped by applying an overly high blankholder pressure. This manner resulted in the material not being able to be drawn into the die as well as subsequent cup wall stretching and generation of fractures. In addition, the material properties, especially plastic strain ratio (R-value), also affect fractures. As is well known, the R-value is the anisotropy property of the material and depends upon the direction along the plane, at 45°, and at 90° to the rolling direction. Namely, the anisotropy property of the material causes the different material flow and formability in each direction along the plane, at 45°, and at 90° to the rolling direction. As this characteristic causes non-axisymmetric material flow on the flange during the deep drawing process, cup wall stretching can easily be generated, as well as that flange wrinkles can be easily formed. Based on these reasons, to prevent the fracture and increase in deep drawing formability, the prevention of cup wall stretching and flange wrinkle should be strictly considered. As per previous studies [11,19,20], the working process parameters on deep drawing process, i.e., blankholder pressure, lubricant, and draw radius, were investigated. However, as aforementioned, the anisotropy property of the material in each direction along the plane, at 45°, and at 90° to the rolling direction related to draw radius die design for reducing the non-axisymmetric material flow during deep drawing process has not yet been investigated. In the present research, the multi draw radius die (or so-called MDR die) is proposed to reduce the non-axisymmetric material flow during deep drawing process and prevent fracture as well as to increase in LDR. The schematic of MDR die was shown in Figure 1. The draw radius was designed related to the anisotropy property of the material in each direction along the plane, at 45°, and at 90° to the rolling direction to reduce the non-axisymmetric material flow during the deep drawing process. As per deep drawing theory, the LDR should be treated as the baseline. It gives only estimated values and does not take into account many factors. Therefore, in the present research, to compare the formability based on the LDR between conventional die and MDR die applications, the other process parameters affecting LDR, excluding die types, were set with the same conditions for both cases of conventional and MDR die applications.

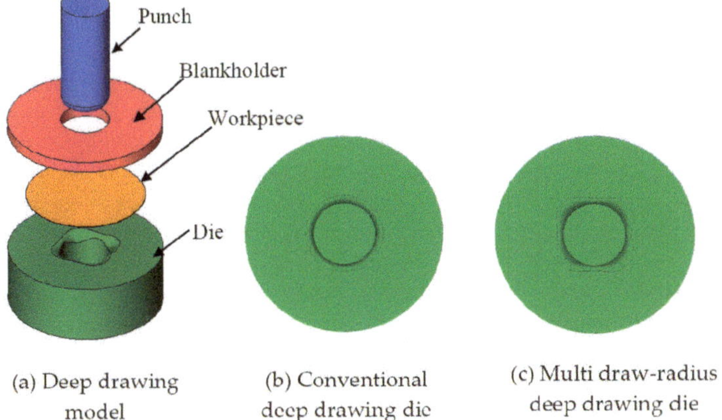

Figure 1. Schematic of deep drawing model and types of deep drawing die: (**a**) Deep drawing model; (**b**) Conventional die; (**c**) Multi draw-radius die.

Figure 2 shows the principle of MDR die based on the material flow characteristic. The comparison of schematic of material flow characteristic between conventional and MDR dies was illustrated. In the case of conventional die as shown in Figure 2a, owing to the effects of the anisotropy property of the material in each direction along the plane, at 45°, and at 90° to the rolling direction, the non-axisymmetric material flow characteristic was formed. By contrast, to encounter these effects, the MDR die was proposed. The different draw radius positioned in each direction along the plane, at 45°, and at 90°

to the rolling direction was designed. Based on this MDR die, the non-axisymmetric material flow characteristic due to the effects of the anisotropy property of the material could be reduced by using a different draw radius in each direction along the plane, at 45°, and at 90° to the rolling direction, as shown in Figure 2b. Therefore, the cup wall stretching and fracture prevention could be achieved, and the LDR could also be increased.

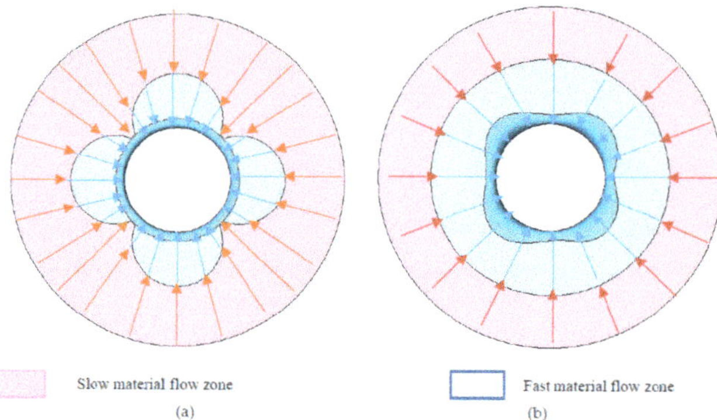

Figure 2. Principle of multi draw radius die based on the material flow characteristic: (**a**) Conventional die; (**b**) Multi draw-radius die.

3. The FEM Simulation and Experimental Procedures

In the present research, the FEM simulation was used as a tool to clarify the deep drawing mechanism of MDR die based on the material flow. The nonlinear FEM commercial code HyperForm 14.0 with RADIOSS script (Altair Engineering Inc., Troy, MI, USA) as the solver was used for FEM simulation of the deep drawing process. The investigated model of MDR die application is shown in Figure 3a. In addition, to clearly understand the deep drawing mechanism of MDR die application, the deep drawing mechanism of conventional die application was also investigated as the model shown in Figure 3b. These 3-D deep drawing models were created by Cimatron 3 (3D Systems Inc., Givat Shmuel, Israel) and then imported as IGES file into HyperForm. The HyperMesh preprocessor was used to create the mesh. The initial blank was set as elastic–plastic and meshed into finite elements of "shell" type. The 4 node quadrilateral shape elements of approximately of 3500 elements were generated. The adaptive remeshing was also set. After remeshing, to lead rather smooth meshes, the combination of 4 node quadrilateral shape elements and triangular shape elements were generated. The tool including punch, die, and blankholder were meshed with the rigid mesh type to prevent their deformations during the deep drawing process. The blankholder pressure was set as gap type. The gap of material thickness was applied. In the present research, the forming limit diagram (FLD) was used to clarify the forming characteristics as well as to predict the fracture zone on deep drawn parts. The workpiece used in this present research was medium carbon steel grade SPCC (JIS standard) with the thickness of 0.5 mm. The material properties of flow curve equation and plastic strain ratio (R-value) were prepared as input parameters for FEM simulation. The workpiece material was described with an elastoplastic, power exponent, isotropic plasticity model of Hollomon power law hardening model, with the constitutive equation determined from the stress–strain curve using the tensile test data. The other necessary material properties, such as the Young's modulus, Poisson's ratio, and ultimate tensile strength are given in Table 1. As per the literature [30–32], for the static compression test, the friction coefficient was set to be from 0.1 to 0.3. In the present research, based on the deep drawing process with lubricant used, the contact surface model was defined by a Coulomb friction law, and friction coefficient (μ) of 0.10 was applied. It was applied for both cases of conventional and MDR deep drawing processes to ignore the effect of lubricant use. Next, the diameters of punch and die were 40.0 and 41.0 mm, respectively, in which the clearance of 0.5 mm was set. The tool radius for conventional die was set following deep drawing theory [1]. Namely, a punch radius of 8.0 mm and die radius of 3.5 mm were set. The MDR draw radii of 3.5–5, 3.5–7, and 3.5–9 mm were investigated. On the basis of deep drawing theory [1], the LDR for this material was approximately of 2.25. Three levels of initial blank diameter of 90, 110, and 115 mm in which the drawing ratios of 2.25, 2.75, and 2.88 were investigated.

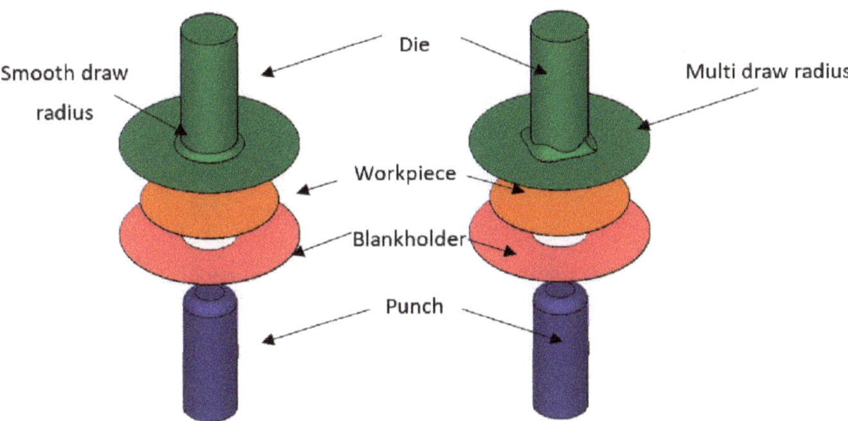

Figure 3. Deep drawing simulation models: (**a**) Conventional deep drawing model; (**b**) Multi draw-radius deep drawing model.

Table 1. FEM simulation and experimental conditions.

Object type	Sheet material: elastic–plastic Tool (punch, die, blankholder): rigid
Sheet material	Medium carbon steel (SPCC, JIS), thickness: 0.5 mm Ultimate tensile strength: 317 MPa Young's modulus: 208 GPa %Elongation: 51 Poisson's ratio: 0.33
Constitutive equation	$\bar{\sigma} = 554.43\bar{\varepsilon}^{0.23} + 208$
Blankholder force	Gap type
Plastic strain ratio (R value)	0° to rolling direction 2.1 45° to rolling direction 1.9 90° to rolling direction 2.6
Blank diameter	90, 100, 110, and 115 mm
Tool geometry	Punch radius 8 mm Punch diameter 40 mm Conventional die radius 3.5 mm MDR die radius 3.5–5, 3.5–7, 3.5–9 mm
Punch velocity	5 mm/s
Clearance	0.5 mm
Friction coefficient (μ)	0.10

The laboratory experiments were performed to validate the FEM simulation results. Figure 4 shows the press machine, which includes a universal sheet metal testing machine (Model SAS-350D, JT Toshi

Inc., Minato-ku, Japan) and the sets of conventional die and MDR die applications. The initial blanks were prepared using a wire electrical discharge machine (Wire-EDM) (Model AQ325L, Sodick Co., Ltd., Ishikawa, Japan). The obtained deep drawn parts were sectioned by wire electrical discharge machine for cup wall thickness examination. The cup wall thickness was measured. Five samples from each deep drawing condition were used to inspect the obtained deep drawing parts. The cup wall thickness was calculated based on these obtained deep drawing parts and the average cup wall thickness values were reported and compared with those analyzed by the FEM simulation.

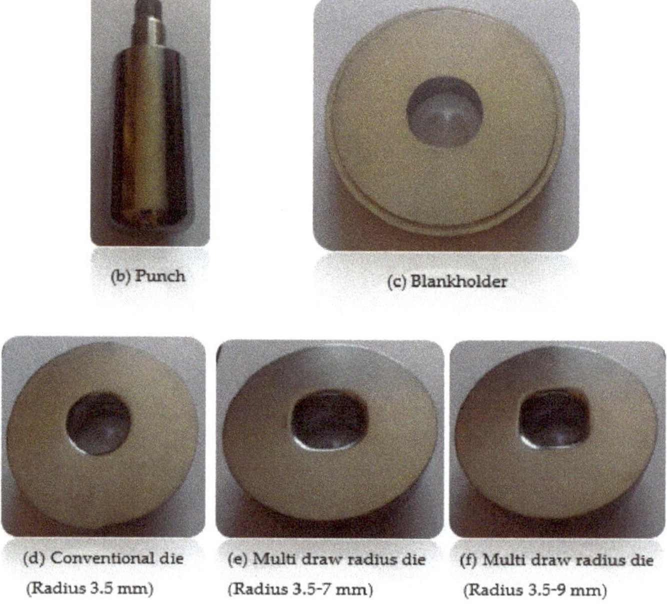

Figure 4. Press machine and set of punch and die: (**a**) Press machine (Universal sheet metal testing machine); (**b**) Punch; (**c**) Blankholder; (**d**) Conventional die (radius 3.5 mm); (**e**) Multi draw-radius die (radius 3.5–7mm); (**f**) Multi draw-radius die (radius 3.5–9mm).

4. Results and Discussion

4.1. The Validation of FEM Simulation Use

The FEM simulation was used, in the present research, as a tool for characterization of the deep drawing mechanism and prediction of the obtained deep drawn parts. Therefore, although the commercial finite element code HyperForm was used, the accuracy of the FEM simulation results should be again validated before starting the discussion section of FEM simulation results. As the validation of the FEM simulation results shows in Figure 5, by comparing with the laboratory experiments, the FEM simulation results showed the successful deep drawn parts and unsuccessful deep drawn parts which corresponded well with the experiments. In addition, the FEM simulation results also showed the earing defects and fracture which corresponded well with the experimental results. The unsuccessful deep drawn part in which the fracture generated as shown in Figure 5b, the FEM simulation result corresponded well with the experiments. On the basis of FLD, the fracture was generated on corner zone and a circumferential character was formed which agreed well with the fracture generated on deep drawn part obtained by experiment. The cup wall thickness was also examined. The comparisons of cup wall thickness distribution between FEM simulation and experimental results are illustrated in Figure 6. The FEM simulation results showed that the predicted cup wall thickness distributions corresponded well with the experiments, in which the errors in the analyzed cup wall thickness were approximately 3% compared with the experimental results. Finally, the deep drawing force was also recorded during experiments to validate the deep drawing force analyzed by FEM simulation. Figure 7 shows the comparison of deep drawing force obtained by FEM simulation and experiment. Again, the analyzed deep drawing force by FEM simulation corresponded well with the experiment, in which the error in the analyzed deep drawing force was approximately 1% compared with the experimental result.

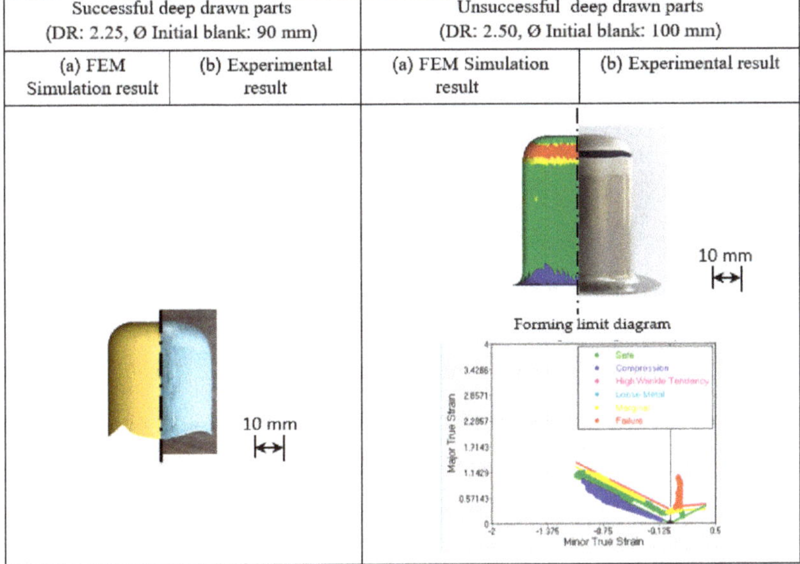

Figure 5. Comparison of deep drawn parts obtained by FEM simulation and experimental results: (**a**) FEM Simulation result; (**b**) Experimental result.

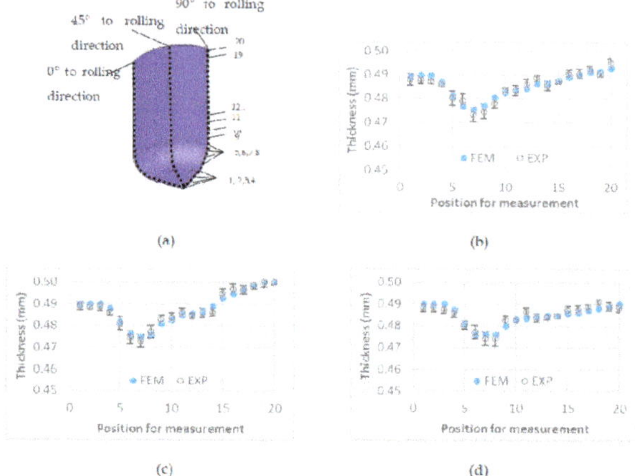

(a)

(b)

(c)

(d)

Figure 6. Cup wall thickness distribution between FEM simulation and experimental results: (**a**) Position for measurement; (**b**) 0° to rolling direction; (**c**) 45° to rolling direction; (**d**) 90° to rolling direction.

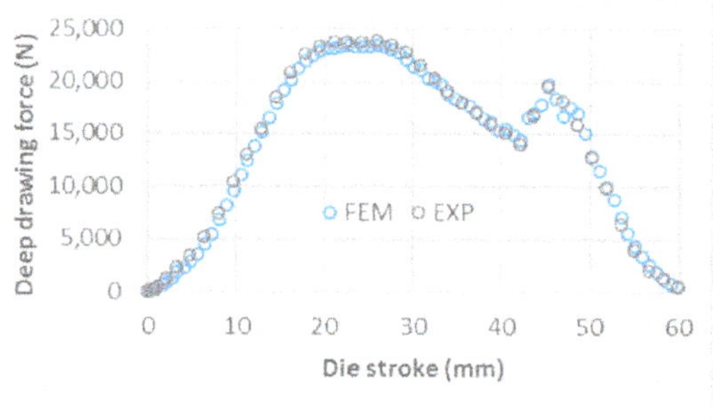

Figure 7. Comparison of deep drawing force between FEM simulation and experimental results (DR: 2.25; Ø initial blank: 90 mm).

4.2. Comparison of Material Flow Analysis between Conventional Die and MDR Die Applications

The principle of MDR die application, as aforementioned, was clearly characterized based on the material flow obtained by FEM simulation. As shown in Figure 8, the comparison of material flow between conventional and MDR die applications was illustrated. For the deep drawing stroke of approximately 17 mm, the material flow showed that the same manner in both cases of conventional and MDR die applications could be observed as shown in Figure 8a. These results corresponded well with deep drawing theory [1]. For the deep drawing stroke of approximately 25 mm, in the case of conventional die application as shown in Figure 8b-1, the effects of the anisotropy property of the material on material flow were clearly illustrated. The non-axisymmetric material flow on flange was clearly illustrated as depicted by dashed lines. These results corresponded well with deep drawing theory [1]. By contrast, in the case of MDR die application as shown in Figure 8b-2, the effects of the anisotropy property on material flow were compensated by multi draw radius especially on the large draw radius zone. However, owing to that there were large radius zones formed on MDR die, it was observed that the material flow velocity in the case of MDR die application was larger than that in the case of conventional die application especially on the large radius zone of MDR die. Moreover, the reduction of the non-axisymmetric material flow characteristic on flange could be obtained and clearly illustrated as depicted by dashed lines. Next, the deep drawing stroke was increased to approximately 37 mm, as shown in Figure 8c, and the effects of the anisotropy property of the material on material flow were clearly evidenced in the case of conventional die application as shown in Figure 8c-1. It was clearly observed that the non-axisymmetric material flow on flange was clearly illustrated as depicted by dashed lines. The flange shape was not a circular but had become square. By contrast, in the case of MDR die application as shown in Figure 8c-2, the effects of the anisotropy property of the material on material flow were continuously compensated by multi draw radius. The reduction of the non-axisymmetric material flow characteristic on flange and the asymmetry of the flange were clearly illustrated. The flange was in a more circular shape. Finally, in the case of conventional die application, the effects of the anisotropy property of the material on material flow were stronger as the deep drawing stroke increased as shown in Figure 8d-1. Vice versa, in the case of MDR die application, the effects of the anisotropy property of the material on material flow were continuously compensated by multi draw radius. The reduction of the non-axisymmetric material flow characteristic on flange were clearly illustrated as well as that the asymmetry of the flange could be continuously reduced and the flange was more circular in shape as shown in Figure 8d-2. These results revealed that by using MDR die application, the effects of the anisotropy property of the material on material flow could be compensated during the deep drawing process. The non-axisymmetric material flow characteristic on the flange and the asymmetry of the flange could be reduced during the deep drawing process. This resulted in preventing cup wall stretching and fracture as well as that the LDR could be increased.

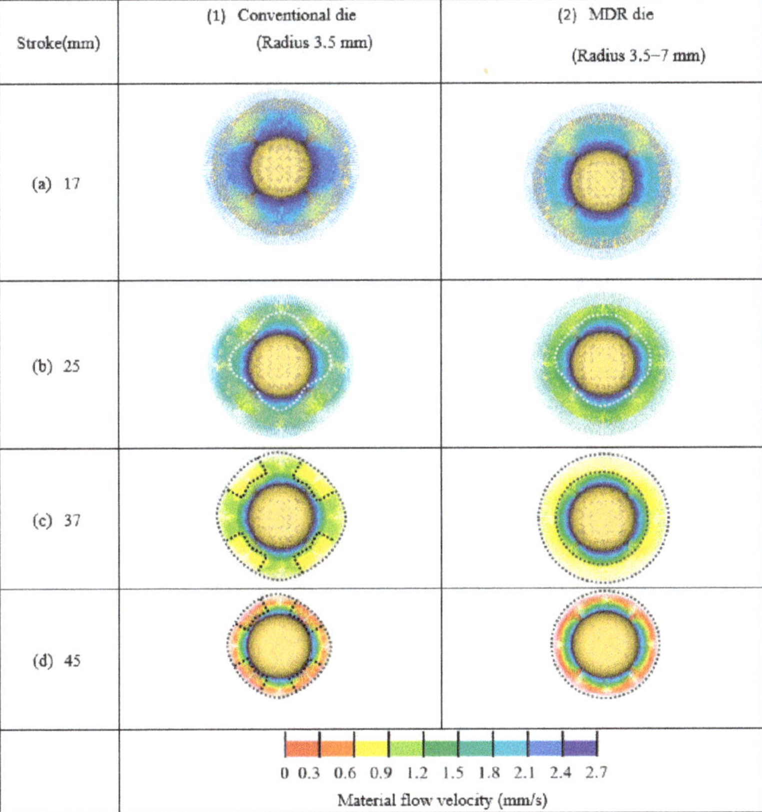

Figure 8. Comparison of the deep drawing mechanism between conventional and multi draw radius (MDR) die applications: (**a**) Die stroke 17 mm; (**b**) Die stroke 25 mm; (**c**) Die stroke 37 mm; (**d**) Die stroke 45 mm.

4.3. MDR Die Design Related to the Anisotropy Property of the Material

As mentioned in the previous section, the results illustrated that the effects of the anisotropy property of the material on material flow were related to the draw radius. Therefore, the MDR die should be strictly designed by relating to the anisotropy property of the material in each direction along the plane, at 45°, and at 90° to the rolling direction. Figure 9b,c show the deep drawn parts obtained by MDR die application related to the anisotropy property of the material. Figure 9b shows the MDR die application by designing the larger draw radius positioned for at 45° to the rolling direction and the small draw radius positioned for along the plane and at 90° to the rolling (named MDR type I). By contrast, Figure 9c shows the MDR die application by designing the larger draw radius positioned for along the plane and at 90° to the rolling direction and the small draw radius positioned for at 45° to the rolling (named MDR type II). The results showed that, in the case of LDR 2.25 (initial blank diameter of 90 mm), the deep drawn parts could be formed by conventional die application as shown in Figure 9a-1. This result corresponded well with the deep drawing theory and literature that by using conventional die application, the deep drawn part could be formed with LDR [1]. The results also showed that the deep drawn parts could be formed by using MDR die application in both cases of MDR die designs as shown in Figure 9b-1,c-1. However, it was observed that in terms of earing defect, the MDR type II showed a larger earing defect than that of MDR type I as well as than that of conventional die application. In addition, it was also observed that the earing defect obtained by using MDR type I was smaller than that of conventional die application. As these results show, to deep draw with LDR, the deep drawn parts could be achieved by using MDR die application. In addition, the quality of deep drawn parts obtained by MDR type I in terms of earing defects was better than that obtained by conventional die. The results illustrate that the MDR die design should be strictly considered as related to the anisotropy property of the material in each direction along the plane, at 45°, and at 90° to the rolling direction. Next, to deep draw over LDR, the result showed that by using conventional die application, the deep drawn part could not be achieved, and a fracture was generated as shown in Figure 9a-2. Based on the FLD, the FEM simulation result clearly showed that the fracture characteristic is formed as a circumferential character. This result agreed well with deep drawing theory [1]. By contrast, in using the MDR die application, the results illustrated that the deep drawn part could be achieved by MDR die type I application as shown in Figure 9b-2. However, using MDR die type II, the deep drawn part could not be achieved. The six ears were also characterized on the basis of FLD, and a fracture was also generated on the top of the deep drawn part as shown in Figure 9c-2. Owing to the anisotropy property of the material in each direction along the plane, at 45°, and at 90° to the rolling direction related to the formability [1], therefore, the design of the larger draw radius positioned at 45° to the rolling direction and the smaller draw radius positioned along the plane and at 90° to the rolling direction was suggested. As shown in Figure 10, the MDR die type II resulted in that the larger non-axisymmetric material flow on flange was formed compared with those in the cases of conventional and MDR die type I applications. As these material flow analyses show, as aforementioned, the non-axisymmetric material flow characteristic on flange and the asymmetry of the flange could be increased and cup wall stretching and fracture were then easier to generate.

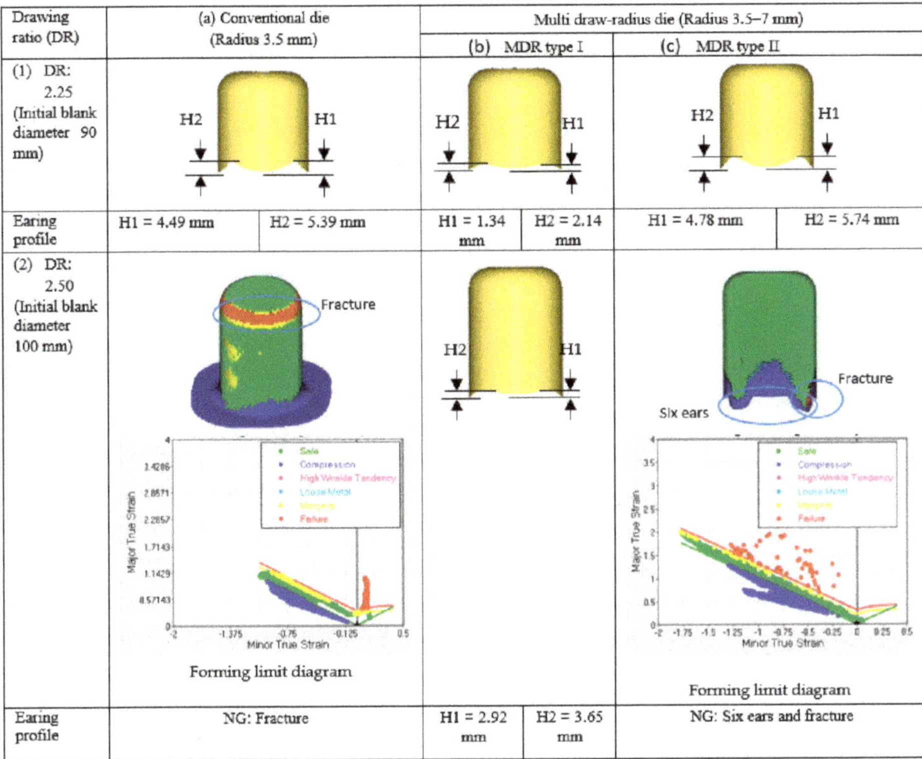

Drawing ratio (DR)	(a) Conventional die (Radius 3.5 mm)		Multi draw-radius die (Radius 3.5–7 mm)			
			(b) MDR type I		(c) MDR type II	
(1) DR: 2.25 (Initial blank diameter 90 mm)	H2 / H1		H2 / H1		H2 / H1	
Earing profile	H1 = 4.49 mm	H2 = 5.39 mm	H1 = 1.34 mm	H2 = 2.14 mm	H1 = 4.78 mm	H2 = 5.74 mm
(2) DR: 2.50 (Initial blank diameter 100 mm)	Fracture / Forming limit diagram		H2 / H1		Six ears / Fracture / Forming limit diagram	
Earing profile	NG: Fracture		H1 = 2.92 mm	H2 = 3.65 mm	NG: Six ears and fracture	

Figure 9. MDR die design related to the anisotropy property of the material: (**a**) Conventional die (Radius 3.5 mm); (**b**) MDR type I; (**c**) MDR type II.

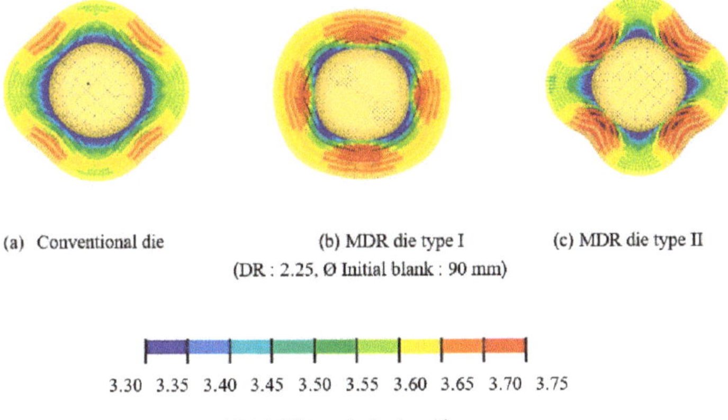

Figure 10. Material flow with respect to deep drawing dies: (**a**) Conventional die; (**b**) MDR die type I; (**c**) MDR die type II.

4.4. Examination of Drawing Ratio with Respect to Multi Draw Radius Dies

Figure 11 shows the obtained deep drawn parts with respect to various MDR dies and drawing ratios. The drawing ratios of 2.75 and 2.88 which were larger than LDR were investigated, as respectively shown in Figure 11a,b. On the basis of deep drawing theory [1], the draw radius of 3.5 mm was recommended and then it was set as a small draw radius in MDR die. Next, the large draw radius values of 5, 7, and 9 mm were set as the large draw radius. With the drawing ratio of 2.75 as shown in Figure 11a, the results showed that the deep drawn parts could not be achieved when the large radius of 5 mm was set, as shown in Figure 11a-1. This result could be explained by the draw radius of 5 mm, set as the large radius, was too small to reduce the non-axisymmetric material flow characteristic during the deep drawing process. Conversely, as the large radius was increased, the greater reduction of non-axisymmetric material flow characteristic could be achieved, and then the deep drawn parts could be achieved as for the large radius values set as 7 and 9 mm, as shown in Figure 11a-2,a-3, respectively. The increases in large radius resulted in that, as aforementioned, the non-axisymmetric material flow characteristic on the flange was reduced, and the asymmetry of flange could also be reduced. This resulted in the more circular flange shape. However, it was also observed that owing to the large radius of 9 mm, the non-axisymmetric material flow characteristic on the flange could be reduced, getting a more circular flange shape than that for the large radius of 7 mm during the deep drawing process. The earing defect in the case of large radius 9 mm was smaller than that obtained in the case of the large radius of 7 mm. Next, with the drawing ratio of 2.88 as shown in Figure 11b, the results showed that the deep drawn parts could not be achieved. Namely, owing to the overly large drawing ratio (overly large initial blank diameter) applied, the non-axisymmetric material flow on the flange could not be effectively reduced during a whole deep drawing process. These results revealed that in the present research, the LDR could be increased by approximately 22.22% using MDR die application. In addition to the increases in LDR, the quality of deep drawn parts in terms of cup wall thickness and earing defects were also increased. Specifically, the earing defect could be reduced approximately 40% compared with the use of conventional die as shown in Figures 9 and 11. Next, the more uniform cup wall thickness in each direction along the plane, at 45°, and at 90° to the rolling direction could be obtained by comparing with the use of conventional die as shown in Figure 12. However, the MDR die should be strictly design related to the anisotropy property of the material in each direction along the plane, at 45°, and at 90° to the rolling direction which was suggested in the previous section.

4.5. Confirmation of MDR Die Application

To validate the accuracy of the MDR die application obtained by FEM simulation, the FEM simulation results were compared with those obtained by experimental results, as shown in Figure 13. The FEM simulation results showed that the predicted deep drawn parts corresponded well with the experiments as shown in Figure 13a-1,b-1 in the cases of MDR draw radius of 3.5–7 and 3.5–9 mm, respectively. In terms of cup wall thickness, the FEM simulation results showed that the predicted cup wall thickness corresponded well with the experiments as shown in Figure 13a-2,b-2 in the cases of MDR draw radius of 3.5–7 and 3.5–9 mm, in which the errors in the analyzed cup wall thickness were approximately 3% compared with the experimental results.

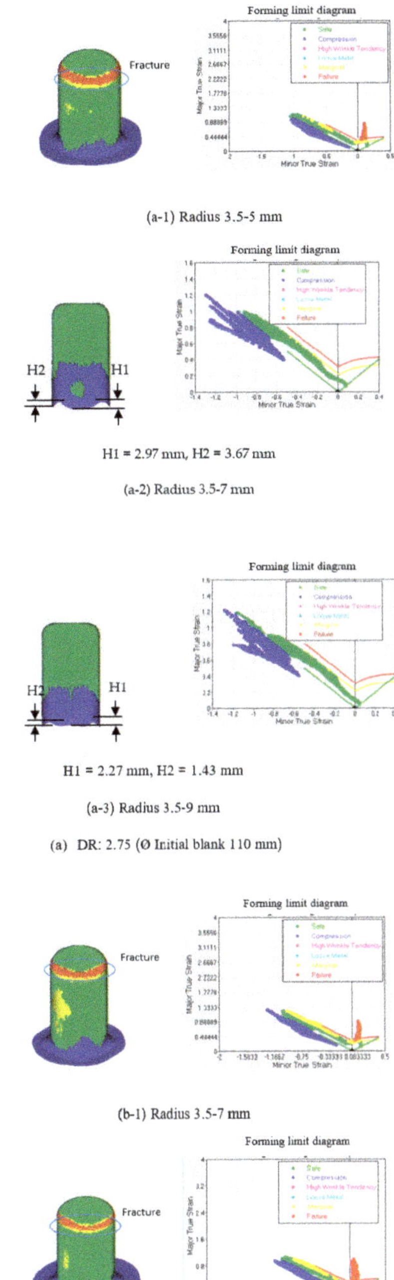

Figure 11. Multi draw radius die application with respect to various DRs: (**a**) DR: 2.75 (Ø Initial blank 110 mm); (**b**) DR: 2.88 (Ø Initial blank 115 mm).

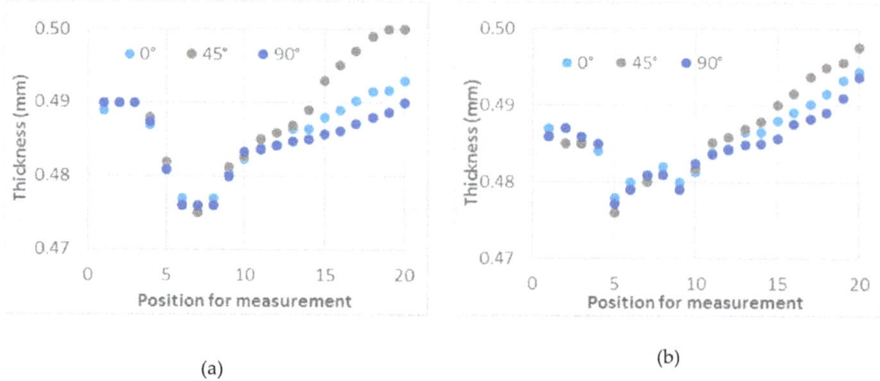

Figure 12. Comparison of cup wall thickness between conventional and MDR die applications obtained by FEM simulation (DR: 2.25; initial blank diameter: 90 mm). (**a**) Conventional die (Radius 3.5 mm); (**b**) Multi draw radius die type I (Radius 3.5–7 mm).

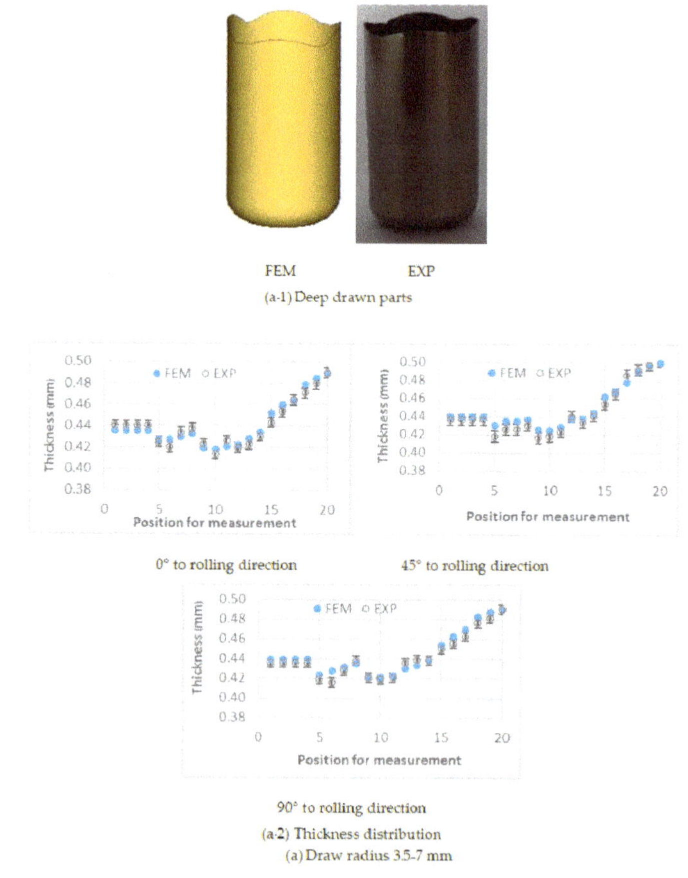

Figure 13. *Cont.*

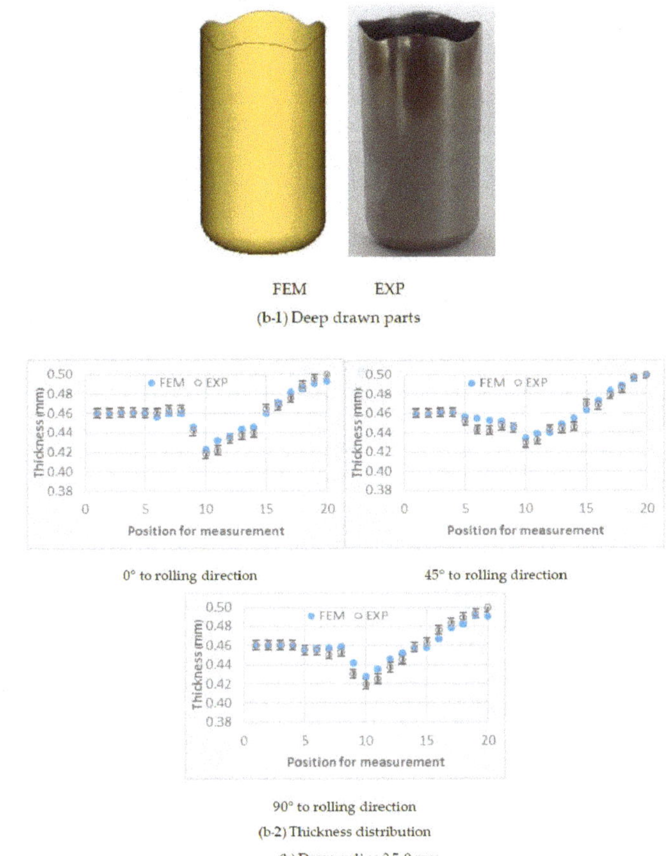

FEM   EXP

(b-1) Deep drawn parts

0° to rolling direction

45° to rolling direction

90° to rolling direction

(b-2) Thickness distribution

(b) Draw radius 3.5-9 mm

Figure 13. Comparison of theoretical and experimental results of multi draw radius die. (DR: 2.75; Ø initial blank: 110 mm): (**a**) Draw radius 3.5-7 mm; (**b**) Draw radius 3.5-9 mm.

5. Conclusions

To increase the drawing ratio and overcome the LDR, the MDR die application was proposed in the present research. First, the conceptual design of MDR die was proposed, and its principle was also clearly elucidated in the present research by FEM simulation based on the material flow. The absolute validation of FEM simulation use was also performed. By using MDR die application, it was revealed that the non-axisymmetric material flow characteristic on the flange as well as the asymmetry of flange shape could be reduced. Specifically, during deep drawing process, the MDR could compensate the effects of the anisotropy property of the material on material flow characteristics in each direction along the plane, at 45°, and at 90° to the rolling direction as well as that the non-axisymmetric material flow characteristic on flange could be effectively reduced and, by reducing the asymmetry of flange, a more circular flange shape could be obtained. Therefore, wall cup stretching could be reduced as well as the delay of fracture. Based on this principle, the LDR could be increased. However, in the present research, the proper MDR die design related to the anisotropy property of the material in each direction along the plane, at 45°, and at 90° to the rolling direction was suggested. Specifically, the larger draw radius positioned for at 45° to the rolling direction and the smaller draw radius positioned for along the plane and at 90° to the rolling direction were recommended. Again, the experiments were also carried out to validate the FEM simulation results in the case of MDR die application. The FEM simulation results showed that the predicted deep drawn parts corresponded well with the experimental results. In addition, the FEM simulation results showed that the predicted cup wall thickness corresponded well with the experiments, in which the errors in the analyzed cup wall thickness were approximately 3% compared with the experimental results. The results of the present research reveal that the LDR could be increased approximately 22.22% using MDR die application. In addition to the increases in LDR, in terms of quality of deep drawn part, the deep drawn parts obtained by MDR die application showed a smaller earing defect compared with those obtained by conventional die. The decreases in earing defect of approximately 40% could be achieved as well as the more uniform cup wall thickness in each direction along the plane, at 45°, and at 90° to the rolling direction could be obtained. However, the MDR die should be strictly design related to the anisotropy property of the material in each direction along the plane, at 45°, and at 90° to the rolling direction as suggested in the present research.

Author Contributions: Conceptualization, W.P. and S.T.; data curation, W.P.; funding acquisition, W.P.; investigation, W.P.; methodology, W.P. and S.T.; project administration, W.P.; supervision, S.T.; writing—original draft, W.P.; writing—review & editing, S.T. All authors have read and agreed to the published version of the manuscript.

Funding: This research was supported by a grant from The Thailand Research Fund (TRF) and Rajamangala University of Technology Rattanakosin under Grant No. MRG6280205.

Acknowledgments: The authors would especially like to thank Rudeemas Jankree and Jaksawat Sriborwornmongkol, for their help in this research. The authors would like to express our gratitude to Pravitr Paramaputi, Srisahawattanakij Co. Ltd., for his support in the Cimatron v.3 program.

References

1. Lange, K. *Handbook of Metal Forming*; McGraw-Hill Inc.: New York, NY, USA, 1985.
2. Ahmadi, M.; Sadeghi, B.M.; Arabi, H. Experimental and numerical investigation of V-bent anisotropic 304L SS sheet with spring-forward considering deformation-induced martensitic transformation. *Mater. Des.* **2017**, *123*, 211–222. [CrossRef]
3. Thipprakmas, S. Finite element analysis of sided coined-bead technique in precision V-bending process. *Int. J. Adv. Manuf. Technol.* **2012**, *65*, 679–688. [CrossRef]
4. Lin, B.-T.; Yang, C.-Y. Using a punch with micro-ridges to shorten the multistage deep drawing process for stainless steels. *Int. J. Adv. Manuf. Technol.* **2016**, *88*, 2693–2703. [CrossRef]
5. Thipprakmas, S.; Komolruji, P. Analysis of bending mechanism and spring-back characteristics in the offset Z-bending process. *Int. J. Adv. Manuf. Technol.* **2015**, *85*, 2589–2596. [CrossRef]
6. Liu, Y.; Tang, B.; Hua, L.; Mao, H. Investigation of a novel modified die design for fine-blanking process to reduce the die-roll size. *J. Mater. Process. Technol.* **2018**, *260*, 30–37. [CrossRef]
7. Thipprakmas, S.; Phanitwong, W. Finite element analysis of flange-forming direction in the hole flanging process. *Int. J. Adv. Manuf. Technol.* **2011**, *61*, 609–620. [CrossRef]
8. Zhang, Z.; Cheng, L.; Sun, H.; Bao, H. Finite element simulation of the punch with inclined edge in the sheet metal blanking process. *Int. J. Comput. Sci. Math.* **2018**, *9*, 377–389. [CrossRef]
9. Thipprakmas, S.; Komolruji, P.K.; Phanitwong, W. Comparison of Offset and Wiping Z-Die Designs for Precision Z-Bent Part Fabrication. *J. Manuf. Sci. Eng.* **2018**, *140*, 021015. [CrossRef]
10. Zhang, C.-B.; Gong, F. Deep drawing of cylindrical cups using polymer powder medium based flexible forming. *Int. J. Precis. Eng. Manuf. Technol.* **2018**, *5*, 63–70. [CrossRef]
11. Modi, B.; Kumar, D.R. Optimization of process parameters to enhance formability of AA 5182 alloy in deep drawing of square cups by hydroforming. *J. Mech. Sci. Technol.* **2019**, *33*, 5337–5346. [CrossRef]
12. Cha, W.-G.; Müller, S.; Albers, A.; Volk, W. Formability consideration during bead optimisation to stiffen deep drawn parts. *Prod. Eng.* **2018**, *12*, 691–702. [CrossRef]
13. Abe, Y.; Mori, K.-I.; Maeno, T.; Ishihara, S.; Kato, Y. Improvement of sheet metal formability by local work-hardening with punch indentation. *Prod. Eng.* **2019**, *13*, 589–597. [CrossRef]

14. Lee, M.S.; Kim, S.J.; Seo, H.Y.; Kang, C.-G. Investigation of Formability and Fiber Orientation in the Square Deep Drawing Process with Steel/CFRP Hybrid Composites. *Int. J. Precis. Eng. Manuf.* **2019**, *20*, 2019–2031. [CrossRef]

15. Liu, J.; Zhuang, L. Cylindrical cup-drawing characteristics of aluminum-polymer sandwich sheet. *Int. J. Adv. Manuf. Technol.* **2018**, *97*, 1885–1896. [CrossRef]

16. Wang, C.; Cheng, L.D.; Liu, Y.; Zhang, H.; Wang, Y.; Shan, D.B.; Guo, B. Research on micro-deep drawing process of concial part with ultra-thin copper foil using multi-layered DLC film-coated die. *Int. J. Adv. Manuf. Technol.* **2018**, *100*, 569–575. [CrossRef]

17. Li, W.; Fu, M.; Wang, J.L.; Meng, B. Grain size effect on multi-stage micro deep drawing of micro cup with domed bottom. *Int. J. Precis. Eng. Manuf.* **2016**, *17*, 765–773. [CrossRef]

18. Gong, F.; Yang, Z.; Chen, Q.; Xie, Z.; Shu, D.; Yang, J. Influences of lubrication conditions and blank holder force on micro deep drawing of C1100 micro conical–cylindrical cup. *Precis. Eng.* **2015**, *42*, 224–230. [CrossRef]

19. Ma, J.; Gong, F.; Yang, Z.; Zeng, W. Micro deep drawing of C1100 micro square cups using microforming technology. *Int. J. Adv. Manuf. Technol.* **2016**, *82*, 1363–1369. [CrossRef]

20. Sen, N.; Karaağaç, I.; Kurgan, N. Experimental research on warm deep drawing of HC420LA grade sheet material. *Int. J. Adv. Manuf. Technol.* **2016**, *87*, 3359–3371. [CrossRef]

21. Wang, W.; Chen, S.; Tao, K.; Gao, K.; Wei, X. Experimental investigation of limit drawing ratio for AZ31B magnesium alloy sheet in warm stamping. *Int. J. Adv. Manuf. Technol.* **2017**, *92*, 723–731. [CrossRef]

22. Singh, A.; Basak, S.; Lin Prakash, P.S.; Roy, G.G.; Jha, M.N.; Mascarenhas, M.; Panda, S.K. Prediction of earing defect and deep drawing behavior of commercially pure titanium sheets using CPB06 anisotropy yield theory. *J. Manuf. Process.* **2018**, *33*, 256–267. [CrossRef]

23. Saxena, R.K.; Dixit, P.M. Finite element simulation of earing defect in deep drawing. *Int. J. Adv. Manuf. Technol.* **2009**, *45*, 219–233. [CrossRef]

24. Dhaiban, A.A.; Soliman, M.-E.S.; El-Sebaie, M. Finite element modeling and experimental results of brass elliptic cups using a new deep drawing process through conical dies. *J. Mater. Process. Technol.* **2014**, *214*, 828–838. [CrossRef]

25. Dehghani, F.; Salimi, M. Analytical and experimental analysis of the formability of copper-stainless-steel 304L clad metal sheets in deep drawing. *Int. J. Adv. Manuf. Technol.* **2015**, *82*, 163–177. [CrossRef]

26. Kesharwani, R.K.; Basak, S.; Panda, S.; Pal, S. Improvement in limiting drawing ratio of aluminum tailored friction stir welded blanks using modified conical tractrix die. *J. Manuf. Process.* **2017**, *28*, 137–155. [CrossRef]

27. Bandyopadhyay, K.; Panda, S.; Saha, P.; Padmanabham, G. Limiting drawing ratio and deep drawing behavior of dual phase steel tailor welded blanks: FE simulation and experimental validation. *J. Mater. Process. Technol.* **2015**, *217*, 48–64. [CrossRef]

28. Faraji, G.; Mashhadi, M.M.; Hashemi, R. Using the finite element method for achieving an extra high limiting drawing ratio (LDR) of 9 for cylindrical components. *CIRP J. Manuf. Sci. Technol.* **2010**, *3*, 262–267. [CrossRef]

29. Halkaci, H.S.; Turkoz, M.; Dilmec, M. Enhancing formability in hydromechanical deep drawing process adding a shallow draw bead to the blank holder. *J. Mater. Proc. Technol.* **2014**, *214*, 1638–1646. [CrossRef]

30. Karagiozova, D.; Shu, D.; Lu, G.; Xiang, X. On the energy absorption of tube reinforced foam materials under quasi-static and dynamic compression. *Int. J. Mech. Sci.* **2016**, *105*, 102–116. [CrossRef]

31. Ha, N.S.; Lu, G.; Xiang, X. High energy absorption efficiency of thin-walled conical corrugation tubes mimicking coconut tree configuration. *Int. J. Mech. Sci.* **2018**, *148*, 409–421. [CrossRef]

32. Ha, N.S.; Lu, G.; Xiang, X. Energy absorption of a bio-inspired honeycomb sandwich panel. *J. Mater. Sci.* **2019**, *54*, 6286–6300. [CrossRef]

Article

Prediction and Control Technology of Stainless Steel Quarter Buckle in Hot Rolling

Hui Li [1], Chihuan Yao [1], Jian Shao [1,*], Anrui He [1], Zhou Zhou [2] and Weigang Li [3]

[1] Institute of Engineering Technology, University of Science and Technology Beijing, Beijing 100083, China; 2012lihui@sina.com (H.L.); yaochihuan@xs.ustb.edu.cn (C.Y.); harui@ustb.edu.cn (A.H.)

[2] Guangxi Beibu Gulf New Materials Company Limited, Beihai 536000, China; zhou_zhou306@sina.com

[3] Engineering Research Centre for Metallurgical Automation and Measurement Technology of Ministry of Education, Wuhan University of Science and Technology, Wuhan 430081, China; liweigang.luck@foxmail.com

* Correspondence: jianshao@ustb.edu.cn; Tel.: +86-10-62336320

Received: 12 July 2020; Accepted: 4 August 2020; Published: 6 August 2020

Abstract: To obtain the good flatness of hot-rolled stainless-steel strips, high-precision shape control has always been the focus of research. In hot rolling, the quadratic wave (centre buckle and edge wave) can usually be controlled effectively. However, the quarter buckle of the strip is still a challenge to solve. In this paper, a prediction model of roll deflection and material flow is established to study the change rule and control technology of the quarter buckle. The effects of the process parameters on the quarter buckle are analysed quantitatively. The process parameters affect the quarter buckle and the quadratic wave simultaneously. This coupling makes the control of the quarter buckle difficult. The distribution of lateral temperature and the quartic crown of the strip have less effect on the quadratic wave but have a great effect on the quarter buckle. Finally, a new technology for work roll contour is developed to improve the quarter buckle. Through industrial application, the model and the new contour are proved to be effective.

Keywords: quarter buckle; roll stack deflection; strip material flow; roll contour optimisation; hot-rolled stainless steel

---

1. Introduction

Hot-rolled strips, especially of stainless steel, are commonly used directly in product processing. The strip should have a good flatness to meet the requirement of users. He et al. [1] divided strip flatness defects into three types: the linear wave, the quadratic wave, and the high-order wave, according to the patterns of these waves in the width direction of the strip. The linear wave is also called the single-edge wave, which is adjusted by roll tilting. The quadratic wave refers to the centre buckle and the edge wave, which can be controlled by the roll bending system. The high-order wave cannot be characterised by linear or quadratic equations. In the production of hot-rolled stainless steel, high-order wave primarily include the flatness defect of the bilateral symmetrical quarter buckle. Figure 1 shows the quarter buckle of the strip after rolling detected by the flatness meter in a 1450 mm hot-rolled stainless steel line. It not only causes the strip edge to easily crack during hot rolling, or even pile-up accidents, but also affects the surface quality of the products in subsequent processes such as pickling and annealing. It is a challenge for researchers to overcome this problem.

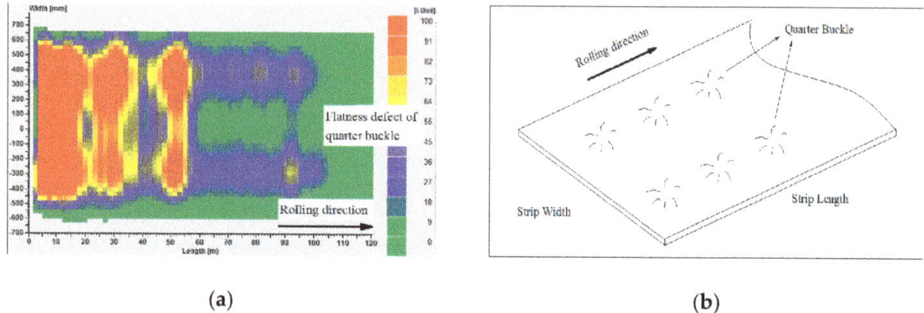

Figure 1. Quarter buckle of the strip after rolling: (**a**) flatness measured by on-line meter; (**b**) diagram of the quarter buckle.

The essence of the flatness problem has been widely recognised. Zhang et al. [2] described the generation mechanism of flatness when studying the shape control of a temper mill. In the process of the plastic deformation of the strip, when the distribution of the longitudinal extension difference along the strip width is uneven, residual internal stress is generated, including positive tensile stress and negative compressive stress. When the difference of the internal stresses exceeds the ultimate load of the strip buckling, a visible wave will appear. This can also explain the generation of the quarter buckle of hot-rolled stainless steel.

Some researchers believe that the roll wear and the grinding error of the roll contour cause the quarter buckle. Du et al. [3] heated the local position of the roll to compensate for the uneven roll wear to eliminate the quarter buckle. Li et al. [4] believed the uneven wear of the back-up roll and the low wear of the work roll to have little effect on the strip flatness, but that the large grinding error of the roll contour caused the quarter buckle. In production, we find that within the working period of the work roll, the quarter buckle appears at all stages. Meanwhile, with the use of high-precision roll grinders and the improvement of roll grinding technology, the grinding accuracy of the roll contour can usually be well controlled. He et al. [5] proposed intelligent roll precision grinding and management technology. Hence, the effects of these on the quarter buckle are ignored in this paper. Under some process parameters, the longitudinal strain distribution of the quarter-buckle after strip rolling is related to the high-order deflection of the loaded roll gap profile. By calculating the deformations of the roll stack and the strip, the distribution of the strip longitudinal strain can be obtained to help study how the changes of process parameters affect the quarter buckle.

Nowadays, with the development of computer performance, the finite element method (FEM) has become popular. Jiang et al. [6] used rigid-plastic FEM to study the friction in thin strip rolling. The study was based on the strip deformation, and the effect of the roll stack was simplified. Chandra et al. [7] analysed the temper rolling process using rigid-plastic FEM. The calculation of roll stack deformation is optimised. Wang et al. [8] studied the effect of roll contour configuration on the flatness of hot-rolled thin strip using two-dimensional varying thickness FEM. The elastic deformation of the roll stack is calculated by this method, which is a simplification of the three-dimensional FEM. The optimisation of the roll contour reduced the concentration of contact stress between the back-up roll and the work roll and improved the flatness control of the bending force. Kong et al. [9] built an elastic-plastic FEM model for the calculation of roll thermal expansion, which improved the calculation accuracy. Kim et al. [10] established a three-dimensional FEM model of thermal–mechanical coupling for the research of lubrication rolling. Hao et al. [11] built an FEM model of aluminium alloy rolling by using the ANSYS software and calculated the deformations of the roll stack and the strip. The calculation accuracy of the FEM has been recognised by many researchers. However, it has the shortcomings of longer calculation and difficulty of attaining convergence. The modelling process with multiple constraints and iterative calculations is complex and difficult to achieve for the commercial finite element software.

The influence function method has high accuracy and short calculation time, which is convenient for on-line applications. Jiang et al. [12] analysed the contact of roll edge during cold rolling of the thin strip using the influence function method. Hao et al. [13] used the influence function method to calculate the deformation of the roll stack for aluminium alloys. Wang et al. [14] described the distributions of the contact pressure between the rolls and the rolling force using a high-order polynomial and achieved a fast calculation of the influence function method. In the influence function method, the stress and displacement fields are obtained from the influence function superposition, so it is not suitable for calculating the plastic deformation of the strip. Wang et al. [15] solved the lateral metal flow of the strip in the hot rolling process using the variational method. Lian et al. [16] believed that the accuracy of the variational method depended too much on assumptions. The three-dimensional difference method was adopted to calculate the strip deformation. Zhang et al. [17] built a rigid-plastic FEM model to calculate the strip deformation, which was combined with the influence function method to get the lateral distribution of the strip thickness. Shao et al. [18] analysed the limitations of the finite difference method, and used the finite volume method to solve the plastic deformation of the strip. Analysis of the quarter buckle with different process parameters requires a large number of simulations. Considering the time cost of the FEM model, in this paper, the rapid influence function method and the finite volume method are used to calculate the distribution of the strip's longitudinal strain after rolling.

The key to solving the problem of the quarter buckle is to make up the high-order loaded roll gap profile and improve the distribution of the strip's longitudinal strain after rolling. In cold rolling, Ma et al. [19] added the roll bending system of the intermediate and work rolls for the researched universal crown mill. In this way, the high-order deflection of the loaded roll gap profile can be adjusted. Guo et al. [20] studied the relationship between the spray cooling of the work roll and the strip cross section. Wang et al. [21] explored the local heat transfer characteristic of the spray cooling of the work roll. Although spray cooling is a solution for local higher-order waves, the effect of changing the local crown of the work roll by heat transfer is slow. Li et al. [22] designed the Baosteel universal roll contour for the intermediate rolls, which was used to adjust the quarter buckle in cooperation with roll shifting. Li et al. [23] upgraded the original roll contour of the continually variable crown to a new roll contour of the quintic curve, which can improve the quarter buckle. Seilinger et al. [24] reduced the quarter buckle by adjusting the sinusoidal component of the roll contour of Smart Crown. In addition, Hara et al. [25] showed the quarter buckle could be corrected by using a partial concave profile to the first intermediate rolls for the Sendzimir mill. Kubo et al. [26] improved the quarter buckle for the Sendzimir mill by using the flexible shaft backing assemblies and the concave roll contour. In the hot rolling, there have been limited reports about the applications of the continually variable crown plus and the Smart Crown. When the quarter buckle is adjusted, the quadratic crown of the loaded roll gap profile is also changed. Hence, it is difficult for them to control the quarter buckle individually, which makes their application more difficult. Therefore, it is of great significance to develop a new control technology for the quarter buckle in the hot rolling.

In this paper, to study the change rule and control technology of the quarter buckle of hot-rolled stainless steel, a prediction model of roll deflection and material flow (RDMF) is established. The rapid influence function method and the finite volume method are used to achieve the iterative computations between the elastic deformation of the roll stack and the three-dimensional plastic deformation of the strip. The model is used to quantitatively analyse the effect of the shape process parameters on the quarter buckle, such as the bending force, the rolling force, the lateral temperature distribution of the strip, the quartic crown of the strip before rolling and so forth. Finally, we develop a new work roll contour to improve the quarter buckle, which is called the middle variable crown (MVC).

2. Methods and Materials

The RDMF model that is built to research the change rule of the quarter buckle incorporates the roll stack deflection and the strip material flow. The former is used to solve the loaded roll gap profile, while the latter can obtain the metal flow and stress distribution. The iterative calculations are performed between them. The calculation process of the RDMF model is shown in Figure 2.

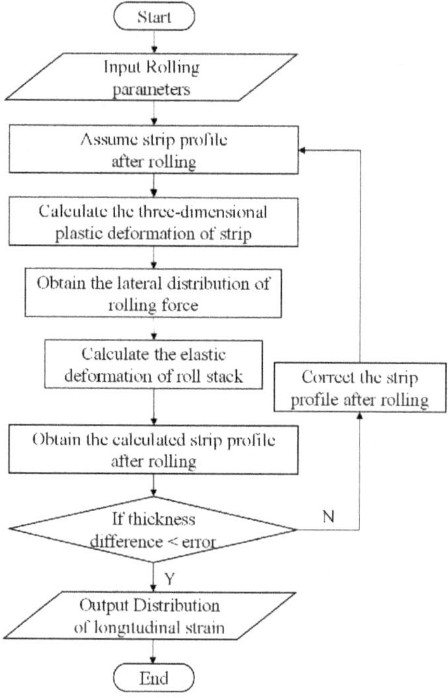

Figure 2. Flow chart of the RDMF model.

2.1. Roll Bending

Because of the symmetry of the rolling system, the calculation process involves the upper rolls. According to the load characteristic of the rolls, both the work roll and the back-up roll can be abstracted into a simply supported beam, respectively. The simplified mechanical model of the upper rolls is shown in Figure 3.

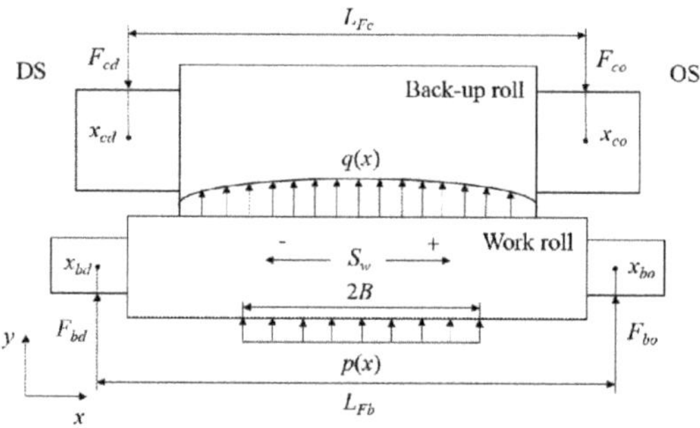

Figure 3. Simplified mechanical model of the upper rolls.

In Figure 3, DS represents the driving side of the mill and OS represents the operating side. x_{cd}, x_{bd} are the positions of the rolling force and bending force of DS. x_{co}, x_{bo} are those of OS. The equilibrium relation in the y direction is as follows:

$$\begin{cases} F_{cd} + F_{co} + \int_{x_{cd}}^{x_{co}} -q(x)dx = 0 \\ \int_{x_{cd}}^{x_{co}} \int_{x_{cd}}^{x_{co}} -q(x)dxdx - F_{co}L_{Fc} + M_{cd} + M_{co} = 0 \end{cases} \tag{1}$$

where $q(x)$ is the distribution of the contact pressure between the rolls, F_{cd} is the rolling force of DS and F_{co} is that of OS, L_{Fc} is the distance between the rolling forces, M_{cd} is the bending moment of DS for back-up roll and M_{co} is that of OS.

For the work roll:

$$\begin{cases} F_{bd} + F_{bo} + \int_{x_{bd}}^{x_{bo}} [p(x) - q(x)]dx = 0 \\ \int_{x_{bd}}^{x_{bo}} \int_{x_{bd}}^{x_{bo}} [p(x) - q(x)]dxdx - F_{bo}(L_{Fb} \pm S_w) + M_{bd} + M_{bo} = 0 \end{cases} \tag{2}$$

where $p(x)$ is the distribution of rolling force, F_{bd} is the bending force of DS and F_{bo} is that of OS, L_{Fb} is the distance between the bending forces, S_w is the shifting value of the work roll in the axial direction. We define the shifting in the direction of OS to be positive. M_{bd} is the bending moment of DS for work roll and M_{bo} is that of OS.

The roll barrel is discretely divided into N slab elements with the same spacing Δx, as shown in Figure 4. The coordinate x_i of the roll barrel is:

$$x_i = i \times \Delta x \quad i = 1, 2, \ldots, N \tag{3}$$

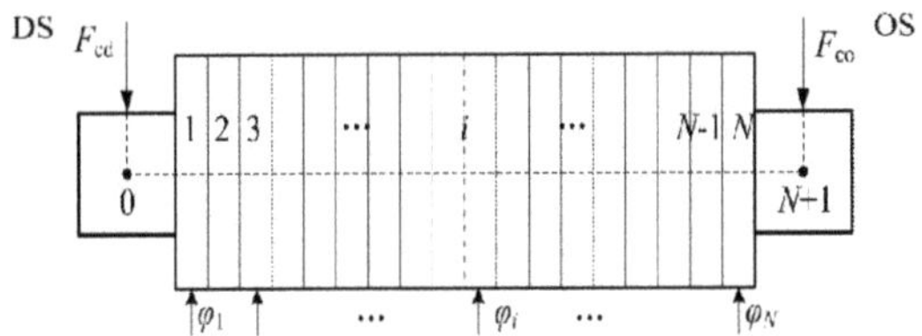

Figure 4. Discretisation of the back-up roll and loads.

According to the bending theory of the beam, the total bending deformation of the roll is expressed by the following equation:

$$y_i = y_M^i + y_Q^i \tag{4}$$

where y_i is the total bending deformation at element i, y_M^i is the deformation at element i caused by the bending moment, y_Q^i is the deformation at element i caused by the shearing force. y_M^i and y_Q^i can be expressed as:

$$\begin{cases} \dfrac{d^4 y_M^i}{dx^4} = \dfrac{\varphi_i}{EI} \\ \dfrac{d^2 y_Q^i}{dx^2} = 2k_{sf}(1 + \mu)\dfrac{\varphi_i}{GS} \end{cases} \tag{5}$$

where φ_i is the concentrated force at element i, E is Young's modulus, I is the polar moment of inertia, k_{sf} is the shear factor of the cylindrical beam, μ is Poisson's ratio, G is the shear elastic modulus, S is the cross-section area of the roll.

According to the boundary conditions of the simplified mechanical model of the rolls and the discretisation of the rolls, the total bending deformation at each element can be derived as:

$$y_{i+1} = y_i + \frac{M_1 + M_2 + \cdots + M_i}{EI}(\Delta x)^2 + \frac{M_{i+1}}{2EI}(\Delta x)^2 + \frac{k_{sf}\varphi_i}{GS}(\Delta x) \tag{6}$$

2.2. Roll Flattening

Based on the Hertz theory, the contact width of an element in the contact zone between the back-up roll and the work roll is assumed to be b:

$$b = \sqrt{\frac{8}{\pi}q\left(\frac{1-\mu^2}{E}\right)\frac{R_1 R_2}{R_1 + R_2}} \tag{7}$$

where q is the specific contact pressure between the rolls at the element, R_1 is the radius of the work roll and R_2 is that of the back-up roll.

As shown in Figure 5, the proximity of the axes of the back-up roll and the work roll caused by the contact flattening is expressed as w:

$$w = \frac{2q\left(1-\mu^2\right)}{\pi E}\left(\frac{2}{3} + \ln\frac{2R_1}{b} + \ln\frac{2R_2}{b}\right) \tag{8}$$

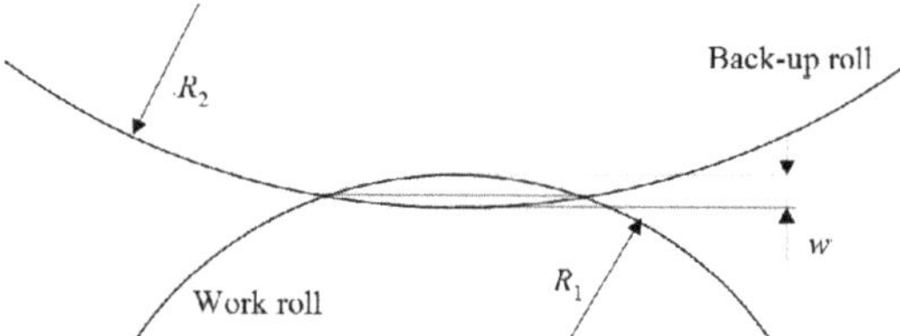

Figure 5. Schematic diagram of contact flattening between the back-up roll and the work roll.

Combined with Equations (7) and (8), this can be obtained as:

$$w = \frac{2q\left(1-\mu^2\right)}{\pi E}\left(\ln\frac{\pi^3 \sqrt{q^2}E}{4(1-\mu^2)} + \ln(2R_1 + 2R_2) - \ln q\right) \tag{9}$$

Because of the coupling relationship between the flattening w and the specific contact pressure q, we fit the functions by the quartic polynomial, respectively. The two expressions are repeatedly iterated and corrected during the calculation of the elastic deformation of the rolls.

$$w = a_0 + a_1 q + a_2 q^2 + a_3 q^3 + a_4 q^4 \tag{10}$$

$$q = b_0 + b_1 w + b_2 w^2 + b_3 w^3 + b_4 w^4 \tag{11}$$

where a_i, b_i are the corresponding polynomial coefficients $i = 0, 1, 2, 3, 4$.

The contact flattening equation of the infinite cylinder is not suitable for the calculation of the flattening between the work roll and the strip. Therefore, the influence function method is used in this paper. The work roll is equidistantly discretised in the direction of the roll barrel according to the above discretisation method. The strip is divided into n slab elements, as shown in Figure 6. The discretised coordinates are expressed as:

$$\begin{cases} n = \frac{B}{\Delta x} \\ x_i = i \cdot \Delta x - \frac{\Delta x}{2} \\ x_j = j \cdot \Delta x - \frac{\Delta x}{2} \end{cases} \tag{12}$$

where B is half the strip width, x_i is the coordinate of the roll barrel at element i, x_j is the coordinate of the strip at element j.

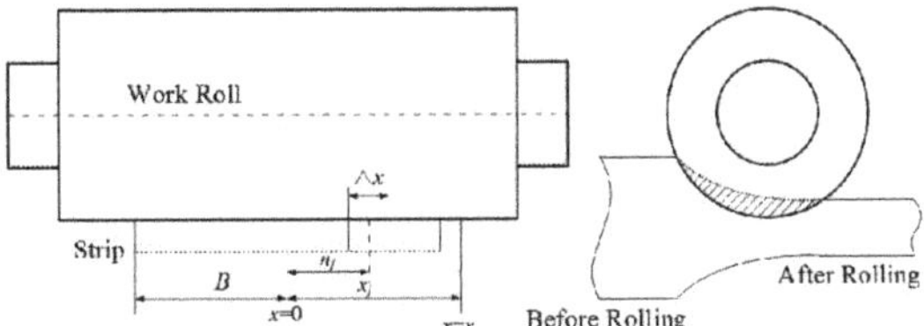

Figure 6. Schematic diagram of contact flattening between the work roll and the strip.

At the coordinate x_i of the roll barrel, the flattening $w_s(x_i)$ between the work roll and the strip is:

$$w_s(x_i) = \frac{(1-\mu^2)}{\pi E} \sum_j^n [a_F]_{ij} p(x_j)$$ (13)

where $p(x_j)$ is the specific rolling force at the coordinate x_j of the strip, $[a_F]_{ij}$ is the influence coefficient matrix of the elastic flattening of the work roll caused by the specific rolling force.

2.3. Strip Material Flow

The strip material flow model is used to calculate the metal flow and stress distribution so as to provide the lateral distribution of the rolling force for the calculation of the roll stack deflection. Figure 7 shows the stress state of the segment of the strip in the deformation zone. The coordinate axes x, y and z correspond to the strip length, thickness and width direction, respectively. The origin o is located at the cross-section centre of the strip passing through the axis of the work roll.

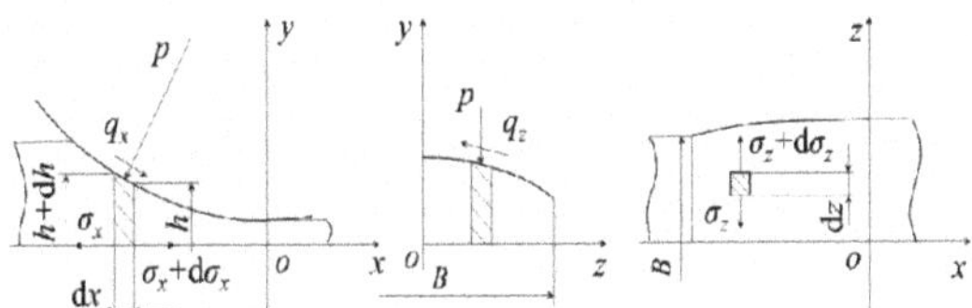

Figure 7. Stress state of the segment of the strip in the deformation zone.

Basic equations include equilibrium differential equation, constitutive equation, the condition of constant volume and the condition of plastic yield. Ignoring the effect of body force, the equilibrium differential equation is:

$$\begin{cases} \dfrac{\partial \hat{\sigma}_x}{\partial \hat{x}} + \dfrac{\partial \hat{\tau}_{xy}}{\partial \hat{y}} + \dfrac{\partial \hat{\tau}_{xz}}{\partial \hat{z}} = 0 \\[2mm] \dfrac{\partial \hat{\tau}_{xy}}{\partial \hat{x}} + \dfrac{\partial \hat{\sigma}_y}{\partial \hat{y}} + \dfrac{\partial \hat{\tau}_{yz}}{\partial \hat{z}} = 0 \\[2mm] \dfrac{\partial \hat{\tau}_{xz}}{\partial \hat{x}} + \dfrac{\partial \hat{\tau}_{yz}}{\partial \hat{y}} + \dfrac{\partial \hat{\sigma}_z}{\partial \hat{z}} = 0 \end{cases}$$ (14)

where $\hat{\sigma}_i$ is the normal stress, $\hat{\tau}_{ij}$ is the shear stress, $i, j = x, y, z$.

For the segment of the contact surface between the strip and the work roll, the equation of its upper surface is:

$$\hat{y} = \frac{1}{2}\hat{h}(\hat{x},\, \hat{z})$$ (15)

where $\hat{h}$ is the strip thickness at the point ($\hat{x},\, \hat{z}$) in the contact surface between the strip and the work roll.

The equilibrium differential equation of the segment of the contact surface is expressed as:

$$\begin{cases} \dfrac{\hat{\sigma}_x}{2}\dfrac{\partial \hat{h}}{\partial \hat{x}} - \hat{\tau}_{xy} + \dfrac{\hat{\tau}_{xz}}{2}\dfrac{\partial \hat{h}}{\partial \hat{z}} + \dfrac{\hat{p}}{2}\dfrac{\partial \hat{h}}{\partial \hat{x}} + \hat{q}_x\dfrac{\left|\overrightarrow{\mathbf{n}}\right|}{\left|\overrightarrow{\mathbf{i}}\right|} = 0 \\[2em] \hat{\sigma}_y - \dfrac{\hat{\tau}_{xy}}{2}\dfrac{\partial \hat{h}}{\partial \hat{x}} - \dfrac{\hat{\tau}_{yz}}{2}\dfrac{\partial \hat{h}}{\partial \hat{z}} + \hat{p} - \dfrac{\hat{q}_x}{2}\dfrac{\partial \hat{h}}{\partial \hat{x}}\dfrac{\left|\overrightarrow{\mathbf{n}}\right|}{\left|\overrightarrow{\mathbf{i}}\right|} - \dfrac{\hat{q}_z}{2}\dfrac{\partial \hat{h}}{\partial \hat{z}}\dfrac{\left|\overrightarrow{\mathbf{n}}\right|}{\left|\overrightarrow{\mathbf{m}}\right|} = 0 \\[2em] \dfrac{\hat{\sigma}_z}{2}\dfrac{\partial \hat{h}}{\partial \hat{z}} - \hat{\tau}_{yz} + \dfrac{\hat{\tau}_{xz}}{2}\dfrac{\partial \hat{h}}{\partial \hat{x}} + \dfrac{\hat{p}}{2}\dfrac{\partial \hat{h}}{\partial \hat{z}} + \hat{q}_z\dfrac{\left|\overrightarrow{\mathbf{n}}\right|}{\left|\overrightarrow{\mathbf{i}}\right|} = 0 \end{cases} \tag{16}$$

where $\overrightarrow{\mathbf{n}}$ is the outward normal vector of the surface, $\overrightarrow{\mathbf{i}}$ is the tangent vector of the friction stress in the x direction and $\overrightarrow{\mathbf{m}}$ is that in the z direction.

The plastic constitutive equation of the strip describes the relationship between plastic stress and strain. The Levy–Mises incremental constitutive theory is used in this paper:

$$\begin{cases} \lambda \hat{s}_{xx} = \dfrac{\partial \hat{u}}{\partial \hat{x}} & \lambda \hat{s}_{xy} = \dfrac{1}{2}\left(\dfrac{\partial \hat{u}}{\partial \hat{y}} + \dfrac{\partial \hat{v}}{\partial \hat{x}}\right) \\[1.2em] \lambda \hat{s}_{yy} = \dfrac{\partial \hat{v}}{\partial \hat{y}} & \lambda \hat{s}_{yz} = \dfrac{1}{2}\left(\dfrac{\partial \hat{v}}{\partial \hat{z}} + \dfrac{\partial \hat{w}}{\partial \hat{y}}\right) \\[1.2em] \lambda \hat{s}_{zz} = \dfrac{\partial \hat{w}}{\partial \hat{z}} & \lambda \hat{s}_{xz} = \dfrac{1}{2}\left(\dfrac{\partial \hat{w}}{\partial \hat{x}} + \dfrac{\partial \hat{u}}{\partial \hat{z}}\right) \end{cases} \tag{17}$$

where $\hat{s}_{ij}$ is the deviator stress, $i, j = x, y, z$; λ is the plastic multiplier factor, which changes with the loading process; $\hat{u}$, $\hat{v}$ and $\hat{w}$ represent the strain rates in the x, y and z directions, respectively.

The equation of constant volume is expressed as:

$$\frac{\partial \hat{u}}{\partial \hat{x}} + \frac{\partial \hat{v}}{\partial \hat{y}} + \frac{\partial \hat{w}}{\partial \hat{z}} = 0 \tag{18}$$

Mises yield criterion is applied in the plastic deformation of the strip:

$$\hat{s}_{xx}^2 + \hat{s}_{yy}^2 + \hat{s}_{zz}^2 + 2\,\hat{s}_{xy}^2 + 2\,\hat{s}_{xz}^2 + 2\,\hat{s}_{yz}^2 = 2\hat{k}_s^2 = \frac{2}{3}\hat{\sigma}_F^2 \tag{19}$$

where $\hat{k}_s$ is the shear strength, $\hat{\sigma}_F$ is the yield strength.

The ratio of the centreline thickness of the strip $\hat{h}_r$ to the length of contact arc $\hat{l}_c$ is defined as the bite ratio δ, as follows:

$$\delta = \frac{\hat{h}_r}{\hat{l}_c} \tag{20}$$

According to the asymptotic analysis method, the variables in the material flow model are normalised:

For the dimension variables:

$$\begin{cases} \hat{x} = x\hat{l}_c \\ \hat{y} = y\hat{h}_r \\ \hat{z} = z\hat{l}_c \end{cases} \tag{21}$$

For the stress variables:

$$\begin{cases} \hat{p} = p\hat{k}_s \\ \hat{s}_{ij} = s_{ij}\delta\hat{k}_s \\ \hat{\sigma}_i = (s_{ii}\delta - p)\hat{k}_s \\ \hat{\tau}_{ij} = \tau_{ij}\delta\hat{k}_s \\ \hat{q}_i = q_i\delta\hat{k}_s \end{cases} \tag{22}$$

For the rate variables:

$$\begin{cases} \hat{u} = u\hat{v}_0 \\ \hat{v} = v\hat{v}_0 \\ \hat{w} = w\hat{v}_0 \end{cases} \tag{23}$$

For the plastic multiplier factor:

$$\hat{\lambda} = \lambda \frac{\hat{v}_0}{\delta \hat{k}_s \hat{l}_c} \tag{24}$$

The equilibrium differential equation after normalisation is:

$$\begin{cases} -\frac{\partial p}{\partial x} + \frac{\partial \tau_{xy}}{\partial y} + \delta \frac{\partial s_{xx}}{\partial x} + \delta \frac{\partial \tau_{xz}}{\partial z} = 0 \\ -\frac{\partial p}{\partial y} + \delta \frac{\partial s_{yy}}{\partial y} + \delta^2 \frac{\partial \tau_{xy}}{\partial x} + +\delta^2 \frac{\partial \tau_{yz}}{\partial z} = 0 \\ -\frac{\partial p}{\partial z} + \frac{\partial \tau_{yz}}{\partial y} + \delta \frac{\partial s_{zz}}{\partial z} + \delta \frac{\partial \tau_{xz}}{\partial x} = 0 \end{cases} \tag{25}$$

The strip thickness is much smaller than the radius of the work roll, so the bite ratio δ is much smaller than one. The first-order asymptotic expansion of the equilibrium equation is deduced, neglecting the small items with δ:

$$\begin{cases} \frac{\partial \tau_{xy}^{(0)}}{\partial y} = \frac{\partial p^{(0)}}{\partial x} \\ \frac{\partial p^{(0)}}{\partial y} = 0 \\ \frac{\partial \tau_{yz}^{(0)}}{\partial y} = \frac{\partial p^{(0)}}{\partial z} \end{cases} \tag{26}$$

Combined with all the basic equations and constraints, the governing equation can be obtained:

$$f_x h^2 \left(\frac{\partial^2 p}{\partial x^2} + \frac{\partial^2 p}{\partial z^2} \right) + 2h f_x \left(\frac{\partial ph}{\partial x^2} + \frac{\partial ph}{\partial z^2} \right) = \frac{(f_x + f_z)v_0}{h^2} \left(\frac{\partial h}{\partial x} + f_z \frac{\partial h}{\partial z} \right) \tag{27}$$

In this paper, the finite volume method is used to solve the partial differential Equation (27). Rectangular element meshing is applied to discretise the deformation zone, as shown in Figure 8.

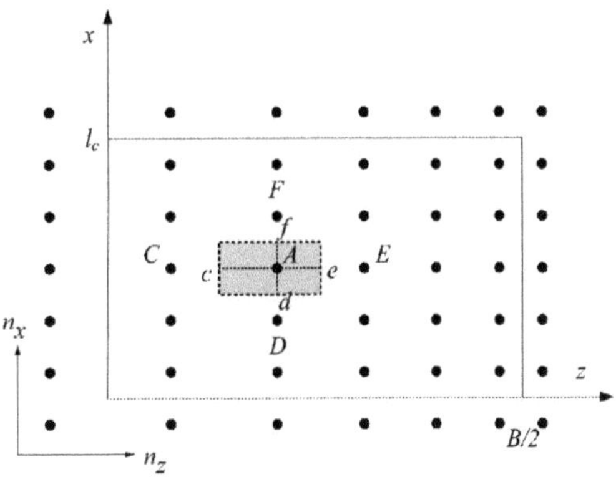

Figure 8. Dispersion of the deformation zone of the strip.

In Figure 8, taking the node A of the shadow region as an example, the points c, d, e and f are the midpoints of the line AC, AD, AE and AF, respectively. The nodes C, D, E and F are the adjacent nodes of the node A in the x and z directions. Based on the finite volume method, Equation (27) is solved in the control volume. According to the Gauss formula, the following equation can be acquired:

$$f_x h^3 \left[S_f \left(\frac{\partial p}{\partial x} \right)_f - S_d \left(\frac{\partial p}{\partial x} \right)_d + S_e \left(\frac{\partial p}{\partial z} \right)_e - S_c \left(\frac{\partial p}{\partial z} \right)_c \right] + 2h^2 f_x \left[S_f \left(\frac{\partial ph}{\partial x} \right)_f \right.$$
$$\left. - S_d \left(\frac{\partial ph}{\partial x} \right)_d + S_e \left(\frac{\partial ph}{\partial z} \right)_e - S_c \left(\frac{\partial ph}{\partial z} \right)_c \right] = (f_x + f_z) v_0 \left(\frac{S_d + S_f}{2} + f_z \frac{S_c + S_e}{2} \right) \tag{28}$$

where S_f is the area of the control volume at the point f in the x direction and S_d is that at the point d in the x direction, S_c is the area of the control volume at the point c in the z direction and Se is that at the point e in the z direction.

Each partial derivative in Equation (28) is obtained by the linear interpolation of stress values at adjacent nodes:

$$\begin{cases} \left(\dfrac{\partial p}{\partial x} \right)_f = \dfrac{P_F - P_A}{dx} & \left(\dfrac{\partial ph}{\partial x} \right)_f = \dfrac{P_F h_F - P_A h_A}{dx} \\[2mm] \left(\dfrac{\partial p}{\partial z} \right)_c = \dfrac{P_A - P_C}{d_{CA}} & \left(\dfrac{\partial ph}{\partial z} \right)_c = \dfrac{P_A h_A - P_C h_C}{d_{CA}} \\[2mm] \left(\dfrac{\partial p}{\partial x} \right)_d = \dfrac{P_A - P_D}{dx} & \left(\dfrac{\partial ph}{\partial x} \right)_d = \dfrac{P_A h_A - P_D h_D}{dx} \\[2mm] \left(\dfrac{\partial p}{\partial z} \right)_e = \dfrac{P_E - P_A}{d_{EA}} & \left(\dfrac{\partial ph}{\partial z} \right)_e = \dfrac{P_E h_E - P_A h_A}{d_{EA}} \end{cases} \tag{29}$$

where P_A, P_C, P_D, P_E and P_F are the rolling forces at nodes A, C, D, E and F, respectively; d_{CA} is the length of the line CA and d_{EA} is that of the line EA.

The large sparse linear equation is built as Equation (30). We used the preprocessing method of the incomplete LU decomposition and the bi-conjugate gradient stabilised method to achieve a fast calculation of Equation (30):

$$\mathbf{Q}_{n \times n} \mathbf{P}_{n \times 1} = \mathbf{R}_{n \times n} \tag{30}$$

where $\mathbf{Q}_{n \times n}$ is the coefficient matrix, $\mathbf{R}_{n \times n}$ is the constant matrix.

2.4. Model Verification

To verify the calculation accuracy of the RDMF model, the last stand of the 1450 mm tandem hot mills in Southwest Stainless Steel Co., Ltd. is taken as the simulation object. The actual parameters of the strip are taken to the RDMF model for numerical simulation and the calculated lateral thickness difference of the strip is compared with the measured one. In addition, considering that the accuracy of FEM is recognised by many researchers, an implicit model of roll-strip coupling is built by using the large-scale commercial finite element software ANSYS [27]. The calculation result of the FEM model is also used to verify the calculation accuracy of the RDMF model. The parameters of the geometrical dimension and the material are listed in Table 1. The FEM model is shown in Figure 9. The element type of the roll stack is SOLID45 and that of the strip is SOLID185. The total number of model elements is 74444, and the number of model nodes is 79185.

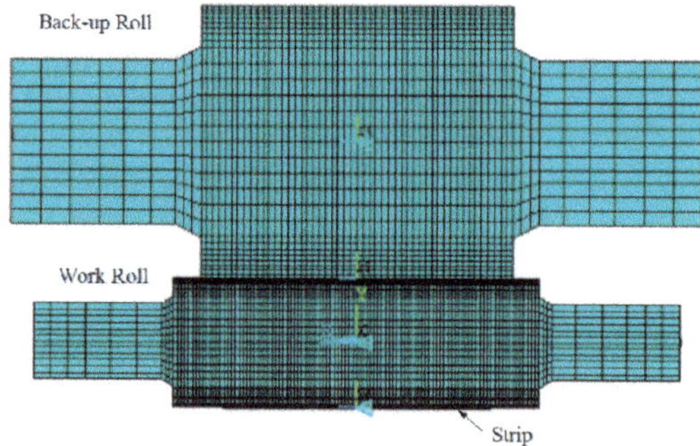

Figure 9. Finite element model of roll-strip coupling.

Table 1. Parameters of the geometrical dimension and the material.

Parameters	Value
Work roll/mm	$\Phi 615 \times 1650$
Distance between the bending forces/mm	2660
Work roll neck diameter/mm	358
Back-up roll/mm	$\Phi 1385 \times 1450$
Distance between the rolling forces/mm	2900
Back-up roll neck diameter/mm	1100
Young modulus of the roll/MPa	2.10×10^5
Poisson's ratio of the roll	0.300
Friction coefficient between the rolls	0.30
Strip width/mm	1240
Strip thickness before rolling/mm	3.1
Alloy code	304-3
Strip temperature/°C	943
Strip yield stress/MPa	269.8
Young modulus of the strip/MPa	1.27×10^5
Poisson's ratio of the strip	0.345
Friction coefficient between roll and strip	0.30

As shown in Figure 10, the distributions of the lateral thickness difference of the strip after rolling are compared. The lateral thickness difference is defined as the difference between the cross section of the strip and the thickness of the strip centre. Taking the midpoint of the strip width as an origin, the normalised coordinate values of the strip width are used as the abscissa axis. It can be found that the calculation results of the RDMF model and the FEM model are basically consistent with the measured distribution of the lateral thickness difference. Without the effect of the edge-drop region, the relative error of the RDMF model is 10.61%, and that of the FEM model is 6.69%. The calculation error of the strip edge may be caused by the friction condition and deformation state in the edge region, but it has a limited effect on the calculation of the strip cross section. Therefore, the calculation accuracy of the RDMF model and the FEM model are able to meet the requirements of actual production. However, the RDMF model has a better solving speed. In the same configuration of calculation (CPU: Intel Core2 Quad Q6600, RAM: 2GB), the RDMF model takes 0.005 s to calculate, while it takes 4.55 h for the FEM model. Because of the complex grade and various specification of the strip in the hot-rolled line, the simulation workload is huge. In the case of ensuring the calculation accuracy and reducing calculation time, the RDMF model proposed in this paper is more practical for the numerical calculation in the hot rolling process.

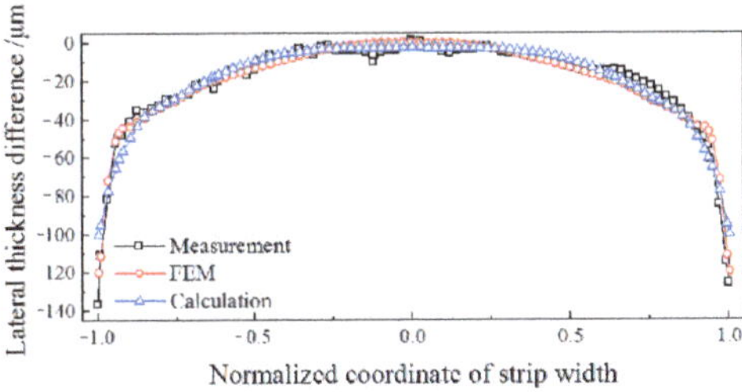

Figure 10. Comparison of the lateral thickness difference.

3. Results and Discussion

Using the proposed RDMF model, a lot of simulated calculations are carried out with different working conditions. Based on the calculation results of the RDMF model, the effects of process parameters on the quarter buckle are analysed. According to the actual production of the last stand of the tandem hot mill, the basic working condition is determined as the control group, as listed in Table 2.

Table 2. Parameters of the basic working condition.

Parameters	Value
Rolling force/kN	10,720
Bending force/kN	184
Back-up roll chamfer/mm	150 × 1.0
Roll contour type of work roll	Quadratic parabola
Roll contour amount of work roll/μm	−50
Strip width/mm	1240
Strip thickness before rolling/mm	3.1
Strip thickness after rolling/mm	2.8
Strip lateral temperature difference/°C	50
Quadratic crown before rolling/μm	44.3
Quartic crown before rolling/μm	0

3.1. Effect of Bending Force

The bending system of the work roll is an on-line adjusting method for controlling the flatness. To study the effect of the change of bending force on the quarter buckle, the distributions of the longitudinal strain differences of the strip are calculated during the increase of the bending force from −16 kN to 384 kN based on the control group, as shown in Figure 11.

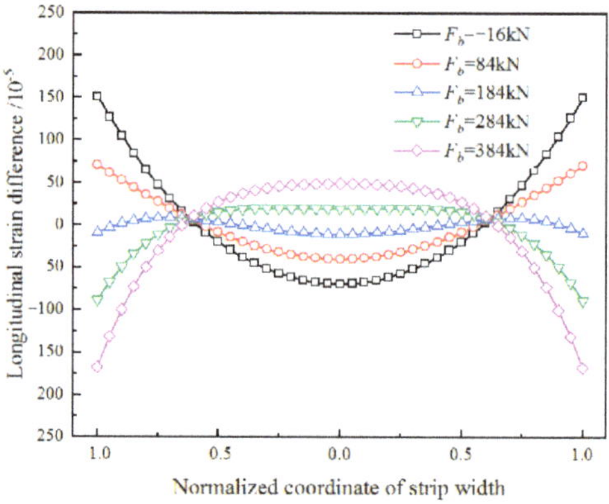

Figure 11. Longitudinal strain difference with different bending forces.

To parameterise the distribution characteristic of the longitudinal strain difference, the curve is fitted to a quartic polynomial, as shown below:

$$\Delta\varepsilon(x) = c_0 + c_2 x^2 + c_4 x^4 \quad x \in [-1,\, 1] \tag{31}$$

where x is the normalised coordinate of the strip width; c_i is the polynomial coefficient, $i = 0, 2, 4$.

According to the Chebyshev polynomial expansion, Equation (31) can also be expressed as:

$$\Delta\varepsilon(x) = d_0 + d_2 T_2(x) + d_4 T_4(x) \tag{32}$$

where d_i is the Chebyshev polynomial coefficient, $i = 0, 2, 4$; $T_2(x)$ is the Chebyshev quadratic component; $T_4(x)$ is the Chebyshev quartic component.

According to Equations (31) and (32), the Chebyshev polynomial coefficients can be expressed as:

$$\begin{cases} d_0 = c_0 + \frac{1}{2}c_2 + \frac{3}{8}c_4 \\ d_2 = \frac{1}{2}(c_2 + c_4) \\ d_4 = \frac{1}{8}c_4 \end{cases} \tag{33}$$

In the distribution of the longitudinal strain difference, the convexity of the quadratic component is defined as $\Delta\varepsilon_2$ and the convexity of the quartic component is $\Delta\varepsilon_4$:

$$\begin{cases} \Delta\varepsilon_2 = d_2 T_2(0) - \frac{1}{2}[d_2 T_2(1) + d_2 T_2(-1)] = -(c_2 + c_4) \\ \Delta\varepsilon_4 = \frac{1}{2}\left[d_4 T_4\left(\frac{\sqrt{2}}{2}\right) + d_4 T_4\left(-\frac{\sqrt{2}}{2}\right)\right] - d_4 T_4(0) = -\frac{c_4}{4} \end{cases} \tag{34}$$

The values of $\Delta\varepsilon_2$ and $\Delta\varepsilon_4$ with different bending forces are shown in Figure 12. As the bending force increases, both of them increase simultaneously. When the value of $\Delta\varepsilon_2$ is close to zero and the value of $\Delta\varepsilon_4$ is large, there is the mode of a quarter buckle. When the bending forces are 284 and 384 kN, the distributions of longitudinal strain difference are in the mode of a centre buckle. The values of $\Delta\varepsilon_2$ are 108.2×10^{-5} and 216.7×10^{-5}, the values of $\Delta\varepsilon_4$ are also large. When the bending forces are -16 and 84 kN, the distributions of longitudinal strain difference are in the mode of an edge wave. In addition, the values of $\Delta\varepsilon_2$ are -219.9×10^{-5} and -110.5×10^{-5}, while the values of $\Delta\varepsilon_4$ are much smaller than that of the centre buckle mode. When the bending force is changed from -16 kN to 384 kN, the quarter buckle mode occurs in the change from the edge wave mode to the centre buckle mode. This is similar to the situation in production: when a quarter buckle appears, increasing the bending force will turn it into a centre buckle, while reducing the bending force will turn it into an edge wave.

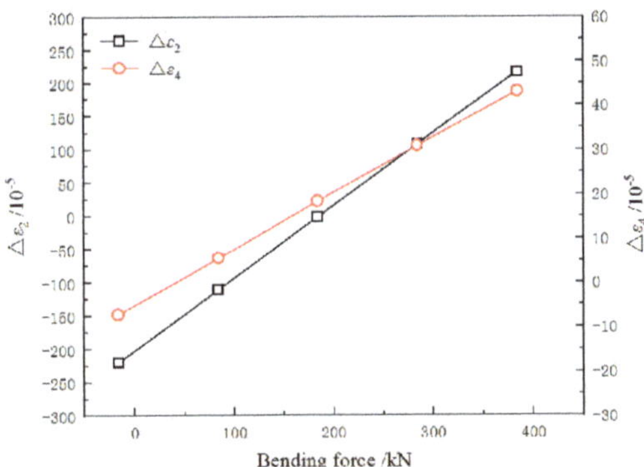

Figure 12. The values of $\Delta\varepsilon_2$ and $\Delta\varepsilon_4$ with different bending forces.

It is difficult to improve the quarter buckle solely by the bending force. When studying the effect of other parameters on the quarter buckle, it is necessary to decouple the quadratic wave and the quarter buckle, namely, by adjusting the bending force to eliminate the quadratic wave, and then studying the change of the $\Delta\varepsilon_4$.

3.2. Effect of Strip Lateral Temperature Difference

In hot rolling, the temperature of the strip edge is lower than the middle. Because the thermal conductivity of stainless steel is poor, the strip lateral temperature difference (ΔT) is usually greater than that of plain carbon steel. The uneven temperature distribution of the strip makes the distribution of the rolling force uneven, which is related to the generation of the quarter buckle. The lateral temperature distribution of the strip in the last stand, detected by a thermal imager, is shown in Figure 13. The temperature difference between the middle and the edge of the strip is 50 °C. By keeping the measured value of the temperature in the middle of the strip constant and decreasing the temperature value at the edge of the strip, the distribution curves with lateral temperature differences of 75 °C and 100 °C were obtained as control groups. In Figure 14, due to the different distribution of the lateral temperature, the deformation resistance of the strip in the width direction is also different.

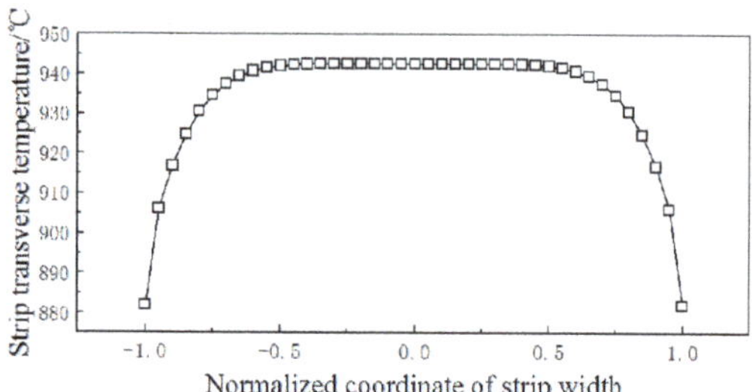

Figure 13. Lateral temperature distribution of the strip in the last stand measured by a thermal imager.

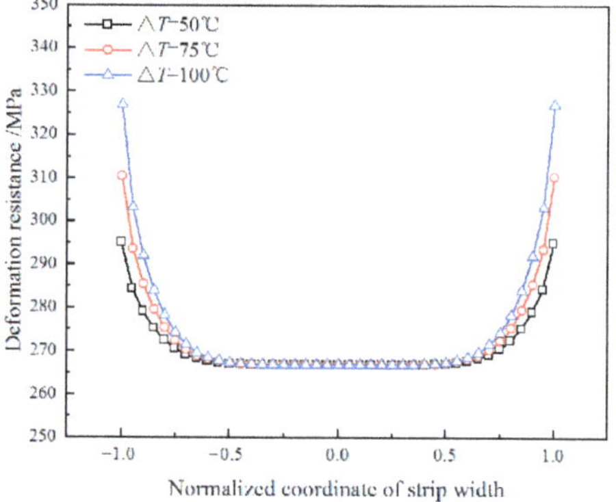

Figure 14. Distribution of the deformation resistance with difference values of ΔT.

In the simulation of the rolling process, the value of $\Delta \varepsilon_2$ is close to zero by changing the bending force. As the ΔT of the strip increases, the bending force required to improve the quadratic wave decreases. The bending force and the $\Delta \varepsilon_4$ are shown in Figure 15. With an increase in ΔT, the value of $\Delta \varepsilon_4$ increases linearly. When ΔT increased by 10 °C, and the value of $\Delta \varepsilon_4$ increased by 3.1×10^{-5}. However, the change of bending force is very small with different temperature differences.

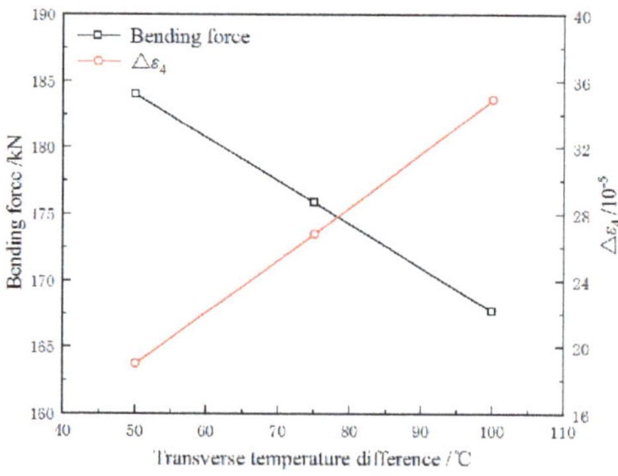

Figure 15. Bending force and $\Delta \varepsilon_4$ of the strip with different values of ΔT.

3.3. Effect of Rolling Force

Research into the effect of rolling force on the quarter buckle can provide a basis for the optimisation of load distribution. With different distributions of lateral temperature, the effect of rolling force on the quarter buckle is considered. The ΔT of the strip is selected to be 50 °C, 75 °C and 100 °C, respectively. Taking 1000 kN as the interval, two rolling force control groups are selected above and below the rolling force of 10,720 kN in Table 2. The calculation results are shown in Figures 16 and 17.

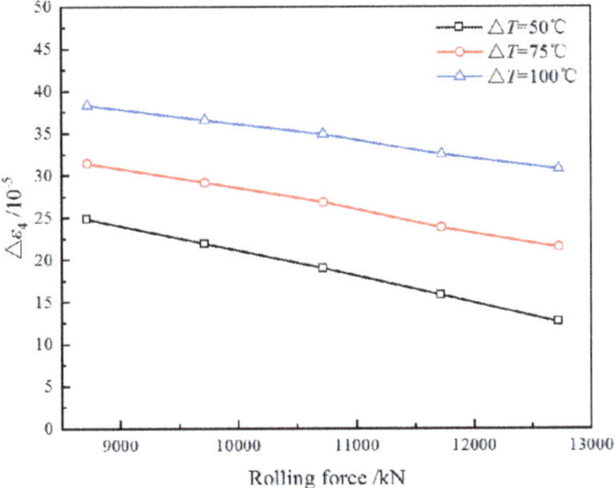

Figure 16. The $\Delta \varepsilon_4$ of the strip with different rolling forces.

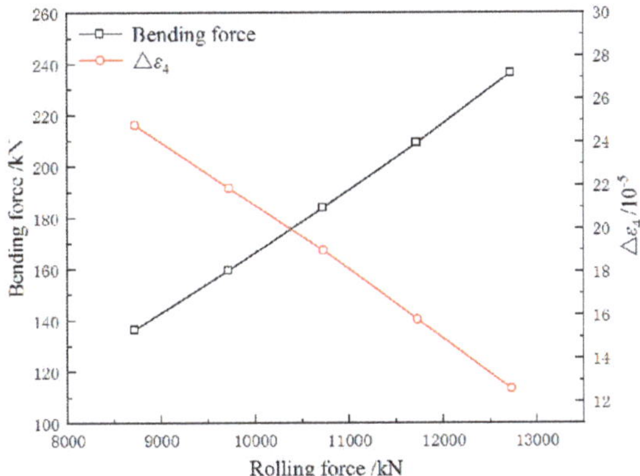

Figure 17. Bending force and the $\Delta \varepsilon_4$ of the strip with different rolling forces (ΔT = 50 °C).

The results show that at the same ΔT the value of $\Delta \varepsilon_4$ decreases linearly with the rolling force increasing, but the bending force required to eliminate the quadratic wave increases gradually. The effect of rolling force on the reduction of the quarter buckle is different with different values of ΔT. When the ΔT is relatively small, as the rolling force increases, the bending force required to improve the quadratic wave is increased, but the effect of reducing the quarter buckle is more obvious as a whole.

3.4. Effect of Strip Quartic Crown before Rolling

The key to the shape control of hot-rolled strips is the change of the strip crown distribution before and after rolling. To study the change rule of the quarter buckle of the strip, the effect of the strip crown before rolling cannot be ignored. The strip crown before rolling can be divided into the quadratic crown and the quartic crown. The former is related to the quadratic wave of the strip, and the latter is related to the high-order wave [23]. In this paper, the effect of the strip quartic crown before rolling on the quarter buckle is studied. The quartic crown is set to 0, 2, 4, 6 and 8 μm. Other parameters remain unchanged. The calculation results are shown in Figures 18 and 19.

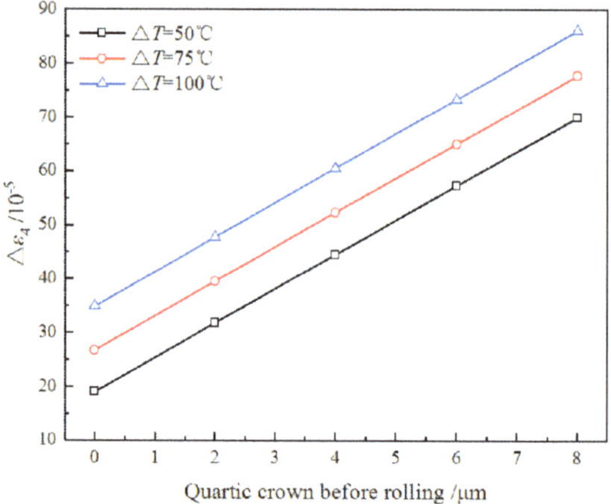

Figure 18. The value of $\Delta\varepsilon_4$ of the strip with different quartic crowns before rolling.

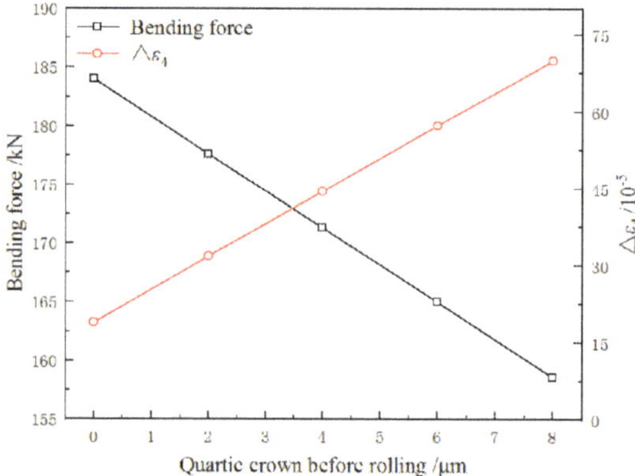

Figure 19. Bending force and $\Delta\varepsilon_4$ of the strip with different quartic crowns before rolling ($\Delta T = 50$ °C).

At the same ΔT, the value of $\Delta\varepsilon_4$ gradually increases with the increase of the quartic crown. With different values of ΔT, as the quartic crown increases, the values of $\Delta\varepsilon_4$ increase at almost the same rate. This indicates that with different values of ΔT, the effect of the quartic crown on the quarter buckle of the strip is only different in the basic value. The magnitude of the quartic crown can reflect the severity of the quarter buckle to a certain extent. According to the comparison of the calculation results, it is found that when the quartic crown changes, the bending force does not change significantly. It has a great impact on the quarter buckle.

3.5. Effect of Back-Up Roll Chamfer Length

The chamfer of the back-up roll helps to reduce the harmful contact zone between the rolls, and affects the deflection of the roll stack. The chamfer length is an important parameter for the back-up roll contour, which affects the range of the contact zone between the rolls. The chamfer length of the back-up roll is selected to be 50, 100, 150, 200 and 250 mm. The effect on the quarter buckle are studied through simulation.

It can be seen in Figure 20 that at the same ΔT, the $\Delta \varepsilon_4$ decreases in a parabola with the increase of the chamfer length, which reduces the possibility of a quarter buckle. At the same time, increasing the chamfer length will reduce the deflection of the roll stack, thereby reducing the strip crown. To compensate for this effect and avoid the quadratic wave, the bending force is significantly reduced, as shown in Figure 21. Therefore, the reduction of $\Delta \varepsilon_4$ in the calculation results from the combined effects of increasing the chamfer length and reducing the bending force.

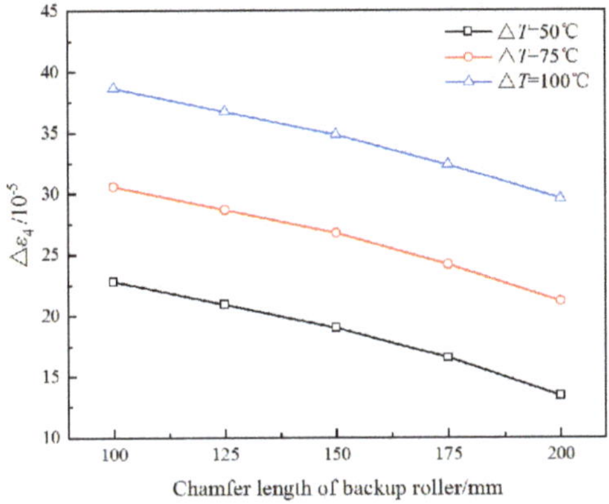

Figure 20. The value of $\Delta \varepsilon_4$ of the strip with different chamfer lengths of back-up roll.

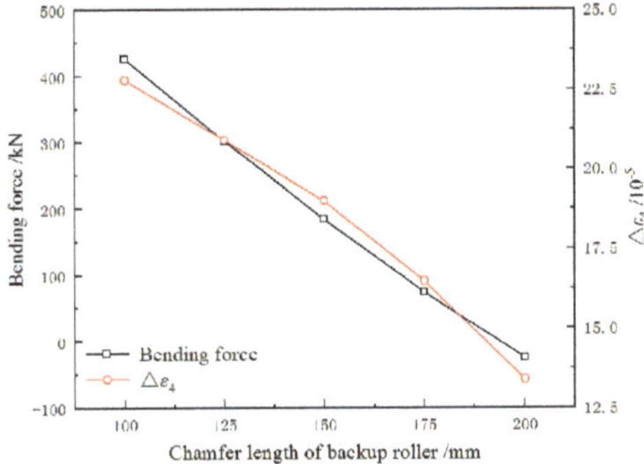

Figure 21. Bending force and $\Delta \varepsilon_4$ of the strip with different chamfer lengths of back-up roll ($\Delta T = 50\ ^\circ$C).

4. Design of MVC and Industrial Application

Aiming at the quarter buckle of hot-rolled stainless steel, a new technology of work roll contour is developed in this paper. The MVC designed by the superposition of the quadratic curve and the sextic curve not only causes the work roll to have the ability to control the quadratic wave, but also improves the quarter buckle. The coefficient of the quadratic curve is designated by the magnitude of the quadratic wave. The coefficients of the sextic curve are determined by the position and magnitude of the quarter buckle. Then, they will be superimposed in different regions. Assuming the strip width is $2B$ and the roll barrel is $2L$, the curve equation of MVC is:

$$y(x) = \begin{cases} e_2 x^2 + \left(f_2 x^2 + f_4 x^4 + f_6 x^6 \right) & x \in [-B,\ B] \\ e_2 x^2 & x \in [-L,\ B) \cup (B,\ L] \end{cases} \tag{35}$$

where e_2 is the coefficient of the quadratic curve; f_i is the coefficient of the sextic curve, $i = 2, 4, 6$.

Figure 22a is a diagram of the compensation for the quarter buckle by the roll contour of MVC. To determine the coefficients of the sextic curve, Equation (36) needs to be satisfied. The contour of the sextic curve is shown in Figure 22b. The key is to determine the values of the extreme point x_0 and the compensation y_0. The position x_0 is generally determined by the position of the quarter buckle, and the compensation y_0 is optimised based on the RDMF model.

$$\begin{cases} f_2 x_0{}^2 + f_4 x_0{}^6 + f_6 x_0{}^6 = y_0 \\ 2f_2 + 4f_4 x_0{}^2 + 6f_6 x_0{}^4 = 0 \\ f_2 B^2 + f_4 B^4 + f_6 B^6 = 0 \end{cases} \tag{36}$$

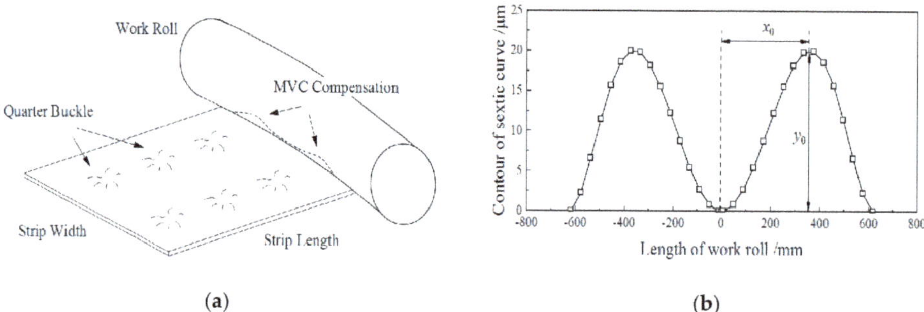

Figure 22. Compensation for the quarter buckle by the MVC: (**a**) schematic diagram; (**b**) contour of the sextic curve.

Taking the quarter buckle problem (Figure 1) of the 1450 mm hot rolling line in Southwest Stainless Steel as an example, we use the MVC technology to optimise the original contour of the work roll (quadratic parabola, −200 μm). The roll contours before and after the optimisation are depicted in Figure 23.

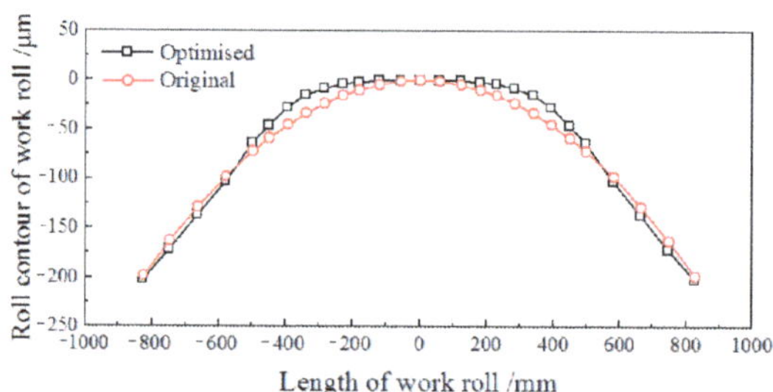

Figure 23. Roll contours before and after optimisation.

According to the actual production, the changes of the $\Delta\varepsilon_4$ with different values of the quartic crown before and after the optimisation are listed in Table 3. When the original contour is used, the $\Delta\varepsilon_4$ increases from 34.9 to 86.2. However, when the optimised contour is used, as the quartic crown of the strip increases, the bending force for adjusting the quadratic wave is little changed, and the $\Delta\varepsilon_4$ can be kept relatively small. To improve the quarter buckle problem, the calculation results show that the optimised contour needs the compensation value of 2.71 to 6.74 μm.

Table 3. Changes of the $\Delta\varepsilon_4$ under different values of the quartic crown before and after the optimisation ($\Delta T = 100\ °C$).

Quartic Crown/µm	0	2	4	6	8
$y_0/\mu m$	2.71	3.75	4.74	5.74	6.74
Bending force/kN	175	172	168	164	161
$\Delta\varepsilon_4$ (Optimised)/10^{-5}	−0.1	−0.7	−0.7	−0.7	−0.6
$\Delta\varepsilon_4$ (Original)/10^{-5}	34.9	47.8	60.6	73.4	86.2

In Figure 24a, the cross section of the strip before and after the optimisation is compared. Using the optimised contour, the cross section of the strip is compensated in the region of the quarter buckle, thereby reducing the relative longitudinal strain of the strip in the region. Correspondingly, in Figure 24b, the relative rolling force of the strip in the region is reduced, and the overall distribution of the rolling force is more uniform, which is beneficial for achieving a more uniform reduction.

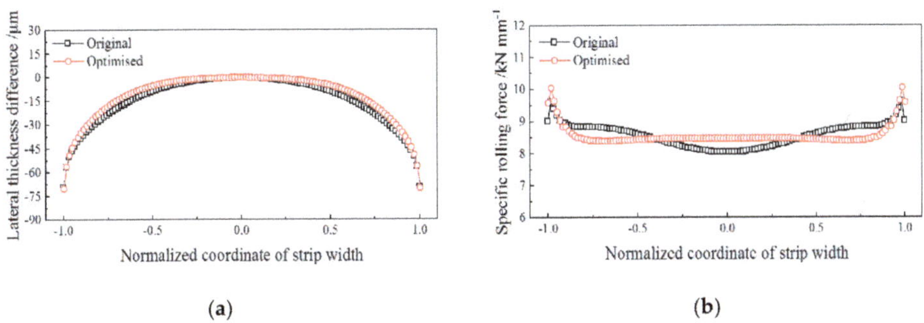

(a)　　　　　(b)

Figure 24. Comparison of calculation results before and after the optimisation ($\Delta T = 100\ °C$, quartic crown is 8 µm): (**a**) lateral thickness difference of the strip; (**b**) specific rolling force.

By comparing the strip flatness detected by the flatness meter before and after the optimisation, the effectiveness of the MVC technology is verified. Figure 25a shows the quarter buckle before solving the problem. With the optimised contour, the quarter buckle of the strip is effectively improved, as shown in Figure 25b. In addition, the research in this paper was successfully applied to other hot rolling lines, as listed in Table 4.

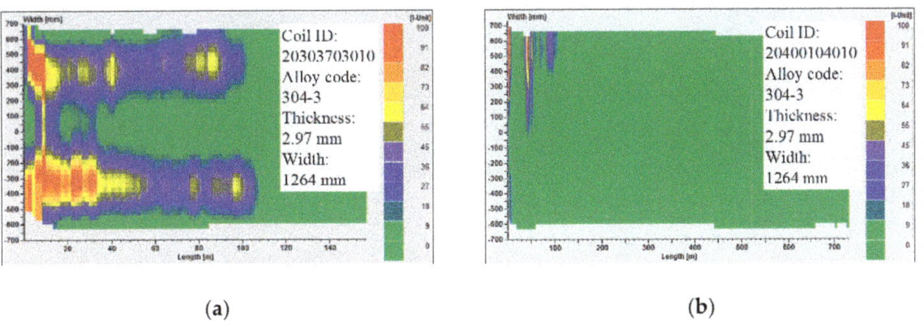

(a)　　　　　(b)

Figure 25. Comparison of the strip flatness detected by the on-line meter: (**a**) before optimisation; (**b**) after optimisation.

Table 4. Other hot rolling lines with a successful application.

Case	Hot Rolling Line
Beibu Gulf New Material Co., Ltd., Guangxi, China	1700 mm
Delong Nickel Industry Co., Ltd., Jiangsu, China	1450 mm
Dingxin Technology Co., Ltd., Fujian China	1780 mm
Guangqing Metal Rolling Co., Ltd., Guangdong, China	1780 mm
Qingshan Stainless Steel Co., Ltd., Indonesia	1780 mm

5. Conclusions

To solve the problem of the quarter buckle in the production of hot-rolled strips, the change rule and control technology of the quarter buckle are studied in this paper. The conclusions are as follows:

(1) A model of roll deflection and material flow for predicting the quarter buckle is established. Compared with the finite element method and the measured data, the accuracy of the RDMF model is verified, and is able to meet the requirements of actual production. However, the RDMF model is much faster in terms of calculation speed than the finite element model, which is suitable for online application and performing a large number of offline calculations.

(2) Quantitative analyses of the effects of the shape process parameters on the quarter buckle are carried out using the RDMF model. The coupling relationship between the quarter buckle and the quadratic wave increases the difficulty of the quarter buckle control. By decoupling analysis, it is found that the lateral temperature difference of the strip and the quartic crown of the strip before rolling have a great impact on the quarter buckle, but their effects on the quadratic wave are small. In addition, increasing the rolling force and the chamfer length of the back-up roll could reduce the quarter buckle, but this would also obviously change the bending force and affect the control of the quadratic wave.

(3) A new control technology for the work roll contour, MVC, is developed for the quarter buckle. The roll contour of MVC improves the strip shape by compensating the loaded roll gap profile at the position of the quarter buckle. It not only has the ability to control the quadratic wave, but also effectively improves the quarter buckle. The roll contour of MVC proved to be effective in industrial applications.

Author Contributions: H.L. and C.Y. carried out the simulation work; A.H. and Z.Z. conceived and designed the industrial verification of the model; J.S. and A.H. developed the work roll contour; H.L. and C.Y. analysed the data. J.S., W.L. and Z.Z. performed the industrial applications. H.L. wrote the paper. All authors have read and agreed to the published version of the manuscript.

Funding: This research was funded by National Natural Science Fund of China, grant number 51674028; Innovative Method Project of Ministry of Science and Technology of China, grant number 2016IM010300; and Guangxi Special Funding Program for Innovation-Driven Development, grant number GKAA17202008.

Acknowledgments: This work was supported by the National Natural Science Fund of China (No.51674028); the Innovative Method Project of Ministry of Science and Technology of China (2016IM010300); and the Guangxi Special Funding Program for Innovation-Driven Development (GKAA17202008).

References

1. He, A.R.; Shao, J.; Sun, W.Q. *Theory and Practice of Shape Control*, 1st ed.; Metall Ind Press: Beijing, China, 2016; pp. 11–15, ISBN 9787502473679.
2. Zhang, Q.D.; Li, B.; Zhang, X.F. Research on the behavior and effects of flatness control in strip temper rolling process. *J. Mech. Eng.* **2014**, *50*, 45–52. [CrossRef]
3. Du, F.S.; Liu, W.W.; Feng, Y.F.; Sun, J.N. Roll profile electromagnetic control process parameters in precision rolling mill. *J. Univ. Sci. Technol. Beijing* **2017**, *39*, 1874–1880. [CrossRef]
4. Li, X.Y.; Zhang, J.; Cheng, X.L.; Zhao, X.M.; Jia, S.H.; Huang, T. Wear of rolls in single-stand temper mill and its effect on the strip shape. *J. Univ. Sci. Technol. Beijing* **2002**, *24*, 326–328. [CrossRef]

5. He, A.R.; Shao, J.; Sun, W.Q.; Song, Y. Key precise control technologies of rolling for smart manufacturing. *Metall. Ind. Autom.* **2016**, *40*, 1–8. [CrossRef]

6. Jiang, Z.Y.; Tieu, A.K.; Zhang, X.M. Finite element simulation of cold rolling of thin strip. *J. Master. Process. Technol.* **2003**, *140*, 542–547. [CrossRef]

7. Chandra, S.; Dixit, U.S. A rigid-plastic finite element analysis of temper rolling process. *J. Mater. Process. Technol.* **2004**, *152*, 9–16. [CrossRef]

8. Wang, X.D.; Li, F.; Wang, L. Development and application of roll contour configuration in temper rolling mill for hot rolled thin gauge steel strip. *Ironmak. Steelmak.* **2012**, *39*, 163–170. [CrossRef]

9. Kong, F.F.; He, A.R.; Shao, J. Finite element model for rapidly evaluating the thermal expansion of rolls in hot strip mills. *J. Univ. Sci. Technol. Beijing* **2014**, *36*, 674–679. [CrossRef]

10. Kim, K.S.; Hong, W.K.; Frédéric, B. Effect of rolling parameters on surface strain variation in hot strip rolling. *Steel Res. Int.* **2017**, *88*, 1600492. [CrossRef]

11. Hao, P.J.; He, A.R.; Sun, W.Q. Formation mechanism and control methods of inhomogeneous deformation during hot rough rolling of aluminum alloy plate. *Arch. Civ. Mech. Eng.* **2018**, *18*, 245–255. [CrossRef]

12. Jiang, Z.Y.; Wei, D.T.; Tieu, A.K. Analysis of cold rolling of ultra-thin strip. *J. Mater. Process. Technol.* **2009**, *209*, 4584–4589. [CrossRef]

13. Hao, P.J.; He, A.R.; Sun, W.Q. Predicting model of thickness distribution and rolling force in angular rolling process based on influence function method. *Mech. Ind.* **2018**, *19*, 302. [CrossRef]

14. Wang, D.C.; Wu, Y.L.; Liu, H.M. High-efficiency calculation method for roll stack elastic deformation of four-high mill. *Iron Steel.* **2015**, *50*, 69–74. [CrossRef]

15. Wang, T.; Xiao, H.; Zhao, T.Y. Improvement of 3-d fem coupled model on strip crown in hot rolling. *J. Iron Steel Res. Int.* **2012**, *3*, 17–22. [CrossRef]

16. Lian, J.C.; Qi, X.D. *Theory of Strip Rolling and Shape Control*, 1st ed.; China Mach. Press: Beijing, China, 2013; pp. 88–95, ISBN 9787111409724.

17. Zhang, G.; Xiao, H.; Wang, C. Three-dimensional model for strip hot rolling. *J. Iron Steel Res. Int.* **2006**, *13*, 23–26. [CrossRef]

18. Shao, J.; Yao, C.; Chen, C.; Sun, W.Q.; He, A.R. A Rapid online calculation method of three-dimensional plastic deformation in strip rolling. *Int. J. U E-Serv. Sci. Technol.* **2016**, *9*, 151–162. [CrossRef]

19. Ma, X.B.; Wang, D.C.; Liu, H.M. Coupling mechanism of control on strip profile and flatness in single stand universal crown reversible rolling mill. *Steel Res. Int.* **2017**, *88*, 1600495. [CrossRef]

20. Guo, X.Y.; He, A.R.; Shao, J.; Zhou, B.; Li, Q.L. Modeling and simulation of subsectional cooling system during hot aluminum rolling. *J. Mech. Eng.* **2013**, *4*, 74–78. [CrossRef]

21. Wang, Z.P. Research on the Local Heat Transfer Based on the Stepped Cooling of the Rolls. Master's Dissertation, Yanshan University, Qinhuangdao, China, 2017.

22. Li, Q.S.; Xu, J.Y.; Zhou, J.G. BURS roll designed for non-quadratic waves. In Proceedings of the 2005 China Iron and Steel Annual Conference, Beijing, China, 1 October 2005; pp. 330–336.

23. Li, H.B.; Zhang, J.; Cao, J.G.; Cheng, F.W.; Hu, W.D.; Zhang, Y. Roll contour and strip profile control characteristics for quantic CVC work roll. *J. Mech. Eng.* **2012**, *48*, 24–30. [CrossRef]

24. Seilinger, A.; Mayrhofer, A.; Kainz, A. SmartCrown—A new system for improved profile and flatness control in strip mills. *Steel Times Int.* **2003**, *26*, 11–12.

25. Hara, K.; Yamada, T.; Takagi, K. Shape controllability for quarter buckles of strip in 20-high Sendzimir mills. *ISIJ Int.* **1991**, *31*, 607–613. [CrossRef]

26. Kubo, T.; Aizawa, A.; Hara, K.; Uchihata, O. Development of high-precise shape control technology in 20-high Sendzimir mills. *Metall. Res. Technol.* **2006**, *103*, 507–513. [CrossRef]

27. *ANSYS Version 14.0*, ANSYS Inc.: Pittsburgh, PA, USA.

Article

Finite Element Analysis on Ultrasonic Drawing Process of Fine Titanium Wire

Shen Liu, Xiaobiao Shan, Hengqiang Cao and Tao Xie *

School of Mechatronics Engineering, Harbin Institute of Technology, Harbin 150001, China;
joseliu2013@outlook.com (S.L.); shanxiaobiao@hit.edu.cn (X.S.); yzhinv@outlook.com (H.C.)
* Correspondence: xietao@hit.edu.cn; Tel.: +86-451-8641-7891; Fax: +86-451-8641-6119

Received: 31 March 2020; Accepted: 23 April 2020; Published: 28 April 2020

Abstract: Ultrasonic drawing is a new technology to reduce the cross-section of a metallic tube, wire or rod by pulling through vibrating dies. The addition of ultrasound is beneficial for reducing the drawing force and enhancing the surface finish of the drawn wire, but the underlying mechanism has not been fully understood. In this paper, an axisymmetric finite element model of the single-pass ultrasonic drawing was established in commercial FEM software based on actual wire length. The multi-linear kinematic hardening (MKINH) model was used to define the elastic and plastic characteristics of titanium. Influences of ultrasonic vibration on the drawing process were investigated in terms of four factors: location of the die, ultrasonic amplitude, drawing velocity, and friction coefficient within the wire-die contact zone. Mises stresses, as well as contact and friction stress, in conventional and ultrasonic drawing conditions, were compared. The results show that larger ultrasonic amplitude and lower drawing velocity contribute to greater drawing force reduction, which agrees with former research. However, their effectiveness is further influenced by the location of the die. When ultrasonic amplitude and drawing speed remain unchanged, the drawing force is minimized when the die locates at the half-wavelength position, while maximized at the quarter-wavelength position.

Keywords: ultrasonic drawing; titanium wire; drawing force; finite element analysis; Mises stress; contact stress; work hardening; numerical simulation

---

1. Introduction

Titanium wires are widely used in various industries, including aerospace, automobile, biomedicine, petrochemistry, fishery, due to their exceptional characteristics, such as high strength, lightweight, good corrosion resistance, excellent biocompatibility, etc. [1–4]. However, unlike many other metallic wires, the manufacturing of titanium wire is usually conducted at elevated temperature, as their cold workability is degraded by high yield stress to tensile strength (Y/T) ratio and the strain hardening phenomenon [5]. The conventional drawing process involves complicated heat preparation, lubrication, and surface treatment, which is neither eco-friendly nor energy-saving. What is more, the wire products often have low dimensional accuracy, poor surface finish, and high breakage ratio [6]. Ultrasonic drawing is a metalworking process to reduce the diameter of a wire, tube, or rod by pulling them through oscillating dies with converging cross-section shapes [7]. Compared with the conventional drawing process, the new technology has the following advantages: reduced drawing force, enhanced surface finish, lower breakage ratio, prolonged tool life, greater area reduction per pass, etc. Therefore, ultrasonic drawing is a promising technology to increase productivity, lower costs, and improve the quality of the products in the industrial manufacturing of metallic wires [8]. The benefits associated with ultrasound are not limited to titanium and the drawing process, but to varied machining processes, such as welding, forging, rolling, and cutting of a wide range of metallic and non-metallic materials like aluminum, steel, and carbon fiber [9–11].

The idea of introducing ultrasonic vibrations into the drawing process was first reported by Blaha et al. in 1955 [12]. They discovered the acoustic softening phenomenon, which is similar to thermal softening effects, with an abrupt 40% reduction in drawing force when ultrasound was superimposed. The observations were explained as the result of activation and increased mobility of dislocations at the lattice defects. In 1968, Winsper et al. reviewed various experimental results on ultrasonic-assisted metal forming processes and pointed out that all these phenomena could be explained by the superposition of an alternating load on a static load, which is later known as stress superposition hypothesis [13]. The above theories triggered numerous studies on the influence of ultrasound in metal forming processes, which remains active to the present day. In 2003, Hayashi et al. compared the influence of longitudinal and radial ultrasonic vibrations on the aluminum wire drawing process using the finite element method (FEM) [14]. It was found that radial vibration yielded superior results than longitudinal vibration with respect to drawing force reduction, surface quality improvement, and critical drawing speed. In 1999, Susan et al. conducted the ultrasonic drawing experiment of steel ball-bearing steel wire. They attributed the drawing force reduction to the surface effect of ultrasound utilizing the friction reversion mechanism [15]. The friction coefficient and the drawing force were calculated according to the relations between three factors: drawing velocity, ultrasonic frequency, and amplitude of the die. In 2004, the proposed calculation method was further revised to calculate the drawing force in the ultrasonic-assisted tube drawing process [16]. They also pointed out that ultrasonic vibration contributes to weakening the strain hardening effect and helps to increase the plasticity of the specimen. In 2009, Qi et al. experimentally examined the effects of longitudinal ultrasonic vibrations in the industrial production of brass wire [17]. Results proved the advantages of ultrasonic drawing over the conventional drawing with a 17% increment in drawing speed, 7% decrease in drawing force, prolonged tension regulation period from 0.5 s to about 1.5 s, and improved surface finish of the drawn wire. In 2012, Shan et al. proposed a new mathematic model to describe the ultrasonic drawing process based on the nonlocal friction theorem and obtained the stress distribution, contact pressure, and drawing force [18]. The ultrasonic anti-friction effect of stainless-steel wire, copper wire, and titanium wires at room temperature were analyzed. In 2016, Yang et al. compared the influences of longitudinal-torsional composite vibration and longitudinal vibration on the titanium wire drawing process through numerical and experimental methods [19]. The results indicated that longitudinal vibration is more beneficial for reducing the friction force and improving the surface finish of the drawn wire than composite vibration. In 2018, Liu et al. performed an experimental study on two-pass titanium wire drawing with two oscillating dies and achieved a drawing force reduction of over 50% [20]. Results showed that increased vibration amplitude leads to greater drawing force decrement, whereas increased drawing velocity brings about the reverse effect. Higher drawing speed with moderate ultrasonic amplitude is more preferable in removing surface defects.

Although the feasibility and advantages of ultrasonic wire drawing have been verified by previous researchers, the underlying mechanisms were explained in different ways. The correctness of presuppositions in quantitative calculations remains to be verified. The recent works mainly focused on experimental research and theoretical studies, whereas little attention was paid to the finite element analysis of the ultrasonic-assisted drawing process. In this paper, the ultrasonic drawing process was described as contact and friction problems between the elastic-plastic traveling string and the rigid vibrating die. The axisymmetric finite element (FE) model was established in commercial FEM software Abaqus, considering the actual length of the wire and the strain hardening characteristics of the material. Influences of ultrasound on the drawing force were discussed and compared in terms of four factors: vibration amplitude, drawing speed, location of the die, and the friction coefficient. Mises stress distribution inside the wire, as well as contact and friction stress within the contact region, was investigated.

2. Finite Element (FE) Model Establishment and Simulation Procedure

The titanium wire drawing process could be explained as the frictional contact problem between a certain length of deformable wire and the internal surfaces of the rigid oscillating die. Considering the inherent symmetry of the geometry and boundary condition, an axisymmetric finite element analysis model was established in commercial software Abaqus (Dassault Systèmes Simulia Corp., Providence, RI, USA) to minimize computational costs, as shown in Figure 1. In this particular problem, the diameter of the raw titanium wire, Ø 0.4 mm, is drawn into Ø 0.36 mm, with a cross-section area reduction of 19%. The generation of heat due to plastic dissipation inside the wire and frictional heat generation within the contact region are not considered, as this paper emphasizes on the independent effects caused by the addition of ultrasound in the general drawing process of long thin metallic wires. The drawn wire is wound up around the drum reel driven by an electrical motor. The length of unwound wire, or the distance between the die and the reel, denoted as viable L in the figure, is usually 100~200 mm and should be larger than the radius of the drum, 40 mm, in practice. The flexibility and elastic deformation of slightness titanium wire would exert a remarkable influence on the ultrasonic drawing process, therefore, the uncoiled drawn wire should be modeled with the full length. In contrast, the length of the raw wire, denoted as l in the figure, has little influence on simulation results and therefore is assigned as a constant of 10 mm. In the discretization process, titanium wire is defined as a deformable part with an overall meshing size of 0.02 mm, whereas the die is modeled as a rigid body with an element size of 0.03 mm. The two parts are all modeled using the 4-node reduced axisymmetric continuum quadrilateral solid element. Penalty and kinematic formulations are employed in the definition of contact interactions. The contact type between the die and the wire is specified as surface-to-surface contact, and the contact behavior is assumed to abide by Coulomb friction law, with an initial coefficient of 0.1. The simulation is also performed with Arbitrary Lagrangian-Eulerian (ALE) adaptive meshing and enhanced hourglass control.

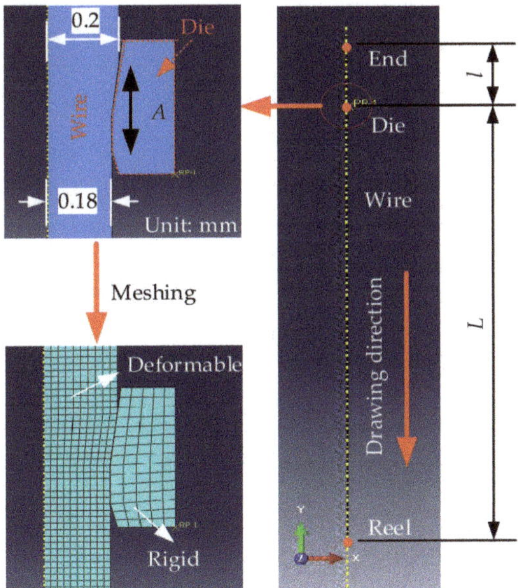

Figure 1. Simulation settings and FE model of single-die ultrasonic wire drawing.

In the meshed model, all the nodes situated on the central axis are constrained at the radial (X) and circumferential (Z) direction. Velocity constraints along the wire drawing direction (−Y) are prescribed to the bottom nodes of the wire near the reel. A reference point NREF is assigned to control the movement of the rigid die. The whole simulation process, which lasts for 5 ms, can be divided into three steps. The duration for the first and the third step is 2 ms, which are separately divided into 100 substeps. In these two steps, the amplitude of the die is assigned as 0 to simulate the conventional drawing process. The second step lasts for 1 ms, which equals to 20 oscillation cycles for, and is divided into 400 substeps, during which ultrasonic vibrations of 20 kHz is applied to the die to stimulate the ultrasonic drawing process. In the study, the frequency of the die remains the same, whereas its amplitude changes from 1 to 10 μm. The variation range of the drawing speed and the distance from the die to the reel are 100~1200 mm/s and 15.8~126.4 mm, respectively. The friction coefficient varies from 0.1 to 0.5. The wire drawing forces could be captured by extracting the reaction force along the Y direction at the NREF point.

In the simulation, the material of wire is selected as TA2 commercial pure titanium, the density, elastic modulus, and Poisson ratio of which are 4500 kg/m^3, 115 GPa, and 0.37, respectively. The wire drawing die mainly consists of two sections: a stainless casing and the diamond nib [21]. In the FE model, only the nib section, which directly contacts the wire, is considered. The diamond mandrel, with 3520 kg/m^3, 1100 GPa, 0.07, respectively, in density, elastic modulus, and Poisson ratio, could be divided into three areas: the reduction area with the semi-cone angle of 7°, the bearing area, and the exit area. Plastic deformation mainly occurs in the reduction area; however, the bearing area is essential for maintaining the dimensional accuracy of the wire products. The stress-strain curve of TA2, as well as the geometry and dimension of the diamond nib, is shown in Figure 2. The plastic deformation of the titanium is assumed to follow the Mises yield criterion, and a multi-linear kinematic hardening (MKINH) model was employed to describe the plastic behavior of the material [22]. Considering the changing stress state at the wire-die interface caused by the oscillation of the die, the simulation process involves not only material nonlinearity, but also nonlinear geometry and boundary conditions; therefore, the transient analysis is conduct in the Abaqus/Explicit module to improve the efficiency and convergence of the calculation.

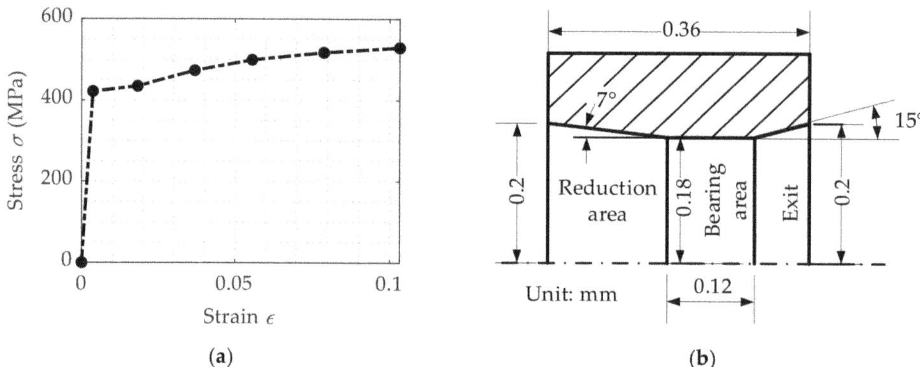

Figure 2. Material property and geometry definition of the die specified in the simulation program: (**a**) Stress-strain curve of TA2 commercial pure titanium in FE simulation; (**b**) Structure and dimensions about the diamond nib of the wire drawing die.

3. Results and Discussion

Based on the above finite element analysis (FEA) model, influences of ultrasonic vibration on titanium wire drawing process are investigated in terms of four factors: the location of the die, the vibration amplitude of the die, the wire drawing velocity, and the friction coefficient. The relationships between each factor and the averaged drawing force are analyzed using the control variates technique. The internal equivalent stress distribution of titanium wire, as well as the contact and friction stress distribution at the die-wire interface, is also displayed to understand the underlying mechanism of the ultrasonic drawing process.

3.1. Influence of Die Location on Drawing Force

In previous research, Hayashi et al. built an FEA model and discussed the influence of vibration amplitude and direction on drawing force and Mises stress distribution inside the wire when drawn with the assistance of axial and radial ultrasonic vibrations [14]. However, the established model was confined to a very narrow region of 1~2 mm in length adjacent to the die. The flexibility and stretching of the elastic slender wire are neglected. In industrial production, the wire drawing velocity is provided by the reel drum, which rotates at a constant speed. For the conventional drawing process, the velocity of the wire adjacent to the reel drum is equal to that near the die. However, for the ultrasonic wire drawing process, when the die oscillates periodically, they are not equal, as the uncoiled drawn wire would be tightened and relaxed intermittently along with the die. The influence of this neglected factor on the drawing process is determined by the length of uncoiled drawn wire, i.e., the distance between the reel drum and the wire.

To investigate the influence of reel-die distance L on the ultrasonic drawing process, the other factors maintain constant, with ultrasonic amplitude, drawing speed, and friction factor specified as 10 μm, 300 mm/s, and 0.1, respectively. As the drawn wire is forced to vibrate and its natural frequency is determined by the length, simulations are conducted when the wire length amounts to $\lambda/8$, $\lambda/4$, $3\lambda/8$, and $\lambda/2$, respectively, where λ denotes the wavelength of ultrasonic vibration when prorogating in the titanium wire. The wavelength can be calculated as 252.8 mm, based on Young's modulus, density of the TA2, and the oscillation frequency of the die, 20 kHz. The variations of drawing force with time under different uncoiled wire lengths are illustrated in Figure 3.

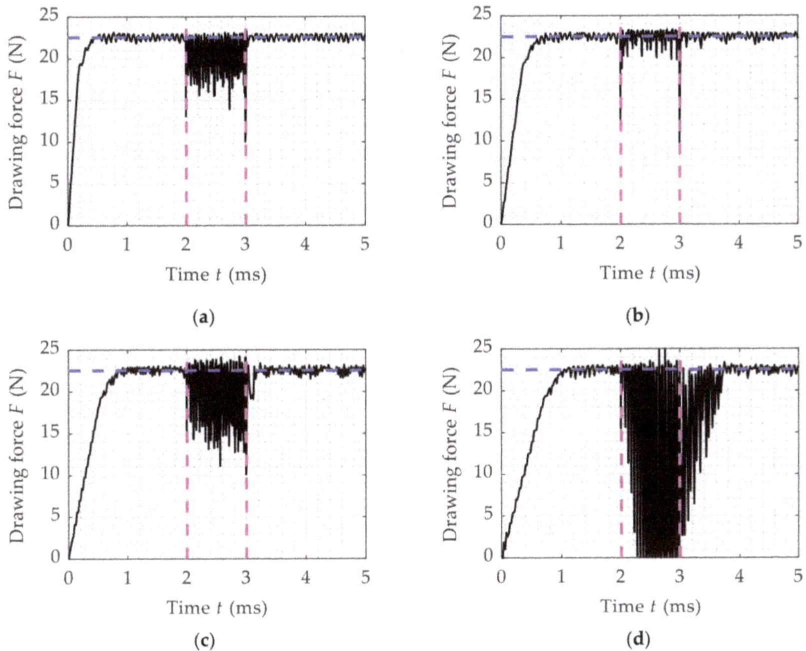

Figure 3. Time-variation of the drawing force for different distances between the die and the reel: (**a**) L = 31.6 mm; (**b**) L = 63.2 mm; (**c**) L = 94.8 mm; (**d**) L = 126.4 mm.

It can be seen that the whole drawing process lasts for 5 ms. Ultrasonic vibrations were imposed during the period of 2~3 ms and then were removed. At the beginning of the first 2 ms, the drawing force ramped up from 0 N to around 20 N within 1 ms, which represents the elastic deformation phase of the wire drawing process. Afterward, the growth of the drawing force gradually slowed down, which means the start of plastic deformation. The drawing force finally stabilized at around 22.5 N, and wire products were steadily pulled through the die. Comparing Figure 3a,d, we can easily find that the distance between the die and the wire does not influence the conventional drawing force, however, it would affect the duration of the elastic deformation period. When ultrasonic vibrations were applied, the drawing force began to fluctuate in a certain range, with the maximum values slightly larger than or equal to the normal drawing force, whereas the minimum values are significantly lower than that, which results in a decrease in averaged drawing forces. The fluctuation range of ultrasonic drawing force, or the reduction ratio of averaged ultrasonic drawing forces compared to the conventional drawing value, is heavily influenced by the location of the die. Specifically, at the same ultrasonic amplitude and drawing velocity, the drawing force fluctuates most violently when the reel locates 126.4 mm ($\lambda/2$) from the die, while the fluctuation is the minimum when the reel-die distance equals 63.2 mm ($\lambda/4$). Besides, there is a difference between the four subfigures with respect to the envelope curve shapes under the ultrasonic drawing condition. When the die is distributed $\lambda/2$ from the reel, the envelope of the ultrasonic drawing force is shaped like a trapezoid, instead of rectangular when located at other positions. For these positions, the amplitudes of drawing force fluctuation reach the maximum as soon as ultrasounds are imposed, and the drawing forces immediately restored to around 22.5 N when vibrations are removed. At the half-wavelength position, however, the minimum drawing force drops steadily in the first five oscillating cycles before reaching a steady state. When ultrasound is removed, the drawing force remains lower than 22.5 N in about 0.7 ms, instead of going back to the original value right away. Finally, even in the half-wavelength condition and even without the consideration of the pliability of the metallic wire, the ultrasonic drawing forces remains above 0, which means no separations occur between the wire and die at their contact interface, and the validity of the presuppositions for reversed friction mechanism should be further verified. Unlike a bar, a long thin wire could not bear the pressure, which is another reason why the drawing force stays above 0 and the separation could not happen.

Figure 4 shows the variation of averaged drawing force with the distance between the die and the reel. It can be seen that, as the reel-die distance increases from 15.8 mm ($\lambda/16$) to 126.4 mm, the averaged drawing force first goes up and then drops, peaking at the quarter-wavelength position, however, remains below the conventional drawing force value of 22.5 N. At the half-wavelength position, the averaged drawing force reaches the minimum value of 10.28 N, with a reduction of more than 50% compared with conventional drawing force. The influence of the reel-die distance on the ultrasonic drawing force can be attributed to the stretching vibration of the uncoiled drawn wire. This section of titanium wire resonates under the drive of vibrating die when its length approximates half-wavelength, as the frequency of the die is approaching the first-order longitudinal resonant frequency of the wire.

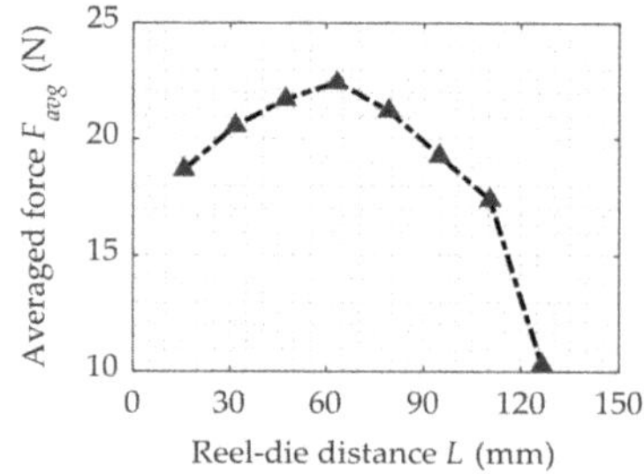

Figure 4. FEM calculation results of average ultrasonic drawing forces at different distances between the die and the reel drum.

3.2. Influence of Ultrasonic Amplitude on Drawing Force

It can be seen from the above analysis, with the same ultrasonic amplitude and drawing speed, maximum drawing force reduction can be achieved when the distance between the reel and the die approximates half-wavelength of ultrasound traveling in the titanium wire. Therefore, the influences of other factors, including ultrasonic amplitude A, wire drawing speed V, and friction coefficient μ, on the ultrasonic drawing force will be further considered under this condition. From previous studies, it can be known that the effects of ultrasonic amplitude and drawing speed on ultrasonic drawing force are correlated [14–20]. Drawing force reduction can only be realized when the prerequisite $V < 2\pi fA$ is satisfied.

Figure 5 illustrates the changing of wire drawing force along with time at the drawing velocity of 300 mm/s and with the friction coefficient of 0.1. It can be found that the upper limit of the drawing force remains around 22.5 N, which will not be affected by the intensity of ultrasonic vibrations. In Figure 5a, where ultrasound amplitude is 2 μm, there is almost no change in drawing force compared with the conventional drawing value, because the drawing speed, 300 mm/s, exceeds the threshold value $2\pi fA$, which is determined by the oscillation amplitude and frequency of the die. In other subfigures, where pre-condition is satisfied, an obvious decrease in drawing force can be observed. The envelope lines of ultrasonic drawing force present the shape of the triangle or trapezoid instead of rectangular, as the die locates half-wavelength off the reel drum, which is consistent with previous analysis. The influence of ultrasound will also last for a short time after the die stops vibrating, instead of disappearing instantly. However, the fluctuation range of the ultrasonic drawing force will be widened with the increment of ultrasonic amplitude. When ultrasound is turned on, the minimum values of the drawing force go down consistently until arriving at the stable phase. The time consumption for this period is shortened with larger ultrasonic amplitude, and the decrement of minimum drawing force for each oscillation period will be enlarged correspondingly.

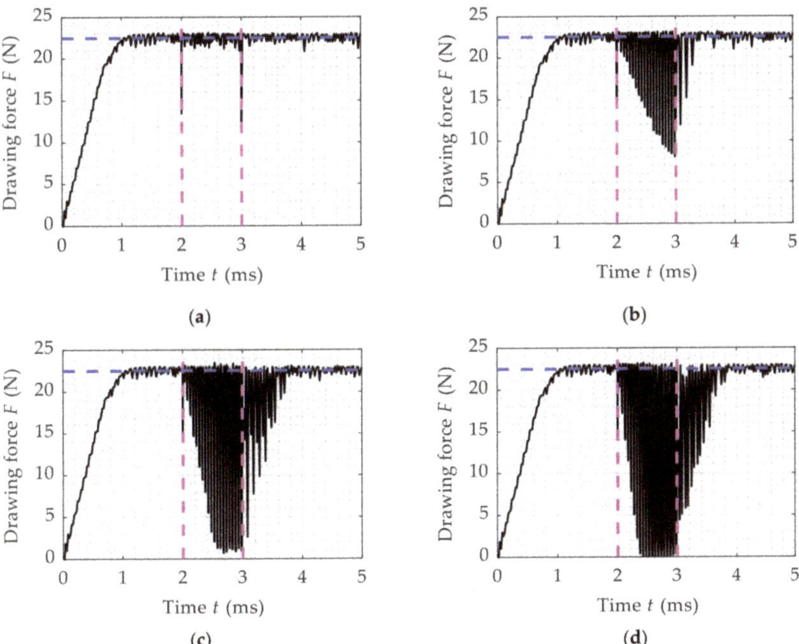

Figure 5. Influence of ultrasonic amplitude on time-variation of the drawing force: (**a**) $A = 2$ μm; (**b**) $A = 4$ μm; (**c**) $A = 6$ μm; (**d**) $A = 8$ μm.

Figure 6 shows the numerical calculation results of averaged ultrasonic drawing force under different ultrasonic amplitudes at drawing speed of 300 mm/s and 600 mm/s, respectively. Overall, the two variation curves show the same trend, that is, the averaged ultrasonic drawing force goes down with the increment of ultrasonic amplitude. However, the drawing force declines faster at a relatively lower drawing speed of 300 mm/s. A flat section could be found at the initial segment of the two curves, where there is almost no reduction in drawing force compared with 22.5 N. Therefore, at a certain drawing velocity, to achieve drawing force reduction, the intensity of ultrasonic vibration has to be large enough. The flat region expands at a relatively higher drawing velocity because the threshold drawing speed ($V = 2\pi fA$) increases with ultrasonic intensity.

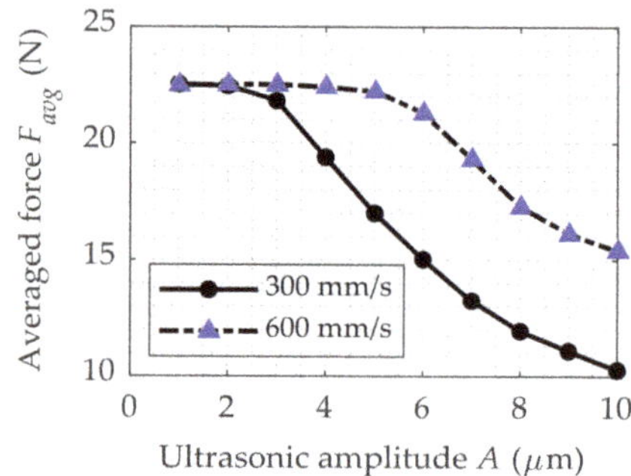

Figure 6. FEM calculation results of average ultrasonic drawing force with different ultrasonic amplitudes at drawing speed of 300 mm/s and 600 mm/s.

Based on the above simulation results, the following hypothesis could be put forward. With the addition of ultrasonic vibration, the conventional continuous drawing process is turned into an intermittent ultrasonic drawing process. A vibration cycle of the die can be divided into two phases: the plastic deformation phase, and the stretching vibration phase. For the former, titanium wire is drawn through the die and the drawing force equals the conventional drawing force. For the latter, however, the wire vibrates along with the die. The drawing force is determined by the stretching force of the wire and should be lower than the conventional drawing force. The time allocation between the two phases is determined by both the amplitude of the die and the wire drawing velocity. With the increment of drawing speed, the proportion of the plastic deformation stage will rise correspondingly, therefore, averaged ultrasonic drawing force will be closer to normal drawing force 22.5 N. On the contrary, when ultrasonic amplitude increases, the time duration of the deformation stage is decreased, and the stretching force vibration is aggravated simultaneously; therefore, the overall drawing force goes down.

3.3. Influence of Drawing Speed on Drawing Force

Figure 7 displays the influence of wire drawing velocity on the time variation of drawing forces when the distance between the die and reel drum equals half-wavelength and the ultrasonic amplitude remains 10 µm. By comparison, it can be found that the steady-state value of conventional drawing force is not affected by drawing speed, which fluctuates slightly around 22.5 N, because strain rate dependence of the flow stress is not included in the material model. However, the time duration of the elastic deformation stage is prolonged when drawing speed increases, with 1.8 ms for 200 mm/s to 0.5 ms for 800 mm/s. For the ultrasonic drawing stage, the fluctuation range of drawing forces is narrowed with the increment of drawing velocity. And the time consumption for ultrasonic drawing forces to reach the steady value increases correspondingly from 5 oscillation cycles for 200 mm/s to 13 oscillation cycles for 800 mm/s. After the die stops vibrating, the fluctuation of the drawing force continues for a while, but the time duration is shortened with the increment of drawing velocity. According to the hypothesis proposed in Section 3.2, with the increment of drawing speed, the plastic deformation phase takes up a greater proportion in a vibration period, therefore the overall drawing force is increased. Meanwhile, the stretching vibration time is reduced, more cycles are undergone before the minimum drawing force gets stabilized. In addition, as the drawing speed approaches the threshold value $2\pi fA$, the decrement of minimum drawing force in each cycle drops, making the total drawing force reduction decreased.

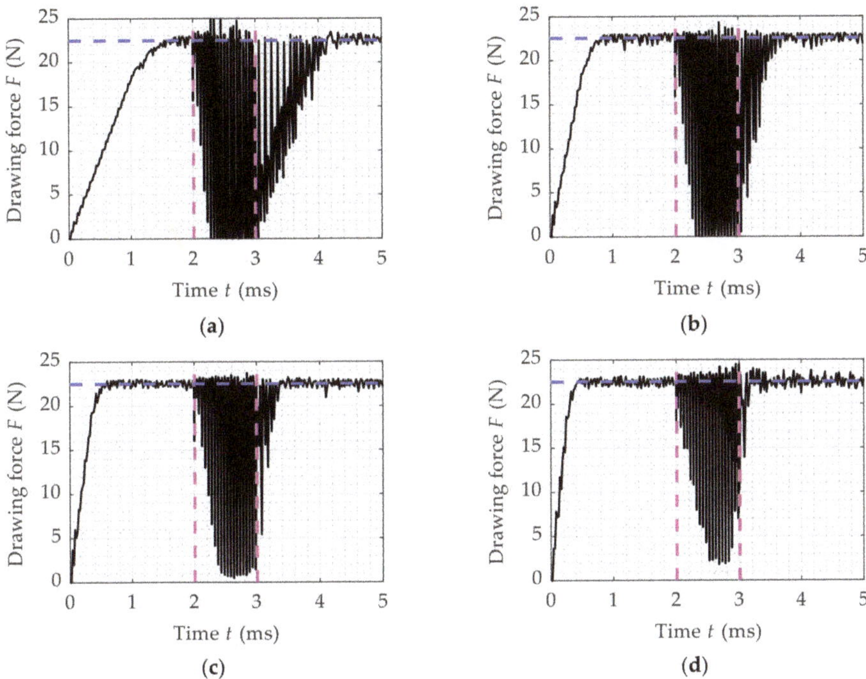

Figure 7. Influence of drawing speed on time-variation of the drawing force with reel-die distance of 126.4 mm: (**a**) $V = 200$ mm/s; (**b**) $V = 400$ mm/s; (**c**) $V = 600$ mm/s; (**d**) $V = 800$ mm/s.

Figure 8 illustrates the time variation of drawing forces at different drawing speed when the die locates 7.9 mm ($\lambda/32$) to the reel drum and the ultrasonic amplitude remains 10 μm. In general, the drawing forces present a similar upward trend as the drawing speed rises. However, the envelope lines of the drawing forces are shaped as rectangular instead of the trapezoid. An abrupt decline in drawing force could be observed in these figures when ultrasonic vibration is imposed. The drawing force will restore to the initial value of 22.5 N immediately when ultrasonic vibrations are removed. Besides, compared with the half-wavelength conditions, the fluctuation range of the corresponding drawing force appears to be more sensitive to drawing speed, which is narrowed down sharply, especially at higher drawing speeds. This phenomenon might be caused by the weakened stretching vibration of uncoiled drawn wire, as the oscillating frequency of the die is far below the first-order longitudinal frequency of the wire.

Figure 9 compares the influence of drawing speed on averaged ultrasonic drawing force when the reel-die distance equals 126.4 mm and 7.9 mm, respectively. Although the two curves present a similar upward trend with the increment of drawing speed, the drawing force at the half-wavelength position is lower than the $\lambda/32$ position. The gap between the two curves is gradually narrowed as the drawing speed rises. When approaching the critical drawing speed 1256 mm/s, which is calculated at the amplitude of 10 μm, and frequency of 20 kHz, the two curves all converge to the conventional drawing force of 22.5 N. The difference between the two curves might be attributed to the stretching vibration of the wire. At the 7.9 mm condition, drawing force improvement caused by increased drawing velocity could not be compensated by the strengthened oscillation of the wire, as occurs in the half-wavelength condition.

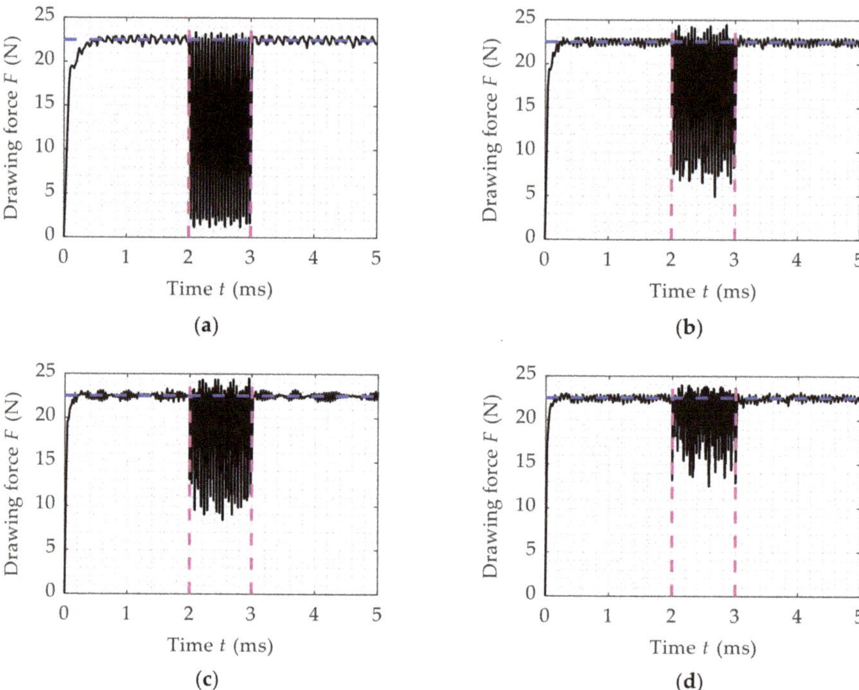

Figure 8. Influence of drawing speed on time-variation of the drawing force with reel-die distance of 7.9 mm: (**a**) V = 200 mm/s; (**b**) V = 400 mm/s; (**c**) V = 600 mm/s; (**d**) V = 800 mm/s.

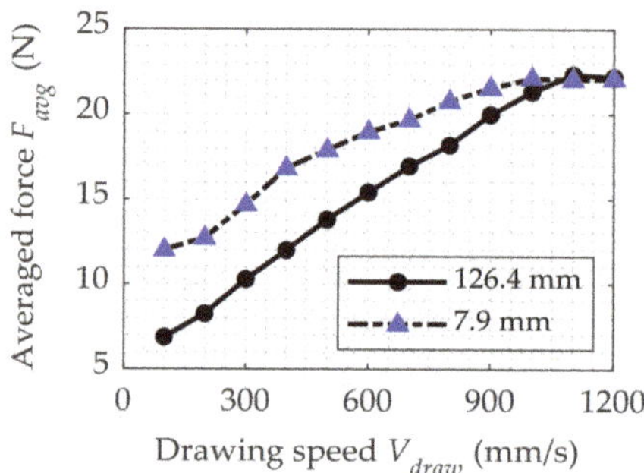

Figure 9. FEM calculation results of average ultrasonic drawing force with different drawing speeds with reel-die distance of 126.4 mm and 7.9 mm.

3.4. Influence of Friction Coefficient on Drawing Force

To further investigate the influence of friction coefficient on the variation of drawing force, ultrasonic amplitude and drawing speed are specified as 10 µm and 300 mm/s, and remain unchanged. Figure 10 shows the time variation of drawing forces when the reel-die distance equals 15.8 mm (λ/16) with friction coefficient assigned as 0.3 and 0.5, respectively. Compared with simulation results in

Figures 3 and 4, in the two subfigures of Figure 10, conventional drawing forces climb up to 36.13 N and 45 N from 22.5 N, respectively, and the averaged ultrasonic drawing forces correspondingly ascend to 32.07 N and 41 N from 18.68 N. Therefore, it is suggested that the friction coefficient influences the drawing force no matter whether ultrasonic vibrations are imposed. In addition, the increments of conventional drawing force, when friction coefficient rises, are very close to that of averaged ultrasonic drawing forces. In other words, ultrasonic vibration does not affect the wire-die friction coefficient. However, with excessive friction, plastic deformation would occur to the drawn wire between the die and the reel, resulting in wire breakage, as plotted in Figure 10b.

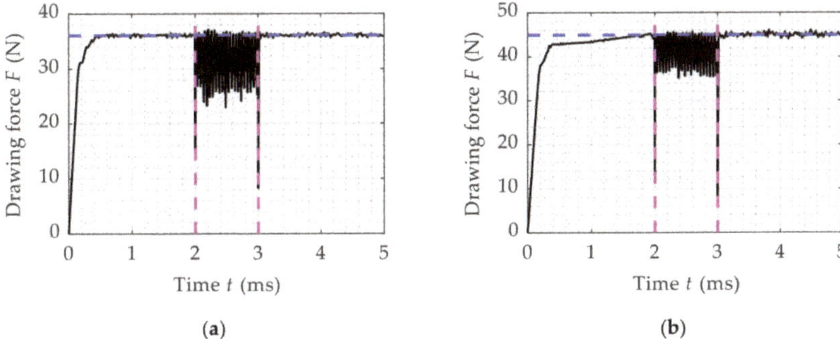

Figure 10. Time variation of drawing force with different friction coefficient when the distance between the reel and the die is 15.8 mm: (**a**) $\mu = 0.3$; (**b**) $\mu = 0.5$.

Figure 11 shows the time variation of drawing forces with friction coefficient of 0.3 and 0.5, respectively, when the die locates 126.4 mm from the reel. Conventional drawing forces are 36.1 N and 42.8 N, which are consistent with the results in Figure 10. Averaged ultrasonic drawing forces increase to 17.15 N and 20.83 N from 10.28 N. Similarly, ultrasonic vibration exerts influence on both the conventional and the ultrasonic drawing forces. However, on this occasion, the increments of conventional drawing force caused by increased friction are obviously larger than that of averaged ultrasonic forces. Therefore, at the half-wavelength condition, ultrasonic vibration helps to decrease the equivalent friction with the contacting region. Compared with Figure 3d, severe distortion could be found in the two subfigures. In the conventional drawing phase, time-consuming to reach the steady-state increases to 1.6 ms and 1.8 ms from 1 ms, respectively. This might be caused by the plastic deformation of the uncoiled drawn wire. In the ultrasonic drawing phase, many negative values appear in the drawing force curve. As the flexibility of the titanium is not fully considered in the FE model, the negative section of the curves will be chopped off in the data post-processing procedure.

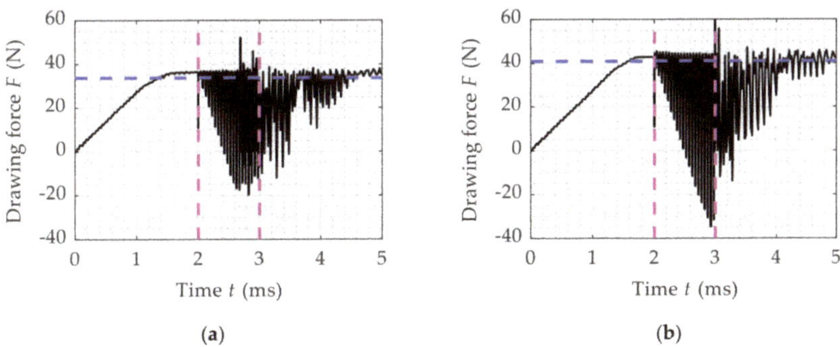

Figure 11. Time variation of drawing force with different friction coefficient when the distance between the reel and the die is 126.4 mm: (**a**) $\mu = 0.3$; (**b**) $\mu = 0.5$.

Figure 12 illustrates the influences of friction coefficient on conventional and ultrasonic drawing forces. In both subfigures, the drawing forces ascend with an increased friction coefficient. For conventional drawing, the simulation results are in good agreement with a friction coefficient below 0.4 and are consistent with theoretical results calculated according to Avitzur's theory, when the friction coefficient is lower than 0.2 [23]. For ultrasonic drawing, the drawing force curve has the same shape with the conventional drawing force curve when reel-die distance equals λ/16. However, at the half-wavelength condition, the increment of ultrasonic drawing force is much smaller than that of conventional drawing force, which means a decreased equivalent friction.

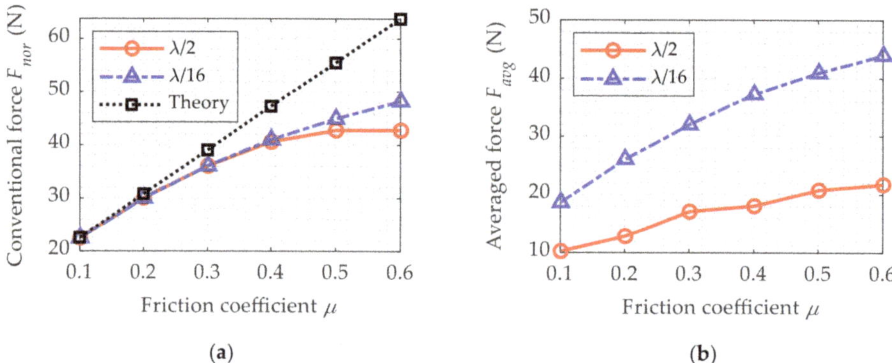

Figure 12. Influence of friction coefficient on drawing forces when the die locates at different positions: (**a**) Influence of friction on conventional drawing force; (**b**) Influence of friction coefficient on averaged ultrasonic drawing forces.

3.5. Influence of Ultrasonic Vibration on Stress Distribution

Figure 13 depicts the Mises equivalent stress distribution of titanium wire under conventional drawing conditions with drawing velocity of 300 mm/s and friction coefficient of 0.1. In the contour, the equivalent stress remains 0~105.9 MPa in a large portion of the feeding area, which locates to the left of the die, when the influence of the back-pull force is neglected. At the entrance region, the Mises stress surged to around 423 MPa. The maximum stress 423.6~529.5 MPa occurs at the reduction area and bearing area, where the die is in direct contact with the wire. However, at the bearing area and the right adjacent area to it, the maximum stress only concentrates on the outer layer of the wire. These are mainly residual stresses caused by the inhomogeneous deformation between the surface and core section of the metallic wire, which declines sharply along the radial direction (−X) of the wire. At the reduction area, however, the equivalent stress maintains 423.6~529.5 MPa throughout the whole internal region of the wire. This is where plastic deformation mainly occurs. For the wires far from the right side of the die, the drawing stress is kept between 211.8 MPa to 311.7 MPa.

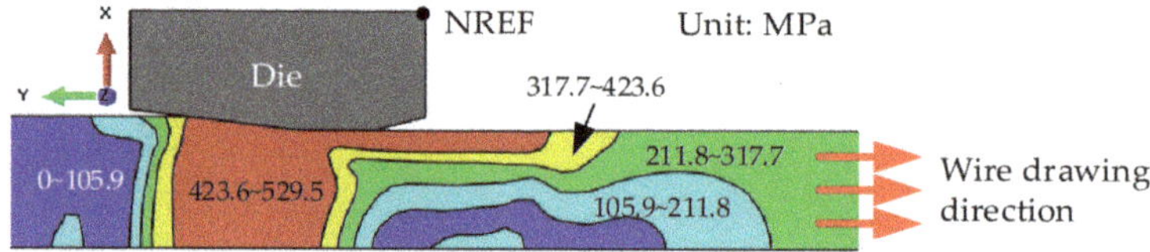

Figure 13. Mises equivalent stress distribution of the wire at the contact region under conventional drawing conditions.

To investigate the influence of ultrasonic vibration on the stress distribution of the wire, the simulation is carried out with an ultrasonic amplitude of 10 μm and a wire-die distance of 15.8 mm. The equivalent contours around the contact region for different movement states of the die are demonstrated in Figure 14. It can be seen that, when the die moves, along the reversed wire drawing direction, from the equilibrium position to the leftmost position, the overall stress distribution is similar to that of conventional drawing, except for slight difference at the core segment of the drawn wire, with a maximum stress of 525.9 MPa. In contrast, when the die moves, along the drawing direction, from the central position to the right extreme position, more significant changes could be observed with the maximum equivalent stress dropping to 515 MPa and shrinkage in its area. In addition, the equivalent stress rises to 206 MPa at the feeding area, 308.8 MPa at the core of the drawn wire, and reduced to below 411.8 MPa at the outer surface layer. The more uniform distribution of the equivalent stress is beneficial for eliminating the defects and residual stress caused by inhomogeneous deformation between the core and surface layer of the titanium wire.

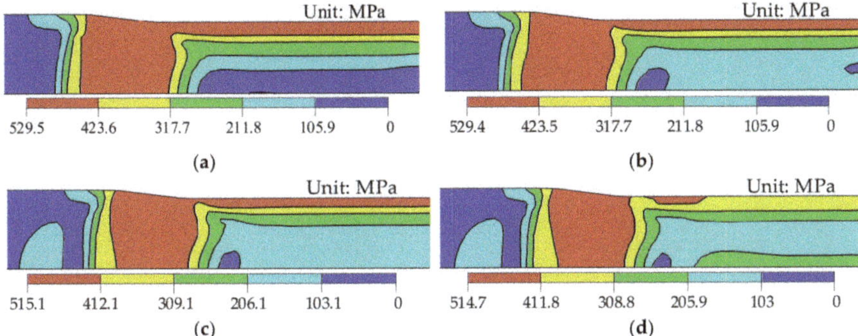

Figure 14. Mises equivalent stress distribution of the wire at the contact region under ultrasonic drawing conditions when the die locates 15.8 mm from the reel and in various motion states: (**a**) Equilibrium position and moves to left; (**b**) Left extreme position; (**c**) Equilibrium position and moves to the right; (**d**) Right extreme position.

Figure 15 illustrates the contact and friction stress distribution at the wire-die interface under conventional drawing conditions. From Figure 15a, it can be seen that the maximum value of contact stress occurs at the two sides of the reduction area, adjacent to the entrance and the bearing areas. This is where the surface plastic deformation mainly happens, as the contact stress range, 513.1~641.4 MPa, is apparently above the yield stress of titanium, which is calculated as around 460 MPa based on theoretical and simulation results. In the middle of the reduction area, contact stress falls into the range of 384.8~513.1 MPa, both elastic and plastic deformations coexist. In the entrance and bearing regions, which locate at the two sides of the reduction region, the contact stress plummets to below 128.3 MPa, which means no plastic deformation happens. In Figure 15b, we can find that the friction stress has the same distribution with the contact stress, but is 10% of the latter in value, as the Coulomb friction model is adopted in the setup of the FEA model with a coefficient of 0.1. Therefore, no additional simulation result about the friction stress will be separately presented. In general, the contact and friction stress distribution is compatible with the equivalent stress distribution.

Figure 16 shows the contact stress contours for the ultrasonic drawing process with the die in different motion states. Compared with traditional drawing, there is almost no change in the distribution of contact stress, except for a slight change in the values. When the die locates at the equilibrium position and moves to the left, the maximum contact stress goes up to 660.4 MPa, which is slightly above 641.4 MPa, as shown in Figure 16a. In other positions, the contact stresses are slightly lower than conventional drawing values. In the whole ultrasonic drawing process, the contact stress at the reduction region is apparently above 0. In other words, no separations would occur at the wire-die interface, as presupposed in the reverse friction mechanism theory. Although with shortened reel-die distance, the minimum contact force could be further decreased, to the point of achieving an intermittent separation between the two parts, which has no practical meaning.

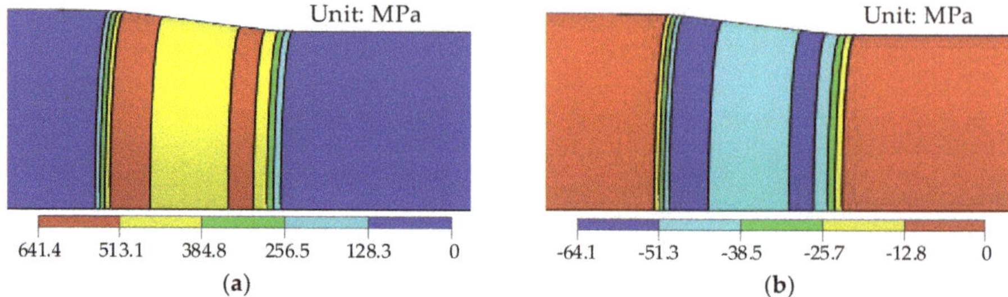

Figure 15. Contact stress and friction stress distribution at the contact interface under conventional drawing conditions: (**a**) Contact stress distribution; (**b**) Friction stress distribution.

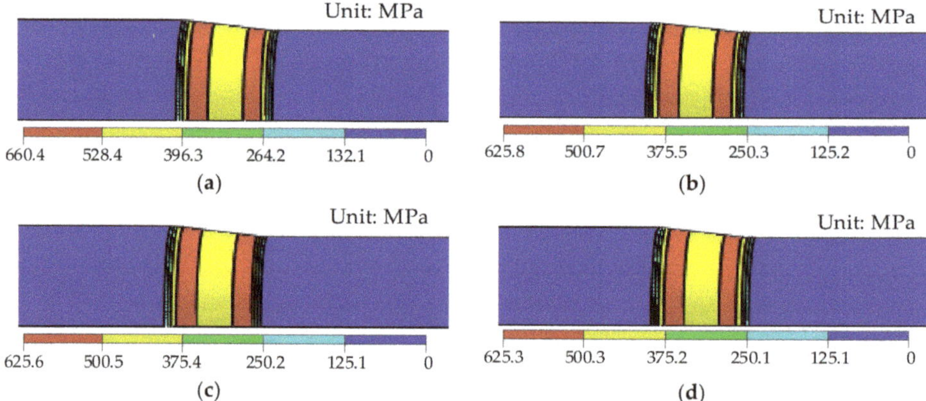

Figure 16. Contact stress distribution at contact interface under ultrasonic drawing conditions when the die locates 15.8 mm from the reel and in various motion states: (**a**) Equilibrium position and moves to left; (**b**) Left extreme position; (**c**) Equilibrium position and moves to the right; (**d**) Right extreme position.

4. Conclusions

An axisymmetric FE model for a single-pass titanium wire drawing process was built in the commercial software Abaqus based on the practical length of wire. Then, the influence of ultrasonic vibrations on the wire drawing process was investigated with respect to four factors: the location of the die, ultrasonic amplitude, drawing velocity, friction coefficient. At last, the stresses distribution contours at the contact region with and without ultrasonic vibrations were compared. The main conclusions of this study can arrive as follows:

(1) Ultrasonic metal wire drawing process can be interpreted as the frictional contact problem between a certain length of deformable metallic string and the rigid die. The drawing force reduction caused by the addition of ultrasonic vibration could be attributed to the stretching vibration of the wire, which locates between the reel and the die. When the reel-die distance approximates the half-wavelength of ultrasonic vibration when propagating in the metallic material, the drawing force saw the greatest decline.

(2) In the ultrasonic drawing process, an oscillation cycle of the die could be divided into two phases: the plastic deformation phase and the stretching vibration phase. The time distribution between

them is determined by the ultrasonic amplitude and wire drawing speed. The drawing force reduction increases with the increment of ultrasonic amplitude and decrease in drawing speed.

(3)   Ultrasonic vibration has almost no effect on the overall friction coefficient, except for the half-wavelength condition or when the reel-die distance is very small.

(4)   There is no separation between the wire and the die at the reduction area for single-pass ultrasonic wire drawing when the back-pull force is not considered. Ultrasonic vibrations do have influences on the internal equivalent stress of the wire and the contact stress at the wire-die interface, however, only slightly.

Author Contributions: S.L. wrote the paper and completed the numerical simulation; T.X. and X.S. provided the funding and supervised the project; H.C. revised the manuscript and helped in the simulation part. All authors have read and agreed to the published version of the manuscript.

Funding: This research was funded by the National Natural Science Foundation of China, grant number 51575130.

Conflicts of Interest: The authors declare no conflict of interest.

References

1.   Baudana, G.; Biamino, S.; Ugues, D.; Lombardi, M.; Fino, P.; Pavese, M.; Badini, C. Titanium aluminides for aerospace and automotive applications processed by Electron Beam Melting: Contribution of Politecnico di Torino. *Met. Powder Rep.* **2016**, *71*, 193–199. [CrossRef]

2.   Soro, N.; Brassart, L.; Chen, Y.; Veidt, M.; Attar, H.; Dargusch, M.S. Finite Element Analysis of Porous Commercially Pure Titanium for Biomedical Implant Application STZ-based modelling of plasticity in Li-Si alloys View project Mechanics of Interfaces: Evolving microstructure and conduction properties View project Finite element analysis of porous commercially pure titanium for biomedical implant application. *Mater. Sci. Eng. A* **2018**, *725*, 43–50.

3.   Attar, H.; Bermingham, M.J.; Ehtemam-Haghighi, S.; Dehghan-Manshadi, A.; Kent, D.; Dargusch, M.S. Evaluation of the mechanical and wear properties of titanium produced by three different additive manufacturing methods for biomedical application. *Mater. Sci. Eng. A* **2019**, *760*, 339–345. [CrossRef]

4.   Xiang, C.; Guo, J.; Chen, Y.; Hao, L.; Davis, S. Development of a SMA-Fishing-Line-McKibben Bending Actuator. *IEEE Access* **2018**, *6*, 27183–27189. [CrossRef]

5.   Arikatla, S.P.; Tamil Mannan, K.; Krishnaiah, A. Surface Integrity Characteristics in Wire Electrical Discharge Machining of Titanium Alloy during Main cut and Trim cuts. In Proceedings of the Materials Today: Proceedings, Hyderabad, India, 31 December 2017; Volume 4, pp. 1500–1509.

6.   Liu, S.; Shan, X.; Guo, K.; Yang, Y.; Xie, T. Experimental study on titanium wire drawing with ultrasonic vibration. *Ultrasonics* **2018**, *83*, 60–67. [CrossRef] [PubMed]

7.   Xie, Z.; Guan, Y.; Zhu, L.; Zhai, J.; Lin, J.; Yu, X. Investigations on the surface effect of ultrasonic vibration-assisted 6063 aluminum alloy ring upsetting. *Int. J. Adv. Manuf. Technol.* **2018**, *96*, 4407–4421. [CrossRef]

8.   Chen, C.Y.; Anok Melo Cristino, V.; Hung, C. Effects of ultrasonic vibration on SUS304 stainless steel subjected to uniaxial plastic deformation. *J. Chin. Inst. Eng. Trans. Chin. Inst. Eng. A* **2018**, *41*, 327–332. [CrossRef]

9.   Narkevich, N.A.; Tolmachev, A.I.; Vlasov, I.V.; Surikova, N.S. Structure and Mechanical Properties of Nitrogen Austenitic Steel after Ultrasonic Forging. *Phys. Met. Metallogr.* **2016**, *117*, 288–294. [CrossRef]

10.   Wang, K.; Shriver, D.; Li, Y.; Banu, M.; Hu, S.J.; Xiao, G.; Arinez, J.; Fan, H.T. Characterization of weld attributes in ultrasonic welding of short carbon fiber reinforced thermoplastic composites. *J. Manuf. Process.* **2017**, *29*, 124–132. [CrossRef]

11.   Xiang, D.H.; Wu, B.F.; Yao, Y.L.; Zhao, B.; Tang, J.Y. Ultrasonic Vibration Assisted Cutting of Nomex Honeycomb Core Materials. *Int. J. Precis. Eng. Manuf.* **2019**, *20*, 27–36. [CrossRef]

12.   Blaha, F.; Langenecker, B. Tensile deformation of zinc crystal under ultrasonic vibration. *Naturwissenschaften* **1955**, *42*, 556. [CrossRef]

13.   Winsper, C.E.; Sansome, D.H. The superposition of longitudinal sonic oscillations on the wire drawing process. *Proc. Inst. Mech. Eng.* **1968**, *183*, 545–562. [CrossRef]

14. Hayashi, M.; Jin, M.; Thipprakmas, S.; Murakawa, M.; Hung, J.-C.; Tsai, Y.-C.; Hung, C.-H. Simulation of ultrasonic-vibration drawing using the finite element method (FEM). *J. Mater. Process. Technol.* **2003**, *140*, 30–35. [CrossRef]

15. Susan, M.; Bujoreanu, L. The metal-tool contact friction at the ultrasonic vibration drawing of ball-bearing steel wires. *Rev. Metal.* **1999**, *35*, 379–383. [CrossRef]

16. Susan, M.; Bujoreanu, L.; Găluşcă, D.; Munteanu, C. Influence of the relative deformation rate on tube processing by ultrasonic vibration drawing. *Rev. Met.* **2004**, *40*, 109–117. [CrossRef]

17. Qi, H.; Shan, X.; Xie, T. Design and Experiment of the High Speed Wire Drawing with Ultrasound. *Chin. J. Mech. Eng.* **2009**, *22*, 580–586. [CrossRef]

18. Shan, X.; Qi, H.; Wang, L.; Xie, T. A new model of the antifriction effect on wiredrawing with ultrasound. *Int. J. Adv. Manuf. Technol.* **2012**, *63*, 1047–1056. [CrossRef]

19. Yang, C.; Shan, X.; Xie, T. Titanium wire drawing with longitudinal-torsional composite ultrasonic vibration. *Int. J. Adv. Manuf. Technol.* **2016**, *83*, 645–655. [CrossRef]

20. Liu, S.; Yang, Y.; Xie, T.; Shan, X. Experimental Study on Fine Titanium Wire Drawing with Two Ultrasonically Oscillating Dies. *IEEE Access* **2018**, *6*, 16576–16587. [CrossRef]

21. Wang, X.; Wang, C.; Sun, F.; Ding, C. Simulation and experimental researches on HFCVD diamond film growth on small inner-hole surface of wire-drawing die with no filament through the hole. *Surf. Coat. Technol.* **2018**, *339*, 1–13. [CrossRef]

22. Achard, V.; Daidie, A.; Paredes, M.; Chirol, C. Optimization of the Cold Expansion Process for Titanium Holes. *Adv. Eng. Mater.* **2017**, *19*, 1500626. [CrossRef]

23. Avitzur, B.; Narayan, C.; Chou, Y.T. Upper-bound solutions for flow through conical converging dies. *Int. J. Mach. Tool Des. Res.* **1982**, *22*, 197–214. [CrossRef]

Article

Optimization of the Continuous Galvanizing Heat Treatment Process in Ultra-High Strength Dual Phase Steels Using a Multivariate Model

Patricia Costa [1,*], **Gerardo Altamirano** [2], **Armando Salinas** [1], **David S. González-González** [3] and **Frank Goodwin** [4]

1   Centro de Investigación y de Estudios Avanzados del Instituto Politécnico Nacional (CINVESTAV), Unidad Saltillo, Zona Industrial, Ramos Arizpe C.P. 25903, Mexico; armando.salinas@cinvestav.edu.mx
2   Instituto Tecnológico de Saltillo. Blvd, Venustiano Carranza #2400, Col. Tecnológico, Saltillo C.P. 25280, Mexico; galtamirano@itsaltillo.edu.mx
3   Facultad de Sistemas, Universidad Autónoma de Coahuila, Ciudad Universitaria, Carretera a México Km 13, Arteaga, Coahuila, Mexico; david.gonzalez@uadec.edu.mx
4   International Zinc Association, 2530 Meridian Parkway, Durham, NC 27713, USA; fgoodwin@zinc.org
*   Correspondence: patricia_sheilla@hotmail.com; Tel.: +52 844-438-9600 (ext. 8549)

Received: 26 April 2019; Accepted: 4 May 2019; Published: 21 June 2019

Abstract: The main process variables to produce galvanized dual phase (DP) steel sheets in continuous galvanizing lines are time and temperature of intercritical austenitizing (t_{IA} and T_{IA}), cooling rate (CR_1) after intercritical austenitizing, holding time at the galvanizing temperature (t_G) and finally the cooling rate (CR_2) to room temperature. In this research work, the effects of CR_1, t_G and CR_2 on the ultimate tensile strength (*UTS*), yield strength (*YS*), and elongation (*EL*) of cold rolled low carbon steel were investigated by applying an experimental central composite design and a multivariate regression model. A multi-objective optimization and the Pareto Front were used for the optimization of the continuous galvanizing heat treatments. Typical thermal cycles applied for the production of continuous galvanized AHSS-DP strips were simulated in a quenching dilatometer using miniature tensile specimens. The experimental results of *UTS*, *YS* and *EL* were used to fit the multivariate regression model for the prediction of these mechanical properties from the processing parameters (CR_1, t_G and CR_2). In general, the results show that the proposed multivariate model correctly predicts the mechanical properties of *UTS*, *YS* and *%EL* for DP steels processed under continuous galvanizing conditions. Furthermore, it is demonstrated that the phase transformations that take place during the optimized t_G (galvanizing time) play a dominant role in determining the values of the mechanical properties of the DP steel. The production of hot-dip galvanized DP steels with a minimum tensile strength of 1100 MPa is possible by applying the proposed methodology. The results provide important scientific and technological knowledge about the annealing/galvanizing thermal cycle effects on the microstructure and mechanical properties of DP steels.

Keywords: dual phase steel; hot dip galvanizing line; multivariate analysis; multi-objective optimization; dilatometry

---

1. Introduction

The thinner gauge advanced high-strength steels (AHSS) are used in the automotive industry for the manufacture of structural components for the car body. The objective of using these materials is a reduction of weight, increase passenger safety and reducing fuel consumption and emission of greenhouse effect gases [1].

AHSS dual phase (DP) steels, with a microstructure consisting of a ferrite matrix, responsible for their good ductility, and martensite and/or bainite islands, which provide high tensile strength, exhibit the greatest rate of production [1]. In addition to mechanical strength and formability, DP steel for automotive and other applications require high corrosion resistance. Therefore, they are usually galvanized to increase the useful life of the components and galvannealed to improved their weldability [2].

Galvanized DP steel strips can be produced from cold rolled sheets using available continuous hot dip galvanizing lines (CGL). However, careful selection and control of processing parameters to generate the dual phase microstructures (ferrite + martensite) are needed to achieve the desired mechanical properties. Basically, a CGL processing line can be divided into five sections as illustrated schematically in Figure 1:

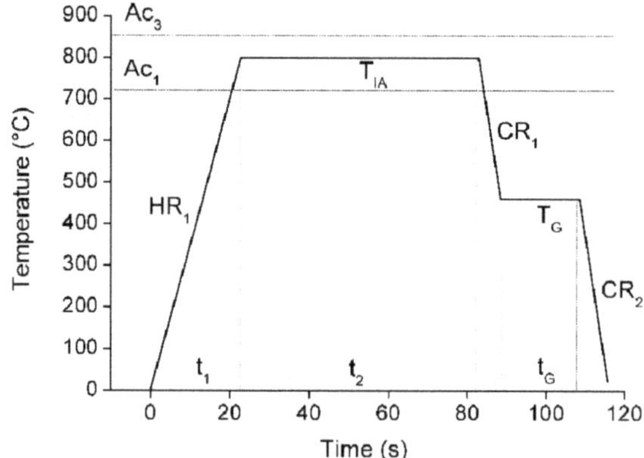

Figure 1. Schematic representation of an idealized continuous galvanizing cycle for the production of galvanized AHSS- dual phase steel strips.

Heating section, where the strip is heated rapidly to temperatures appropriate to produce the microstructures required for further processing of the strip.

Soaking section, where typical processing times can be as long as 60 s.

Primary cooling section, where the strip is rapidly cooled to a temperature as close as possible to the temperature of the galvanizing bath.

Galvanizing section, where the Zn coating is formed by an initial reaction between the steel and the liquid zinc and rapid solidification of a liquid zinc film as the strip leaves the zinc pot and passes through air-knives.

Secondary cooling section, where the strip is cooled to room temperature for further processing.

It should be evident that heating and cooling rates, as well as the time of interruption of cooling at the galvanizing temperature all depend on the size (length) characteristics of each section, the heating and cooling capacities of each section, the line speed and the thickness and width of the strip.

In contrast to the processing parameters used in the fabrication of other types of galvanized products, for example formable steel strips, manufacture of DP-galvanized strips requires the cold worked strip to be heated to temperatures (T_{IA}) within the intercritical (ferrite + austenite) range (i.e., between Ac_1 and Ac_3). The initial heating rate (HR_1), T_{IA} and ($t_1 + t_2$) allow conditioning of the intercritical austenite and ferrite (initial microstructure).

Metallurgically, the most important characteristics of the intercritical austenite are its volume fraction and hardenability which depend on: (a) T_{IA}, ($t_1 + t_2$), (b) grain size, and (c) chemical composition. Also, the metallurgical state of the intercritical ferrite is important (degree of recrystallization, solute content (C, N) and second particle precipitate size and distribution).

The hardenability of the intercritical austenite is the most important metallurgical property in the cooling and galvanizing sections of the process. If the intercritical austenite can be maintained in the metastable form at $T < T_G$ and the secondary cooling rate (CR_2) is fast enough, the intercritical austenite will transform to martensite in this section and a galvanized dual phase steel strip will be produced. It is noteworthy that, in this case, the Zn coating will be formed on an $(\alpha + \gamma)$ microstructure. The effect of the volume change associated with the martensite transformation on the adherence of the Zn coating has never been investigated. On the other hand, if the metastable intercritical austenite transforms to non-martensite products (pro-eutectoid ferrite, perlite or bainite) at $T > T_G$, due to slow CR_1 cooling rates, the Zn coating will be formed on a complex microstructure consisting of intercritical ferrite, pro-eutectoid ferrite, perlite, bainite and residual metastable austenite. This residual austenite may transform to martensite on further cooling to $T < M_s$ or remain in a metastable form at room temperature. Of course, this will depend on the dynamic changes in hardenability that take place during continuous cooling and the interrupted cooling of the strip at T_G.

It is evident from the above discussion that process and product development in CGL's to produce galvanized DP-AHSS strips appears quite a complex task. In general, the chemical compositions of the DP-steels are adjusted to increase their hardenability by adding alloy elements, such as Cr, Mo, Si, B etc. However, the hardenability of intercritical and residual austenite changes dynamically during the cooling stages of the process and makes very difficult to predict their transformation behavior under actual industrial processing conditions.

Several authors have investigated the behavior of dual phase steels during processing in continuous annealing and galvanizing lines [3–10]. For example, Calcagnotto et al. [3] studied the effect of temperature and time of intercritical austenitization on the microstructure and properties produced in dual phase C-Mn steels with an ultra-fine initial microstructure composed of ferrite and perlite. The results showed that the amount of martensite and the ferritic grain size increase when the intercritical austenitization time and temperature are both increased. Also, they reported that the heating rate does not have a significant influence on the phase transformation kinetics of intercritical austenite.

Recent studies [11] indicate that the interruption of cooling at the temperature of the zinc bath (450–460 °C) can cause the formation of bainite or stabilize the austenite. In both cases the result is an undesirable microstructure with limited mechanical properties.

Due the importance of the mechanical properties control in these kind of advanced steels, in the scientific literature have been reported several works to predict the mechanical properties of DP steels using numerical methods based on microstructural characteristics, such as the morphology, phases ratio and diffusion equations for dual phase steel with maximum tensile strength of 1000 MPa [12–20]. In this context, Pernach et al. [18] used numerical methods to model the decomposition of austenite in DP600 steel based on the equations that govern the kinetics and diffusion in steels. Bzowski et al. [19] also used numerical methods to predict the morphology and carbon distribution in austenite. Kim et al. [20] used an orthogonal statistic design to predict the *UTS* and *%EL* in DP-50Kg-grade employing as input variables the annealing and galvanizing temperature.

Van et al. [21], studied the effect of the process parameters on the mechanical properties in TRIP Steels using univariate statistical analysis. However, it is not common to find in the literature the direct relationship of mechanical properties with processing parameters of DP steels using multivariate statistical modelling and optimization.

According to Ray et al. [22] process modelling and optimization are tools for obtaining the best combination of process parameters to maximize or minimize a given product property. There are several ways to model and optimize processes. Some researchers choose to perform experiments and statistical analysis through graphs and thus decide the best parameters for a given process. For example, Lombardi et al. [23] optimized the heat treatment of an aluminum alloy by correlating its mechanical properties with results of metallographic analysis and statistical analysis of the data.

Other authors prefer to use empirical mathematical models based on statistical analysis methods, such as response surface methodology (RSM) and the Taguchi method, as well as advanced analysis methods, such as neural networks, genetic algorithms and finite element modeling. Cavaliere et al. [24] optimized different parameters of the nitriding of a steel by combining RSM with a genetic algorithm. Their models can predict the mechanical properties, the microstructural evolution and the phase transformations as a function of the chemical composition of the steel and the parameters of the nitriding process. Chaouch et al. [25] studied the effect of heat treatment on the mechanical properties of an AISI 4140 steel. Among all the mentioned techniques, RSM is frequently used to model and optimize processes, due to its low computational time and its high level of precision in optimization [26–28].

This methodology is frequently used in the manufacturing process area [26], such as welding processes [29], but can be used for other processes where the response variable is influenced by different input variables [30]. Reisgen et al. [31] developed a statistical model for predicting the heat input and weld bead geometry in laser welding. Cruz et al. [32] obtained the GTAW welding parameters to produce quality joints on Ti6Al4V plates. With these results, the adequate amperage, voltage and advance speed are selected so that the union had the maximum toughness. Miguel et al. [33] also used this methodology to optimize the weld bead geometry using the GMAW welding process for aluminum alloy plates (AA 6063-T5). Singla et al. [26] used the factorial design for investigating the effect of cerium oxide addition on the dry sliding adhesive wear behavior of hardfaced Fe-18Cr-1.1Nb-2.1C alloy and a regression equation of the model was developed and validated with experimental tests. García Nieto et al. [34] used multivariate regression to predict the segregation in continuous cast steel slabs and the results allowed to determinate the most important variables that impact in the industrial process directly.

In the stage of isothermal holding for the galvanizing process, the chemical composition of the residual austenite changes dynamically and consequently the microstructure and mechanical properties of the steel. Due to these dynamic changes in the austenite it is difficult to obtain a general mathematical model that adequately describes this behavior, making necessary the determination of empirical models. There are several studies on the influence of the temperature of intercritical austenitizing on the mechanical properties [2,35–37]. However, the impact of interrupted cooling time at galvanizing temperature and the effects of the cooling rates prior and after galvanizing on the microstructure and mechanical properties of dual phase steel strips processed in CGL's have not been investigated previously. In addition, multivariate modeling in conjunction with multi-objective optimization using genetic algorithms is not habitually applied in the area of heat treatments to model thermal processes, since when the response variables are correlated the statistical analysis becomes more complex.

Therefore, the aim of this work is to optimize the continuous galvanizing cycle to produce DP steels with a minimum ultimate tensile strength of 1100 MPa, yield strength between 550–750 MPa and a minimum elongation of 10% from an advanced cold rolled steel.

2. Materials and Methods

2.1. Theoretical Aspects of Multivariate Model

In this work the response variables to be optimized are the ultimate tensile strength (*UTS*), the yield strength (*YS*) and the elongation (*EL*). From a metallurgical point of view, these mechanical properties are correlated with each other. Therefore, the type of analysis indicated for this case is the multivariate analysis which is appropriate when several measurements are obtained from each experimental sample subjected to an experiment with several input variables [38]. With the use of the statistical modeling it is possible to jointly establish the correlation of the galvanizing thermal cycle parameters (CR_1, t_G and CR_2) with the response variables (*UTS*, *YS* and *EL*). In the regression model of Equation (1) each response variable y in a sample of n observations is represented as a linear function of the process variables x plus a random error ε.

$$\begin{aligned}
y_1 &= \beta_0 + \beta_1 x_{11} + \beta_2 x_{12} + \ldots + \beta_q x_{1q} + \varepsilon_1 \\
y_n &= \beta_0 + \beta_1 x_{n1} + \beta_2 x_{n2} + \ldots + \beta_q x_{nq} + \varepsilon_n
\end{aligned} \tag{1}$$

In this equation the β_i coefficients represent the "weights" that the magnitudes of the input variables (x), by themselves or due to their interactions, have on the magnitude of the output variables

(y). In other words, they represent the magnitude of the influence of the values of the process variables (CR_1, t_G and CR_2) on the ultimate tensile strength (UTS), yield strength (YS) and elongation (EL) of the material. It is noteworthy that, although the UTS and other mechanical properties depend on the final microstructure of the steel and the final microstructure depends strongly on the evolution of microstructure from T_{IA} to room temperature. Therefore, the final mechanical properties depend on the phase transformation behavior of the intercritical austenite during the cooling cycle.

Thereby, the multivariate model is constructed as follows [39]:

a) Selection of the input variables according to the objective of the investigation;
b) Selection of experimental design and generation of the experimental matrix;
c) Perform the experiments according to the experimental matrix designed;
d) Statistical analysis of the experimental data to obtain the fit of the polynomial function; i.e., obtain the β_i coefficients in Equation (1).
e) Statistical evaluation of the fitted model using multivariate variance analysis (MANOVA) and analysis of determination coefficients (R^2);

Once it is ensured that the model meets with the statistical assumptions and presents good prediction, the regression equations are used as functions in the optimization stage. In this investigation the multivariate model is developed using three dependent variables (UTS, YS and EL) and three independent variables (CR_1, t_G and CR_2). The multivariate model can be written in its general form as follows [38]:

$$Y = XB + E. \tag{2}$$

Therefore, the coefficients B can be estimated by the following equation [38]:

$$\hat{B} = (X'X)^{-1}X'Y. \tag{3}$$

When the process variables have an influence on the response variable, hypothesis tests for the analysis of variance should be performed. The hypothesis test is used to determine which hypothesis is best supported by the data. There are two hypotheses: The null hypothesis (H_0) and the alternative hypothesis (H_1). In this investigation the test of Wilks (Λ) was used to determine if the process variables have a significant influence on the response variables. This test is based on the calculation of Λ:

$$\Lambda = \frac{|Y'Y - \hat{B}'X'Y|}{|Y'Y - n\overline{yy'}|}. \tag{4}$$

The null hypothesis is H_0: B1 = 0, where B1 is the matrix X without the first column. The null hypothesis is rejected if $\Lambda \leq$ table value and it means that the response variables are influenced by the process variables.

The performance of a model can be expressed by the goodness coefficient or determination coefficient R^2. For multivariate modeling there are several measurements of association between the y's and the x's. One of these measurements is based on Wilks test:

$$R^2 = 1 - \Lambda. \tag{5}$$

In the present investigation a composite central design with three center points is used for obtaining the experimental matrix. The experimental data obtained are then used to evaluate the feasibility of manufacturing cold-rolled galvanizing dual phase (ferrite + martensite) steel strip with a $UTS > 1100$ MPa, YS between 550–750 MPa and a minimum elongation of 10% in an idealized continuous galvanizing line where the thermal profile to which the strip is subjected is, as shown in Figure 1. In practice, the heating and cooling rates in each step are defined by the speed and the actual temperature of the strip along the processing line. It is noteworthy that the galvanizing stage effectively represents an interruption of cooling at temperatures between 450 and 460 °C. Therefore, phase transformations that take place on the cooling of the strip may be affected significantly.

The mechanical properties investigated are important in advanced steels because they determine the potential of weight savings of the components used in the automotive industry [1]. The goal in this work is to evaluate the effect of CR_1, t_G and CR_2 on the mechanical properties of the steel with chemical composition listed in Table 1.

Table 1. Chemical composition of the experimental steel.

Element	C	Si	Mn	P	S	Cr	Mo	Ni	B
wt. %	0.154	0.260	1.906	0.013	0.0009	0.413	0.108	0.048	0.0010
Element	Al	Cu	Nb	Ti	V	Ca	N	Fe + Impurities	
wt. %	0.036	0.018	0.004	0.044	0.008	0.001	0.0036	Balance	

The experimental region of the process variables (CR_1, t_G and CR_2) is presented in Table 2 and the 17 conditions produced by the experimental design are shown in Table 3. The experimental matrix represents a composite center design with 3 central points for investigating the relationship of the mechanical properties with the process variables.

Table 2. Independent process variables and experimental design levels.

Notation	Process Variable	Unit	Level	
			Low −1	High +1
x_1	Cooling rate (CR_1)	°C/s	10	110
x_2	Hold time (t_G)	s	3	20
x_3	Cooling rate (CR_2)	°C/s	10	110

Table 3. Design matrix with independent process variables.

	Process Variables		
Run	CR_1 °C/s	t_G s	CR_2 °C/s
1	30	17	90
2	10	11	60
3	110	11	60
4	60	11	60
5	90	6	30
6	60	11	10
7	30	17	30
8	30	6	90
9	30	6	30
10	90	6	90
11	60	20	60
12	60	11	60
13	60	11	110
14	90	17	90
15	60	11	60
16	90	17	30
17	60	3	60

The models obtained after going through the inferential statistical analysis can be used as objective functions to perform the optimization of input parameters (CR_1, t_G and CR_2) looking for the results of optimal mechanical properties for the manufacture of a DP steel with the desired properties. For this a genetic algorithm NSGAII (Non-Dominated Sorting Genetic Algorithm) was used, it allows to optimize several responses at the same time, that is, a multi-objective optimization.

The use of the genetic algorithm NSGAII is based on the mechanisms of genetics and natural selection, combining the survival of the fittest in the form of structured representations of solutions to the problem, called individuals, with mechanisms of information exchange for the generation of new solutions. The search is based on the information obtained from the problem and as the space of possible solutions is traversed, the search is directed towards the best values [40]. Thereby, to optimize the parameters of the thermal cycle presented in Figure 1, a solution is searched that maximizes the *UTS*, as well as elongation and minimizes the yield strength (*YS*) in order to obtain a DP steel with minimum *UTS* of 1100 MPa, yield strength between 550 and 750 MPa and a minimum elongation of 10%. Five solutions obtained from optimization were selected to perform experimental tests. Furthermore, the hypothesis of a normal distribution of the data was verified.

2.2. Experimental Methods

Figure 2 illustrates the time-temperature-transformation (TTT) and continuous cooling transformation (CCT) diagrams calculated using JMatPro (Java-based Materials Properties, 9.0 version, Sente Software Ltd., Guildford, UK) for the steel investigated. The calculation was performed assuming $T_{IA} = 800\ °C$. Under these conditions of intercritical austenitizing, the software predicts an initial microstructure consisting of 11.3% ferrite and 88.7% austenite, an intercritical austenite grain size of 8.5 µm and an M_s of 365.6 °C.

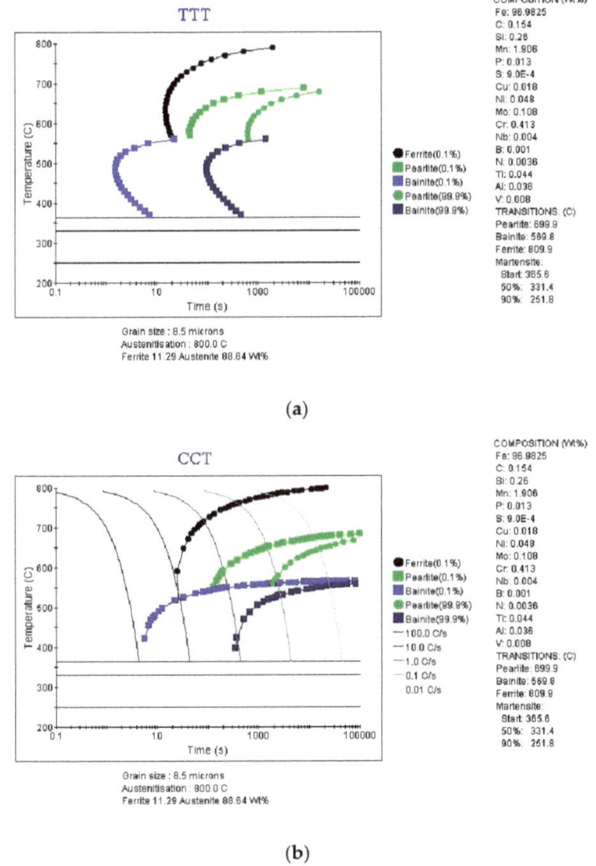

(a)

(b)

Figure 2. Calculated TTT (**a**) and CCT (**b**) diagrams for the steel investigated. The calculation was performed assuming a $T_{IA} = 800\ °C$.

As can be appreciated in the CCT diagram, cooling rates faster than 100 °C/s are needed to produce ferrite + martensite microstructures while cooling rates between 30 and 100 °C/s result in ferrite

+ bainite + martensite microstructures. Cooling rates slower than 1 °C/s produce microstructures consisting of ferrite+pearlite and martensite can not be formed under these cooling conditions. It is noteworthy that, according to the TTT diagram, any isothermal holding at temperatures between 450 and 460 °C will produce a transformation of any metastable residual austenite to bainite. Another important observation is that, for cooling rate slower than 10 °C/s, the ferrite volume fraction in the final microstructure has contributions from the intercritical ferrite present at the beginning of cooling and the pro-eutectoid ferrite formed on cooling. Ferrite + perlite + bainite microstructures are produced when the cooling rates are slower than 30 °C/s. It is therefore evident that, with the chemical composition of the steel and using a T_{IA} = 800 °C, a wide variety of microstructures, and consequently of properties can be produced depending on the actual cooling conditions during processing. In this work, the effect of introducing an interrupted cooling stage (isothermal holding) at 460 °C to simulate the thermal effects of galvanizing on the microstructure and mechanical properties after final cooling to room temperature is investigated.

The thermal cycles (Figure 1) were simulated using a quenching dilatometer Linseis model L78 RITA (Selb, Germany). The experiments were carried out in both miniature tension test samples with dimensions presented in Figure 3, and standard test samples of 5 mm × 11 mm × 1.1 mm for microstructure and microhardness analysis. The tensile tests were performed using a miniature extensometer (MTS System Corporation, Eden Prairie, MN, USA) MTS model 632.29 with a 5 mm calibrated gauge length. This extensometer has a strain measuring the range of −10% to 30%, which is perfectly adequate for the tensile test specimens designed for dilatometry. The tensile tests carried out using an electromechanical universal testing system MTS model QTEST/100 (MTS System Corporation, Eden Prairie, MN, USA) at a crosshead speed as 1.5 mm/min. Experimental data were obtained to provide ultimate tensile strength (*UTS*), yield strength (0.2% proof stress) and elongation to fracture (*EL*). Al experiments were performed by duplicate.

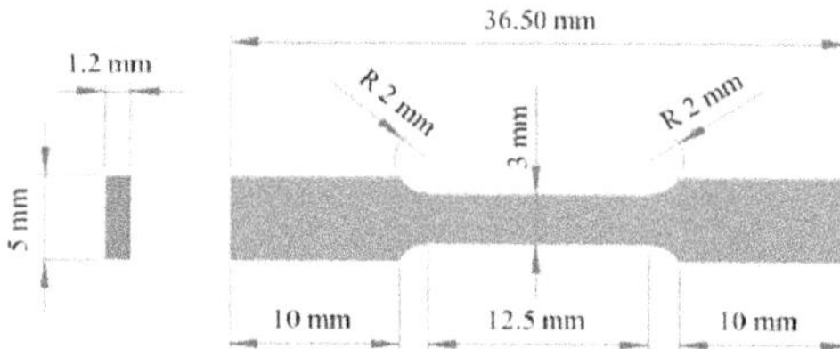

Figure 3. Dimensions of miniature tensile test specimens used in dilatometry experiments.

For metallographic analysis, the samples were cut in the longitudinal section, prepared using conventional metallographic techniques and analyzed by scanning electron microscopy (SEM) model JSM-7800F (JEOL Ltd., Akishima, Japan). Microhardness tests were carried out using a Vickers Tukon 300-FM/10kg microdurometer (Wilson Instruments®, Binghamton, NY, USA) with a 500 gf load applied during 12 s. The actual initial microstructure of the steel at 800 °C for all dilatometry experiments consists of 35% intercritical ferrite and 65% intercritical austenite with an average intercritical austenite grain size of 2.8 μm. This later value is about 3 times smaller than the one predicted by JMatPro. CR_1 and CR_2 are varied between 10–110 °C/s, while T_G, the galvanizing temperature, is kept constant at 460 °C. Finally, the interrupted cooling time (t_G) is varied from 3 to 20 s.

3. Results and Discussion

3.1. Effect of Thermal Cycle on Mechanical Properties

The mechanical properties resulting from the 17 experiments performed (different thermal cycles) included in the experimental design matrix (Table 3) are listed in Table 4. With these results, the model coefficients (β) of Equation (1) were determined using the least squares method described by

Rencher [38]. The multivariate variance analysis (MANOVA) was used to evaluate the fit of the model and identifying which factors and interactions among them were more significant.

Table 4. Mechanical properties for each of the experimental combination (standard deviation in brackets).

	Response Variables				Response Variables		
Run	UTS	YS	EL	Run	UTS	YS	EL
	MPa	MPa	%		MPa	MPa	%
1	1142 (3)	729 (13)	11.3 (3.3)	10	1274 (13)	959 (30)	8.6 (0.6)
2	1174 (34)	754 (21)	12.1 (3.1)	11	1123 (4)	730 (28)	9.9 (1.2)
3	1237 (1)	829 (7)	10.3 (0.5)	12	1187 (9)	828 (17)	10.5 (1.8)
4	1245 (4)	853 (9)	10.8 (0.6)	13	1203 (20)	781 (13)	10.7 (0.3)
5	1264 (6)	890 (19)	8.3 (0.7)	14	1145 (13)	779 (24)	9.4 (1.6)
6	1141 (18)	745 (31)	9.9 (1.6)	15	1199 (8)	844 (33)	9.5 (1.1)
7	1131 (32)	725 (18)	10.7 (0.9)	16	1166 (12)	777 (16)	10.1 (0.8)
8	1226 (3)	841 (20)	8.6 (1.3)	17	1294 (31)	1015 (11)	8.0 (0.6)
9	1196 (8)	791 (11)	9.8 (0.8)	-	-	-	-

3.2. Development of Statistical Model

Before making the estimation of the parameters of the model, it is important to determine if the response variables are correlated. The answer to this question allows to select the method that should be employed to estimate the β parameters, that is, whether the calculations should be done using univariate or multivariate methods. To this end, the Pearson correlation test was used. This test calculates the correlation coefficient between the response variables considered. If one variable tends to increase while the others decrease, the correlation coefficient is negative. On the other hand, if the two variables tend to increase at the same time, the correlation coefficient is positive. The hypothesis test performed is as follows [41], H_0: $r = 0$ versus H_1: $r \neq 0$ where r is the correlation between a pair of variables.

The null hypothesis H_0 indicates that there is no correlation between the variables analyzed. Thus, the p-value is used to reject or not the null hypothesis, that is, if the P-value is less than the value of alpha ($\alpha = 0.05$) the null hypothesis is rejected and therefore the variables could be correlated. The level of alpha often used is 0.05 which means that the possibility of finding an effect that does not really exist is only 5%. Then, when the alpha value is equal to 0.05, the results can be accepted with 95% confidence [41].

Table 5 presents the results of the Pearson correlation test applied to the y responses showed a Table 4. As can be seen at Table 5, all the p-values are less than 0.05 ($\alpha = 0.05$), which means that, with 95% confidence, the responses variables (*UTS*, *YS* and *EL*) are correlated with each other.

Table 5. Pearson correlation test between the response variables.

Variable	*UTS*	*YS*
YS	0.926 0.000	-
EL	−0.570 0.017	−0.717 0.001
Contents of the cell: Pearson Correlation p-value		

Pearson correlation values demonstrate that the correlation between *UTS* and *YS* is positive, that is, when the *UTS* increases the *YS* tends to increase. However, when these two variables increase the elongation decreases since the EL presents a negative correlation with *YS* and *UTS*. From physical metallurgy point of view, these results are completely coherent. Since the response variables are correlated and several input variables (*x*'s) are considered, the linear regression model must be calculated using the equations for multivariate modeling (Equations (3) and (4)). The resulting models are:

$$UTS \ (MPa) = 1241.7 + 0.6368CR_1 - 9.0971t_G + 0.3278CR_2 \tag{6}$$

$$YS \ (MPa) = 868.05 + 1.0918CR_1 - 12.836t_G + 0.4549CR_2 \tag{7}$$

$$E \ (\%) = 9.772 - 0.0204CR_1 + 0.1099t_G + 0.001639CR_2 \tag{8}$$

Analyzing the values of the β coefficients of the model for UTS (Equation (6)), it becomes evident that cooling rates CR_1 and CR_2 have a positive linear relationship with the UTS. In contrast, the cooling interruption time (t_G) at 460 °C has a negative linear relationship with the UTS. The variable that has the greatest impact on UTS is t_G at 460 °C, where increasing t_G causes a decrease in UTS. The same behavior is noted for YS, but for EL the cooling rate CR_1 and the time t_G have opposite effects than those presented by UTS and YS.

As can be seen in the calculated TTT or CCT diagrams of Figure 2, interruption of cooling at 460 °C causes an isothermal transformation of metastable austenite to bainite. Therefore, the UTS and YS at room temperature of the steel will decrease with increasing t_G since smaller amounts of martensite would be present at room temperature after the final cooling.

The multivariate determination coefficient (R^2) was calculated according to Equation (5): $R^2 = 91\%$. This result indicates that the model has a good fit since the coefficient of determination is greater than 80% [41]. It can be that 91% of the variability of the process is explained. The models in Equations (6)–(8) allow optimizing the heat treatment process using the optimization algorithm NSGA II and helps to predict the behavior of the process considering input variables changes.

After estimating the coefficients, it is important to verify if the data satisfy some assumptions, which according to Montgomery [41] are:

a) The variance of the errors (residuals) must be homogeneous;
b) Errors must be independent;
c) Errors must have a normal distribution.

Using the Henze-Zirkler multivariate normality test [42], the normality of the data can be verified. The p-value of the Henze-Zirkler test for the calculated models was 0.0829 indicating that the data follows a normal distribution and therefore comply with part of the assumptions for multivariate variance analysis. To verify the homogeneity of the variance, another statistical test is performed called the Breusch-Pagan test. The null hypothesis of this test is that the variance is homogeneous over the residuals of the model. If this is not the case (rejection of H_0), the model can lose its efficiency and makes erroneous predictions [43]. The variance of the models (Equations (6)–(8)) is homogeneous since the p-values of the tests were greater than $\alpha = 0.05$ according to the results in Table 6.

Table 6. Breusch-Pagan test for verification of the homogeneity of the variance of the residuals for each response variable.

Studentized Breusch-Pagan Test	
Model	**p-Value**
$(UTS \sim V1 + t2 + V2)$	0.71
$(YS \sim V1 + t2 + V2)$	0.2588
$(EL \sim V1 + t2 + V2)$	0.6142

Once the model data satisfies the assumptions for statistical analysis, the next step is to verify the effect of the input variables and their interactions over the UTS, YS, and EL. Thus, the multivariate variance analysis (MANOVA) was carried out using the Wilks test. The results of the p-values for each input variable and the first order interactions between them are presented in Table 7.

Table 7. *p*-Values for multivariate analysis of variance (MANOVA).

Terms	Valor–P
V_1	0.0432031
t_2	0.0003479
V_2	0.5830329
$V_1 t_2$	0.7387451
$t_2 V_2$	0.8638424
$V_1 V_2$	0.6460354

In this case, the null hypothesis (H_0) is that the variable under analysis is not significant. According to the results in Table 7, the interactions between the input variables are not significant (p-value $> \alpha = 0.05$). The results also show that CR_2 is not significant. However, it was assumed that this variable is significant considering that, from physical metallurgy point of view, CR_2 does play an important role in the transformation to martensite of any residual austenite at the end of the interruption of cooling at 460 °C. Of course, this is important when considering the thermal cycles involved in continuous galvanizing.

The hypothesis tests carried out as part of the inferential analysis indicated that the models obtained have a good performance ($R^2 = 91\%$). Thus, Equations 6-8 were used as objective functions in the optimization of input variables, as will be shown in the next sections.

3.3. Optimization

Process optimization techniques allow the selecting of the best parameters combination from a set of available alternatives. The problem to be optimized is defined by an objective function that returns as a result a real value or a vector of real values. Multi-objective optimization is used when it is necessary to optimize more than one objective function simultaneously, where the solution is composed of a set of optimal elements and, generally, it needs a decision maker to select one of them [40].

In this research it is sought to minimize the resistance to yield (YS) and to maximize both the ultimate tensile strength (UTS) and elongation (EL). The objective functions used in the optimization process are the statistical models presented in Equations (6)–(8).

The results of the optimization are presented in the form of a Pareto Front graph, as illustrated in Figure 4. The graph presents the set of optimal solutions in the target space where all the objectives in play are considered. All the options presented in the Pareto Front are equally valid [40]. The next step is to perform experimental tests with solutions presented in the Pareto Front.

Of the 21 optimal solutions presented in Figure 4, five conditions were selected to corroborate de optimization experimentally and the mechanical properties results are presented in Table 8. The criterion for selecting the conditions for the experimental test was that they should comply with a minimum UTS of 1100 MPa, YS between 550 and 750 MPa and a minimum elongation of 10%.

The five conditions selected were run experimentally using the miniature tensile test specimens designed to perform dilatometry thermal cycles. The mechanical properties were measured, and their results compared with those calculated by the model (Figure 5). The error in the prediction of the UTS is less than 3.4%, in the YS it is less than 7.7% and that of the elongation to fracture is less than 7.6%. These results indicate that the models obtained have a very good predictive capacity since they present errors of less than 10% [31].

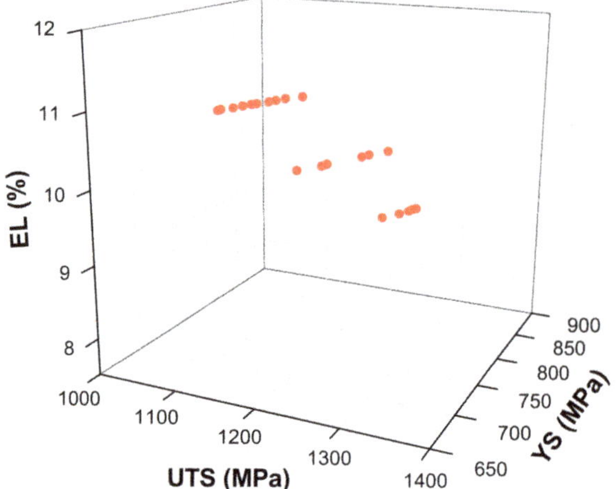

Figure 4. Pareto front for the results of multi-objective optimization.

Table 8. Results of the multi-objective optimization using the models.

Test	CR_1 (°C/s)	t_G (s)	CR_2 (°C/s)	Model Results			Experimental Results		
				UTS (MPa)	YS (MPa)	EL (%)	UTS (MPa)	YS (MPa)	EL (%)
TVF-1	10	15	26	1120	698	11.3	1140	749	10.5
TVF-2	10	15	13	1116	692	11.2	1109	683	10.7
TVF-3	13	13	25	1144	732	10.9	1116	734	11.3
TVF-4	17	14	15	1126	708	11.0	1129	767	11.6
TVF-5	15	15	24	1123	704	11.1	1162	757	10.6

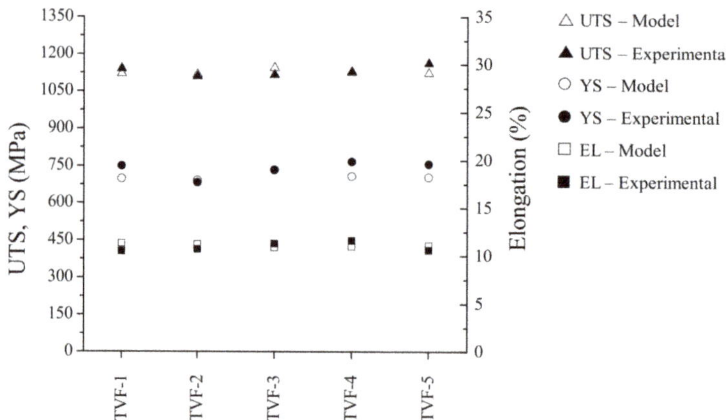

Figure 5. Results of mechanical properties for the experimental tests with optimization results.

In addition, the thermal TVF-3 presented in Table 8 can be used to produce dual phase steel (DP1100) that is desired in this investigation. It is worth mentioning that this condition results in a *YS/UTS* ratio of 0.66. This value is important because dual phase steels must have very good formability for the manufacture of components for the automotive industry, so it is important to minimize the *YS/UTS* ratio. Figure 6 presents the microstructure of this condition where the thermal cycle is optimized.

the microstructure was quantified by metallography and is formed by 13% martensite (α'), 32% bainite (B) and 55% ferrite (α).

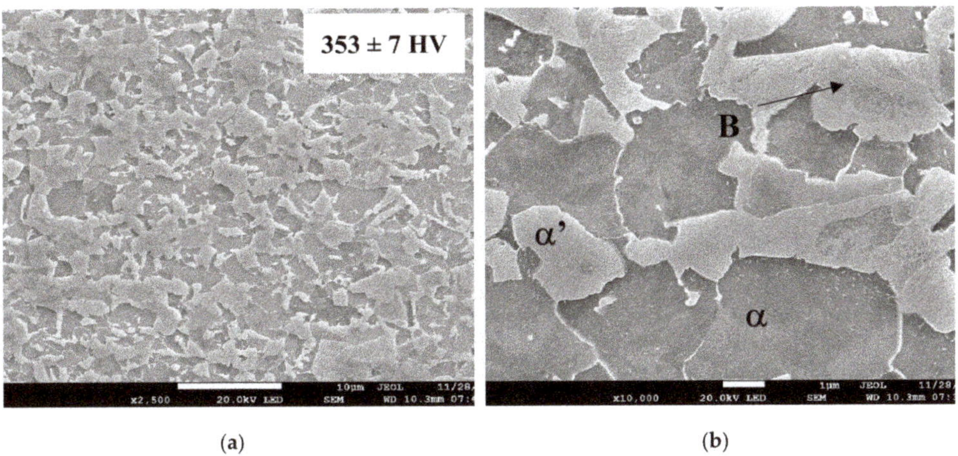

(a)　　　　　　　　　　　　　　　(b)

Figure 6. Microstructure of TVF-3 heat treated sample: (**a**) 2500× and (**b**) 10,000×; B: Cainite; α: Ferrite and α': Martensite.

3.4. Effects of Process Parameters

In the manufacturing of galvanizing DP-AHSS steels, the cooling rate (CR_1) must be fast enough to avoid the formation of perlite or bainite and retain carbon in solution in the metastable austenite until the transformation to martensite takes place at $T < M_s$. With the statistical models developed in this work (Equations (6)–(8)) it is possible to predict the effect of CR_1 over the mechanical properties to DP steel, as shown in Figure 7. To plot this graph, the secondary cooling rate (CR_2) was kept constant at 25°C/s and t_G at 13 s. As can be seen, both the *UTS* and *YS* tend to decrease with slower initial cooling rates (CR_1).

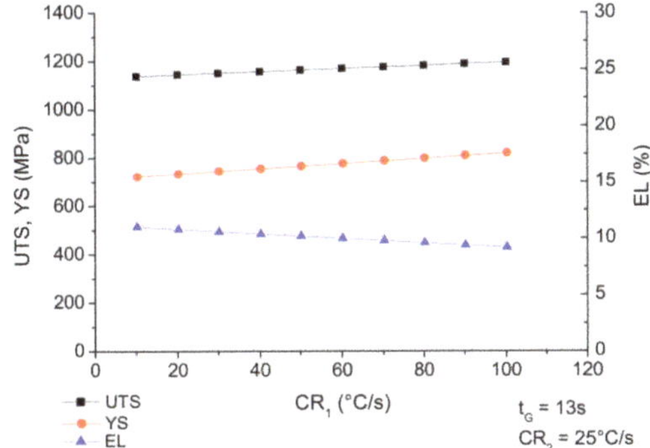

Figure 7. Effect of CR_1 on mechanical properties of dual phase (DP) steel. Calculations with statistical models.

According to these results, a galvanizing DP steel with a minimum *UTS* of 1100 MPa, *YS* between 550 and 750 MPa and a minimum elongation to fracture of 10% can be produced using values of CR_1 between 10 °C/s and 30 °C/s with the other parameters constants at: $HR_1 = 35$ °C/s; $T_{IA} = 800$ °C, $t_2 = 60$ s; $t_G = 13$s, $T_G = 460$ °C and $CR_2 = 25$ °C/s.

During the time of interruption of cooling at 460 °C, a certain amount of metastable residual austenite is available which, during the isothermal holding transforms to bainite as can be seen in both the TTT and CTT diagrams presented in Figure 2. These transformations have a significant impact on the mechanical properties of a galvanizing DP steel, as can be seen in Figure 8. With the increase of t_G at 460 °C the *UTS* and the *YS* decreases, and the elongation increases. Thus, in order to ensure the desired mechanical properties (*UTS* > 1100 MPa, *YS* between 550 and 750 MPa and elongation to fracture greater than 10%), the time of interruption of cooling must be between 13 and 17 s when the other process variables are set to: $HR_1 = 35$ °C/s; $T_{IA} = 800$ °C, $t_2 = 60$ s; $CR_1 = 13$ °C/s, $T_G = 460$ °C and $CR_2 = 25$ °C/s.

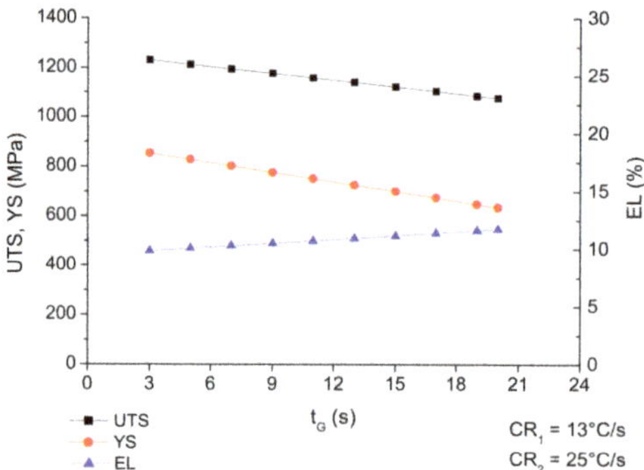

Figure 8. Effect of interruption of cooling (t_G) on the mechanical properties of DP steel. Calculations with statistical models.

Figure 8 shows that as t_G increases, the *UTS* decreases and this behavior is associated with the formation of lower strength microstructures. The formation of bainite and consequently the decrease in the fraction of martensite in the microstructure causes the observed decrease in tensile strength. According to the TTT and CCT diagrams (Figure 2), the transformation of metastable austenite to bainite occurs during the interruption of cooling. Fonstein [1] reported that there is little information about the effect of the presence of bainite in DP steels. However, some data indicate that the hardening generated by bainite in DP steels is weaker than that generated by the presence of martensite. Also, Fonstein et.al [44] report that a 10% replacement of martensite by bainite decreases the *UTS* in 40 MPa.

The transformation of austenite during the interruption of cooling at 460 °C can be of vital importance since, in general, the martensite start-temperature (M_s) of most industrial DP-steel grades are lower than 450 °C. Therefore, the formation of bainite in processes that involve interruption of cooling, such continuous galvanizing, can inhibit the formation of martensite and, consequently, the desired strength in the steel will not be obtained.

It can be seen in Figure 9 that the final cooling rate (CR_2) does not significantly influence the final mechanical properties. This result is similar to that reported by other researchers [1], who suggest that this final cooling rate (CR_2) does not have a significant influence on the strength of the steel. The reason for this suggestion is that the amount of martensite generated is determined by the initial cooling rate in thermal cycles to obtain DP-steels without interrupted cooling at 460 °C. However, in this work, the isothermal holding at 460 °C prior to the final cooling produces residual metastable austenite with more carbon compared to the intercritical austenite and then the final rate for the transformation of the residual austenite has a significant effect in the *UTS*. Furthermore, using $HR_1 = 35$ °C/s; $T_{IA} = 800$ °C, $t_2 = 60$ s; $CR_1 = 13$ °C/s, $T_G = 460$ °C and $t_G = 13$ s, it is possible to obtain a galvanizing DP steel with minimum *UTS* of 1100 MPa, *YS* between 550 and 750 MPa and a minimum elongation of 10% with the final cooling rate (CR_2) between 10 and 70 °C/s.

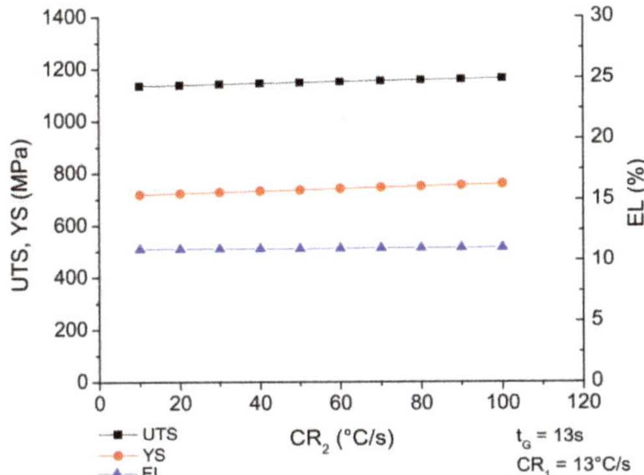

Figure 9. Effect of CR$_2$ on UTS mechanical properties in DP steel. Calculations with statistical models.

Analysis of the effect of the process variables of the thermal cycle presented in Figure 1 on the mechanical properties of the steel using and the statistical models developed, allow defining a process window in which the process parameters can be combined, so that the DP1100 steel can be produced. Of course, the applicability of the models developed is limited to the specific chemical composition and amount of deformation by cold rolling of the steel for which they were developed. In the present case, processing with $HR_1 = 35$ °C/s; $T_{IA} = 800$ °C, $t_2 = 60$ s; $CR_1 = 12$–30 °C/s, $T_G = 460$ °C, $t_G = 13$–17 s and $CR_2 = 15$–30 °C/s, allow to obtain galvanizing DP steel with a minimum UTS of 1100 MPa, a maximum YS of 750 MPa and a minimum elongation to fracture of 10%.

It is important to emphasize that, from the results of the present investigation, it is clear that the holding time at 460 °C has the greatest influence on the resulting mechanical properties. The decrease in UTS when t_G is increased is related to the lower amount of martensite produced as a result of the final cooling, due to the formation of bainite during the holding period at 460 °C. These results are in accordance with those reported by Bellhouse [2] that studied the effect of the holding time at 465 °C on the mechanical properties and microstructure of TRIP steels using a thermal cycle very similar to the one used in this work.

4. Conclusions

The models developed using the multivariate analysis combined with multi-objective optimization shown good prediction capabilities for the mechanical properties of DP steels processed under continuous galvanizing conditions with a prediction error less than 10%. With this model it is possible to design, from very specific chemical composition, different grades of galvanized DP steels with UTS higher than 1100 MPa.

It is possible to obtain dual phase microstructures with the steel chemistry selected in this investigation and the intercritical austenitization temperature of 800 °C. For this, an initial cooling rate to 460 °C of 13 °C/s, an interruption of cooling of 13 s at 460 °C and a final cooling rate of 25 °C/s seem to be most appropriate.

In relation to the effect of the heat treatment parameters, the results show that the holding time at the temperature of the zinc bath (460 °C) is the most significant variable for the processing of galvanized DP steels. The evolution of the microstructure and the final mechanical properties of these steel grades are directly controlled by this process parameter. The t_G has a strong impact on the UTS, due to the transformation of the metastable austenite to bainite that causes that the UTS and YS decrease as the time of interruption increases.

Furthermore, the multivariate model results are agreed with the metallurgist knowledge about the phenomenon discussed in this study and the methodology applied can be used to another dataset like the variables used in this investigation, but it is important to consider the specific characteristics of each industrial process.

Author Contributions: P.C. performed the experiments, characterization and analysis of the data as part of her PhD. activities; G.A support in the metallurgical analysis and characterization; A.S. and F.G. conceived and design the experiments; G.A. and P.C. coordinated and planned the activities proposed in the methodology; A.S. and F.G. provided the materials, computer programs and other analysis tools; A.S. supervised the development of the methodology; P.C. and G.A. discussed the results and wrote the first draft of the manuscript; D.S.G.-G. support in the statistical analysis; All authors revised and approved the final version of the manuscript.

Funding: This research was funded by Consejo Nacional de Ciencia y Tecnología (CONACYT), grant number 379955.

Acknowledgments: The authors gratefully acknowledge the support and guidance provided by the International Zinc Association (IZA). A.S. and P.C. are indebted to CONACYT for support in the form at a graduate studies fellowship.

Conflicts of Interest: The authors declare no conflict of interest.

References

1. Fonstein, N. *Advanced High Strength Sheet Steels*; Springer International: Chicago, IN, USA, 2015. [CrossRef]
2. Bellhouse, E.M.; McDermid, J.R. Effect of continuous galvanizing heat treatments on the microstructure and mechanical properties of high Al-low Si transformation induced plasticity steels. *Metall. Mater. Trans. A* **2010**, *41*, 1460–1473. [CrossRef]
3. Calcagnotto, M.; Ponge, D.; Raabe, D. Microstructure control during fabrication of ultrafine grained dual-phase steel: Characterization and effect of intercritical annealing parameters. *ISIJ Int.* **2012**, *52*, 874–883. [CrossRef]
4. Park, I.J.; Kim, S.T.; Lee, I.S.; Park, Y.S.; Moon, M.B. A Study on Corrosion Behavior of DP-Type and TRIP-Type Cold Rolled Steel Sheet. *Mater. Trans.* **2009**, *50*, 1440–1447. [CrossRef]
5. Sodjit, S.; Uthaisangsuk, V. Microstructure based prediction of strain hardening behavior of dual phase steels. *Mater. Des.* **2012**, *41*, 370–379. [CrossRef]
6. Aslam, I.; Li, B.; Martens, R.L.; Goodwin, J.R.; Rhee, H.J.; Goodwin, F. Transmission electron microscopy characterization of the interfacial structure of a galvanized dual-phase steel. *Mater. Charact.* **2016**, *120*, 63–68. [CrossRef]
7. Ebrahimian, A.; Banadkouki, S.G. Effect of alloying element partitioning on ferrite hardening in a low alloy ferrite-martensite dual phase steel. *Mater. Sci. Eng. A* **2016**, *677*, 281–289. [CrossRef]
8. Tasan, C.C.; Diehl, M.; Yan, D.; Bechtold, M.; Roters, F.; Schemmann, L.; Zheng, C.; Peranio, N.; Ponge, D.; Koyama, M.; et al. An Overview of Dual-Phase Steels: Advances in Microstructure-Oriented Processing and Micromechanically Guided Design. *Annu. Rev. Mater. Res.* **2014**, *45*, 391–431. [CrossRef]
9. Garcia, C.I.; Hua, M.; Cho, K.; Redkin, K.; DeArdo, A.J. Metallurgy and continuous galvanizing line processing of high-strength dual-phase steels microalloyed with Niobium and Vanadium. *Metall. Ital.* **2012**, *104*, 3–8.
10. Niakan, H.; Najafizadeh, A. Effect of niobium and rolling parameters on the mechanical properties and microstructure of dual phase steels. *Mater. Sci. Eng. A.* **2010**, *527*, 5410–5414. [CrossRef]
11. Wu, R.; Wang, L.; Jin, X. Thermal Stability of Austenite and Properties of Quenching & Partitioning (Q&P) Treated AHSS. *Phys. Procedia* **2013**, *50*, 8–12. [CrossRef]
12. Ramazani, A.; Pinard, P.; Richter, S.; Schwedt, A.; Prahl, U. Characterisation of microstructure and modelling of flow behaviour of bainite-aided dual-phase steel. *Comput. Mater. Sci.* **2013**, *80*, 134–141. [CrossRef]
13. Ramazani, A.; Mukherjee, K.; Quade, H.; Prahl, U.; Bleck, W. Correlation between 2D and 3D flow curve modelling of DP steels using a microstructure-based RVE approach. *Mater. Sci. Eng. A* **2013**, *560*, 129–139. [CrossRef]
14. Ramazani, A.; Mukherjee, K.; Prahl, U.; Bleck, W. Modelling the effect of microstructural banding on the flow curve behaviour of dual-phase (DP) steels. *Comput. Mater. Sci.* **2012**, *52*, 46–54. [CrossRef]
15. Ramazani, A.; Mukherjee, K.; Prahl, U.; Bleck, W. Transformation-Induced, Geometrically Necessary, Dislocation-Based Flow Curve Modeling of Dual-Phase Steels: Effect of Grain Size. *Met. Mater. Trans. A* **2012**, *43*, 3850–3869. [CrossRef]

16. Schoof, E.; Schneider, D.; Streichhan, N.; Mittnacht, T.; Selzer, M.; Nestler, B. Multiphase-field modeling of martensitic phase transformation in a dual-phase microstructure. *Int. J. Solids Struct.* **2017**, *134*, 181–194. [CrossRef]

17. Wei, X.; Asgari, S.; Wang, J.; Rolfe, B.; Zhu, H.; Hodgson, P. Micromechanical modelling of bending under tension forming behaviour of dual phase steel 600. *Comput. Mater. Sci.* **2015**, *108*, 72–79. [CrossRef]

18. Pernach, M.; Bzowski, K.; Pietrzyk, M. Numerical modeling of phase transformation in dual phase (DP) steel after hot rolling and laminar cooling. *Int. J. Multiscale Comput. Eng.* **2014**, *12*, 397–410. [CrossRef]

19. Bzowski, K.; Rauch, L.; Pietrzyk, M. Application of statistical representation of the microstructure to modeling of phase transformations in DP steels by solution of the diffusion equation. *Procedia Manuf.* **2018**, *15*, 1847–1855. [CrossRef]

20. Kim, S.-J.; Cho, Y.-G.; Oh, C.-S.; Kim, D.E.; Moon, M.B.; Han, H.N. Development of a dual phase steel using orthogonal design method. *Mater. Des.* **2009**, *30*, 1251–1257. [CrossRef]

21. Van, H.D.; Van, C.N.; Ngoc, T.T.; Manh, T.S. Influence of heat treatment on microstructure and mechanical properties of a CMnSi TRIP steel using design of experiment. *Mater. Today Proc.* **2018**, *5*, 24664–24674. [CrossRef]

22. Ray, P.; Ganguly, R.; Panda, A. Optimization of mechanical properties of an HSLA-100 steel through control of heat treatment variables. *Mater. Sci. Eng. A* **2003**, *346*, 122–131. [CrossRef]

23. Lombardi, A.; Ravindran, C.; Mackay, R. Optimization of the solution heat treatment process to improve mechanical properties of 319 Al alloy engine blocks using the billet casting method. *Mater. Sci. Eng. A* **2015**, *633*, 125–135. [CrossRef]

24. Cavaliere, P.; Perrone, A.; Silvello, A. Multi-objective optimization of steel nitriding. *Eng. Sci. Technol. Int. J.* **2016**, *19*, 292–312. [CrossRef]

25. Chaouch, D.; Guessasma, S.; Sadok, A. Finite Element simulation coupled to optimisation stochastic process to assess the effect of heat treatment on the mechanical properties of 42CrMo4 steel. *Mater. Des.* **2012**, *34*, 679–684. [CrossRef]

26. Cartuyvels, R.; Booth, R.; Dupas, L.; De Meyer, K. Process technology optimization using an integrated process and device simulation sequencing system. *Microelectron. Eng.* **1992**, *19*, 507–510. [CrossRef]

27. Singla, Y.K.; Arora, N.; Dwivedi, D. Dry sliding adhesive wear characteristics of Fe-based hardfacing alloys with different CeO2 additives—A statistical analysis. *Tribol. Int.* **2017**, *105*, 229–240. [CrossRef]

28. Correia, D.S.; Gonçalves, C.V.; Da Cunha, S.S.; Ferraresi, V.A.; da Cunha, S.S., Jr. Comparison between genetic algorithms and response surface methodology in GMAW welding optimization. *J. Mater. Process. Technol.* **2005**, *160*, 70–76. [CrossRef]

29. Benyounis, K.; Olabi, A. Optimization of different welding processes using statistical and numerical approaches—A reference guide. *Adv. Eng. Softw.* **2008**, *39*, 483–496. [CrossRef]

30. Šumić, Z.; Vakula, A.; Tepić, A.; Čakarević, J.; Vitas, J.; Pavlić, B. Modeling and optimization of red currants vacuum drying process by response surface methodology (RSM). *Food Chem.* **2016**, *203*, 465–475. [CrossRef]

31. Reisgen, U.; Schleser, M.; Mokrov, O.; Ahmed, E. Statistical modeling of laser welding of DP/TRIP steel sheets. *Opt. Laser Technol.* **2012**, *44*, 92–101. [CrossRef]

32. Cruz, C.; Hiyane, G.; Mosquera-Artamonov, J.D.; L, J.M.S. Optimización del proceso de soldadura GTAW en placas de Ti6Al4V. *Soldag. Insp.* **2014**, *19*, 2–9. [CrossRef]

33. Miguel, V.; Martínez-Conesa, E.J.; Segura, F.; Manjabacas, M.C.; Abellan, E. Optimización del proceso de soldadura GMAW de uniones a tope de la aleación AA 6063-T5 basada en la metodología de superficie de respuesta y en la geometría del cordón de soldadura. *Rev. Met.* **2012**, *48*, 333–350. [CrossRef]

34. Nieto, P.J.G.; Suárez, V.M.G.; Antón, J.C.A.; Bayón, R.M.; Blanco, J.A.S.; María, A.; Fernández, A. A New Predictive Model of Centerline Segregation in Continuous Cast Steel Slabs by Using Multivariate Adaptive Regression Splines Approach. *Materials* **2015**, *8*, 3562–3583. [CrossRef]

35. Chang, L.; Chen, T.R.; Pan, Y.T.; Yang, K.C. Practical method for producing galvanised Dual Phase steels with superior strength-ductility combination. *Mater. Sci. Technol.* **2009**, *25*, 1265–1270. [CrossRef]

36. Colla, V.; DeSanctis, M.; DiMatteo, A.; Lovicu, G.; Valentini, R. Prediction of Continuous Cooling Transformation Diagrams for Dual-Phase Steels from the Intercritical Region. *Met. Mater. Trans. A* **2011**, *42*, 2781–2793. [CrossRef]

37. Ding, W.; Hedström, P.; Li, Y. Heat treatment, microstructure and mechanical properties of a C–Mn–Al–P hot dip galvanizing TRIP steel. *Mater. Sci. Eng. A* **2016**, *674*, 151–157. [CrossRef]

38. Rencher, A.C. *Methods of Multivariate Analysis*, 2nd ed.; John Wiley & Sons Inc.: Danvers, MA, USA, 2002. [CrossRef]

39. Bezerra, M.A.; Santelli, R.E.; Oliveira, E.P.; Villar, L.S.; Escaleira, L.A. Response surface methodology (RSM) as a tool for optimization in analytical chemistry. *Talanta* **2008**, *76*, 965–977. [CrossRef]

40. López, J. *Optimización Multi-objetivo: Aplicaciones a problemas del mundo real*, 1st ed.; Editorial de la Universidad Nacional de La Plata: Buenos Aires, Argentina, 2014. [CrossRef]

41. Montgomery, D.C. *Diseño y análisis de experimentos*, 2nd ed.; Editorial Limusa, S.A. de C.V.: Distrito Federal, Mexico, 2004.

42. Delgado, S.C.; Palacio, S.R.; Barajas, F.H. Comparación de Pruebas de Normalidad Multivariada. In Proceedings of the XXVI Simposio Internacional de Estadística, Sincelejo, Sucre, Colombia, 8–12 August 2016; pp. 1–4.

43. Breusch, T.S.; Pagan, A.R. A Simple Test for Heteroscedasticity and Random Coefficient Variation. *Econometrica* **1979**, *47*, 1287–1294. [CrossRef]

44. Fonstein, N.; Jun, H.; Huang, G.; Sadagopan, S.; Yan, B. Effect of Bainite on Mechanical Properties of Multiphase Ferrite-Bainite-Martensite Steels. *Mater. Sci. Technol.* **2011**, *1*, 634–641.

MDPI

St. Alban-Anlage 66

4052 Basel

Switzerland

Tel. +41 61 683 77 34

Fax +41 61 302 89 18

www.mdpi.com

Metals Editorial Office

E-mail: metals@mdpi.com

www.mdpi.com/journal/metals